PYRIDINE AND ITS DERIVATIVES

SUPPLEMENT IN FOUR PARTS

PART ONE

This is the fourteenth volume in the series

THE CHEMISTRY OF HETEROCYCLIC COMPOUNDS

THE CHEMISTRY OF HETEROCYCLIC COMPOUNDS

A SERIES OF MONOGRAPHS

ARNOLD WEISSBERGER and EDWARD C. TAYLOR

Editors

PYRIDINE
AND
ITS DERIVATIVES
SUPPLEMENT
PART ONE

Edited by

R. A. Abramovitch

University of Alabama

AN INTERSCIENCE® PUBLICATION

JOHN WILEY & SONS

NEW YORK • LONDON • SYDNEY • TORONTO

An Interscience ® Publication

***Library of Congress Cataloging in Publication Data*:**

Abramovitch, R. A. 1930–
Pyridine supplement.

(The Chemistry of heterocyclic compounds, v. 14)
"An Interscience publication."
Supplement to E. Klingsberg's Pyridine and its derivatives.
Includes bibliographical references.
1. Pyridine. I. Klingsberg, Erwin, ed. Pyridine and its derivatives. II. Title.

QD401.A22 547'.593 73–9800

ISBN 0–471–37915–8

Printed in the United States of America

10 9 8 7 6 5 4 3 2 1

Contributors

R. A. ABRAMOVITCH, *Department of Chemistry, University of Alabama, University, Alabama*

N. S. BOODMAN, *United States Steel Corporation, Applied Research Laboratory, Monroeville, Pennsylvania*

J. O. HAWTHORNE, *United States Steel Corporation, Applied Research Laboratory, Monroeville, Pennsylvania*

ROBERT E. LYLE, *Department of Chemistry, University of New Hampshire, Durham, New Hampshire*

P. X. MASCIANTONIO, *United States Steel Corporation, Applied Research Laboratory, Monroeville, Pennsylvania*

OSCAR R. RODIG, *Department of Chemistry, University of Virginia, Charlottesville, Virginia*

A. W. SIMON, *United States Steel Corporation, Applied Research Laboratory, Monroeville, Pennsylvania*

G. M. SINGER, *Oak Ridge National Laboratory, Oak Ridge, Tennessee*

TO THE MEMORY OF

Michael

The Chemistry of Heterocyclic Compounds

The chemistry of heterocyclic compounds is one of the most complex branches of organic chemistry. It is equally interesting for its theoretical implications, for the diversity of its synthetic procedures, and for the physiological and industrial significance of heterocyclic compounds.

A field of such importance and intrinsic difficulty should be made as readily accessible as possible, and the lack of a modern detailed and comprehensive presentation of heterocyclic chemistry is therefore keenly felt. It is the intention of the present series to fill this gap by expert presentations of the various branches of heterocyclic chemistry. The subdivisions have been designed to cover the field in its entirety by monographs which reflect the importance and the interrelations of the various compounds, and accommodate the specific interests of the authors.

In order to continue to make heterocyclic chemistry as readily accessible as possible new editions are planned for those areas where the respective volumes in the first edition have become obsolete by overwhelming progress. If, however, the changes are not too great so that the first editions can be brought up-to-date by supplementary volumes, supplements to the respective volumes will be published in the first edition.

Arnold Weissberger

Research Laboratories
Eastman Kodak Company
Rochester, New York

Edward C. Taylor

Princeton University
Princeton, New Jersey

Preface

Four volumes covering the pyridines were originally published under the editorship of Dr. Erwin Klingsberg over a period of four years, Part I appearing in 1960 and Part IV in 1964. The large growth of research in this specialty is attested to by the fact that a supplement is needed so soon and that the four supplementary volumes are larger than the original ones. Pyridine chemistry is coming of age. The tremendous variations from the properties of benzene achieved by the replacement of an annular carbon atom are being appreciated, understood, and utilized.

Progress has been made in all aspects of the field. New instrumental methods have been applied to the pyridine system at an accelerating pace, and the mechanisms of many of the substitution reactions of pyridine and its derivatives have been studied extensively. This has led to many new reactions being developed and, in particular, to an emphasis on the direct substitution of hydrogen in the parent ring system. Moreover, many new and important pharmaceutical and agricultural chemicals are pyridine derivatives (these are usually ecologically acceptable, whereas benzene derivatives usually are not). The modifications of the properties of heteroaromatic systems by *N*-oxide formation are being exploited extensively.

For the convenience of practitioners in this area of chemistry and of the users of these volumes, essentially the same format and the same order of the supplementary chapters are maintained as in the original. Only a few changes have been made. Chapter I is now divided into two parts, Part A on pyridine derivatives and Part B on reduced pyridine derivatives. A new chapter has been added on pharmacologically active pyridine derivatives. It had been hoped to have a chapter on complexes of pyridine and its derivatives. This chapter was never received and it was felt that Volume IV could not be held back any longer.

The decision to publish these chapters in the original order has required sacrifices on the part of the authors, for while some submitted their chapters on time, others were less prompt. I thank the authors who finished their chapters early for their forebearance and understanding. Coverage of the literature starts as of 1959, though in many cases earlier references are also given to present sufficient background and make the articles more readable. The literature is covered until 1970 and in many cases includes material up to 1972.

I express my gratitude to my co-workers for their patience during the course of this undertaking, and to my family, who saw and talked to me even less than usual during this time. In particular, I acknowledge the inspiration given me by the strength and smiling courage of my son, Michael, who will never know how much the time spent away from him cost me. I hope he understood.

R. A. ABRAMOVITCH

University, Alabama

Contents

Part One

Part Two

Part Three

Part Four

PYRIDINE AND ITS DERIVATIVES

SUPPLEMENT IN FOUR PARTS

PART ONE

This is the fourteenth volume in the series

THE CHEMISTRY OF HETEROCYCLIC COMPOUNDS

CHAPTER IA

Properties and Reactions of Pyridines

R. A. ABRAMOVITCH

University of Alabama, University, Alabama

AND

G. M. SINGER

Oak Ridge National Laboratory, Oak Ridge, Tennessee

I. Introduction

The first edition of this series surveyed the rapid development of pyridine chemistry through 1959. Since that time, as in all fields of chemistry, the volume of the pyridine literature has expanded tremendously. This review treats some of the major advances in studies of the properties and reactions of pyridine derivatives since 1959. As in the first edition, this chapter emphasizes the differences between pyridine chemistry and benzene chemistry.

In the intervening decade or so, the greatest advances have probably been the development of various spectroscopic methods and the application of these techniques to more detailed studies of the physical properties of organic compounds, tremendously simplifying physical organic chemical investigations, as well as the discovery of new substitution reactions of this ring system. With these analytical techniques available, many more reaction mechanisms have been investigated more thoroughly than was ever possible before. This, in turn, has contributed greatly to the very rapid growth of pyridine literature. Equally significant is the fact that mechanistic and orientation studies are being carried out kinetically and quantitatively at an accelerated rate.

This chapter follows the general pattern set in the first edition, starting with a presentation of the physical properties of the pyridines, followed by a discussion of their chemical reactivity. The effect of the pyridine nucleus on the reactivity of functional groups is treated, and finally a brief discussion of free energy relationships in this system is given. No attempt at exhaustive coverage has been made; instead, references to articles reviewing the various topics discussed are given wherever possible.

The preparation and properties of the reduced pyridines are taken up in Part B of this chapter.

II. Physical Properties

1. Basicity

Most physical properties of pyridine derivatives reflect an interaction between substituents and the pyridine ring, and this is especially so of basicities and dipole moments.

Several groups have studied the effect on p*K*'s of varying the substituent at a particular nuclear position (1–4) or the effect of changing the nature of the substituents (5, 6). The pK_a's of a series of 4-substituted nicotinic acid-1-oxides have also been measured (7). The ionization constants of substituted pyridines have been tabulated, and effects of substituents discussed (8) in terms of their mesomeric, inductive, and steric effects. Substituents at the 2- and 4-positions can influence the basic strength by stabilizing the conjugate acid through mesomeric and/or inductive electron release, or, vice versa, by destabilizing the conjugate acid through a *-M* or *-I* effect. The pK_a's of a series of 3-substituted pyridines were correlated with the σ-constants of the substituents (9) and these studies were extended (5, 10). Correlations of pK_a's of 2- and 4-styrylpyridines (11) and of 4-*R*-pyridinium ions (12) with substituent constants have also been reported. The pK_a's of some alkylpyridines are discussed in Chapter V.3.A.

Clarke and Rothwell (13) found a linear relation between the pK_a of the base and the rate of formation of alkyl pyridinium salts. 2-Substituted pyridines gave a different line but with the same slope as that for the 3- and 4-substituted derivatives. This "*ortho*-effect" was not attributed to a steric effect *per se*, as originally suggested by Brown and Kanner (14) because of the small steric size of the proton. McDaniel and Özcan's reinvestigation (15) of 2,6-di-*tert*-butylpyridine supported this view, finding that the lowered pK_a was due to steric hindrance to solvation of the cations rather than to strain involving the bound proton. A similar conclusion was reached on the basis of the measurement of the heats of neutralization of pyridine, the picolines, and 2,6-lutidine, the values being extrapolated to infinite dilution (16) (*cf.* 17).

Jaffé and Doak (18) and later Jaffé (19) proposed a general equation relating the pK of substituted pyridine-1-oxides to the substituent constant σ^R of the group attached to the ring:

$$pK = (1.89 \pm 0.07)\sigma^R + 0.812$$

where σ^+ is used for mesomerically electron donating substituents, and σ^- for mesomerically electron withdrawing substituents. 2-Substituted pyridine-1-oxides have also received the Hammett treatment (20).

Ionization constants have provided a means of studying tautomeric ratios in substituted pyridines such as the amino- and hydroxypyridines (8, 21). From a consideration of the ionization constants and u.v. spectra of amino-2 (and 4)-pyridones and their *O*- and nuclear *N*-methyl derivatives it has been established that protonation of 3- and 5-amino-2-pyridone and 3,4-diamino-2-pyridone occurs first at the 3-(or 5)-amino group, but 4- and 6-amino-2-pyridone and 2- and 3-amino-4-pyridone are protonated first at the oxygen atom (22). First protonation of dimethylaminopyridines and their *N*-oxides occurs on the ring nitrogen (pyridines) or at the *N*-oxide oxygen (23).

H_2N … N H … O $\overset{H^+}{\rightleftharpoons}$ H_3N^+ … N H … O

$$H_2N\text{-pyridin-2(1H)-one} \underset{}{\overset{H^+}{\rightleftharpoons}} H_2N\text{-2-hydroxypyridinium } (\overset{+}{N}H)$$

Hammett correlations have been used (24) to evaluate the tautomeric equilibria in a number of 2-arylsulfonamidopyridines.

The proton transfer reactions:

$$C_5H_5\overset{+}{N}H + C_5H_4N\text{-}CO_2^- \rightleftharpoons C_5H_5N + C_5H_4\overset{+}{N}H\text{-}CO_2^-$$

have been studied (25), and it was concluded that the entropy change results largely from electrostatic effects, while ΔH° is a result of both electrostatic and conjugative effects.

A correlation was observed between the IR spectral shifts ($\Delta\nu$) due to hydrogen-bonding between methanol and a number of alkylpyridines (determined in an inert solvent) and the pK_a of the base. With other substituted pyridines the correlation was not as good (26). Steric hindrance of the basic site

$$\Delta\nu_{\mathrm{MeOH}} = 19.6\ \mathrm{p}K_a + 169\,(\mathrm{cm}^{-1})$$

may influence pK_a more than $\Delta\nu$, and differences in solvation effects probably contribute to some of the scatter observed. Also, pK_a's were determined in water where resonance and inductive effects play a major role both in the unprotonated (but hydrogen-bonded) base and in its protonated form (also hydrogen-bonded). Since $\Delta\nu$ values were measured in CCl_4 solution, these are influenced by resonance and inductive effects in the hydrogen-bonded base only. If resonance effects differed markedly on going from unprotonated to the protonated form of a base than correlation of pK_a's with $\Delta\nu$ would not have been expected (26).

2. Dipole Moments

Molecular dipole moments are calculated as the vector sums of the bond moments of the various functional groups in a molecule, assuming no mesomeric interaction between the groups. To the extent that there is an interaction, the experimental dipole moment differs from a theoretical moment calculated on the basis of group moments determined from aliphatic compounds. Because the pyridine ring itself possesses a dipole moment–as opposed to benzene, which does not–mesomeric interactions are more common and more complex in this series.

The determination and application of the dipole moment data in the heterocyclic series, including pyridines and pyridine-1-oxides, have been surveyed (27).

It has been shown previously that electron donating substituents interact with the pyridine ring to a greater extent than with the benzene ring (28). More recent and extensive determinations (29–32) tend to support this and strengthen the conclusion (28) that "electron attracting substituents interact only as much as with benzene if at all." In fact, a comparison of the moments of pyridine (**IA-1**), 4-nitropyridine (**IA-2**), and 4-cyanopyridine (**IA-3**) with those of their carbocyclic analogs (**IA-4** and **IA-5**, respectively) suggests that there may be even less interaction between these -M substituents and the ring in the pyridine series than in the benzene series (33). This is in accord with the finding that the π-electron ring current is larger for pyridine than for benzene (34) (*vide infra*); that is, a greater perturbation of the aromatic ring is needed to produce as great a degree of electron withdrawal in pyridine as in benzene.

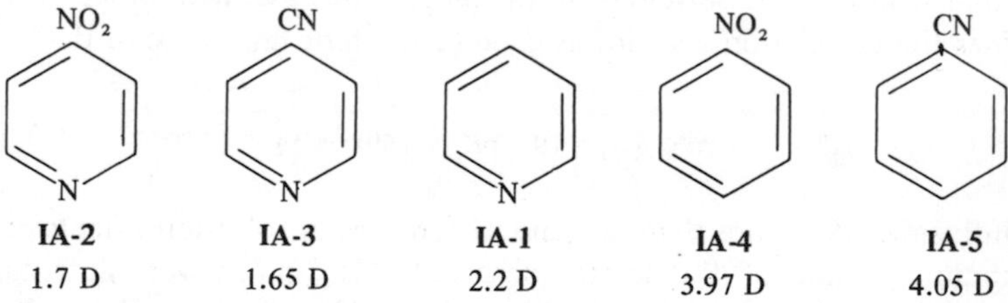

The dipole moments of some pyridine-1-oxides have been measured, and generally, these are found to be about 2 D larger than those of the corresponding amine (35). Dipole moments have been used to compare the electron accepting ability of the pyridine-1-oxide ring with that of nitrobenzene and **IA-1**, and have been correlated with substituent constants (32). The dipole moment of pyridine-1-oxide was calculated to be $\mu_{tot} = 4.78$ D from computed SCF-charge distributions in the ground S_0 state (36). This compares well with the experimentally found value $\mu_{ex} = 4.78 \pm 0.07$ D.

3. Spectroscopic Properties

In the first edition of this series a brief discussion of the use of ultraviolet and infrared spectroscopy in the elucidation of structural details of pyridine derivatives was given. In the last decade, the definition of spectroscopy has been expanded to include techniques which do not involve the absorption of light energy. In particular, nuclear magnetic resonance, mass, and to some degree, electron spin resonance spectrometry have been valuable additions to the pyridine chemist's "tool chest."

A. *Ultraviolet Absorption*

This topic has been discussed extensively by Schofield (8) who has tabulated the data for the free bases and the cation, anion, and zwitterion where these are applicable (see also Ref. 37). The ultraviolet and visible spectra of monocyclic azines contain a weak, but not forbidden, $n \rightarrow \pi^*$ transition which has no counterpart in carbocyclic compounds (38). The importance of this transition is dependent on the separation between the heteroatom and the potential conjugative substituent (39, 40) and on the nature of the substituents; thus the $n \rightarrow \pi^*$ transition is not observed for 2-bromopyridines because of the inductive effect exerted by the bromine atom (41). Systematic studies of the ultraviolet absorption characteristics of numerous pyridine derivatives have been carried out and summarized (8). The UV spectra of alkylpyridines are discussed further in Chapter V.I.3.B., those of pyridine-1-oxides in Chapter IV.II.8.

Numerous MO calculations have been carried out to predict transition probabilities and differences in energy levels in pyridine and pyridinium salts (42–44, and Ref. cited therein), and also to interpret $n \rightarrow \pi^*$ transitions (45).

The application of UV spectroscopy to tautomerism of various pyridine derivatives has been very fruitful, since the electronic spectrum of a compound arises from its π-electron system which, to a first approximation, is unaffected by substitution of an alkyl group for a hydrogen atom. A comparison of the UV spectrum of a potentially tautomeric compound with the spectra of the alkylated forms of both tautomers often shows which is the predominant tautomer in the unalkylated species (for review, see Ref. 46). For example, equilibrium constants were derived from the p*H* dependency of the ultraviolet spectra of pyridoxine and pyridoxyl analogs in aqueous solution, thus providing information about the *in vivo* action of the biologically important but more complex compounds (47). The clear differences in the UV spectra of the thiopyridone-pyridyl sulfide tautomers have been described (48, 49).

The color of some heteroaromatic salts in solution and in the solid state has been attributed to charge-transfer transitions. In these cases, the more polarizable the anion the deeper is the color. The energies of the charge-transfer bands are related to the electron affinity of the cation, but not to the nucleophilic atom localization energies of the carbon atoms in the heterocyclic cation. This is in agreement with a delocalized model of the excited state for the charge-transfer transition (50):

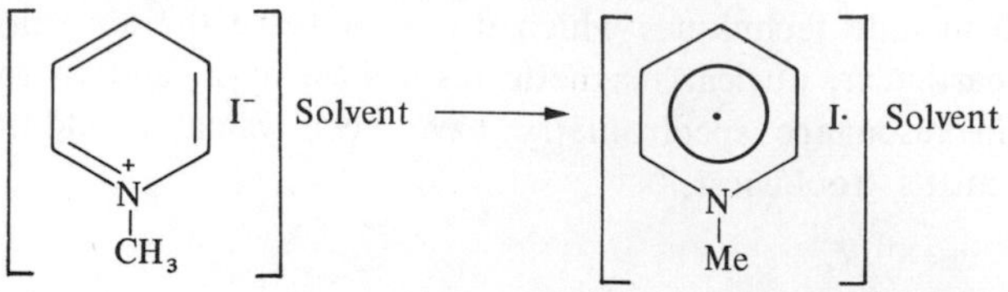

[N-methylpyridinium I^-] Solvent ⟶ [N-Me pyridinyl radical I·] Solvent

The ultraviolet spectra of pyridine-1-oxides have been discussed in terms of hydrogen bonding and donor-acceptor behavior (51).

Since the ultraviolet maximum of pyridine-1-oxide is sensitive to solvent polarity, its use in determining "*Z*-values" for solvents has been advocated (52). Because of intramolecular charge transfer from oximate oxygen to the pyridinium ring, 1-methyl-4-cyanoformylpyridinium oximate (CPO) has two electronic absorption bands whose positions are measures of solvent polarity (53a). Intramolecular charge transfer phenomena have also been observed for several *N*-arylalkylpyridinium ions (54). The effect of solvents on the ultraviolet spectra of *N*-oxides has been discussed (53). Phosphorescence spectra, excitation

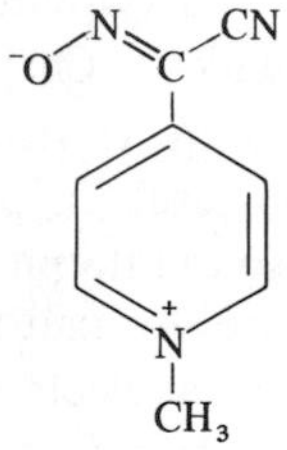

(CPO)

spectra, and the degree of polarization of 4-nitropyridine-1-oxide were recorded under degassed conditions at 77°K (55). The phosphorescence spectra show a blue shift in some alcohol solvents compared with those in ether, indicating hydrogen bonding interaction $\overset{+}{N}–\overset{-}{O} \cdots HOR$ in the protic solvents. The assignments and nature of excited states were compared with the results of PPP-SCF-MO-CI calculations. The π-π^* band appearing in the longest wavelength region has the character of an intramolecular charge transfer-band from the *N*-oxide group to the NO_2 group. The orbital electron distributions at the highest occupied (HOMO) and lowest vacant (LVMO) molecular orbitals were calculated (the charge transfer arises by one-electron transfer from the HOMO to the LVMO). Applications of the use of ultraviolet spectral data on *N*-oxides to the determination of pK_a values, intramolecular hydrogen bonding, charge transfer complexes, tautomerism of hydroxy-, amino-, and mercaptopyridine-1-oxides, among others, have been summarized (56).

B. *Infrared Absorption*

As more and more compounds have been examined, correlations peculiar to pyridine ring substitution have become apparent. For example, correlations have been found for 2- and 3-substituted pyridines in the 600 to 650 cm^{-1} region; 4-substituted derivatives resemble their benzene analogs (57). In terms of the C–H out-of-plane bendings, pyridine may be treated as a monosubstituted benzene. Extending the analogy, a 2,3-disubstituted pyridine may be considered to be a 1,2,3-trisubstituted benzene, and two bands are indeed found at 813 to 769 and 752 to 725 cm^{-1} (58). Similarly, a 2,5-disubstituted pyridine shows the two expected bands at 828 to 813 and 735 to 724 cm^{-1} (58), characteristic of a 1,2,4-trisubstituted benzene. A distinction between 2,3- and 2,5-disubstituted pyridines was found in the C=C and C=N stretching vibration region (59): 2,3-disubstituted derivatives exhibit a band at 1578 to 1588 cm^{-1}, and 2,5-disubstituted derivatives one at 1599 to 1605 cm^{-1}, which is a useful correlation unless absorption by substituents obscures this region. Similarly, 3,5-bis(carbalkoxy)pyridines unsubstituted at C-4 exhibit two bands between 1540 to 1580 and 1590 to 1610 cm^{-1}, but those bearing a 4-substituent have only one band in the range of 1500 to 1700 cm^{-1} (60). Katritzky and Ambler (61) have surveyed and assigned the bands due to pyridine, pyridine complexes, salts, and *N*-oxides, as well as the influence of the nature and position of substituents upon the intensities of the characteristic modes of vibration. The IR spectra of alkylpyridines are discussed further in Chapter V.13.B.

Tautomeric equilibria and hydrogen bonding have been investigated extensively (61). Particular use has been made recently of the variation in the C=O (62) and C=N (63–65) stretching frequencies in tautomeric mixtures.

Infrared spectroscopy has been used to examine the extent of the mesomeric interaction between the *N*-oxide oxygen of pyridine-1-oxides and substituents (66–69). In one case, a lack of correlation was attributed to a coupling of absorption frequencies (70). In another instance, a correlation was found using corrected and derived σ values (69). The infrared spectra of pyridine-1-oxides is discussed further in Chapter IV.II.7.

The infrared spectra of a number of 3-benzenesulfonamidopyridines and *N*-sulfonyliminopyridinium ylides have been measured (71). The 3-sulfonamides exhibited two strong SO_2 bands in the normal range of 1320 to 1340 and 1160 to 1170 cm^{-1}. The NH stretching band appeared at 2880 to 2650 cm^{-1}, however, suggesting that the sulfonamides existed predominantly in the zwitterionic form in the solid state. In contrast to the sulfonamides, the

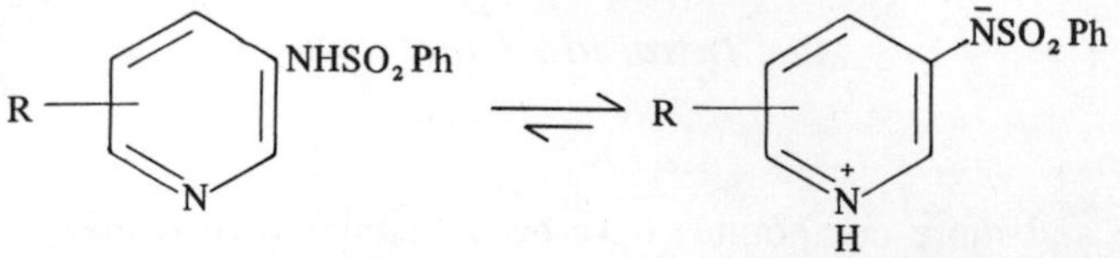

N-sulfonylimino ylides exhibited two strong bands due to SO_2 in the 1270 to 1285 and 1130 to 1140 cm^{-1} regions. The bathochromic shift may be due to the delocalization of the electron pair on the imino nitrogen onto the sulfonyl group, rather than toward the pyridine ring, so that back-donation may not be important in these ylides.

Assignments of the Raman bands have been made for a series of pyridines (72), picolines (73), and their N-oxides.

C. *Nuclear Magnetic Resonance Spectroscopy*

The application of NMR spectroscopy to pyridine chemistry has been discussed extensively (74–76) and is not covered in detail here. Reduced symmetry in pyridine (C_{2v} relative to D_{6h} for benzene) and electronic perturbations of the ring by the nitrogen atom cause unsubstituted pyridine, pyridine-1-oxide, and simple pyridinium salts to show three magnetically nonequivalent groups of protons. The order of absorption is H-2 < H-4 < H-3 for pyridine, H-2 ≪ H-4 < H-3 for pyridinium salts, but H-2 ≪ H-3 < H-4 for pyridine-1-oxide (77) (Table IA-1). The change in order for pyridine-1- oxide is due to mesomeric "back-donation" by the N-oxide function, increasing the electron density at C-4 relative to C-3 which can only be affected by the strong inductive withdrawal by the N-oxide (for more details, see Chapter IV.II.6).

TABLE IA-1. Chemical Shifts (δ) for Proton Resonances[a] of Pyridine, Pyridine-1-Oxide, and Pyridinium Ion in CCl_4 [b]

	Pyridine	Pyridine-1-oxide	Pyridinium ion
H-2	8.53	8.10	8.83
H-3	7.19	7.28	8.19
H-4	7.58	7.08	8.73

[a] Adapted from Ref. 77.
[b] Pyridinium ion values from measurements of pyridines in 18 N D_2SO_4

The substitution pattern of a pyridine derivative can be determined from the observed spin-spin coupling pattern. The individual coupling constants (J_{ij}) fall into a narrow range of values, those for the pyridine-1-oxides generally being somewhat larger than those for the pyridines (Table IA-2). Varying the ring substituents (hence, the electron density at and between the various carbons) affects the various constants differently (79). A number of compilations of NMR

TABLE IA-2. Ranges of Coupling Constants (Hz) of Pyridines and Pyridine-1-Oxides

	Pyridines (78)	Pyridine-1-oxides (77)
$J_{2,3}$	4.5–5.1	6.4–7.4
$J_{3,4}$	7.4–8.2	*ca.* 8.0
$J_{2,4}$	1.5–2.0	*ca.* 1.5
$J_{2,5}$	0.6–0.9	0–1.0
$J_{2,6}$	0–0.3	*ca.* 2.0
$J_{3,5}$	1.0–1.6	2.8–3.4

data of pyridines are available (80–82), including Brügel's work (75) covering over 150 compounds. The NMR spectra of alkylpyridines are discussed briefly in Chapter V.I.3.B.

Many simple monosubstituted pyridines present relatively complex spin-spin coupling patterns (83). In fact, because of the ready availability of these otherwise simple compounds, they often served as experimental models for the derivation and testing of the calculations of the parameters of 3-, 4-, and 5-spin systems (84–86). The 5-spin system in pyridine has been measured and calculated at both 40 MHz (87) and 60 MHz with ^{14}N-decoupling (88).

Chemical shifts vary with the localized electron density. NMR spectroscopy provides one of the best experimental tests of theoretical calculations of such electron densities in pyridines (89). Some examples include local charge densities of the pyridinium ion (90), correlations with a variety of calculated charge density parameters (91–95), and the "degree" of aromaticity of pyridones (96–98).

While one group (99, 100) had concluded that proton chemical shifts appear to be a somewhat unreliable measure of π-electron densities in nitrogen heterocycles, especially in positions adjacent to a nitrogen atom, another (101) has reported observing a simple linear relationship between proton chemical shifts and calculated π-electron densities (by the simple Hückel method): $\delta = 7.97 \pm 9.52(1 - q)$. The high field shift of the nitrogen signal on protonating pyridine was accounted for by the large paramagnetic contribution to the susceptibility of the pyridine nitrogen atom. The low shielding of the α-protons in pyridine (relative to β and γ) is influenced appreciably by both the magnetic anisotropy of the nitrogen atom and by the local dipole moment associated with the nitrogen lone-pair (102).

The NMR spectra of a series of brominated and iodinated alkylpyridines have been reported (103).

A number of investigators have examined the effect of different solvents and of Lewis acids on the chemicals shifts of pyridine protons. The shift of H-2 is affected less than those of H-3 and H-4 (104–108), and steric inhibition of solvation by 2-substituents has been observed (*cf.* Section II.1).

Hydrogen-bonding between the pyridine nitrogen or the pyridine-1-oxide oxygen and chloroform has been demonstrated (77). The chemical shifts for a number of methyl-substituted pyridines have been measured in solvents of varying dielectric constant, including two hydrogen bonding solvents (MeOH and D_2O). The results were interpreted in terms of the electrostatic field effect of the nitrogen lone-pair and the variation in the effective electrostatic dipole with the nature of the solvent (109). The reduction in the effective dipole moment of the nitrogen lone-pair in more polarizable solvents is accompanied by a decrease in the chemical shift difference between the α-protons (where the dipole electrostatic field is most strongly deshielding), and the β- and γ-protons. The additivity of solvent shifts (on going from CCl_4 to C_6H_6) has been discussed (110). The effect of the counterion (Cl^-, Br^-, I^-) on the PMR spectrum of the pyridinium ion has been examined (110a). It has been assumed that the counterion reduces considerably the π-electron polarization in $C_5H_4\overset{+}{N}H$, as calculated for the isolated ion by the VESCF method. The average of all the C-H proton shifts was found to be $\Delta\sigma_2 = 1.53 \pm 0.13$ ppm, $\Delta\sigma_3 = -0.78 \pm 0.13$ ppm, and $\Delta\sigma_4 = -1.30 \pm 0.13$ ppm.

Relative PMR isotropic shifts for the ligands in methyl-substituted pyridine Ni(II) complexes have been determined and compared with values calculated using the INDO-MO method (110b).

The induced proton chemical shifts in pyridine and some simple methylated derivatives by lanthanide (111) and europium complexes (112) have been studied. Tris(dipivaloylmethanoto)europium [$Eu(DPM)_3$] induces shifts (Δ_{Eu}) of the 2,6-protons in pyridine which are much greater than those of the 3,5- and 4-protons. The multiplet from the 3,5-protons, which occurs upfield from that of the 4-proton in the uncomplexed material, appears at lower field than that of the 4-proton peaks when 0.5 mole equivalents of $Eu(DPM)_3$ is added; the same reversal is observed with $Yb(DPM)_3$. In the case of 2-picoline, the signal from the methyl group is shifted slightly further than that of the 6-proton by $Eu(DPM)_3$, and the induced shift of the H-3 signal is larger than that from H-5. Again, in 3-picoline the C-2 proton signal moves further than that from C-6. These observations are in accord with a *pseudo*-contact mechanism (112), as is the observation that the induced isotropic chemical shifts are to a great extent stereospecific (113). The use of $Pr(C_5H_5)_3$ as a shift reagent for nicotine has been reported (114).

NMR data have been used to provide information on the conformation of *ortho*-substituted *N*-aralkylpyridinium ions (115). The shielding of the *ortho* positions of the adjacent nucleus due to the ring current of the ring is a function of molecular conformation.

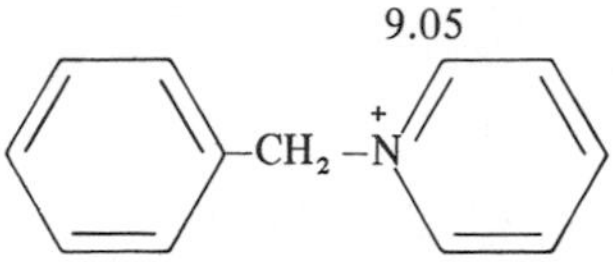

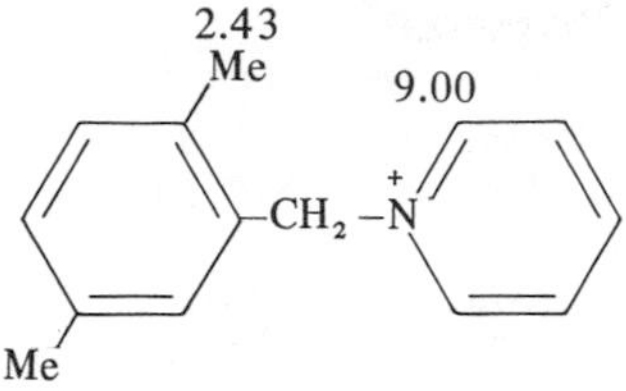

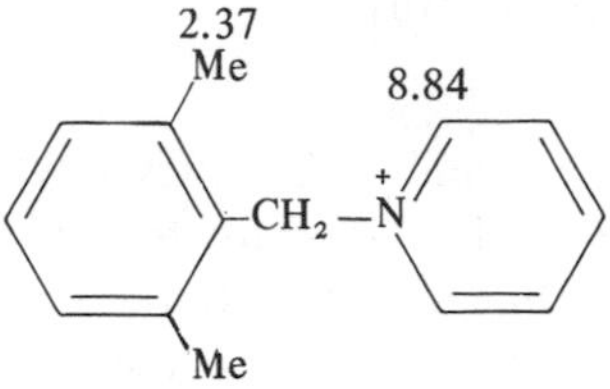

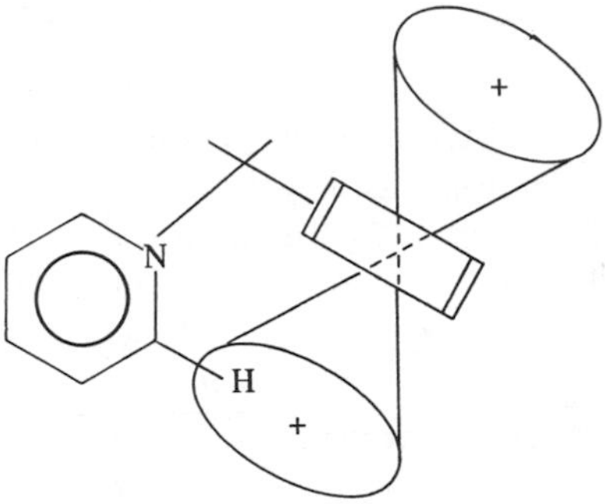

In addition to proton magnetic resonance (PMR), the ^{19}F (116–118), ^{14}N (119), and ^{13}C NMR (120–125) spectra of pyridine derivatives have been studied. It was found that the calculated proton shifts obtained by an SCF treatment nearly reproduced the experimental shifts, but an empirical term had to be added to the calculated ^{13}C shifts to fit the experimental results. This empirical term was directly proportional to the π-electron density on carbon at that position. The ^{13}C shifts were thus nicely accounted for by adding a direct π charge term to the atom anisotropic term (125). An interesting recent application of NMR spectroscopy has been to the determination of the spacial requirements of the pyridine-1-oxide oxygen atom (126). The smallest number of ring members for which the bridge in **IA-6** could still pass by the intraannular group X (X = H, F, Cl, Br, CH_3, C≡N, OCH_3) was determined (temperature-dependent ^1H NMR study). Pyridinophane-1-oxides **(IA-7a-i)** were then examined. Those compounds having many ring members **(IA-7f–i)** exhibited no steric interaction in the NMR (no hindrance to ring inversion). In contrast, **IA-7a–b)** are conformationally rigid. It was thus established that the spatial requirement of the aromatic *N*-oxide oxygen is of the same order of magnitude as that of the aromatically bound fluorine atom.

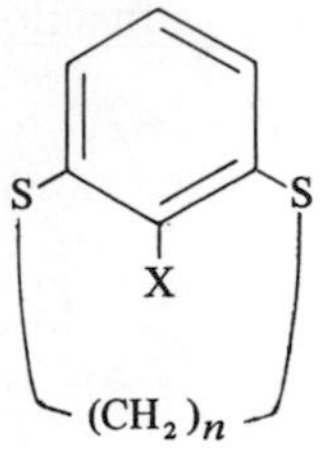

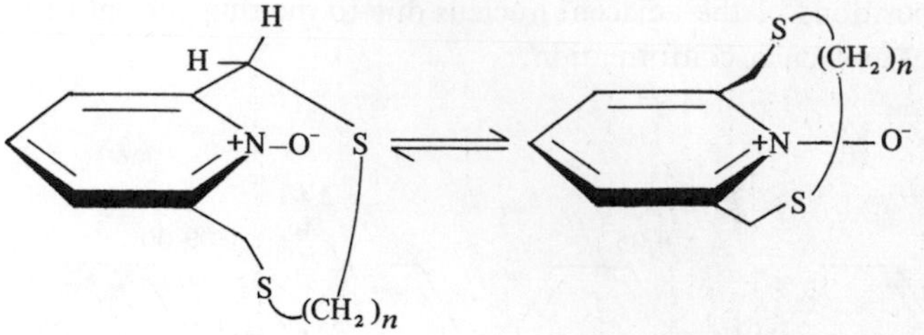

IA-6

IA-7a $n = 1$ IA-7c $n = 3$ IA-7e $n = 5$ IA-7h $n = 7$
IA-7b $n = 2$ IA-7d $n = 4$ IA-7f $n = 6$ IA-7i $n = 8$

D. *Mass Spectroscopy*

Until recently the pyridines had not been examined by mass spectrometry as extensively as might have been expected. The literature to 1967 has been reviewed (127). The mass spectrum of pyridine reflects both the stability of the aromatic ring–the parent ion is the base peak–and the ease, relative to benzene and its derivatives, with which the rings are cleaved (loss of HCN being significant) (128). Deuterium labelling has shown (129) that, even at 14 eV, loss of HCN involves the α-, β-, and γ-hydrogen atoms statistically, so that randomization of the protons occurs (perhaps *via* azaprismane and azabenzvalene radical cation intermediates) prior to fragmentation. Indeed, it has recently been shown (130) that complete and independent randomization of the carbon and hydrogen atoms occurs in the molecular ion of pyridine before

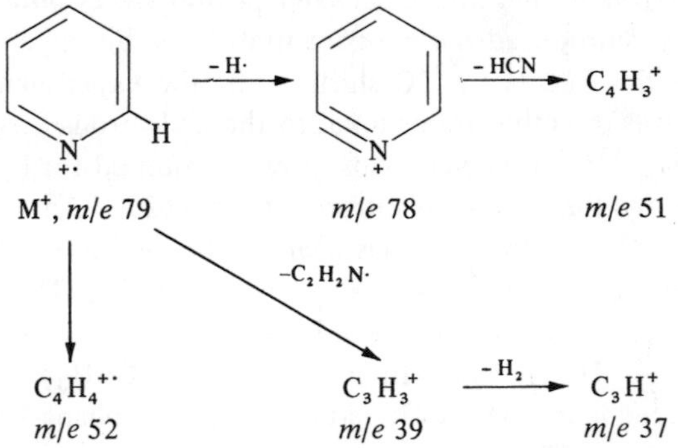

M^+, m/e 79 $\xrightarrow{-H\cdot}$ m/e 78 $\xrightarrow{-HCN}$ $C_4H_3^+$ m/e 51

M^+ → $C_4H_4^{+\cdot}$ m/e 52

M^+ $\xrightarrow{-C_2H_2N\cdot}$ $C_3H_3^+$ m/e 39 $\xrightarrow{-H_2}$ C_3H^+ m/e 37

fragmentation. These observations are fairly general throughout the pyridine family. For example, hydrogen scrambling occurs in the methylpyridines (131). β-Fission is most favorable in 3-alkylpyridines, a reflection of the higher electron density at that position (128). In contrast, γ-cleavage is most favored with 2-alkylpyridines (see also Chapter V.I.3.B.) where the resultant carbonium ion can be stabilized by the ring nitrogen; for example, the base peak for 2-ethylpyridine is at $M^{+}-1$ and is over twice as intense as the parent ion (127).

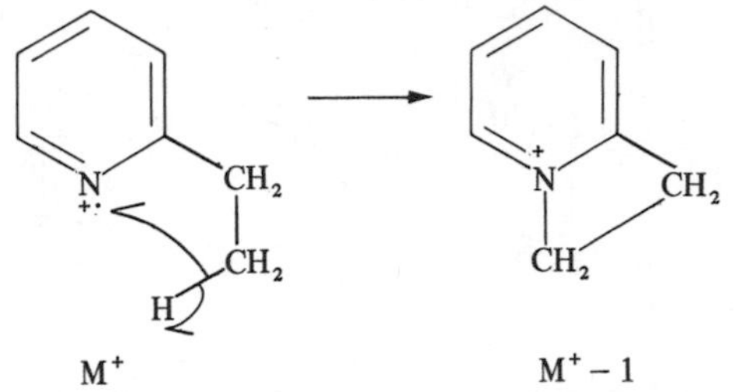

M^{+} $M^{+}-1$

Other *ortho*-effects have also been observed. The base peak for 3-nitropyridine corresponds to loss of the nitro group. Similarly, loss of HCl is an important mode of fragmentation in 2-chloropyridine and 2-chloropicolines (132). A quantum chemical method has been reported which aims at determining the most plausible ion structure and fragmentation mechanisms for pyridine and other nitrogen heteroaromatic molecules upon electron impact (133). The total HMO electronic energies of all possible fragment structures are compared on the hypothesis that the reaction would go, in high probability, through the process that gives the most stable ion. The calculations confirmed that for pyridine and its methyl derivatives, the first bond-scission to the intermediate ion occurs in the β-bond to the nitrogen atom on which the unpaired electron is located.

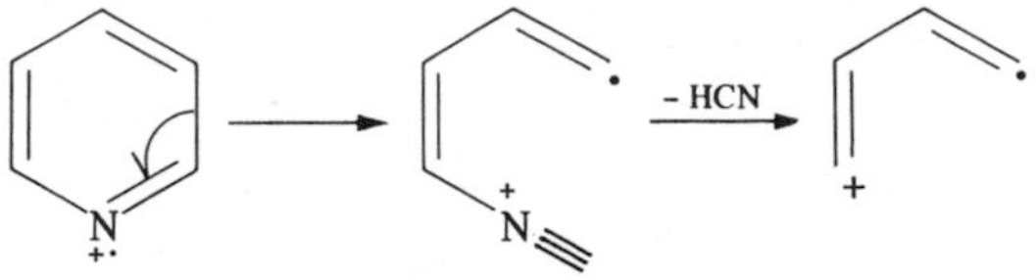

- HCN

Hydroxypyridines eliminate carbon monoxide as an important fragmentation pathway (128, 132, 134–136) but this fragmentation is much less prominent in 3-hydroxypyridines than in 2- and 4-pyridone derivatives (128). Similar

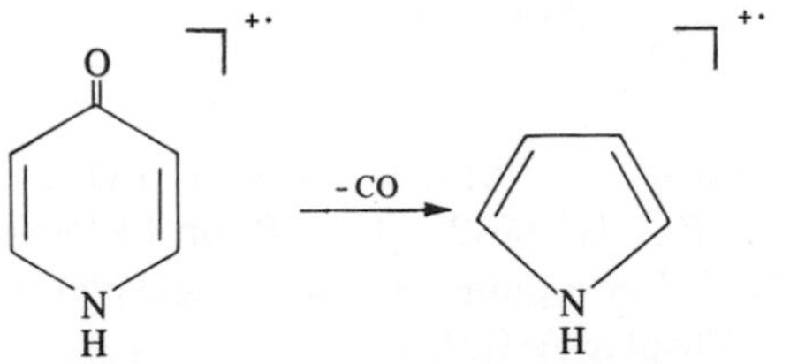

- CO

eliminations occur in pyridinethiones (132, 135, 137). There is often a pronounced *ortho*-effect enhancement of fragmentation from 3-alkyl-2-aminopyridines and 3-alkyl-2-pyridinethiones (132). Some interesting rearrangements occur in the molecular ion of methyl isonicotinate (138):

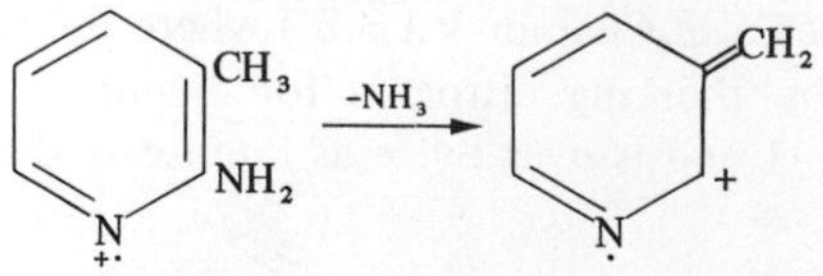

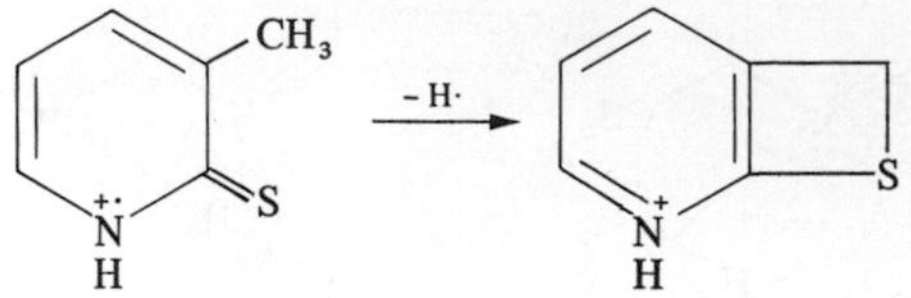

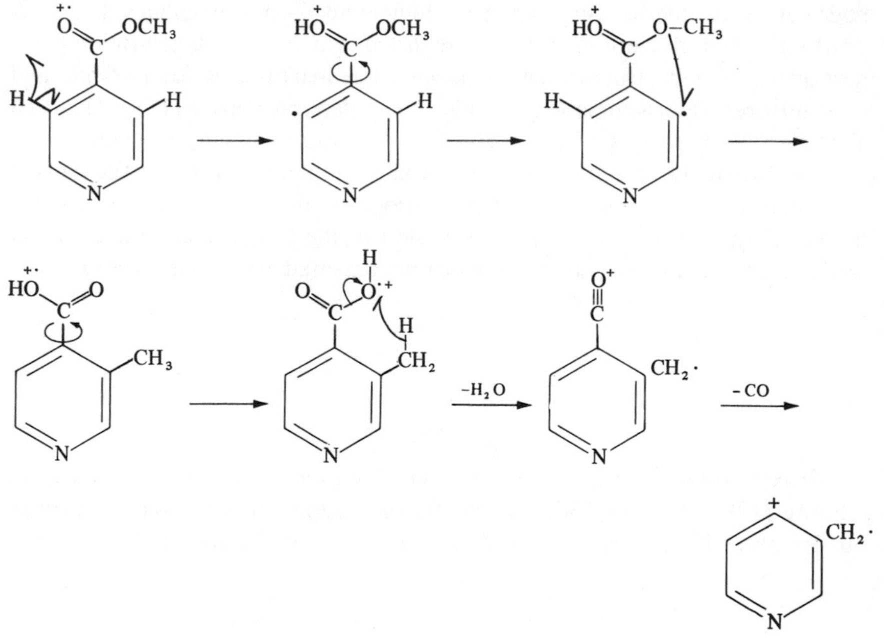

The most prominent feature of the mass spectra of pyridine-1-oxides is the loss of the *N*-oxide oxygen (M^+-16) (132, 139, 140) or the loss of OH (M^+-17) from 2-alkylpyridine-1-oxides. For a more extensive discussion of the mass spectra of pyridine-1-oxides, see Chapter IV.II.9.

Reports on the mass spectra of pyridinium salts are practically nonexistent, probably because of their generally low volatility. The major fragmentations of the interesting phenoxypyridinium salt **(IA-8)**(141) are to the pyridine radical cation and to the *p*-nitrophenoxy radical. Subsequent fragmentations occur according to expected pathways.

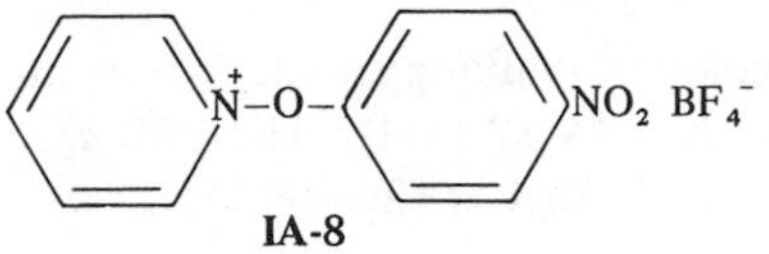

$\overset{+}{N}$–O–C₆H₄–NO_2 BF_4^-

IA-8

E. *Electron Spin Resonance*

Several workers have used electron spin resonance (ESR) spectroscopy to observe the formation of radicals from pyridine by irradiation with an electron beam (142, 143), by γ-irradiation (144), or by reaction with metallic sodium (145). The species produced on γ-irradiation of **IA-1** at 77°K appears to be the 2-pyridyl radical **(IA-9)** which, on warming to −90°C, forms the dimeric radical (144).

Irradiation of pyridinium hydrochloride in solid anhydrous hydrogen chloride produces the 4*H* radical **(IA-10)** (146). The ESR spectrum of the stable free

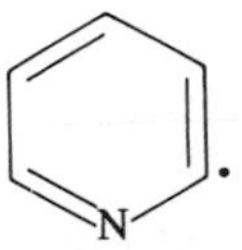

IA-9

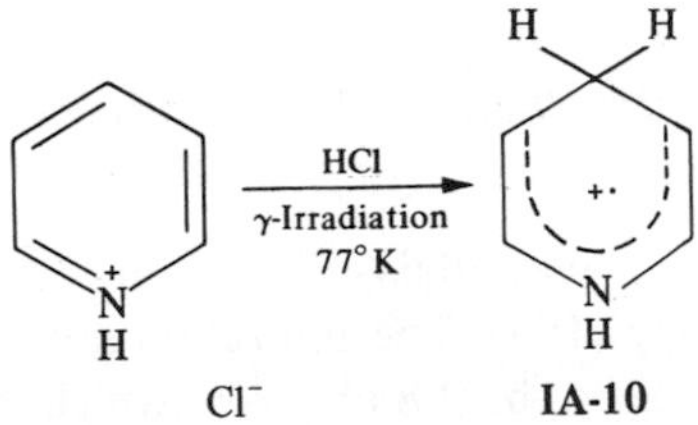

HCl, γ-Irradiation, 77°K

Cl^- **IA-10**

radical, 1-ethyl-4-carbomethoxypyridinyl **(IA-11)**, has been determined in dilute solution (147), as has the ESR spectrum of methylviologen cation radical in

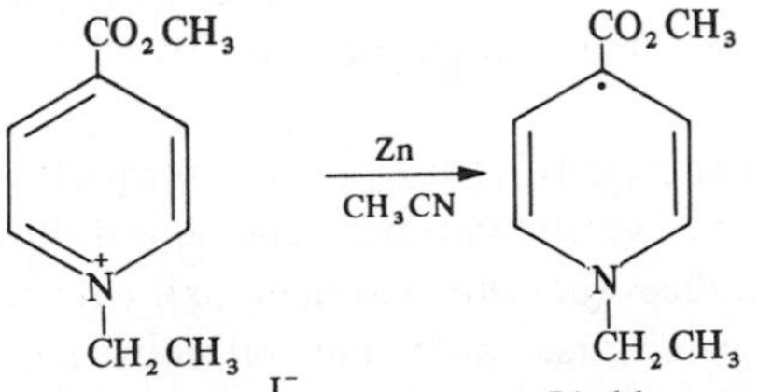

CO_2CH_3 CO_2CH_3

Zn, CH_3CN

CH_2CH_3 I^- CH_2CH_3

IA-11

acetonitrile (148). When solid *N,N'*-diacetyltetrahydro-4,4'-dipyridyl is heated the ESR spectrum observed at about 120° has been attributed to the *N*-acetyl-4-pyridinyl radical (149).

Oxygenation of the anion of 4-alkylpyridine-1-oxides has been shown by ESR to lead to nitroxides such as **IA-12** (150). The ESR spectra of radicals derived from *N*-oxides are discussed further in Chapter IV.II.10.

CH_3 ... t-BuO^-K^+ → Dark red solution → O_2 →

IA-12

F. *Nuclear Quadrupole Resonance*

A few preliminary studies have been carried out of the nuclear quadrupole resonance (NQR) of pyridine derivatives (151–153). One interesting result of an NQR study of variously substituted pyridines has been that "inductive effects do not attenuate in an aromatic ring and this, in turn, implies that σ-bonds are not as localized as had been assumed"(154).

4. Miscellaneous Physical Properties

The first six ionization potentials of pyridine have been measured by photoelectron spectroscopy (155). The ionization potential of the nitrogen lone pair of pyridine was deduced to be 9.8 eV, from which it was concluded that the nitrogen lone pair is appreciably delocalized (156), in agreement with CNDO calculations (157) (Section III.1.A.a). The π-electron ring current of pyridine, calculated from the diamagnetic anisotropy derived from experimental Cotton-Mouton constants, was found to be slightly larger than that of benzene and appreciably larger than that of the five-membered heteroaromatic compounds (34).

As part of an exhaustive study of the physical properties of organic molecules, 25 pyridine derivatives were examined and the following properties were measured: refractive indices (at four wavelengths), densities, surface tension (at four temperatures), parachors, molecular refractivities, molecular refraction coefficients, ultraviolet, and infrared spectra (158). The gas-liquid chromatographic (GLC) behavior of a number of pyridine derivatives has been examined and the trends discussed (159, 160). In addition, GLC separations have been used extensively in a number of other studies (e.g., Ref. 161 and references cited therein).

III. Chemical Reactivity

1. Theoretical Treatment

The development of new methods for the calculation of the electronic properties of organic molecules has more than kept pace with the growth of pyridine chemistry. Most of the new methods have been applied to pyridine or its derivatives with varying results.

A. *Pyridines*

a. ELECTRONIC PROPERTIES. The values of the formal atomic charges shown in **IA-13** were derived from a semi-empirical LCAO calculation using a monoelectronic effective Hartree-Fock Hamiltonian without neglecting σ or π

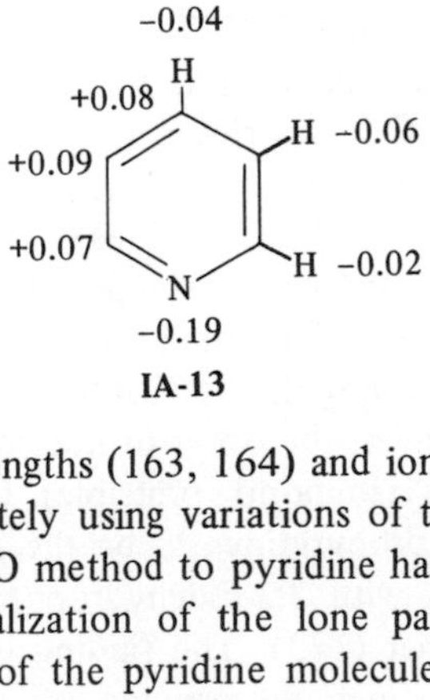

-0.04
H
+0.08
H -0.06
+0.09
+0.07
H -0.02
N
-0.19

IA-13

interactions (162). Bond lengths (163, 164) and ionization potentials (163) can be calculated quite accurately using variations of the SCF-LCAO-MO method.

Application of the CNDO method to pyridine has led to the conclusion that there is significant delocalization of the lone pair of electrons of nitrogen throughout the σ-orbitals of the pyridine molecule. This delocalization occurs predominantly through the $2p_y$ orbitals of the carbon atoms α and β to the nitrogen. Calculations lead to a surprisingly small amount of s character (14%) in the pyridine lone pair orbital (157) (see also Section III.4).

Effects of substituents on transition energies have been calculated by the one-electron approximation (45) and the effect of hyperconjugation was found to be larger than the inductive effect (165). The effect of the various halogens on the transition energies has been calculated by the Pariser-Pople method (166, 167).

Calculations have been presented that would indicate that not only are 2- and 4-hydroxypyridines, but so are also 3-hydroxypyridines, more accurately depicted as the oxo-forms (168, 169), even if this requires invoking the concept of a pentacovalent nitrogen (**IA-14a**) (169).

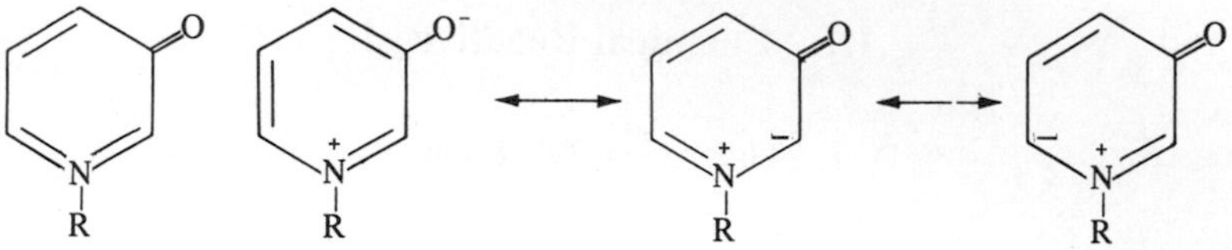

IA-14a **IA-14b** **IA-14c**

The position of $\nu_{C=O}$ (1590 cm^{-1}) for "1-methyl-3-pyridone" is between the positions for the 2- (1659 cm^{-1}) and 4-isomers (1575 cm^{-1}) (168). Structures such as **IA-14b** and **c** could account for the latter observation. More discussion of this point is given in Chapter XII. The inductive and mesomeric effects of substituents in pyridine derivatives have been estimated using the simple Hückel method, and found to be basically no different than in the case of benzene derivatives (170).

b. SUBSTITUTION REACTIONS. Most calculations of the π-electron densities at the nuclear positions of pyridine find little difference between C-2 and C-4. It is well known, however, that nucleophilic substitution by strong nucleophiles occurs almost exclusively at C-2 in pyridine itself, and more often at C-2 than at C-4 in appropriate derivatives, whereas with weaker nucleophiles, the order of reactivities is usually reversed. EHT calculations of the total electron densities at nuclear positions (i.e., both σ- and π- electron densities) find a 6.8% lower charge density at C-2 than at C-4 (171). In general, the pattern of substitution observed for strong nucleophiles will follow the pattern of ground state electron densities while, with weaker nucleophiles, the order or nucleophilic localization energies will usually be followed (Hammond principle) (172). The same type of calculation finds 3,4-didehydropyridines to be the most stable of the various possible didehydropyridines, with 2,3-didehydropyridine being less stable than even a *meta*- or a *para*-isomer (173). The finding is in good accord with the dearth of evidence for a 2,3-didehydropyridine generated by the action of base on a 3-halopyridine. These conditions generally lead to a 3,4-pyridyne.

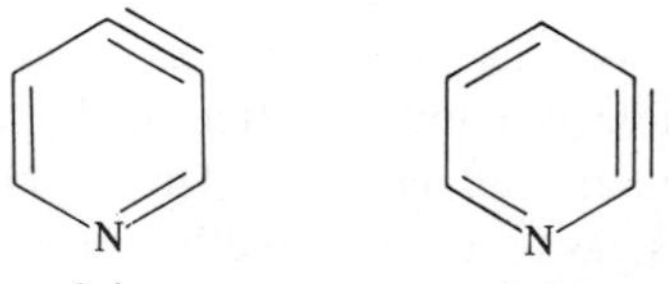

3,4- 2,3-

An unspecified method of calculation predicted that substitution would be favored at C-2 of 3-substituted pyridines rather than at C-4 or C-6 because of the "higher energy level of the intermediate from the 4- or 6-substitution" (174), a result in accord with experimental findings (172).

The susceptibility of C-3 toward electrophilic attack is predicted on the basis of the π-electron densities calculated by the LCAO-MO and PPP methods (175).

Kuthan and his co-workers (176) have attempted to correlate various theoretical parameters with experimental data on the rates of homolytic phenylation as follows:

$$\begin{aligned}\log k_i &= 23.71\, F - 9.39 \\ &= 13.35\, S_r - 11.10 \\ &= -6.37\, L_r - 16.27\end{aligned}$$

where k_i = relative rate constant for homolytic phenylation at C_i
F = free valence
S_r = chemical superdelocalizability
L_r = Wheland atom localization energy for radical substitution

The correlation is reasonable for pyridine, 3-picoline, and 2,6-lutidine, but only S_r correlates well for 2- and 4-picoline. Indices of reactivity and the nature of the transition states in homolytic phenylation and alkylation of pyridine and pyridinium salts have recently been calculated (177).

B. *Pyridinium Salts and Pyridine-1-oxides*

In accord with most experimental evidence, most calculations of susceptibility toward nucleophilic attack [EHT (178), CNDO/2 (179)] find little change to result from quaternization of the ring nitrogen atom, and substitution is still expected to occur at C-2. The Pariser-Pople method has been used to calculate the electronic transition energies and electron densities of methylpyridinium salts. Best fit with the experimental u.v. spectrum, polarographic reduction, and charge-transfer phenomena was obtained by inclusion of the polarization of the σ-skeleton into the π-calculation (44). The net charges on pyridine derivatives and the corresponding $N{\rightarrow}SO_3$ salts have been compared (103), and the electron-attracting influence of the Lewis acid was demonstrated.

+0.020
-0.002
+0.031
N

+0.112
-0.009
+0.169
N
↓
SO_3

The concept of a cyclic π-septet (formed by the transition of a single electron into the lowest antibonding π^*-orbital of pyridine or its salt has been used to explain some reactions of pyridine and its salts) which occur predominantly at the 4-position (180) and the thermal transformations of pyridinium salts (181) (the Ladenburg rearrangement: see Section III.4.A).

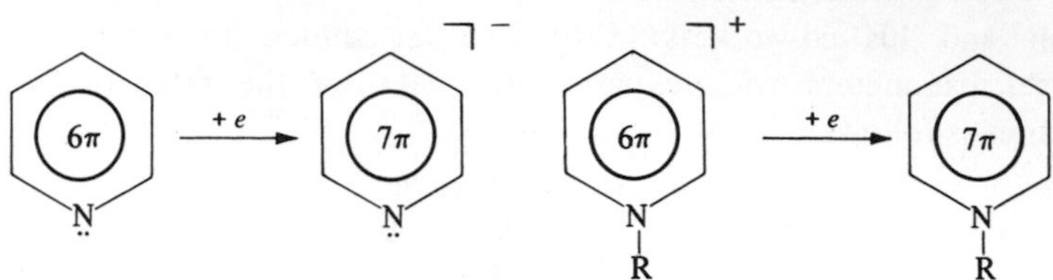

Calculation of the π-electron densities in pyridine-1-oxide using the simple LCAO method predicted nucleophilic attack at either C-2 or C-4, without a preference for either (28). A more recent investigation (PPP-SCF-CI method) (182) did yield the result that nucleophilic attack should occur preferentially at C-2, as is usually observed.

A CNDO-CI analysis of ground and excited state properties of pyridine-1-oxide has been carried out (36). The authors checked the validity of the postulated geometry of pyridine-1-oxide [d(C–N) = 1.40 Å, $d(\overset{+}{N}-\overset{-}{O})$ = 1.28 Å, d(C–C) = 1.40 Å, d(C–H) = 1.08 Å, all angles = 120°] by computing and comparing the results of the total dipole moment and the electronic absorption maxima with the experimental values: dipole moment μ_{theory} = +4.78; μ_{expt} = 4.78; electronic absorption (first excited band): theoretical 3.76 eV; expt. 3.81 eV. The total ground state charge distribution in pyridine-1-oxide was as follows:

0.970
H 4.058
0.963 H 4.004
4.026
0.948 H
$\overset{+}{N}$ 4.526
$\overset{-}{O}$ 6.566

The 3,4-didehydropyridinium ion is calculated to be the most stable of the cationic pyridynes (173), but the predicted stability difference between the 3,4-isomer and the 2,3-isomer is small, apparently because the repulsion between the carbanion at C-2 and the nitrogen lone pair is removed in the quaternary compound, an argument which is also valid for the α-carbanions derived from the *N*-oxides (183).

2. Additions to the Pyridine System

A. *Pyridines*

A number of what appear to be nucleophilic substitution reactions of pyridines and pyridine-1-oxides involve initial complexation at the ring nitrogen, attack at the α-position with the formation of a cyclic intermediate, and aromatization with ring opening of the intermediate (see Section III.5.D for examples). In another class of reactions, the products do not undergo ring opening and the overall result is the formation of a second ring fused to the pyridine ring. The additions may occur in a stepwise manner as described above, or may be concerted as for example, in a 1,3-dipolar addition.

One of the oldest examples of this latter type is the reaction of pyridine with dimethyl acetylenedicarboxylate (DMA), first reported by Diels and Alder. The correct structures of the products have now been shown to be **IA-15** and **IA-16** (Ref. 184, and references cited therein).

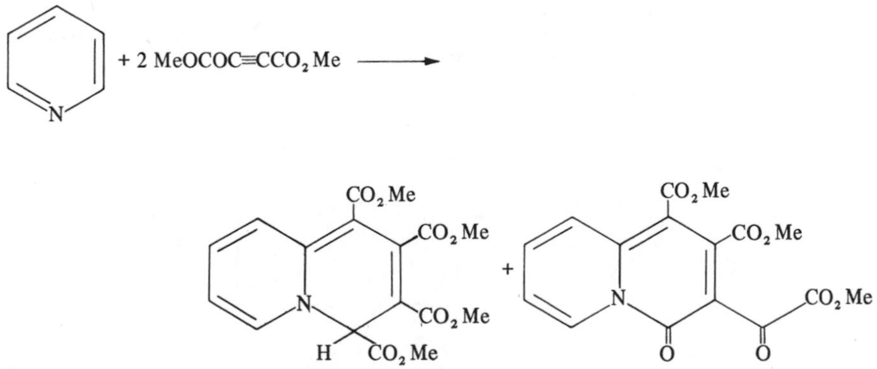

N + 2 MeOCOC≡CCO_2Me ⟶

CO_2Me CO_2Me N CO_2Me H CO_2Me
IA-15

+

CO_2Me CO_2Me N CO_2Me O O
IA-16

On the other hand, the addition of one equivalent of DMA to 2-phenylethynylpyridine (**IA-17**) gave a pyrido[*a*]pyrrole derivative (**IA-18**), possibly by the mechanism shown (185).

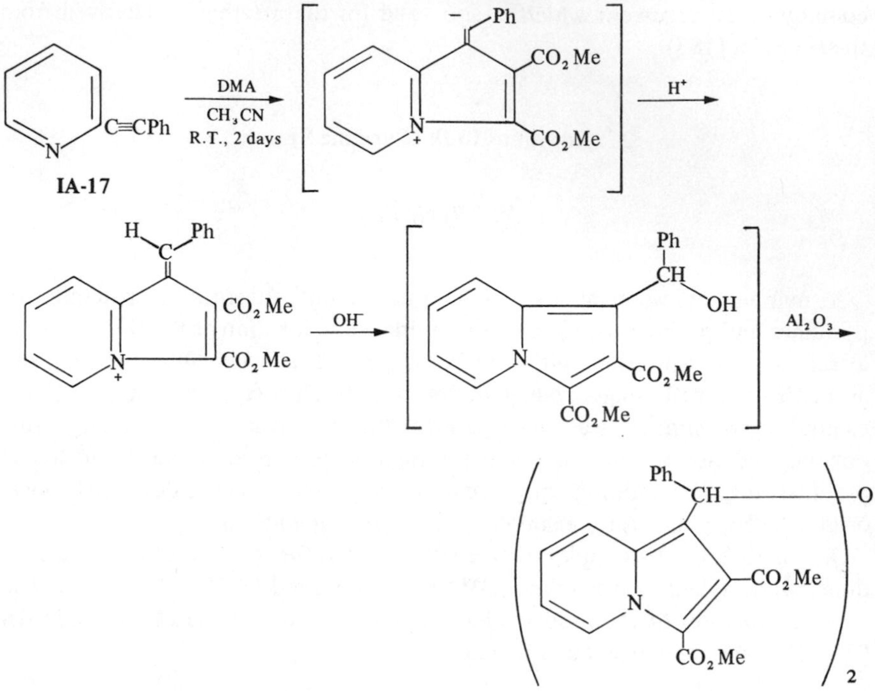

IA-17

DMA, CH_3CN, R.T., 2 days; H^+; OH^-; Al_2O_3

IA-18

Stable dihydroindolizines may be prepared by intramolecular 1,5-cyclization of ylides (186).

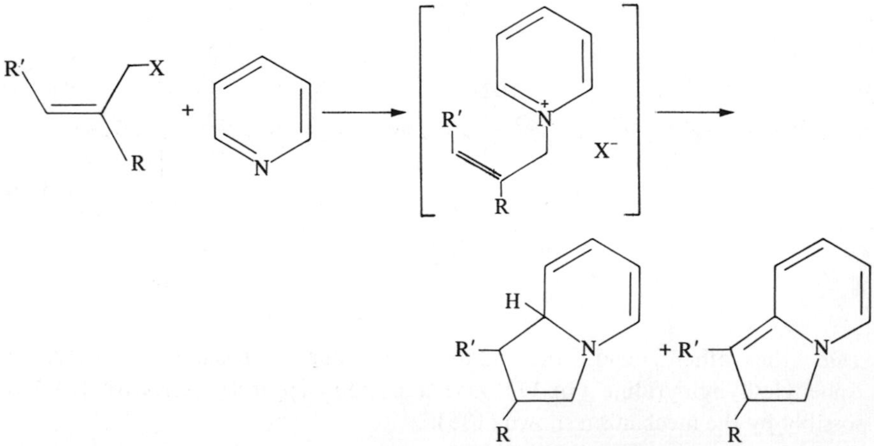

Pyridine, 4-picoline, and 3,5-lutidine react with sulfene to give a bicyclic adduct **(IA-19)** (187), a product analogous to those formed by the reaction between sulfene and enamines. This reaction contrasts with the failure of pyridine to react with the adduct between ketene-sulfur dioxide (188), even though this adduct reacts smoothly with Schiff bases to give thiazolone *S,S*-dioxides **(IA-20)** (189).

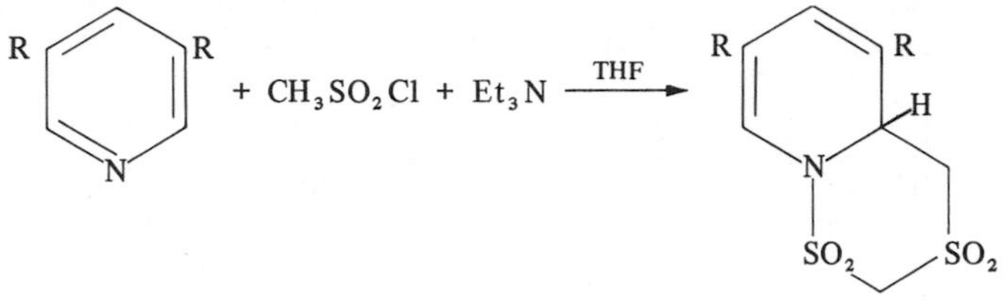

$$\text{R,R-pyridine} + CH_3SO_2Cl + Et_3N \xrightarrow{\text{THF}} \textbf{IA-19}$$

IA-19

$$\text{RCH=NR}' + O{=}\bar{C}{-}CH_2{-}\overset{+}{S}O_2 \longrightarrow \textbf{IA-20}$$

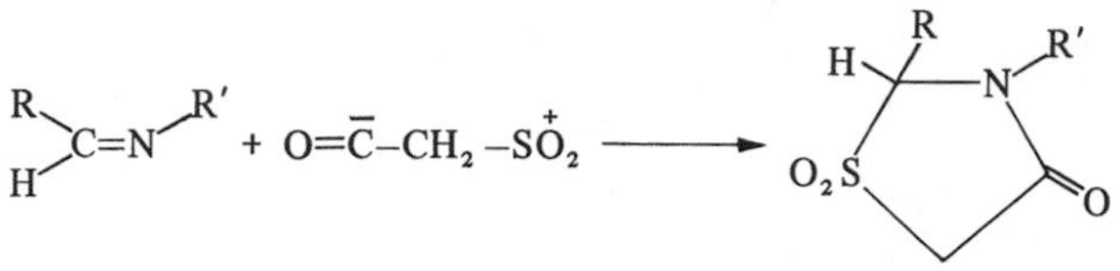

IA-20

Photolysis of a methanolic solution of 5,6,7,8-tetrahydro-2-quinolone **(IA-21)** and diphenylacetylene gives not only the expected photodimer **(IA-22)** but also the tetracyclo-compound **(IA-23)** which must be formed from the product of the initial thermal Diels-Alder reaction (190).

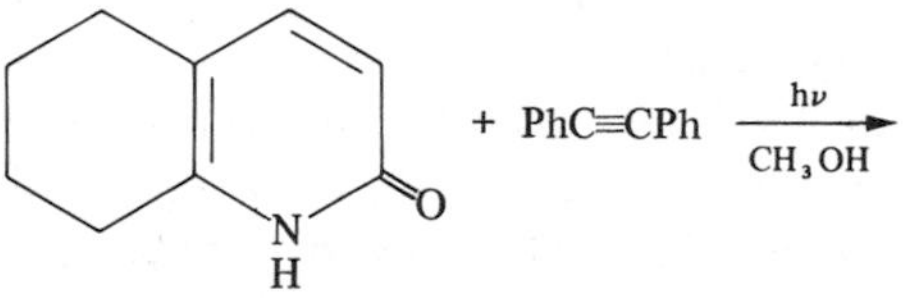

$$\textbf{IA-21} + PhC{\equiv}CPh \xrightarrow[CH_3OH]{h\nu}$$

IA-21

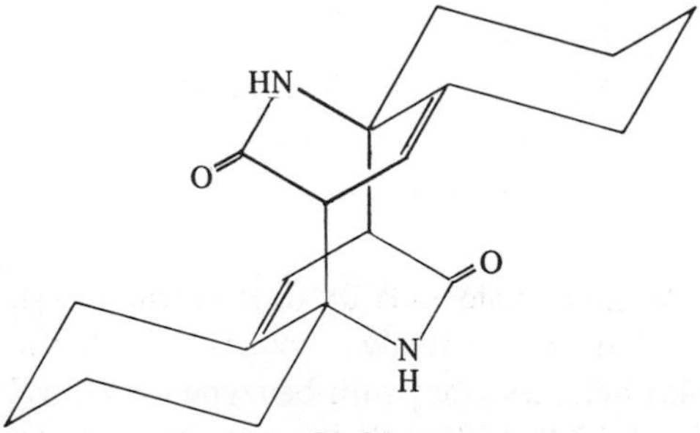

IA-22 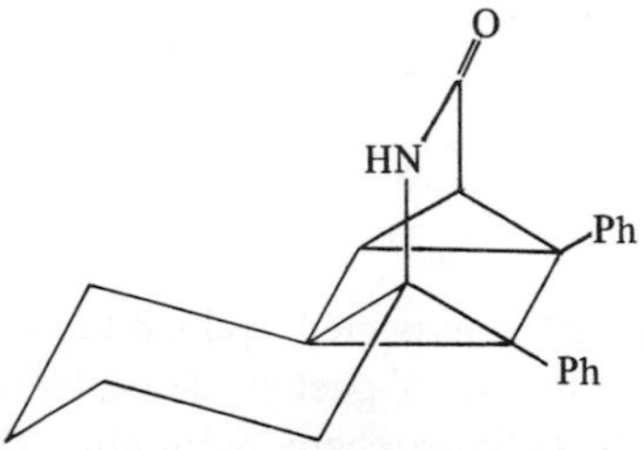 + **IA-23**

1-Methyl-2(1*H*)-pyridone undergoes the Diels-Alder reaction with maleic anhydride in boiling toluene (191) to give the expected *endo*-dicarboxylic anhydride. The reaction with fumaronitrile gave only a 3% yield of adduct.

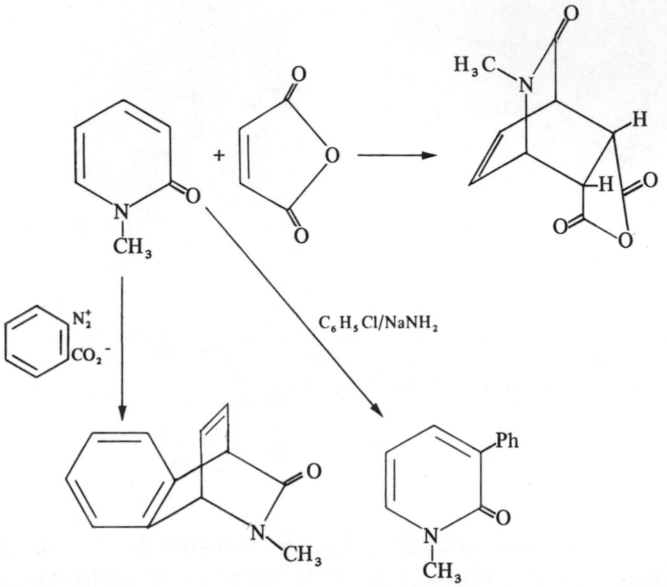

1-Methyl-2-pyridone reacts similarly with benzyne (from diazotized anthranilic acid) to give the Diels-Alder product (192). With chlorobenzene and sodamide, however, it gives 1-methyl-3-phenyl-2-pyridone. On the other hand, 2-pyridone itself reacts with benzyne to give 2-phenoxypyridine (192).

Anhydro-3-hydroxy-1-methylpyridinium hydroxide reacts with electron-deficient olefins such as acrylonitrile or methyl acrylate to yield 8-azabicyclo [3.2.1] octane derivatives (193). The corresponding *N*-phenylpyridinium betaine

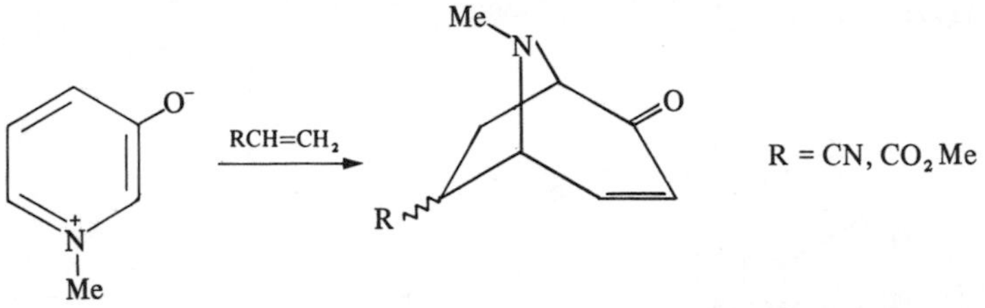

$$\xrightarrow{RCH=CH_2} \qquad R = CN, CO_2Me$$

gave the expected cycloadducts with *N*-phenylmaleimide or acrylonitrile (194). Both the *N*-phenyl- (**IA-24d**) and what was initially thought to be the *N*-2,4-dinitrophenylpyridinium (**IA-24a**) betaines react with benzyne *via* a novel benzyne displacement reaction to give **IA-24b** (195a). It has now been shown

(195b) that **IA-24a** first rearranges to **IA-24c** before undergoing *N*-phenylation followed by elimination of the 2,4-dinitrophenyl group to give **IA-24d**.

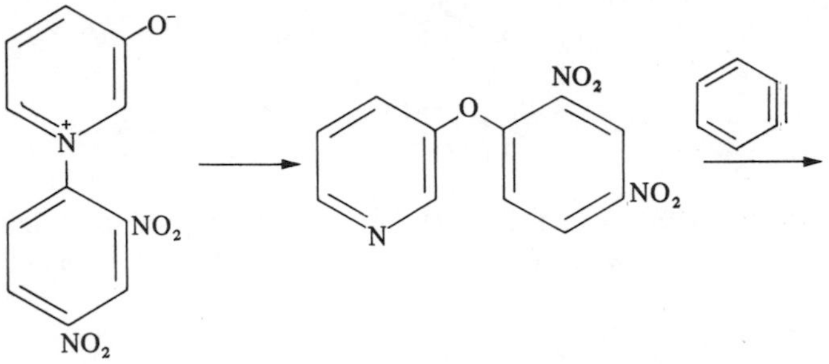

IA-24a **IA-24c**

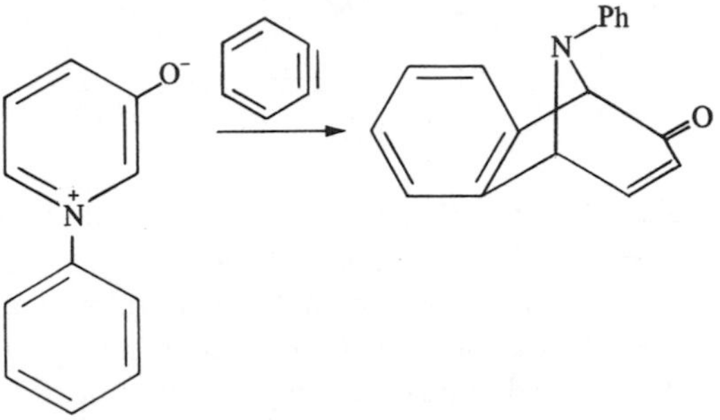

IA-24d **IA-24b**

Reaction of 2,2′-azopyridine with phenylcarbene leads to the 3-(2-pyridyl)pyrido[1,2-*a*] triazoline (196). Treatment of the tosylhydrazone **IA-25a** with base leads to an intramolecular cyclization product (**IA-25b**) (197). Intramolecular cyclization of the alkyl diazo-compound is faster than carbene formation. The equilibrium between 2-azidopyridines (**IA-26**) and the tetrazolopyridine

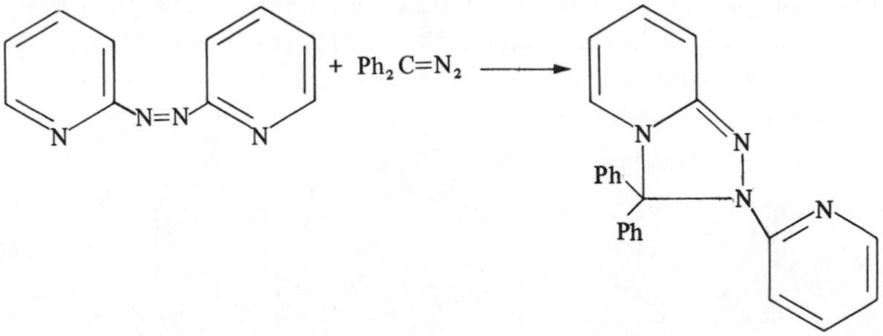

+ $Ph_2C{=}N_2$ →

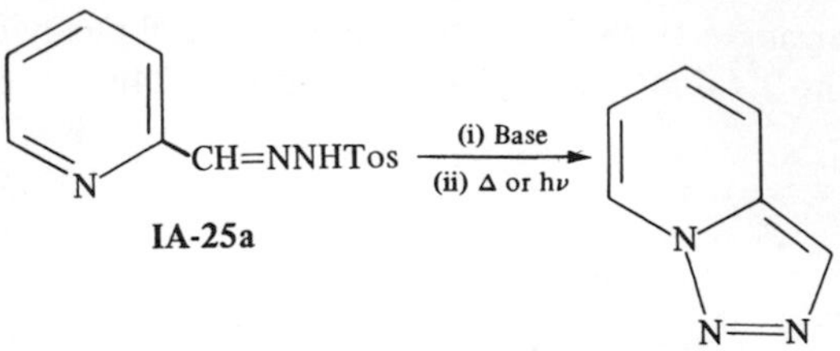

IA-25a

IA-25b

(**IA-27**) has been studied (198). When R = H or 5-Br, the equilibrium lies exclusively toward the tetrazole; when R = 6-Cl, K_T = 2.5–18.0, depending on

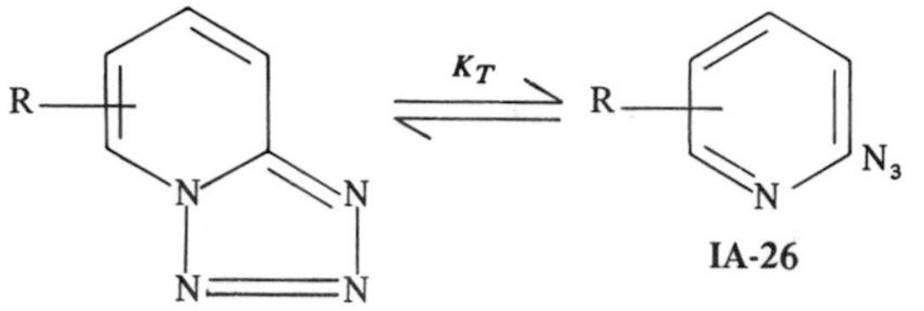

IA-27

IA-26

solvent – in $CHCl_3$ at 40°, the molecule exists exclusively in the azide form. When R = 5-NO_2, K_T = 0.83–3.84.

Photolysis of diazomethane in pyridine leads to a good yield of 2-picoline (199). It was suggested that this reaction proceeds through initial attack on nitrogen and intramolecular rearrangement (or 1,3-dipolar addition?), rather than *via* a free carbene.

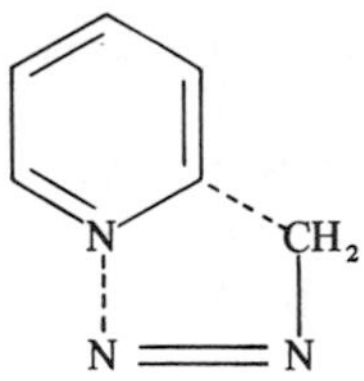

Pyridine and ketene react to give a tricyclic compound (**IA-28**) (200), and 2-aminopyridine (201) and 2-aminolutidines (202) react with diketene to give the pyrido[1,2-*a*]pyrimidone derivatives **IA-29**.

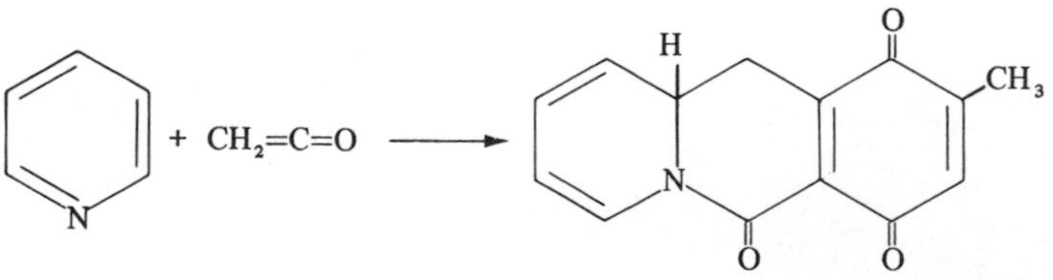

IA-28

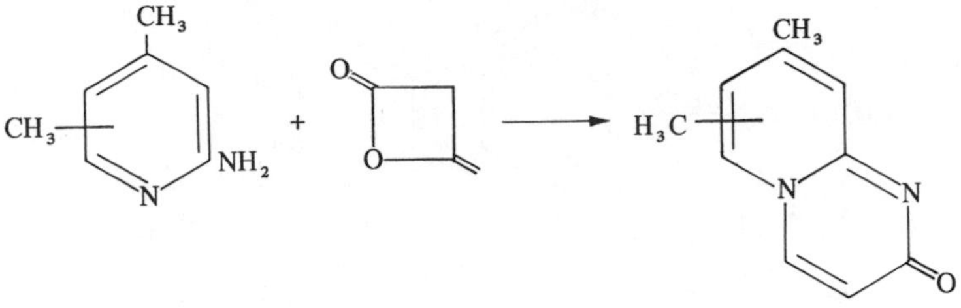

IA-29

The product obtained by the addition of sodium bisulfite to pyridine reacts with benzenediazonium salts to give 3-phenylazopyridine, and undergoes nitration at 20° to give a 15% yield of 3-nitropyridine (203). This constitutes the most convenient method of synthesizing this latter compound.

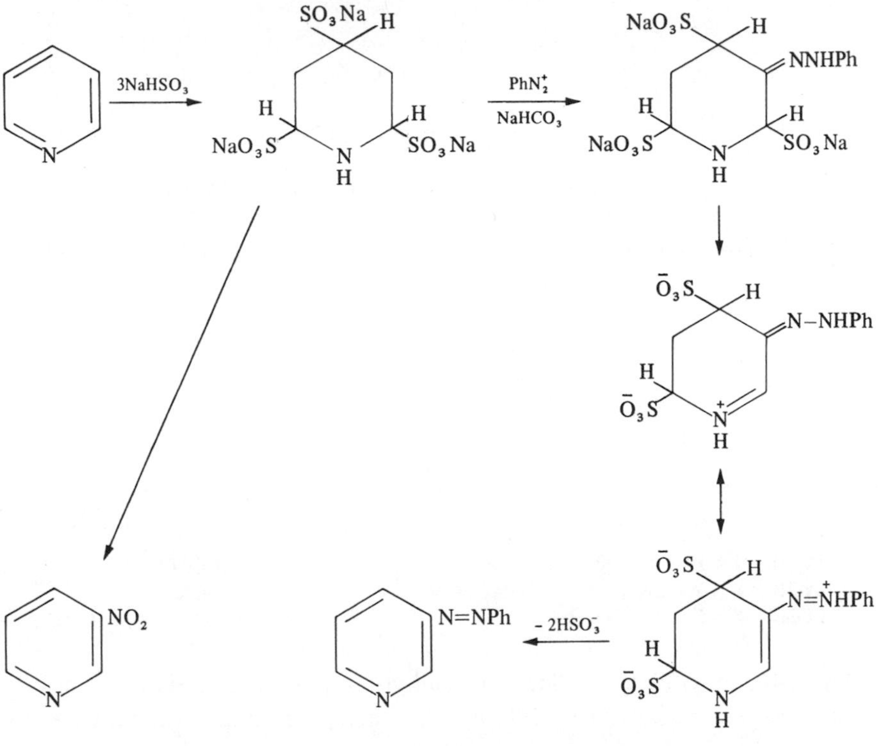

B. *Pyridine-1-oxides and Pyridinium Ylides*

A number of reactions of pyridine-1-oxides have been postulated to involve cyclic intermediates (see Section III.5.D.b.).

In contrast to its reactions with the free bases, addition of ketene to pyridine- or 2,6-lutidine-1-oxide leads only to alcohols and hydroxy derivatives (204).

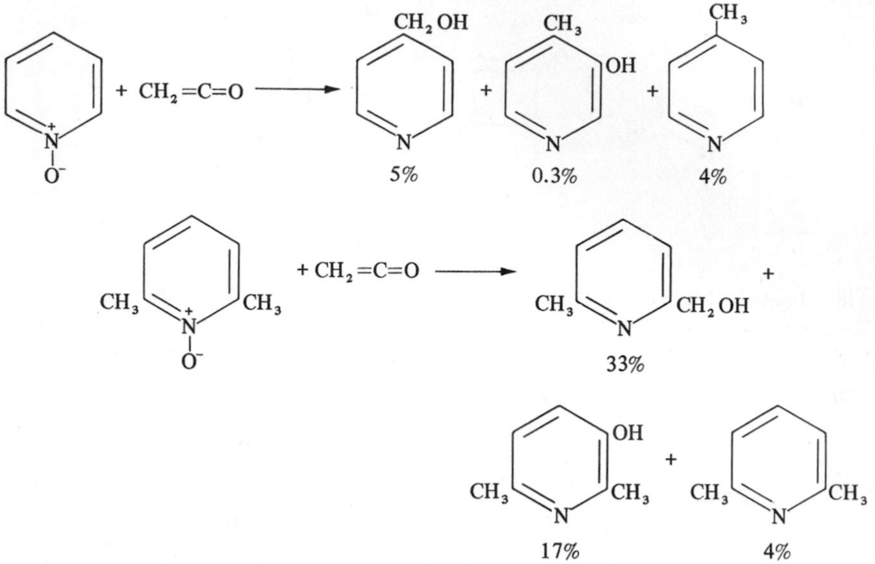

A number of pyridinium ylides undergo 1,3-dipolar cycloadditions with acetylenes such as dimethyl acetylenedicarboxylate (DMA). For example, **IA-30** and DMA give substituted pyrrocolines (Ref. 205, and ref. cited therein; Ref.

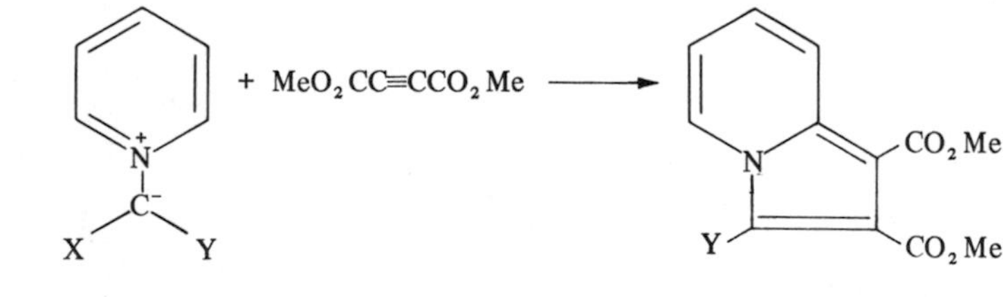

IA-30a X = H, Y = COPh
IA-30b X = Y = CN
IA-30c X = CN, Y = CO_2R

Y = COPh
Y = CN
Y = CN or CO_2R

206). 1-Iminopyridinium ylides and ethyl propiolate give 1-carbethoxy-3-azapyrrocoline (207). The addition of dimethyl acetylenedicarboxylate to

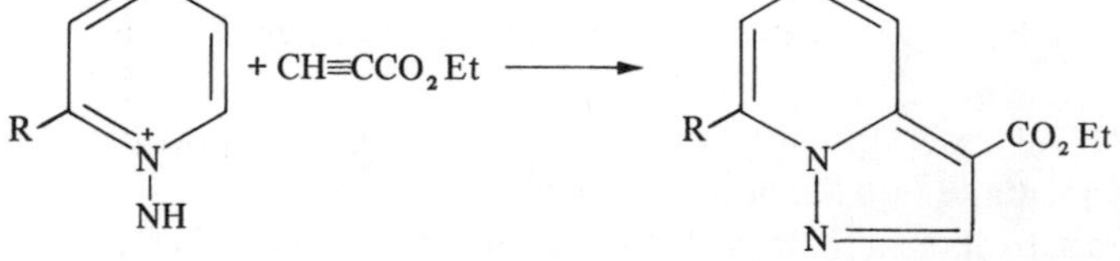

1-alkoxycarbonyliminopyridinium ylides give pyrazolopyridine derivatives (**IA-31**) which are said to rearrange to 3-substituted pyridines (see Section

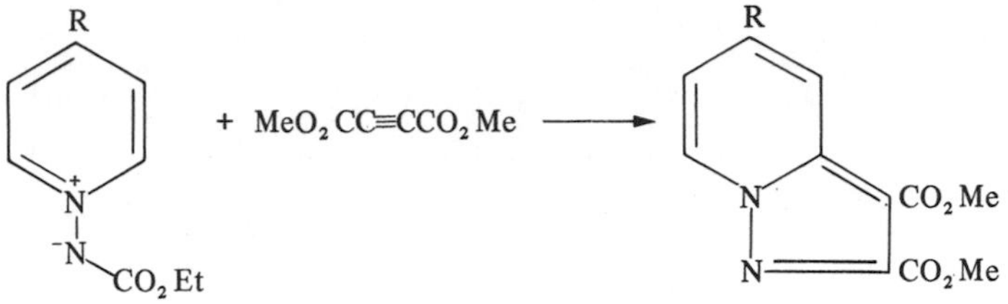

R, N^+, ^-N, CO_2Et + $MeO_2CC{\equiv}CCO_2Me$ ⟶ R, N, N, CO_2Me, CO_2Me

IA-31

III.5.D) (298, 209). Similarly, **IA-32** and dimethyl acetylenedicarboxylate give the pyrido[1,2-*a*] pyrrole **IA-33** (210).

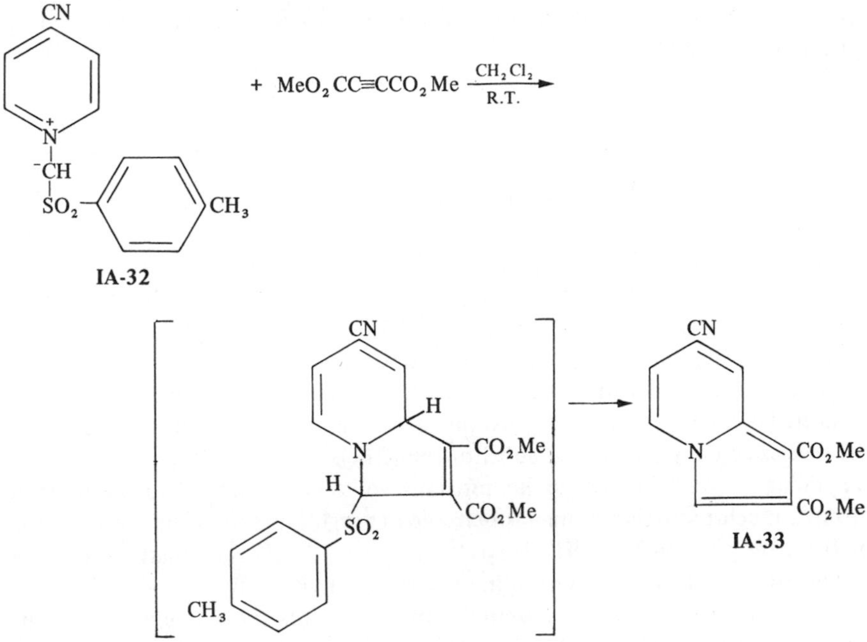

CN, N^+, ^-CH, SO_2, CH_3 + $MeO_2CC{\equiv}CCO_2Me$ $\xrightarrow[\text{R.T.}]{CH_2Cl_2}$

IA-32

[CN, H, N, CO_2Me, H, CO_2Me, SO_2, CH_3] ⟶ CN, N, CO_2Me, CO_2Me

IA-33

4-Cyanopyridine adds to 1-iminopyridinium betaine to give 2-(4-pyridyl)pyrido[1,2-*b*]-1,2,4-triazole, also obtained from 1-aminopyridinium iodide with aqueous potassium cyanide. In the latter case, nucleophilic addition of CN^- to the pyridinium salt first leads to 4-cyanopyridine which reacts further with the ylide (211).

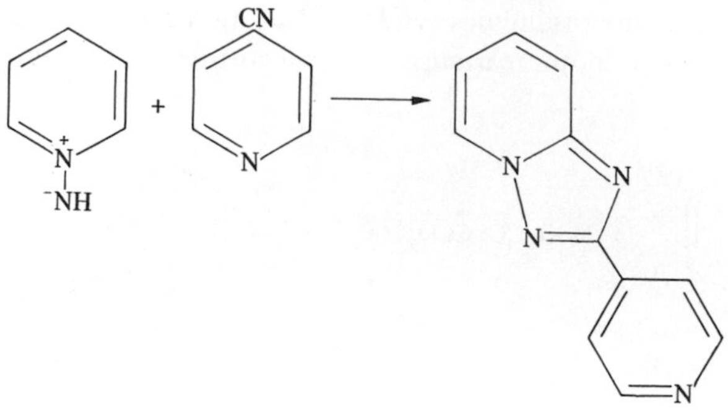

3. Electrophilic Substitutions

The kinetics of many substitution reactions have been studied in detail since the first edition was written and substituent effects put on a much more quantitative basis.

A. *Pyridines*

a. SULFONATION AND NITRATION. All of the picolines are readily sulfonated by a hot mixture of $HgSO_4$ and 20% oleum to give the *x*-methyl-5-pyridinesulfonic acid (212).

The mechanism of the nitration of pyridines and hydroxypyridines has been studied extensively in recent years. The relation between reaction rate and the Hammett acidity function (H_0) of the reaction medium has been examined, and the rate profiles of substitutions of the free bases under investigation have been compared with those of analogous quaternary salts to determine the nature of the species actually undergoing substitution (213).

Pyridine is well known to be resistant to electrophilic substitution. It is nitrated much more slowly than is benzene (a rough estimate of the deactivation of the *meta* position by =NH– is 10^{-12}) and more slowly even than is quinoline or imidazole (214). For the nitration of pyridine in 90% sulfuric acid at 25° $k_2 = <10^{-8}$ 1/(mole)(sec); while the first nitration is on the conjugate acid, the second is on the free base (214). This trend appears to be general. Pyridines with pK_a greater than +1 are nitrated as their cations with the orientation of the entering nitro group depending on the nature of the substituents present. Very weakly basic pyridines (pK_a less than −2.5) are nitrated as their free bases (215, 216).

The same trend appears in the nitration of pyridones (217). The 2-pyridones, which are nitrated as the free bases, are in the oxo-form (217, 218). Nitration of

3-hydroxypyridines (219, 220) provided the first examples of such electrophilic substitution at C-4, in contrast to earlier results (8). The effect of substituents upon orientation in nitration and sulfonation of pyridines has been reviewed (8, 172) (see also Chapters VIII and XV).

b. HALOGENATION. The rate of bromination of 2-aminopyridine derivatives (as their free bases) in aqueous sulfuric acid was found to be dependent on the concentration of bromide ion, high $[Br^-]$ decreasing the concentration of free bromine according to: $Br_2 + Br^- \rightleftharpoons Br_3^-$ (221). The facile β-chlorination of pyridine by SO_2Cl_2 or S_2Cl_2 (222) was later rediscovered (223). The

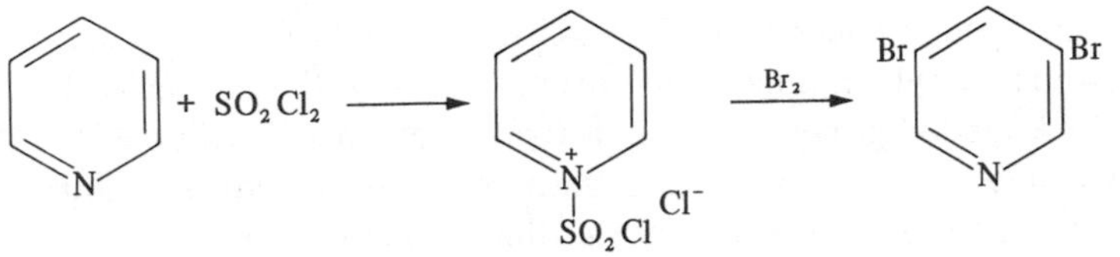

+ SO_2Cl_2 → $\overset{+}{N}$–SO_2Cl Cl^- —Br_2→ Br, Br, N

bromination and iodination of pyridine and alkylpyridines in oleum has been studied. The reactive substrate is the $C_5H_5N \rightarrow SO_3$ salt in these cases. Attack occurs mainly at the β-positions but, in some cases, substitution also occurs at C-2 and C-6. It has been suggested that the latter may be due to activation of the α-positions by hydrogen-bonding of the C-2 proton to the SO_3 group (103):

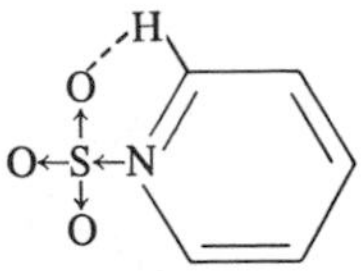

H, O, O←S←N, O

The orientation in the bromination of hydroxypyridines or pyridyl ethers depends on the tautomeric form the species is in. Bromination of 2,4-dihydroxy-, 2-hydroxy-4-ethoxy-, or 2-ethoxy-4-hydroxypyridine gave good yields of the 3-bromo-derivatives, resulting from activation of the 3-position by the adjacent carbonyl moiety (224). On the other hand, 2,4-diethoxypyridine, in which no carbonyl group is present, gave the 5-bromo-derivative; steric hindrance to bromination at C-3 may also play a role here (224). As in the case of bromination, nitration of 2(1*H*)-pyridones also favors the 3-position, while 2-alkoxypyridines are attacked at C-5. Chelation of the nitronium ion with the pyridone oxygen in the transition state [chelation has been suggested for nitration (225) of 2-pyridone] is less likely in the bromination. C-2 is activated similarly toward bromination in 3-hydroxypyridine (226). Halogenation is discussed further in Chapter VI and in reviews (8, 172).

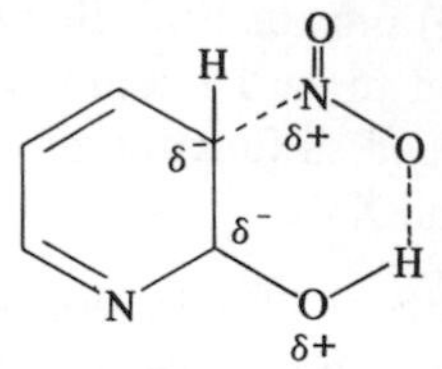

c. MISCELLANEOUS ELECTROPHILIC SUBSTITUTION REACTIONS. Acid-catalyzed hydrogen exchange does not occur with pyridine and 4-picoline but takes place smoothly at C-3 with 2,6-lutidine and 2,4,6-collidine (227). The conjugate acid is the species undergoing H-T exchange as indicated by the fact that (*a*) 1,2,4,6-tetramethylpyridinium ion reacted slightly faster than 2,4,6-collidine since the *N*-methyl group is slightly activating compared with $=\overset{+}{N}H-$; (*b*) log *k* increased linearly with $-H_0$, the magnitude of the straight line slopes being of the order expected for reaction *via* the pyridinium ion; and (*c*) collision frequency factors are inconsistent with the concentration of free base present.

The rate of acid-catalyzed hydrogen-deuterium exchange of 2- and 4-pyridones is dependent on H_0, with the exchange occuring on the free base throughout most of the acidity range (228). The rate-profiles for acid-catalyzed H–D exchange at the 3-position of 2-amino-5-chloropyridine, 2-amino-3-methylpyridine, and 5-chloro-2-dimethylaminopyridine have been reported (229). 2-Amino-3-methylpyridine undergoes exchange as the conjugate acid:

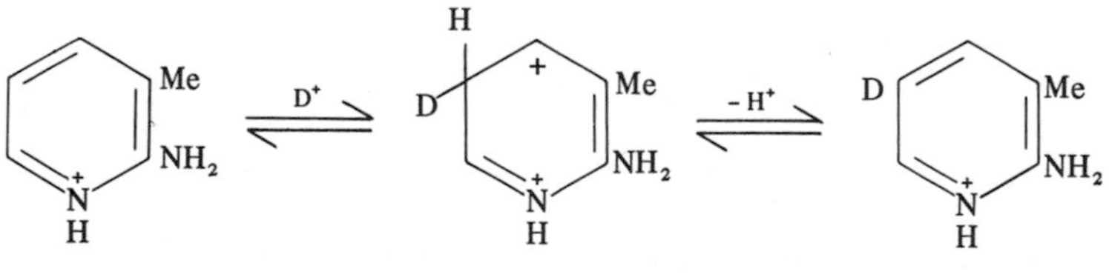

2-Amino-5-chloropyridine and 5-chloro-2-dimethylaminopyridine react as the free base at acidities less than $D_0 = -2$, and as the conjugate acids at higher acidities.

A recent study revised the old finding (230, 231) that only *O*-alkylation occurred in the reaction of 2-pyridone with diazomethane. The *N*-methyl isomer is actually the major product (232). The reaction of diazomethane with 6-hydroxy-2-pyridone gives, in addition to the expected *N*- and *O*-alkylation products, a novel hydrazone **(IA-34)** which results from electrophilic substitution at C-3 (233). Similar substitution occurs in the reaction of benzenediazonium chloride with 6-hydroxy-2-pyridone and with 2-methyl-4-pyridone (234).

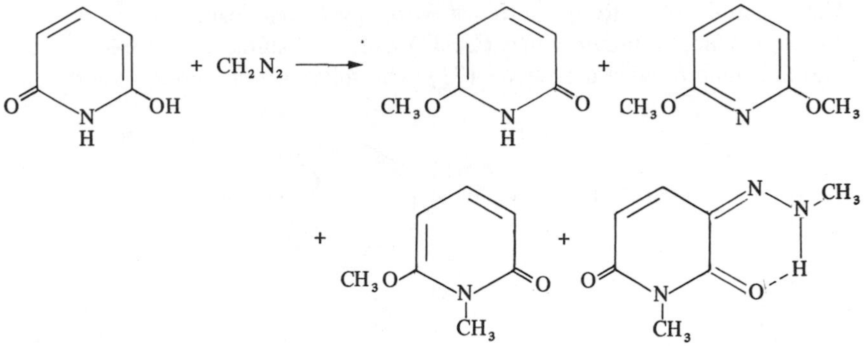

O N(H) OH + CH_2N_2 → CH_3O N(H) O + CH_3O N OCH_3

+ CH_3O N(CH_3) O + O N(CH_3) O N N(CH_3) H

IA-34

2-(Trimethylsilyl)-, 2-(trimethylgermyl)-, and 2-(trimethylstannyl)pyridine are solvolysed by water and by alcohols, as are 2-silylpyridines bearing groups other than methyl on the silicon (235, 236). The effects of ring methyl substituents upon the rates of methanolysis of the 2-(trimethylsilyl) derivatives gave the following relative reactivities: 3-Me < 5-Me < H < 6-Me < 4-Me. For hydrolysis, the relative rates were 5-Me < H < 3-Me < 6-Me < 4-Me (235). A mechanism was proposed involving rate-determining nucleophilic attack on the silicon by the oxygen of a hydrogen-bonded solvent molecule through a five-centered cyclic activated complex (**IA-35**). Support for such a complex comes from the

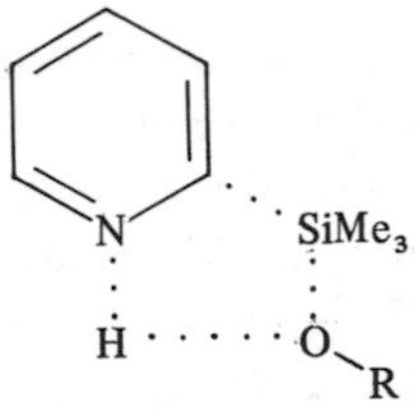

N SiMe$_3$ H O R

IA-35

observation that (+)-2-(methyl-1-naphthylphenylsilyl)pyridine reacts with methanol or water to give pyridine with retention of configuration at silicon (237). Cleavage of the Si–C bond in 2-(trimethylsilyl)pyridine can be effected with benzaldehyde and yields **IA-36** (238).

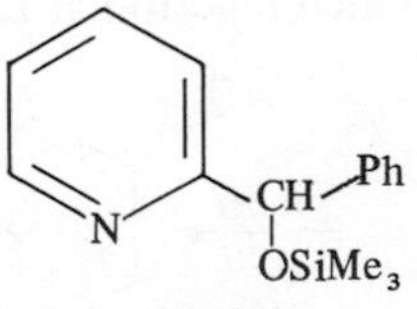

N CH Ph OSiMe$_3$

IA-36

The reaction of sulfonyl nitrenes with pyridines leads to mixtures of *N*-sulfonylaminopyridinium ylides (**IA-37**) and, with suitably activated rings, to 3-sulfonylaminopyridine derivatives (71). The substitution products are thought

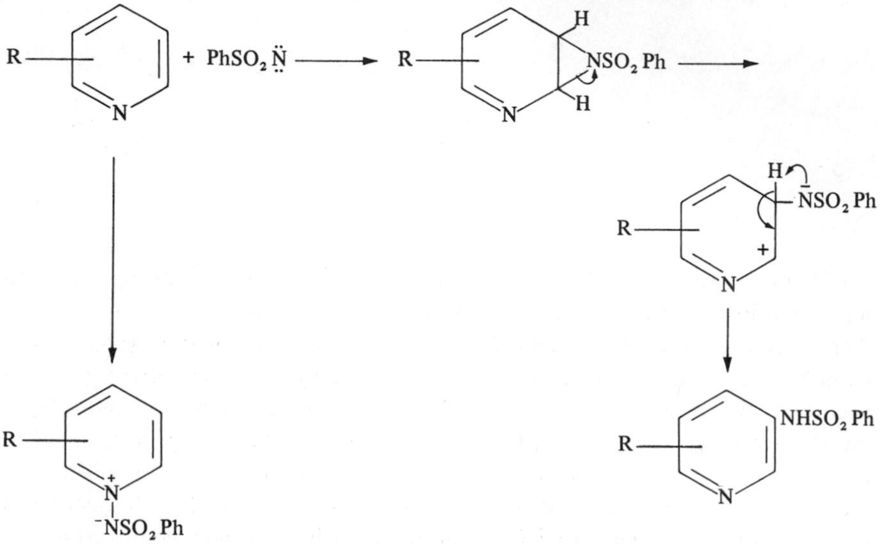

IA-37

to be formed in a rate-determining addition to the "2,3-double bond," followed by fast ring-opening and proton-shift to give the 3-sulfonylaminopyridine.

Photodeboronation of pyridineboronic acids (**IA-38**) can be achieved by irradiation in methanol solution (239). Hydroxyl groups activate the ring

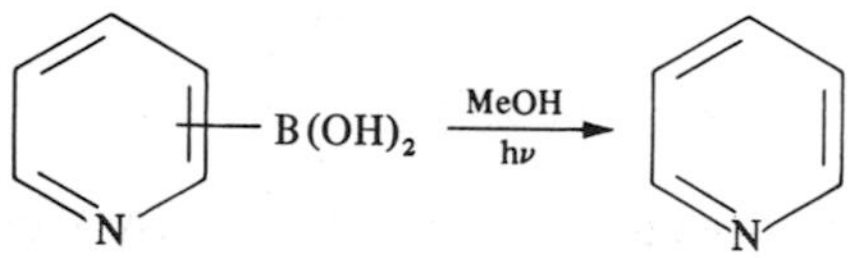

IA-38

sufficiently that 3-hydroxypyridine undergoes hydroxymethylation under basic conditions, attack taking place preferentially at C-2 (240). 3-Hydroxypyridine derivatives also undergo the Mannich reaction at the 2-position (241).

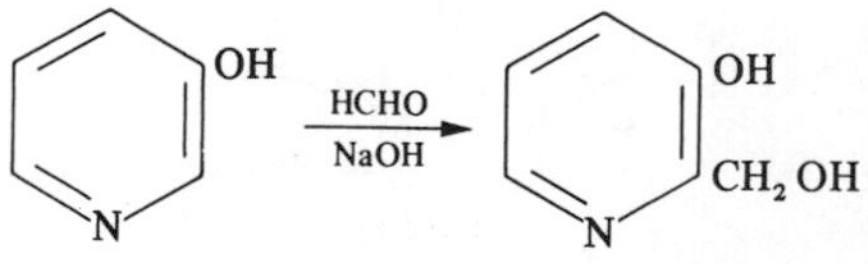

B. *Pyridine-1-Oxides*

Substituent effects on orientation in electrophilic substitution in pyridine-1-oxides have been discussed qualitatively (172). To explain the position of attack of the *free base* by various reagents it has been postulated that very soft reagents (as in mercuration) attack position 2, rather hard reagents (nitration) position 4, and very hard ones (sulfonation) position 3 (242). Attack at C-3 usually takes place in the *O*-protonated or complexed species, however, and not on the free base.

a. SULFONATION AND NITRATION. Sulfonation of pyridine-1-oxide under vigorous conditions occurs mainly at C-3, only small amounts of the other two isomers being formed (α: γ-ratio, 1: 2) (243, 244). It has been suggested that in fuming sulfuric acid a complex is formed between pyridine-1-oxide and SO_3, resulting in greater deactivation of the 2-, 4-, and 6-positions than of the 3-position.

Mechanistic studies of the nitration of pyridine-1-oxide have compared the rate of reaction of the *N*-oxide with an analogous quaternary salt, usually the *N*-methoxy compound. The nitration of pyridine is more than 530 times as fast as the nitration of the *N*-methoxypyridinium ion (245, 246), indicating that it is the free base of pyridine-1-oxide which undergoes nitration. Other *N*-oxides which undergo nitration as the free bases are invariably substituted at C-4 (247, 248). Nitration at C-3 and C-2 involves reaction of the conjugate acid (**IA-39**). Nitration of pyridine-1-oxide at the 3-position was achieved in the case of

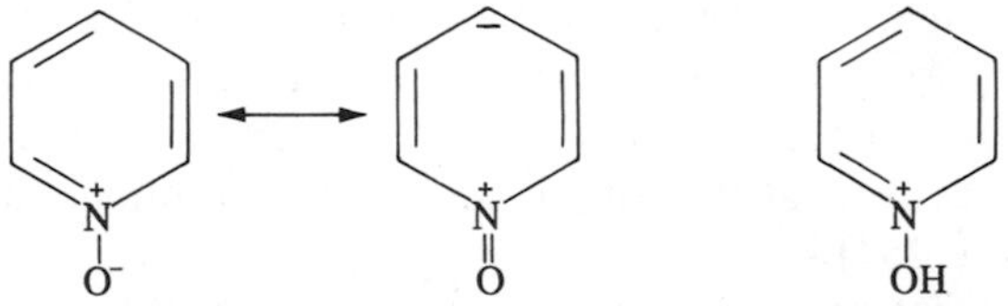

IA-39

2,6-dimethyl-4-methoxypyridine-1-oxide (**IA-40**), and the rate profile indicated that nitration of the conjugate acid was occurring (247). Both the greater ease of

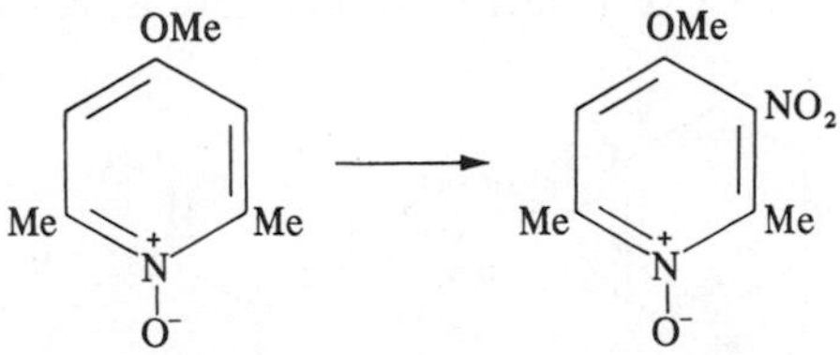

IA-40

nitration of free pyridine-1-oxide than of pyridine and the fact that substitution occurs at C-4 are consistent with the concept of mesomeric "back donation" to C-4 by the *N*-oxide function.

These results correlate well with those of acid-catalyzed hydrogen-deuterium exchange experiments. Exchange at C-2 and C-4 occurs *via* the free base, but the conjugate acid is the reactive species for exchange at C-3 (249). At low acidities, and especially at high p*D* values, exchange is base-catalyzed at all positions, with $k_{2,6} > 100k_4$.

b. HALOGENATION. Bromination of pyridine-1-oxide is a complicated reaction. Low yields of 2- and 4-bromopyridine-1-oxide are obtained in 90% sulfuric acid at 200° in the presence of silver sulfate (250). In 65% fuming sulfuric acid, however, 3-bromopyridine-1-oxide is formed in 55% yield, along with a complicated mixture of *ortho*- and *para*-dibromo products. The β-substitution is thought to occur because of complexing of the *N*-oxide oxygen with SO_3, preventing "back donation" and leading to deactivation of C-2 and

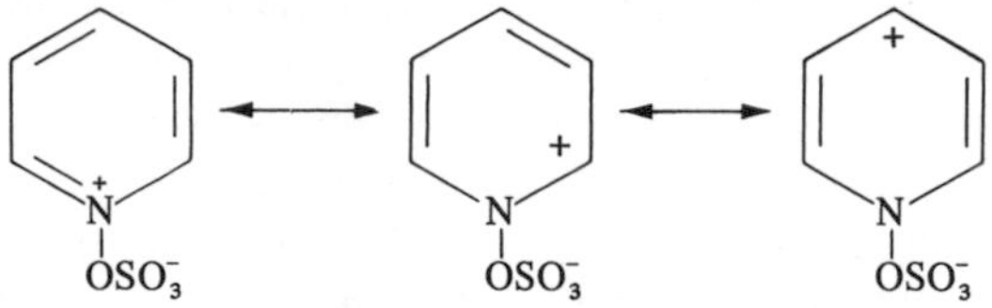

C-4. Subsequent bromination of the monobromo isomers gave only *ortho*- and *para*-dibromo products, indicating either that the directive effect of Br is stronger than that of $=\overset{+}{N}-OSO_3^-$, or that the presence of a bromo substituent reduces the basicity of the *N*-oxide sufficiently that complex formation does not occur. Bromination proceeds more readily in this medium than in 90% sulfuric acid, and this has been attributed to the presence of bromonium ions in high concentration (cf. 251).

Under quite different conditions (Br_2, Ac_2O, and NaOAc in $CHCl_3$), 35% of 3,5-dibromopyridine-1-oxide was obtained (252), indicating that the species undergoing dibromination is the *O*-acetate ($=\overset{+}{N}-OAc$).

c. MANNICH REACTION. 3-Hydroxypyridine-1-oxides undergo aminomethylation preferentially at C-2 unless that position is blocked, in which case

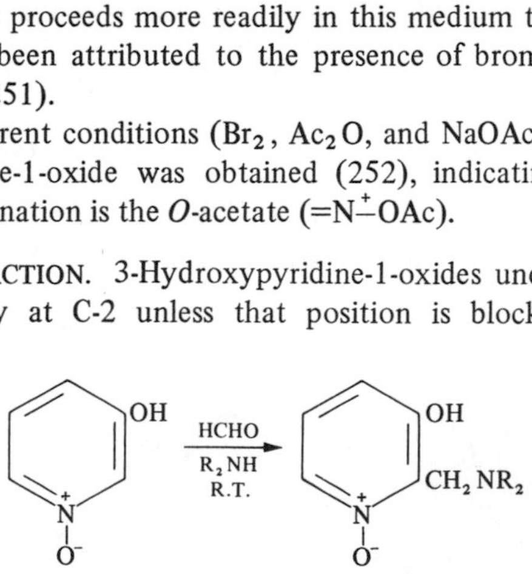

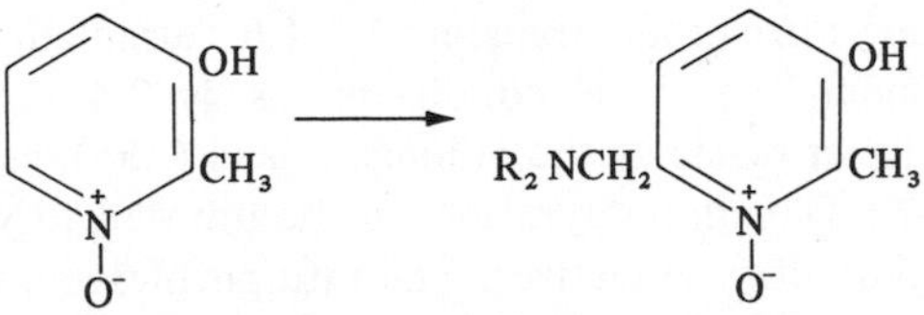

OH CH3 → R2NCH2 OH CH3 (N-oxide structures)

substitution occurs at C-6 (253). If both the 2- and 6-position are blocked, aminomethylation occurs at C-4.

C. *Pyridinium Compounds*

Pyridine is used widely as a catalyst in acetylation reactions. Although *N*-acetylpyridinium acetate **(IA-41)** had long been assumed to be an intermediate in these reactions, its presence was only recently demonstrated by following its formation by u.v. monitoring of the acetylation of anisidine in a stopped-flow apparatus (254). The elusiveness of this intermediate is due to its rapid reverse reaction: $k_1 = 80\ M^{-1}$, $k_{-1} = 900\ M^{-1}$.

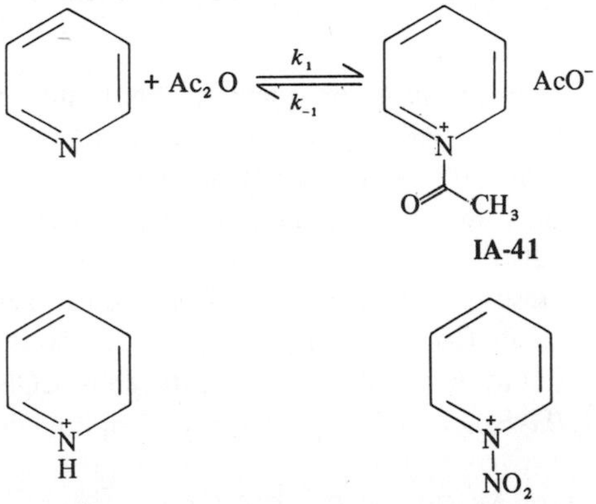

$$\text{Pyridine} + Ac_2O \underset{k_{-1}}{\overset{k_1}{\rightleftharpoons}} \text{N-acetylpyridinium}\ AcO^-$$

IA-41

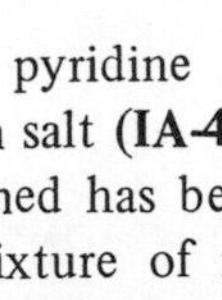

IA-42 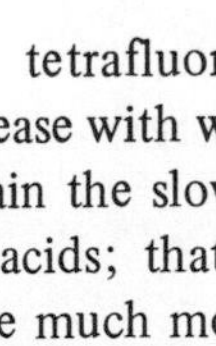 **IA-43**

Nitration of pyridine with nitronium tetrafluoroborate gives the *N*-nitropyridinium salt **(IA-43)** (255, 256). The ease with which this exceedingly stable salt is formed has been invoked to explain the slow rate of nitration of pyridine in a mixture of nitric and sulfuric acids; that is, not only is the pyridinium ion **IA-42** formed, but so also is the much more deactivated **IA-43** (255). 1-Nitro-2-methylpyridinium tetrafluoroborate, but not 1-nitropyridinium tetrafluoroborate, is a mild but powerful agent for the nitration of aromatic hydrocarbons (257). Apparently, the 2-methyl-group sterically inhibits coplanarity of the nitro-group and the pyridine ring, thus weakening the N–N bond, and facilitating nitro-group transfer. 1,2,4,6-Tetramethylpyridinium ion can be nitrated under the same conditions as is 2,4,6-trimethylpyridine, indicating that the latter reacts as the conjugate acid (258), as is confirmed by the kinetic results. 2,6-Dimethoxypyridine can be nitrated to yield successively the 3-mono- and 3,5-di-nitro derivatives. The rate profiles show that the first of these nitrations occurs on the 2,6-dimethoxypyridinium ion, and that the introduction of the first nitro-group lowers the basicity so much that sufficient amounts of the free base are present for it to compete successfully with the corresponding conjugate acid in the second nitration stage (213).

MeO, OMe, $\overset{+}{N}$, H $\xrightarrow{HNO_3}$ MeO, NO_2, OMe, N $\xrightarrow{HNO_3}$ O_2N, NO_2, MeO, OMe, N

4. Free Radical Reactions

A. *Pyridines and Pyridinium Salts*

Most free radical substitutions of pyridine, but not all, give primarily 2-substituted derivatives. Photolytic chlorination of pyridine gives 2-chloropyridine in over 70% yield (259) as opposed to the mixtures and polyhalogenated products obtained from chlorinations at elevated temperatures, reactions which are also temperature dependent (see Chapter VI). Mixed halogens give similar results (260), as does free radical acylation (261) and amidation (262), although C-4 substitution is also observed in the latter case. The orientation observed in the amidations indicates that $\cdot CONH_2$ radicals (from formamide and H_2O_2/H_2SO_4 and $FeSO_4$) have nucleophilic character.

$$HCONH_2 + \cdot OH \text{ (or } t\text{-BuO}\cdot) \rightarrow H_2O + \cdot CONH_2$$

N $+ \cdot CONH_2 \longrightarrow$ N, $CONH_2$ $+$ $CONH_2$, N

(> 80% yield)

The Emmert reaction has been shown to be either a 1,3-dipolar addition or a nucleophilic cyclization following coordination of the ring nitrogen atom with the Mg^{+2} ion (161) and not a free radical reaction as previously suggested (see also Section III.5.D). The variation in the product distribution of the Ladenburg rearrangement (262) when the halide ion is varied from iodide to bromide or chloride has been attributed to the iodide's greater participation in the charge transfer complex proposed as the first intermediate (264) (see also Section III.1.B).

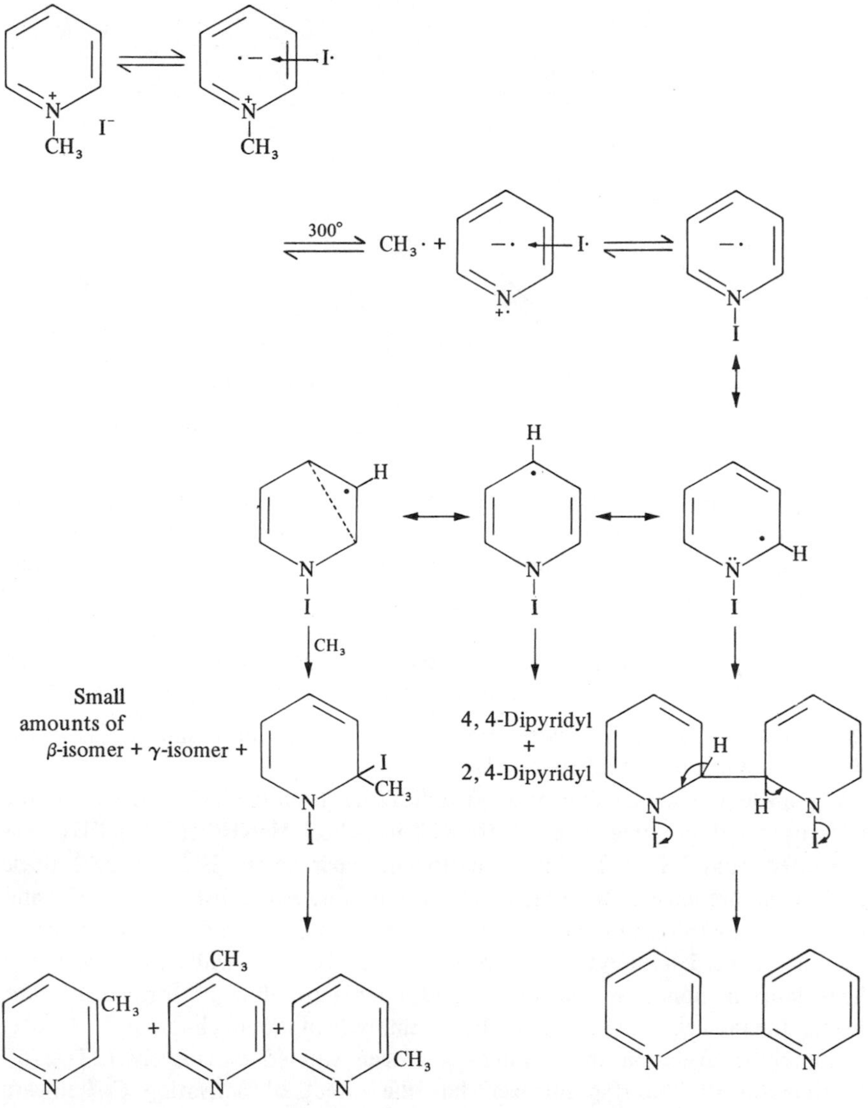

$\overset{+}{N}$ CH_3 I^- ⇌ $\overset{+}{N}$ CH_3 I·

$\underset{\rightleftharpoons}{300°}$ CH_3· + N$^{+\cdot}$ I· ⇌ N–I

H

N–I ⟵⟶ N–I ⟵⟶ N–I H

CH_3

Small amounts of β-isomer + γ-isomer + N–I (I, CH_3)

4, 4-Dipyridyl + 2, 4-Dipyridyl

H H N I N I

CH_3 CH_3 + + CH_3 N N N

N N

Dialkylation products are formed by the radical alkylation of the original charge transfer complex followed by further rearrangement.

The radical coupling product of the diphenyl-3-pyridylmethyl radical was shown to be **IA-44** (265).

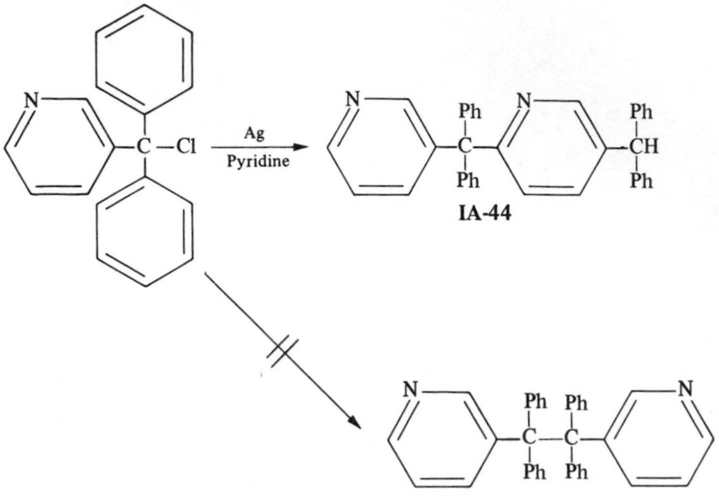

IA-44

Free radical alkylation, for example, methylation, of pyridine and methylpyridines (as the free bases) occurs at C-2 to a greater extent than does free radical arylation (266–268); thus the methyl radical is more nucleophilic than the phenyl radical.

The reactivity of 4-picoline toward the cyclohexyl radical has been investigated. The total rate ratio relative to benzene was surprisingly large: ${}_{C_6H_6}^{Py}K = 29$ and the partial rate factors were: $F_2 = 81.5$; $F_3 = 5.6$, which are *much* larger than other partial rate factors found in homolytic aromatic substitution reactions (269).

The nucleophilic character of alkyl radicals has been further confirmed by use of protonated pyridine as a substrate. The radical $MeNHCO(CH_2)_4CH_2\cdot$ was generated from 2-methyl-3,3-pentamethyleneoxazirane and Fe^{2+} in 20% sulfuric acid in the presence of pyridine to give the products of substitution at C-2 and C-4 (2-: 4- = 34: 66) and some 5,5′-dimer (262). Protonation of pyridine activates the ring toward homolytic alkylation (268, 270). Thus while pyridine is not benzylated in nonacidic solution, 2- and 4-benzylpyridine are formed in acidic media, the radicals being generated by thermolysis of dibenzylmercury.

Homolytic arylation of pyridines has been studied extensively (271-275). Protonation of the ring nitrogen has the effect of activating C-2 toward homolytic arylation relative to C-4 (276-278). Partial rate factors have been determined for the arylation of pyridines. Pyridine is arylated slightly more readily than is benzene by phenyl radicals (271, 279), more readily by

TABLE IA-3. Total Rate Ratios in the Gomberg-Hey Arylation of Pyridine at 40° (281)

$XC_6H_4\cdot$	${}^{Py}_{C_6H_6}K$	$XC_6H_4\cdot$	${}^{Py}_{C_6H_6}K$
Ph·	1.14	*o*-$MeC_6H_4\cdot$	1.72
o-$MeOC_6H_4\cdot$	1.27	*p*-$MeC_6H_4\cdot$	1.44
p-$MeOC_6H_4\cdot$	1.30	*o*-$NO_2C_6H_4\cdot$	0.47
o-$BrC_6H_4\cdot$	0.70	*p*-$NO_2C_6H_4\cdot$	0.78
p-$BrC_6H_4\cdot$	0.87		

nucleophilic aryl radicals, and much less readily by electrophilic aryl radicals (271, 280, 281). 3- And 4-methylpyridines are activated toward homolytic arylation to a greater extent than is pyridine, the positions *ortho* to the substituent undergoing the greatest activation (282). A slight concentration-dependent substitution orientation effect for arylation carried out in nitrobenzene has been attributed to oxidation by a trace of nitrosobenzene in the solvent (273).

The phenylation of the cyanopyridines (using benzoyl peroxide as the source of radicals, in the presence of nitrosobenzene) has been studied quantitatively and has led to the following total rate ratios and partial rate factors (283):

2-Cyanopyridine ${}^{CN}_{H}K = 3.6$

(2-cyanopyridine: 3-position 7.14, 4-position 8.95, 5-position 1.51, 6-position 4)

3-Cyanopyridine ${}^{CN}_{H}K = 6.22$

(3-cyanopyridine: 2-position 12.3, 4-position 9.3, 5-position 2.24, 6-position 13.4)

4-Cyanopyridine ${}^{CN}_{H}K = 5.7$

(4-cyanopyridine: 3-position 13, 2-position 4.1)

The free radical phenylation of PyH^+Cl^- and $MePyH^+Cl^-$ in acetic and formic acids have been studied. The reactivity of the 2- and 4-positions is found to be increased (as predicted by HMO calculations) relative to that in the free base, while the 3-position's reactivity remains unchanged (284).

The free radical halogenation of pyridine has been explored mechanistically further and the following mechanism proposed (285) for gas phase nuclear halogenation:

$$ArH + X_2 \rightleftharpoons (ArH^+X_2^-)$$
$$(ArH^+X_2^-) \rightarrow \text{rearr.} \rightarrow \sigma\text{-complex} \rightarrow ArX + HX$$

which differs from the earlier proposal in which the following steps were proposed (286):

$$(\mathrm{ArH^+})(\mathrm{X^-}) \xrightarrow[\text{determining step}]{\text{Isomer ratio}} \mathrm{Ar\cdot} + \mathrm{HX}$$

$$\mathrm{Ar\cdot} + \mathrm{X_2} \xrightarrow{\text{Fast}} \mathrm{ArX} + \mathrm{X\cdot}$$

$$\mathrm{X\cdot} + \mathrm{ArH} \longrightarrow (\mathrm{ArH^+})(\mathrm{X^-})$$

Halogenations are discussed further in Chapter VI.

Oxidation of 2-pyridone and of nicotinic acid by Fenton's reagent gives oxalic acid (287):

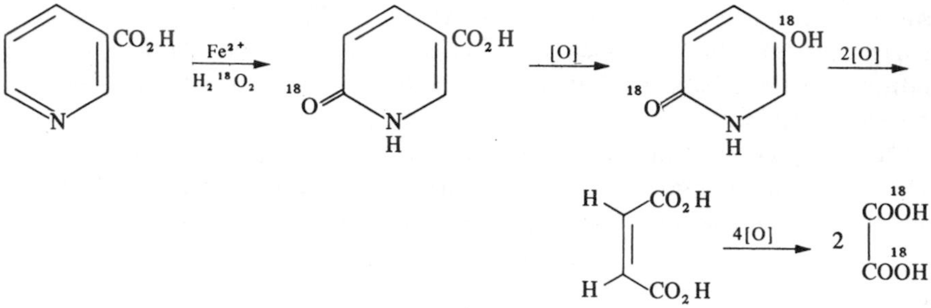

CO_2H (pyridine, N) $\xrightarrow[H_2{}^{18}O_2]{Fe^{2+}}$ ^{18}O, CO_2H (N–H) $\xrightarrow{[O]}$ ^{18}O, ^{18}OH (N–H) $\xrightarrow{2[O]}$

H, CO_2H / H, CO_2H $\xrightarrow{4[O]}$ 2 $CO^{18}OH$ | $CO^{18}OH$

B. *Pyridine-1-oxides*

Appreciably less effort has been expended on studies of the homolytic substitution of pyridine-1-oxides than has been on pyridines. The yields from phenylation from the decomposition of benzoyl peroxide were too low to measure and the yields were variable when radicals were generated by the decomposition of diphenyltriazole (PhN=NNHPh) (288). Phenylation of pyridine-1-oxide at 60° with $PhN_2^+\ BF_4^-$ and pyridine gave 0.9% of the isomeric phenylated oxides and 6% of phenylated pyridines (289). Higher yields (*ca.* 35%) of phenylation products were obtained by the electrolytic reduction at 0° of benzenediazonium tetrafluoroborate in acetonitrile containing the *N*-oxide (290). The α: β: γ isomer ratio was 89: < 1: 10 (determined by reduction of the crude reaction mixture to the corresponding phenylpyridines–some of which may have been present already). The most striking feature of this work was the total rate ratio of ${}^{\mathrm{PyO}}_{\mathrm{C_6H_6}}K = 52$ observed and the resultant partial rate factors: $F_2 = 139$; $F_3 = 1.5$, and $F_4 = 31.2$. Calculated values from localization energies at 0° taking $\beta = 0.95$ were ${}^{\mathrm{PyO}}_{\mathrm{C_6H_6}}K = 48.5$, $F_2 = 134$, $F_3 = 1.2$, and $F_4 = 19.7$. It would be of interest to determine whether selective solvation of the diazonium ion by (or complex formation with) the *N*-oxide in the competitive experiments (relative to anisole) could account for the unusually high value of the total rate ratio.

5. Nucleophilic Substitutions

In carbocyclic aromatic compounds, electrophilic substitution with subsequent elimination of a proton is very much favored over nucleophilic substitution which requires displacement of the very unstable hydride ion. This order is reversed only in cases where strongly electron withdrawing groups, such as nitro groups, activate the ring toward nucleophilic attack, but it is usually a substituent group which is displaced and not a hydride ion.

The ring nitrogen, or protonated or alkylated *N*-oxide function, in pyridine derivatives is a strongly electron withdrawing group, both inductively and mesomerically, so that all ring positions are deactivated toward electrophilic substitution. In fact, as discussed above in Section III.4, electrophilic substitution often requires harsh conditions, and there are no authenticated reported examples of a Friedel-Crafts substitution in the pyridine series. Instead, nucleophilic substitution is facilitated and, as is expected, both from qualitative resonance concepts and from theoretical calculations, nucleophilic substitution is particularly facile at C-2, C-4, and C-6.

A. *Hydrogen-Deuterium Exchange Reactions*

Although hydrogen-deuterium exchange in pyridine derivatives was once assumed to involve intermediates such as **IA-45** (291), it is now generally agreed that intermediates such as **IA-46** are involved (183, 292, 293). This type of reaction is important not only because it provides isotopically labeled

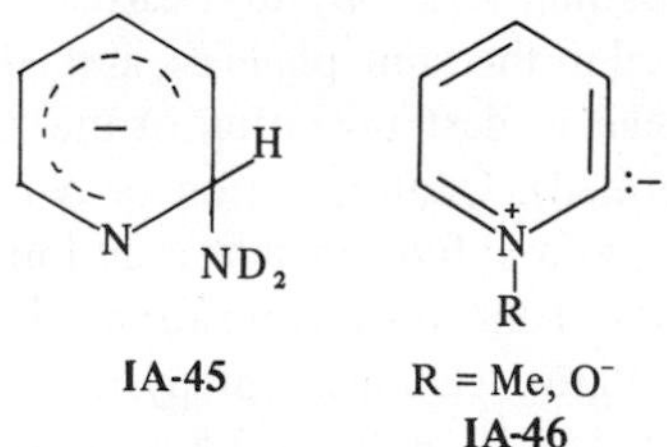

IA-45

R = Me, O^-
IA-46

compounds for other studies (183, 294) [carbanions are also involved in deuterodecarboxylation, a convenient method for the preparation of labeled pyridines (295)], but also because the relative rates of exchange at the various nuclear positions reflect (*a*) the relative electron densities at these positions, which in turn are related to (*b*) the ability of the ring to support the carbanion charge at that position (183). Synthetic applications of this reaction are discussed below. The possible biological importance of the base-catalyzed proton abstraction in pyridinium salts has also been pointed out (296).

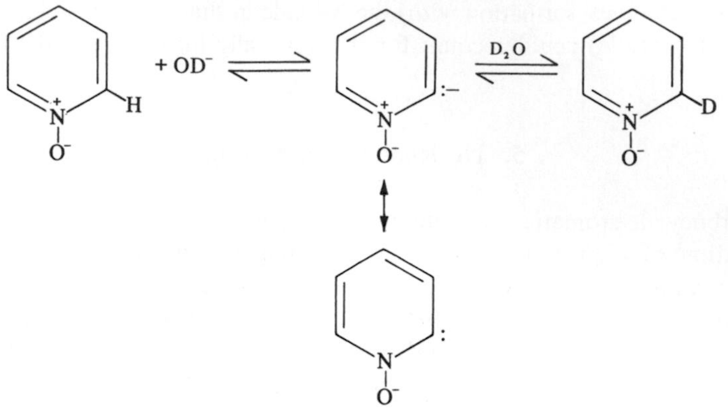

It is well established that the exchange is slowest at H-2 in pyridines (H-4 exchanges fastest) (297-299), but is fastest at H-2 in pyridinium salts (183, 293, 299–301), and *N*-oxides (183, 298, 299, 302). The accepted explanation of this apparent contradiction is that quaternization of the ring nitrogen or *N*-oxide formation prevents a destabilizing C-2-carbanion nitrogen-lone-pair interaction.

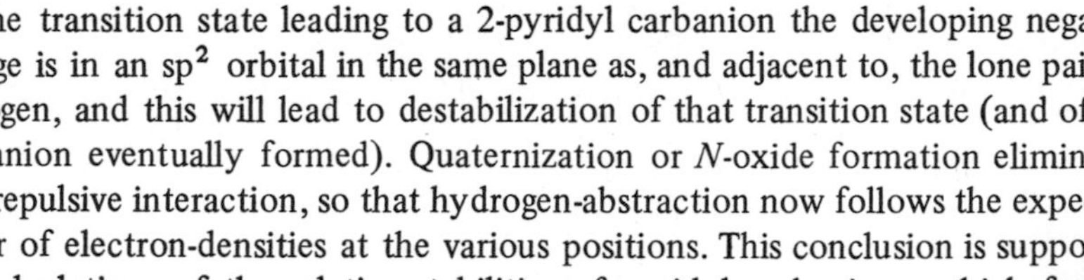

In the transition state leading to a 2-pyridyl carbanion the developing negative charge is in an sp^2 orbital in the same plane as, and adjacent to, the lone pair on nitrogen, and this will lead to destabilization of that transition state (and of the carbanion eventually formed). Quaternization or *N*-oxide formation eliminates this repulsive interaction, so that hydrogen-abstraction now follows the expected order of electron-densities at the various positions. This conclusion is supported by calculations of the relative stabilities of pyridyl carbanions, which further show that 2,3-pyridyne is less stable than 3,4-pyridyne because of the repulsive interaction between the lone pairs of electrons at N-1 and C-2 (173), contrary to an earlier calculation which arrived at the opposite conclusion (303).

The base-catalyzed proton abstraction from pyridine-1-oxides has permitted the direct introduction of a variety of electrophilic reagents at the 2-position of such molecules with the retention of the *N*-oxide function. Thus alkylation (296, 304), acylation (304, 305), halogenation (306, 307), mercuration (307), thiolation (306), and aminoalkylation (308), among others, have been achieved (see Chapter IV for details). The effect of substituents on orientation in some of these reactions has been discussed (296).

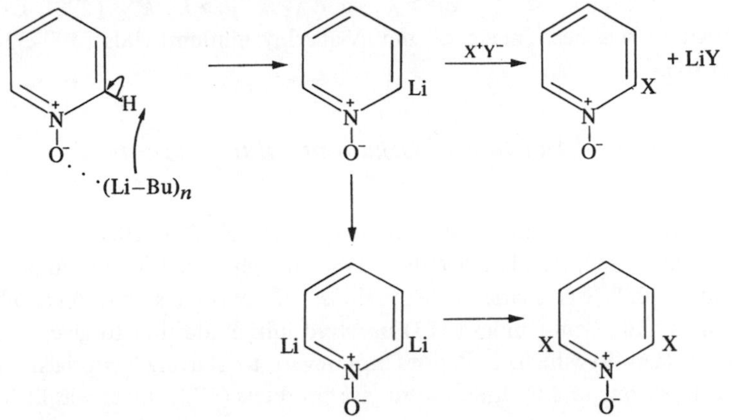

It is interesting that although base-catalyzed hydrogen-deuterium exchange is faster at H-2 than at the *N*-methyl group of *N*-methylpyridinium salts (183,

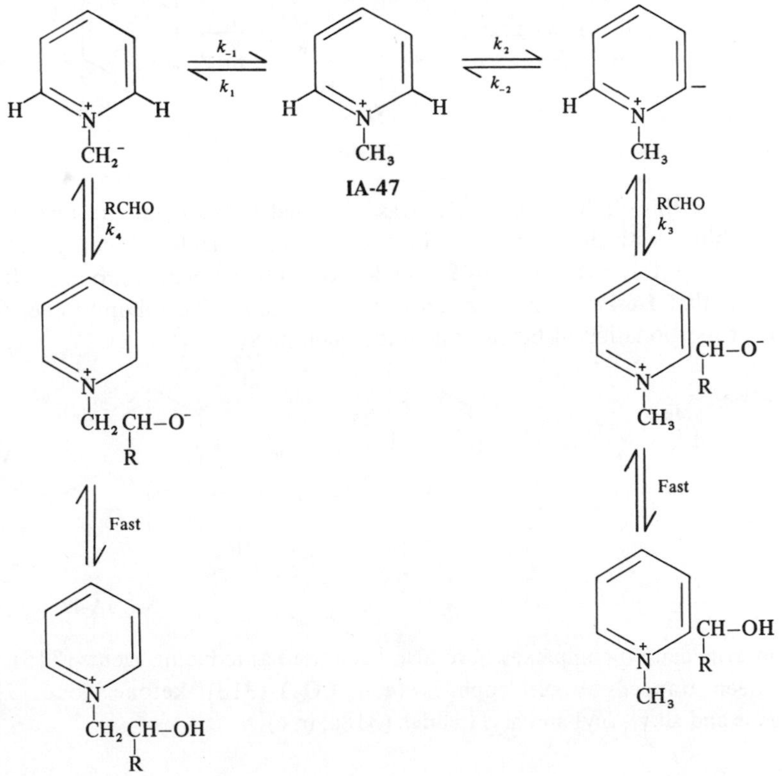

293), treatment of an *N*-alkylpyridinium salt with base and an aldehyde gives substitution only at the *N*-methyl group (293, 309). This has been explained by the equilibria shown from **IA-47**, where the following relative rates were found: $k_1 \ll k_{-1}$, $k_2 \ll k_{-2}, k_2 > k_1$, and $k_4/k_{-1} > k_3/k_{-2}$ (293). Extensive synthetic use has been made of the *N*-alkylpyridinium ylids (309–311) (see Chapter III).

B. *Addition of Organometallic Reagents*

a. PYRIDINES. Pyridines are usually susceptible to attack by strongly nucleophilic organometallic reagents (312). Though it had long been postulated that nucleophilic aromatic substitutions of pyridines by organo-lithium compounds and alkali amides (313) involved initial addition to give a Wheland σ-complex (**IA-48**) which, if it does not revert to starting materials, could be induced to eliminate a hydride ion to give products (172), there was little *direct*

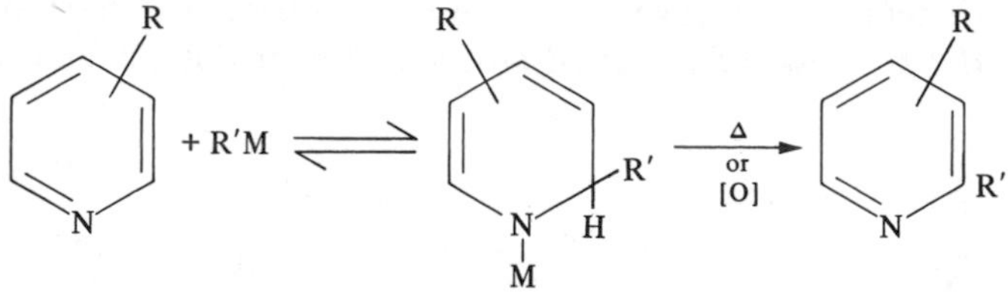

R R R + R'M R' H N M Δ or [O] R' N N

IA-48

evidence of this. Indirect evidence was obtained (314) from the reaction of *o*-tolyllithium with 3-picoline (not in excess) which yielded 3-methyl-1,2,5,6-tetrahydro-2-*o*-tolylpyridine (**IA-49**) in addition to the expected products. It was suggested that **IA-49** resulted, in addition to 3-methyl-2-*o*-tolylpyridine, from the disproportionation of the intermediate σ-complex:

2 Me H N Li Me — 2LiH Me Me N + Me Me N H

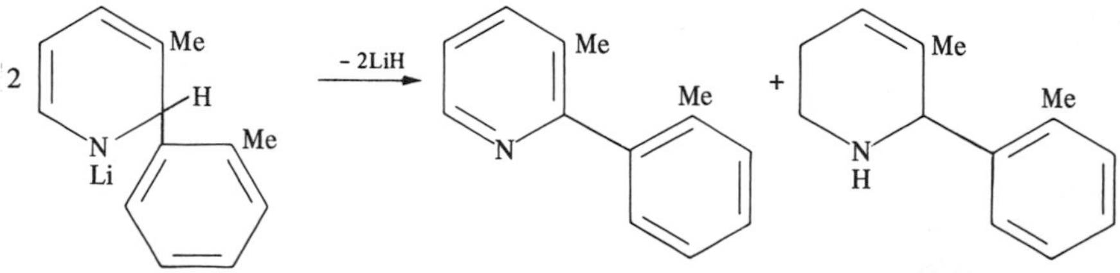

IA-49

The intermediate σ-complexes have also been used as reducing agents (315), and have been trapped by electrophiles (e.g., CO_2) (316), ketones (316, 317), halogens and alkyl, acyl and aryl halides (318a, b, c):

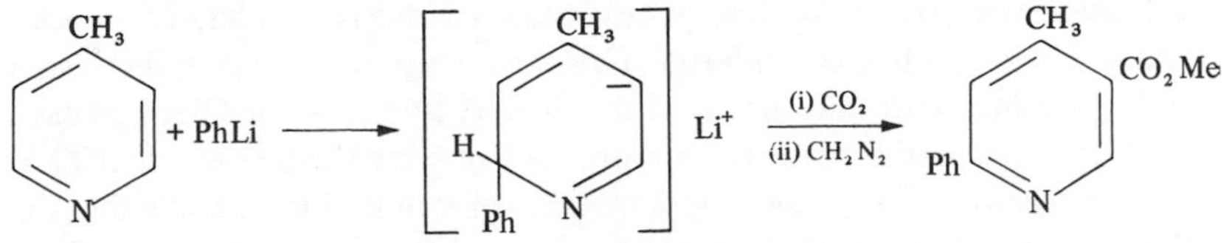

The intermediacy of **IA-48** has now been proved unequivocally by the measurement of the NMR spectra of a series of pyridine-butyllithium adducts in solution (319, 320), and by the actual isolation of these adducts by careful manipulation under a nitrogen atmosphere (321). A tetrahydro-intermediate has been isolated from the reaction of pyridine with two equivalents of *tert*-butyllithium at -70° (321a).

Most studies have been concerned with the very strongly nucleophilic organolithium reagents but some studies with other organometallic compounds have been carried out. Thus the methyl Grignard reagent gives dihydro derivatives on reaction with 3,5-dicyanopyridines (322). The reaction of phenyl Grignard reagent with 3-benzoylpyridine has been shown to give 3-benzoyl-4-phenyl-1,4-dihydropyridine (323), and the reaction of pyridine with phenylcalcium iodide gives 2-phenyl- and 2,6-diphenyl- or 2,5-diphenylpyridine (**IA-50**), depending on reaction conditions (324). The formation of 2,5-diphenylpyridine in this reaction, and of some 3-methyl-5-*o*-tolylpyridine from 3-picoline and *o*-tolyllithium (314), have been explained as follows (172, 314):

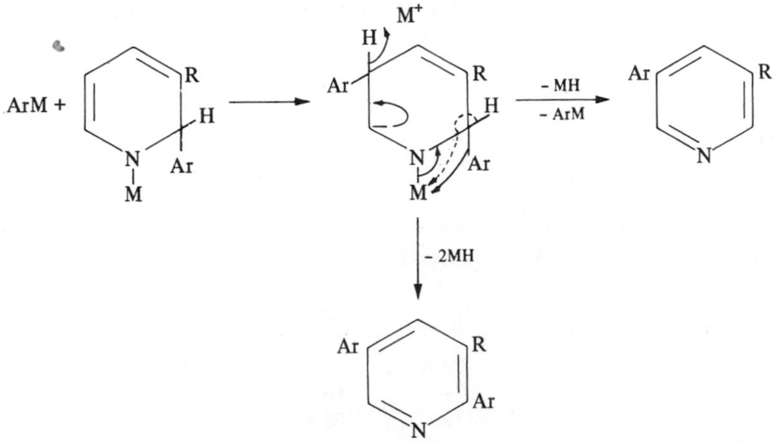

The orientation of substitution is strongly dependent on the nature of the nucleophile. Aryl- and alkyl-lithium reagents will attack the α-position preferentially in a rate- and product-determining step. Benzyllithium, on the other hand, attacks the γ-position preferentially, as expected for a more stable, hence more selective, anion whose orientation will follow the order of localization energies rather than ground state electron densities. The reaction of pyridine with preformed *n*-butyl Grignard reagent or *n*-butyllithium gives 2-butylpyridine, with only traces of the 4-butyl isomer in the Grignard case; but 4-butylpyridine is the predominant product obtained (but only in 10% yield) from the reaction of pyridine with a mixture of *n*-butyl halide and either lithium or magnesium metal (325). On the other hand, Abramovitch and Giam found no differences in the ratios of products formed from reaction of 3-alkylpyridines with preparations of phenyllithium which were either halide free or contained excess lithium bromide (326, 327). A possible explanation of the formation of a 4-butyl product above could be that a 4-pyridyllithium intermediate is formed (proton abstraction from C-4 in pyridine) followed by alkylation with the halide.

The orientation in the addition of aryllithium compounds to a series of 3-substituted pyridines has been studied extensively by Abramovitch and his co-workers (for a review see Ref. 172). Substitution was found, in most cases, to take place preferentially at C-2. Steric effects only became important with a β-substituent much larger than methyl (Table IA-4). On going from phenyllithium to *o*-tolyllithium, the ratio of products from 3-picoline *increases* slightly in favor of the 2,3-isomer, showing clearly that steric effects are not important with the 3-methyl substituent (314). These observations have been interpreted in terms of a transition state in which the phenyl anion is almost

TABLE IA-4. Ratio of Products from the Reaction of 3-Alkylpyridines with Phenyllithium (172)

3-R-pyridine + PhLi ⟶ 2,3- (2-Ph-3-R-pyridine) + 2,5- (2-Ph-5-R-pyridine)

R	% Composition	
	2,3-	2,5-
Methyl	94.6	5.4
Ethyl	84	16
Isopropyl	70	30
t-Butyl	4.5	95.5
2-(*N*-Methyl)pyrrolidyl	30	70
Ph	16.5	83.5
OCH_3	100	0
NH_2	100	0
SO_2NEt_2	100	0

perpendicularly above C-2, and the ground state configuration of the pyridine ring has been only slightly distorted (172). The exclusive formation of the 2,3-isomer in the reaction of PhLi both with 3-amino- and 3-methoxy-pyridine (Table IA-5) has been attributed to the formation of a complex such as **IA-50** between the lone-pair of electrons on the 3-substituent and the lithium atom, in which the phenyl group would be suitably oriented to attack at the 2-position (328).

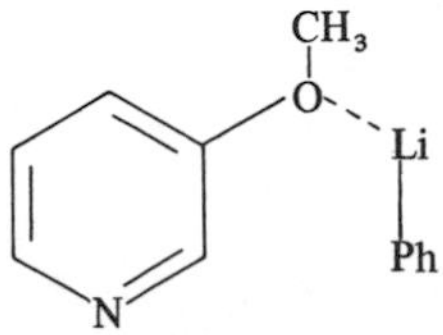

CH_3 O Li Ph N

IA-50

The most startling result of these studies has been the observation that a 3-methyl- or a 3-ethyl-group *activates* the 2-position in pyridine towards nucleophilic attack by phenyllithium (327, 329) (but not by methyllithium) (330), while the 6-position was deactivated normally (Table **IA-5**). The product-determining step in these reactions is the addition to give the σ-complex, and the reactions have been shown not to be rapidly reversible. No deuterium isotope effect was observed in the reaction of 3-picoline-2*d* with phenyllithium (326). To explain the observed activation of C-2 it has been suggested that both an electron-deficient bond *and* London dispersion force interaction between the 3-substitutent and the polarizable phenyllithium reagent come into play. That such an explanation is plausible and that activation of C-2 toward nucleophilic attack by a 3-methyl substituent is not unique to phenyllithium has been established by the finding (331, 332) that the bromine atom in 2-bromo-3-methylpyridine is displaced faster by the polarizable thiophenoxide ion than it is in 2-bromopyridine. With thiomethoxide ion the ease of displacement is reversed; that is, 2-bromopyridine reacts faster than does 2-bromo-3-methylpyridine and thus parallels the behavior of methyllithium with 3-picoline. Here again, the 3-methyl group deactivates C-6 normally, as expected for a nucleophilic attack. Thus with thiomethoxide (and methoxide, as well as phenoxide ion) (332) the greater reactivity of 2-bromo-3-methyl- over 2-bromo-5-methyl-pyridine is attributed to ion-dipole interaction, possible between the substituent and the attacking reagent during attack at C-2, but not at C-6. This is not sufficient to overcome the deactivating influence of the 3-methyl group toward nucleophilic attack, and 2-bromopyridine reacts faster in this case than either of the methyl derivatives. When thiophenoxide (which is

TABLE IA-5. Relative Reactivities of Pyridine, 3-Picoline, 3-Ethylpyridine, and 3-Isopropylpyridine towards Phenyl-, Methyl-, Isopropyl-, and Benzyllithium

	Phenyllithium	Methyllithium	Isopropyllithium	Benzyllithium
3-Picoline	${}^{\mathrm{Me}}_{\mathrm{H}}K = 1.30$	${}^{\mathrm{Me}}_{\mathrm{H}}K = 0.28$	${}^{\mathrm{Me}}_{\mathrm{H}}K = 0.17$	${}^{\mathrm{Me}}_{\mathrm{H}}K = 0.125$
	${}^{\mathrm{Me}}F_2 = 2.4$	${}^{\mathrm{Me}}F_2 = 0.47$	${}^{\mathrm{Me}}F_2 = 0.07$	${}^{\mathrm{Me}}F_4 = 0.125$
	${}^{\mathrm{Me}}F_6 = 0.13$	${}^{\mathrm{Me}}F_6 = 0.09$	${}^{\mathrm{Me}}F_6 = 0.26$	
3-Ethylpyridine	${}^{\mathrm{Et}}_{\mathrm{H}}K = 0.79$			
	${}^{\mathrm{Et}}F_2 = 1.4$			
	${}^{\mathrm{Et}}F_6 = 0.16$			
3-Isopropylpyridine	${}^{\mathrm{isoPr}}_{\mathrm{H}}K = 0.56$			
	${}^{\mathrm{isoPr}}F_2 = 0.82$			
	${}^{\mathrm{isoPr}}F_6 = 0.30$			

much more polarizable than are the other ions used) is the attacking reagent, London dispersion attractive forces can arise between a 3-methyl-(or a 3-bromo-) substituent and the reagent approaching C-2. This, superimposed upon the ion-dipole interactions, is sufficient to overcome the deactivation by the alkyl group and to lead to the observed activation of C-2. A similar explanation has been proposed for the observed reactivity order in the Tschitschibabin reaction (see Section III.5.C.a.i).

The proportion of 2,3-disubstituted (relative to 2,5-disubstituted) product is reported to increase with temperature in the reaction of 3-picoline [or of *N*-(picolyl)pyrrolidine] with methyllithium in boiling toluene (b.p. 110°) [as opposed to the reaction in boiling tetrahydrofuran (b.p. 66°)] (333). This is different from what is observed with 3-picoline and phenyllithium, where the ratio of products appears to be independent of temperature. It may be that the isomeric σ-complexes formed from methyllithium are undergoing side-reactions at different rates at the higher temperature and are thus being removed selectively.

There is a report that the reaction of nicotine with methyllithium gives mainly 6-methylnicotine contaminated with the 4-methyl derivative (334). This is unusual since such reactive anions do not usually attack the γ-position. The complex **IA-51** has been invoked to explain the attack at C-4 (334). In contrast, the reaction of nicotine with phenyllithium gives the 2,3- and 2,5-isomers in the ratio of 30:70 (329, 335), and 3-picoline and methyllithium give only the 2,3-

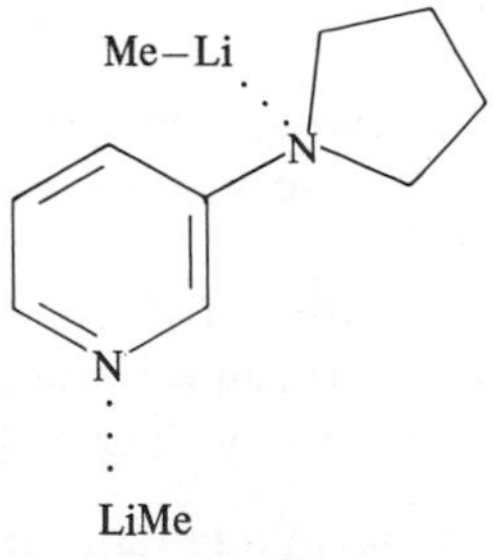

Me–Li
N
N
LiMe

IA-51

and 2,5-isomers in the ratio of 84.4:15.6 (330). Complex **IA-51** is very similar to **IA-50**, which has been proposed to account for substitution at C-2.

2-(4-Pyridyl)-2-methylpropyl metal compounds close to the spirourethane (**IA-52**) with ethyl chloroformate (336). 4-(4-Pyridyl)butyl metal compounds also exist in the 4,4-spirodihydropyridine structure rather than the open form (337).

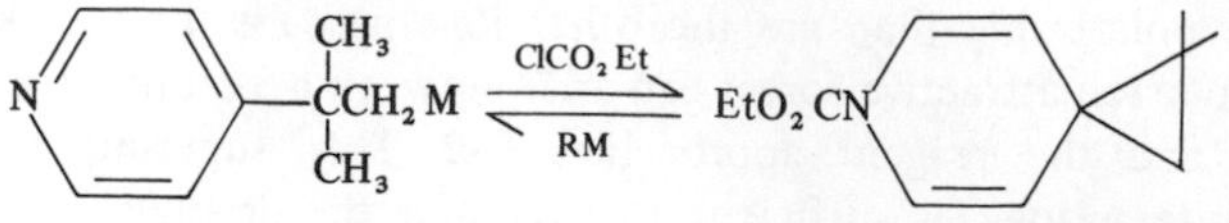

CH_3 | CCH$_2$M | CH_3 ⇌ ($ClCO_2Et$ / RM) EtO_2CN

IA-52

M = Li, Na, K, MgCl

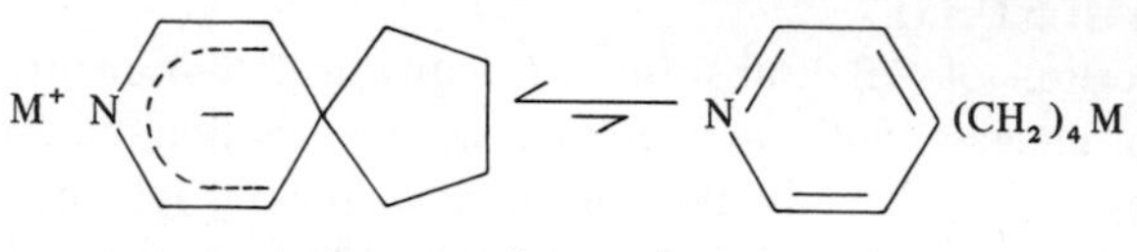

M^+ N ⇌ N $(CH_2)_4M$

b. PYRIDINE-1-OXIDES. As expected, pyridine-1-oxides undergo substitution readily with organometallic reagents. Pyridine-, and 2- and 4-picoline-1-oxide react with phenylmagnesium bromide to give the 2- phenylpyridine (338). The intermediate isolated from this reaction was originally assigned the structure 1-hydroxy-2-phenyl-1,2-dihydropyridine (**IA-53**) (338). More recent NMR data strongly suggest that it is, instead, the open chain *syn*-5-phenyl-*E*,*Z*-9 4-pentadienal oxime (**IA-54**), which recycles and aromatizes on heating (339). In the same way, *N*-methylanabasine *N'*-oxide and the *N*, *N'*-dioxide give

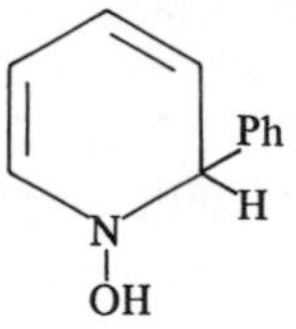

Ph H N OH

IA-53

Ph N OH

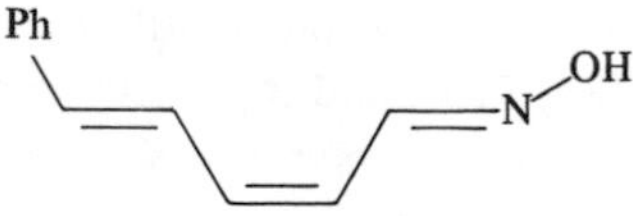

IA-54

6-methyl-3-(*N*-methyl-2-piperidyl)pyridine on treatment with methylmagnesium iodide (340), the observed orientation probably being the result of the steric influence of the substituent at C-3.

Phenyllithium reacts smoothly with pyridine-1-oxide, but gives a plethora of products. Only 2-phenylpyridine has been identified and it is likely that some of the products are the result of pyridine ring cleavage (341).

c. PYRIDINIUM SALTS. *N*-Alkoxypyridinium salts are substituted at C-2 by Grignard reagents with concurrent elimination of the *N*-substituent (342). In some cases, the alkoxy group behaves as a protected aldehyde, and a carbinol is formed as a by-product (343).

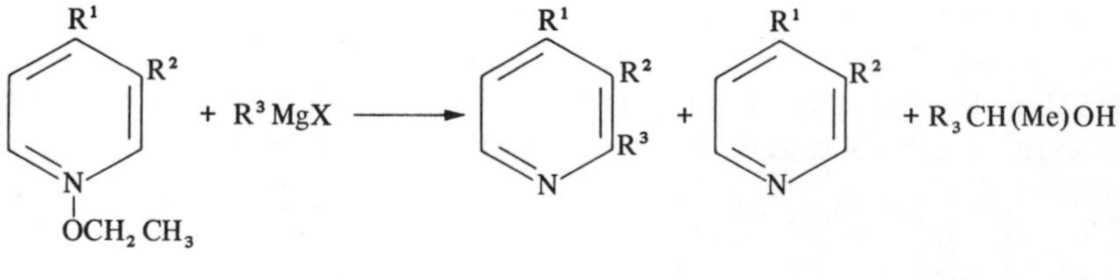

R^1, R^2, N–OCH_2CH_3 + R^3MgX ⟶ R^1, R^2, R^3, N + R^1, R^2, N + $R_3CH(Me)OH$

Powerful nucleophiles add readily to pyridinium salts, but the resulting dihydropyridines are very prone to undergo polymerization in the presence of strong bases, and alot of tars are formed (see Chapter I, Part B). Thus *N*-benzylpyridinium chloride and phenyllithium gave a small amount of 1-benzyl-1,2-dihydro-2-phenylpyridine (344) but mainly tar. On the other hand, the much more stable cyclopentadienyl carbanion (from cyclopentadienyllithium) gave the internally stabilized zwitterion **(IA-55)** (345).

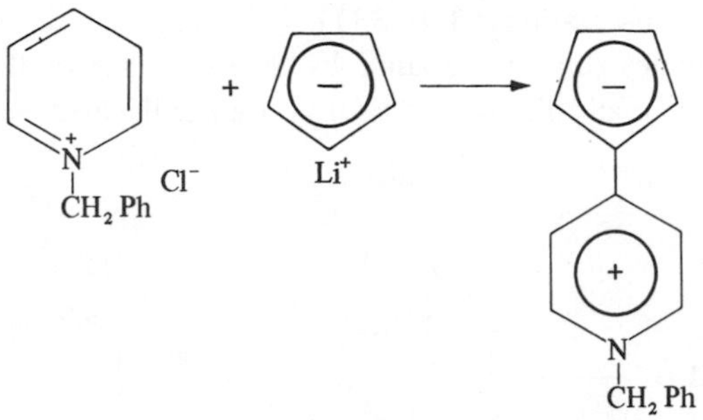

$$\text{C}_5\text{H}_5\overset{+}{\text{N}}\text{CH}_2\text{Ph}\ \text{Cl}^- + \text{C}_5\text{H}_5^-\ \text{Li}^+ \longrightarrow$$

IA-55

Grignard reagents, on the other hand, give readily isolable (though unstable) products with quaternary pyridinium salts. These have usually been reduced further. When the pyridinium salt bears a 3-substituent attack at C-2 is preferred to attack at C-6 (346, 347).

The reaction of pyridinium salts with Grignard and organo-cadmium reagents is described further in Chapter III (Section II.2).

C. *Other Substitutions Reactions*

a. NUCLEAR HYDROGEN

(1) Pyridines. Amination of pyridine with sodium or potassium amide proceeds at an appreciably higher temperature than the reaction with organolithium compounds.

The scope of aminations by the Tschitschibabin reaction and of alkylamination reactions was broadened to include substituted pyridines (348–352). A review has appeared of this reaction (353).

The mechanism of the amination has been elucidated (354). The absence of a deuterium isotope effect in the Tschitschibabin reaction and the effect of a 3-substituent upon orientation showed that, contrary to a suggestion (355), pyridynes are not intermediates and that the initial addition is irreversible. A 3-methyl substituent directs the entering amide ion mainly to the 2-position (2,3-: 2,5-isomer ratio 10.5: 1) as it does phenyllithium. Contrary to what is observed with the latter, however, no overall activation of the pyridine nucleus by the methyl group is observed (${}^{\mathrm{Me}}_{\mathrm{H}}K = 0.22$; ${}^{\mathrm{Me}}F_2 = 0.40$, ${}^{\mathrm{Me}}F_6 = 0.039$). The situation here is then similar to that found for the reaction of methyllithium with 3-picoline and with methoxide or thiomethoxide ion with 2-bromo-3-methyl- and 2-bromo-5-methyl-pyridine, namely overall deactivation by the methyl group, but the rate for 2,3-isomer formation greater than that of the 2,5-isomer (331, 332). An ion-dipole attractive interaction between the approaching amide or methoxide and the 3-methyl group has been proposed to account for these observations (329, 331).

Substituted amines can be prepared by an extension of the Tschitschibabin reaction (356, 357), and the mechanism undoubtedly involves the formation of a σ-complex:

+ RNHNa → [σ-complex: N–Na, H, NHR] —(- NaH)→ [2-NHR-pyridine] —(NaH)→ [2-N(Na)R-pyridine]

Examples of ring-contractions which occur in certain cases where pyridines are treated with potassium amide in liquid ammonia are discussed in Section III.5.C.b.i.

The effect of a 3-methyl-substituent upon the orientation of the entering group in the Emmert reaction has been studied quantitatively (161). 3-Picoline and cyclohexanone in the presence of magnesium gave a mixture of 1-(3-methyl-2-pyridyl)- (**IA-56**; R = H) and 1-(5-methyl-2-pyridyl)cyclohexanol (**IA-57**; R = H) in the ratio of 29.1 : 70.9, together with an aldol condensation product, 2-cyclohex-1-enylcyclohexanone. The total rate ratio from a competitive reaction with pyridine was ${}^{\mathrm{Me}}_{\mathrm{H}}K = 0.28$. When 2-methylcyclohexanone was used the ratio of **IA-56** (R = Me) to **IA-57** (R = Me) obtained was 24.6 : 75.4. These, together with other results including the absence of any influence of atmospheric oxygen, were taken to support either a 1,3-dipolar cycloaddition mechanism or a two-step process involving coordination at nitrogen followed by rate-determining nucleophilic cyclization, rather than a mechanism involving discrete free radicals.

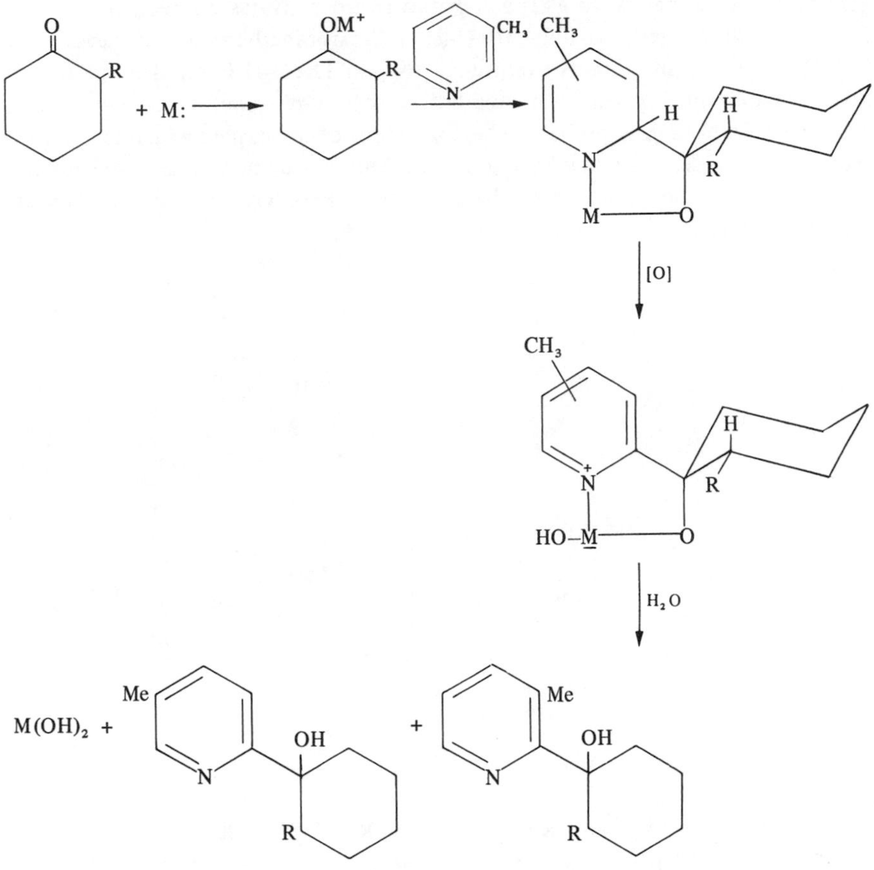

IA-57 **IA-56**

(2) ***Pyridinium Salts.*** The alkaline ferricyanide oxidation of pyridinium salts has been studied extensively (See Chapter III). Effects of substituents have been examined qualitatively (for summary, see Ref. 172) and quantitatively and the mechanism of the reaction appears to be very complex (358). In only one case (R^2 = CN) was a 4-pyridone isolated. A 3-methyl group activates slightly (${}^{Me}_{H}K$ =

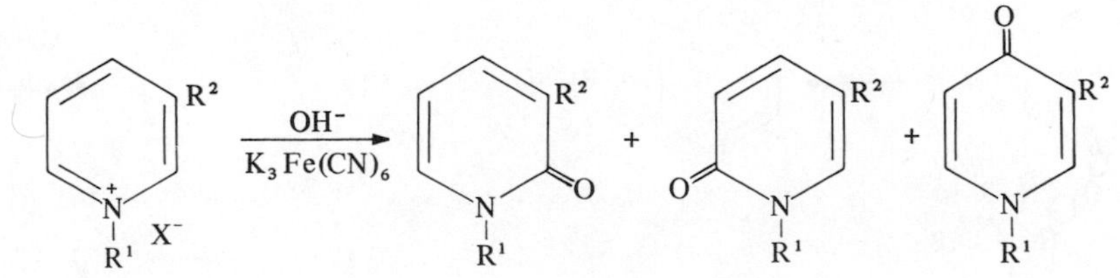

1.1) and directs mainly to C-2; a 3-cyano-group activates appreciably (${}^{CN}_{H}K$ = 20.2) and also directs mainly to C-2; a 3-carbomethoxy-group deactivates (${}^{CO_2Me}_{H}K$ = 0.05) and directs exclusively to C-6. The C–H bond cleavage step is not rate-determining in the oxidation of 3-picoline methiodide. The rate-determining step is probably the formation of a complex which reacts with more ferricyanide, oxidation taking place within a second complex. A 3-methyl group facilitates the approach of the polarizable $Fe(CN)_6^{3-}$. 3-Substituents exert both electronic and steric effects.

R³ R² + OH⁻ ⇌ R³ R² OH H ⇌ Fast N R¹

R³ R² O⁻ H N R¹ $\underset{Slow}{\overset{Fe(CN)_6^{3-}}{\rightleftharpoons}}$ CN⁻ + [R³ R² O–Fe(CN)₅ H N R¹]³⁻ $\xrightarrow[CN^-]{Fe(CN)_6^{3-}}$

[R³ R² $Fe(CN)_6^{4-}$ H O· $Fe(CN)_6^{3-}$ N R¹] $\xrightarrow{Fast}$ R³ R² O N R¹ + $2Fe(CN)_6^{4-}$ + H^+

Attack by nucleophiles on *N*-alkoxypyridinium salts can, in principle, take place in one of five ways: Katritzky's paths *A* through *D* (359), and by nuclear hydrogen abstraction (path *E*):

N⁺ O–CH–R H :B ⟶ N + RCHO (path *A*)

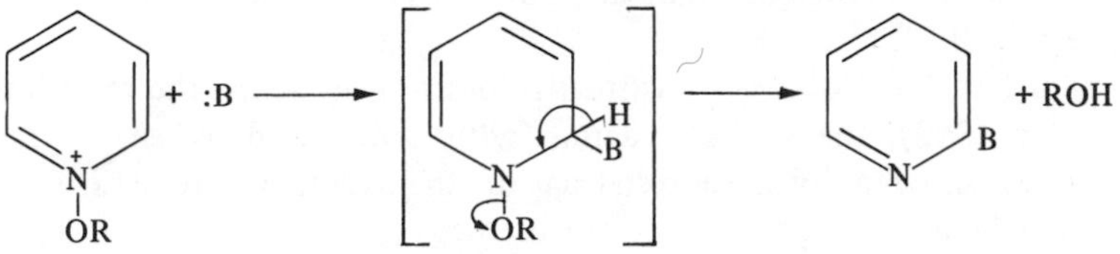

(path *B*)

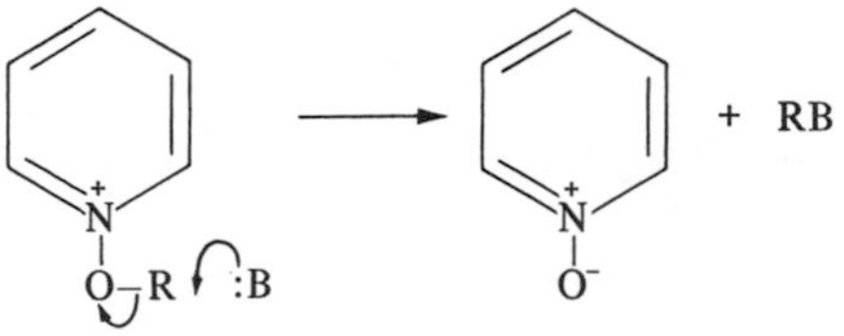

(path *C*)

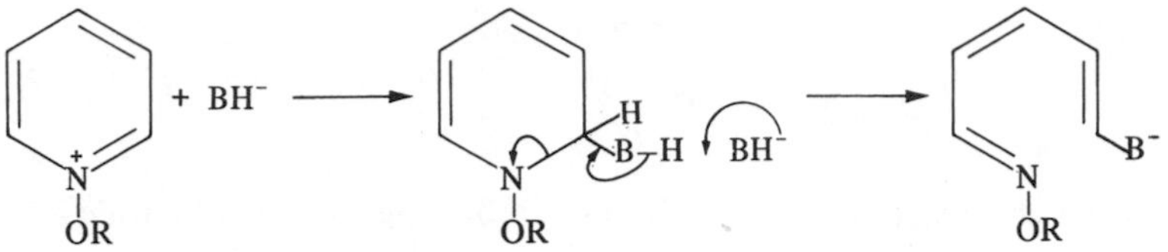

(path *D*)

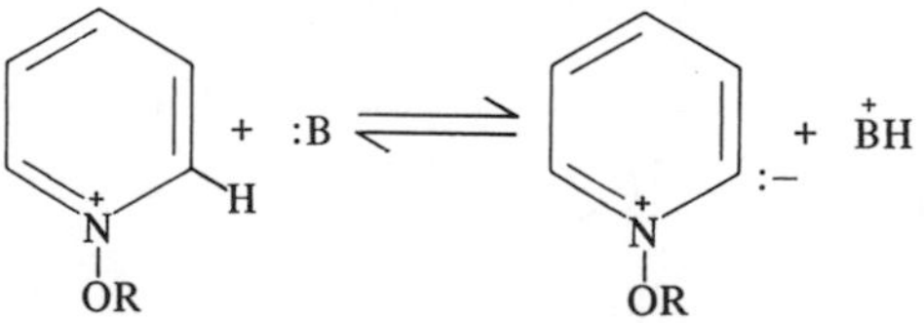

(path *E*)

Cyanide ion does not react with pyridine-1-oxide nor does the latter undergo the Reissert reaction. On the other hand, 4-halopyridine-1-oxides do react with cyanide ion in the presence of benzoyl chloride to give 2-cyano-4-halopyridines (360). In contrast, treatment of *N*-alkoxypyridinium salts with aqueous potassium cyanide gives either 4-cyanopyridine (361) or a mixture of 2- and 4-cyanopyridines, consisting predominantly of the 2-isomer (path *B*) (362). When C-4 is blocked, the reaction conditions determine whether substitution at C-2 is the major pathway (363, 364), or whether a mixture of products including those of displacement of substituents at C-4 is obtained (365).

The nature of the substituent at C-3 has a profound influence on the reaction. There is a linear relationship between ΔH° reaction and the position of the long wavelength absorption band of the substrate (365). The orientation of the products is also strongly dependent on the nature of the 3-substituent. Generally, electron withdrawing groups direct attack to C-4 and C-6 (366, 367) while electron donating groups lead to substitution at all three positions (362, 366). In aqueous solution, cyanide attacks *N*-alkyl-3-carbamoylpyridinium ions at C-4 (368). Aqueous cyanide ion reacts at room temperature with some pyridinium salts to give viologen radical cation derivatives (369) (see also Chapter III, Section II.2).

While the nature of the *N*-alkoxy substituent does not influence the reaction with cyanide ion (362), the similar reaction with substituted *N*-benzylpyridinium salts does show a linear correlation of the reaction rate with σ^*: $\log k_x/k_0 = \sigma\rho^*$ where

$$\sigma_{(XC_6H_4CH_2)} = \frac{0.56}{1.72}\sigma_x^{\,n} + \sigma^*_{(C_6H_5CH_2)}$$

and

$$\rho^* = 2{\cdot}2$$

In contrast to the reactions of cyanide ion, carbanions and borohydride add to pyridinium salts to give 1,4-dihydropyridine derivatives (370, 371). In organic

NO$_2$ + $\bar{C}$HXY ⟶ H CHXY NO$_2$

N$^+$–CH$_3$ N–CH$_3$

X	Y
COOMe	COOMe
CN	COOEt
CN	CN
CN	PhCO

solvents, hydride ion adds initially to the α-positions, but the 1,2- or 1,6-dihydropyridines so formed can rearrange to the thermodynamically more stable 1,4-isomers (372) (see also Chapter IB and Chapter III, Section II.2). Hydride from hydrosulfite ion reduction gives substituted 1,4-dihydropyridines (373).

Alkali metal derivatives of dialkyl phosphonates react with *N*-methoxypyridinium sulfate to give dialkylpyridine-2-phosphonates (path *B*) (374). A

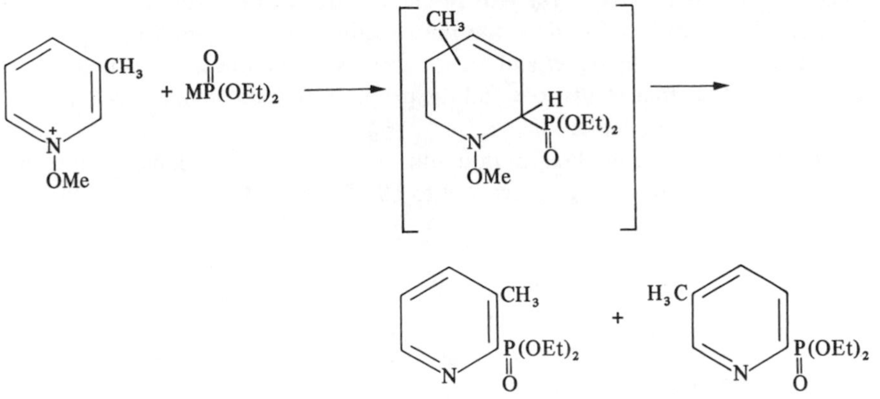

6 : 1

3-methyl-group directs mainly to C-2, as expected. If both the α-positions are blocked, attack takes place at C-4 (path *B*), but paths *A* and *C* compete with this route:

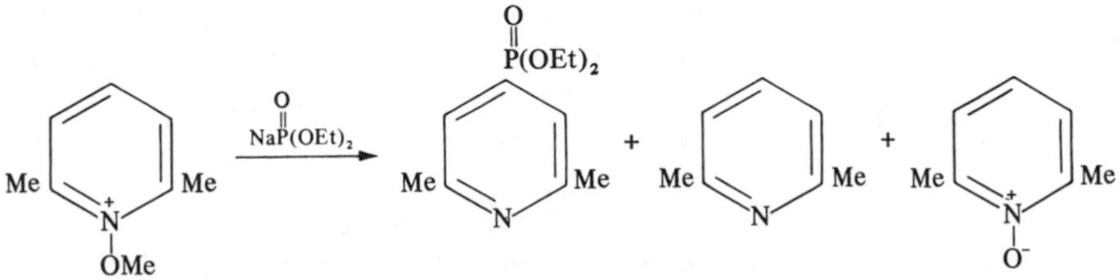

Weaker nucleophiles (NO_2^-, I^-, $S_2O_3^-$, N_3^-, CNS^-, $PhSO^-$) do not displace nuclear hydrogens at all in *N*-methoxypyridinium salts, but instead attack the *O*-methyl group to displace pyridine-1-oxide in an S_N2 reaction (path *C*) (359).

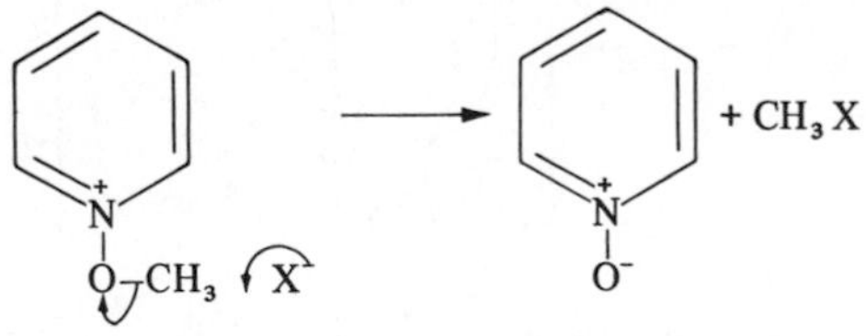

Reactions of suitable *N*-alkyloxypyridinium salts with methoxide ion in methanol and with phenoxide in ethanol proceed by path *A* (path *E* is also likely here but may only be detected by H-D exchange studies, which have not been carried out); attack by aqueous hydroxide ion involves paths *A* and *D*, while reaction with thioalkoxide ions in ethanol in the presence of excess thiol involves paths *A*, *B*, and *C* (359).

The reaction of 1-alkoxy-, 1-acyloxy-, and 1-sulfonyloxypyridinium salts with thiols involves initial addition of the thiol to the α- or the γ-position, and has been studied in detail by Bauer and his co-workers. The initial adducts undergo a variety of further transformations and the products formed are manifold. This is discussed in detail in Chapter IV.

Reaction of the *N*-methoxypyridinium cation with piperidine results in addition at C-2 followed by ring cleavage to **IA-58** (path *D*) (359, 375).

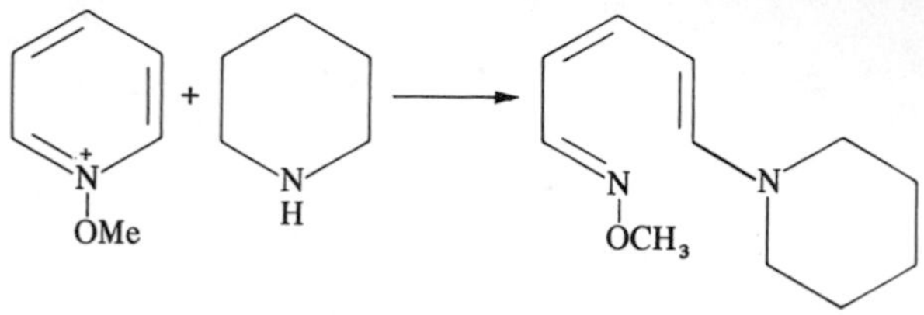

IA-58

It has been suggested (242) that the position of attack by a nucleophile upon the pyridinium ion is determined by whether the nucleophile is hard or soft: a hard nucleophile (BH_4^-, aniline, OH^-) attacks position 2, whereas a soft one (CN^-, $S_2O_4^{2-}$) reacts at C-4. This is an alternative to Kosower's charge-transfer complex theory (376) which could not explain the behavior of CN^- (the latter does not appear to form a charge-transfer complex with pyridinium ion), and is another way of describing strong and weak nucleophiles, which concept has been used earlier (172).

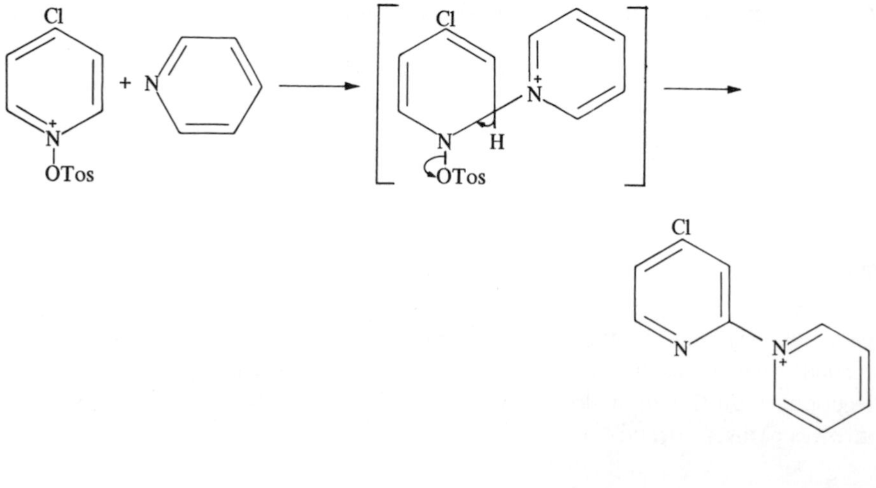

In the presence of tosyl chloride, 2- or 4-chloropyridine-1-oxide (as the *O*-tosylates) react with pyridine to give 4- and 2-*N*-pyridinium salts, respectively, the nuclear chlorine atoms not taking part in the reaction (377). Enamines react with pyridine-1-oxide *O*-benzoates or *O*-tosylates to give ketones (see Chapter IV, Section IV.2.C.) (377a).

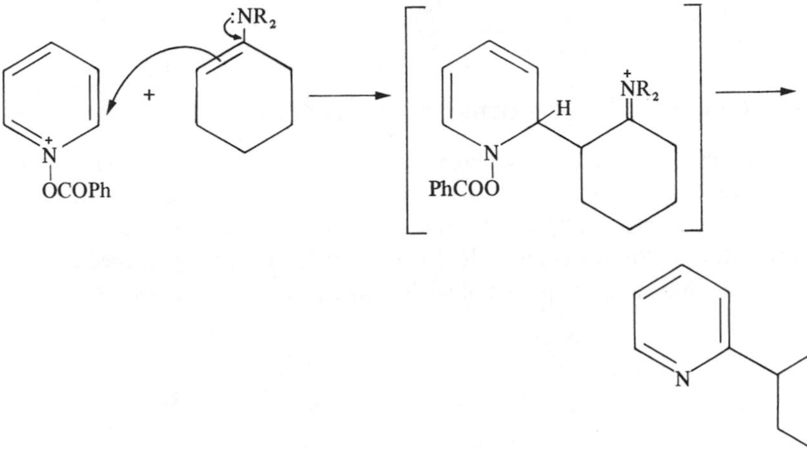

(3) ***Pyridine-1-oxides.*** Direct nucleophilic displacement of hydride in pyridine-1-oxides is usually preceded by attack at the oxygen atom followed by intramolecular nucleophilic attack at the α-position. These reactions are, therefore, discussed in Section III.5.D. In quinoline-1-oxide, however, initial nucleophilic attack by ozone at an α-position leading to 1-hydroxy-2-quinolone

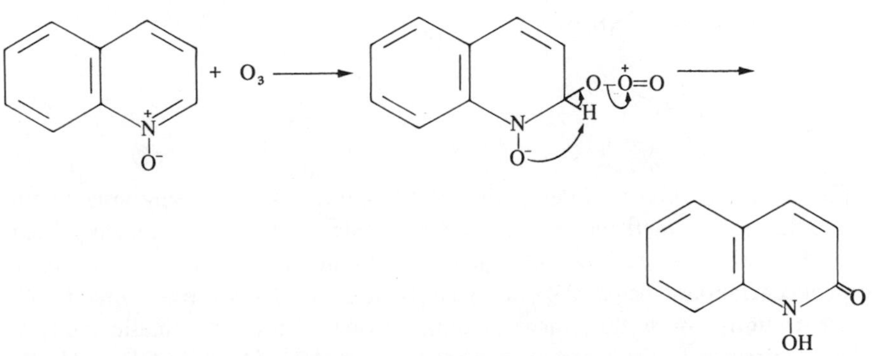

has been suggested (378, 379). Sufficiently activating substituents may lead to direct attack of the pyridine-1-oxide nucleus. Thus 3,5-dinitropyridine-1-oxide probably forms a Meisenheimer-type complex in alkali (380).

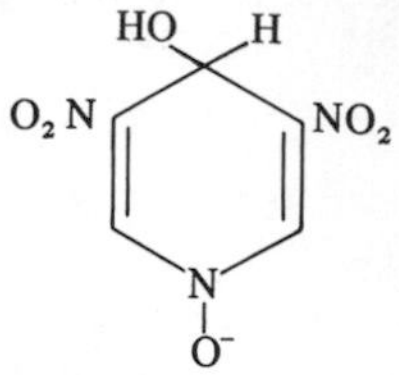

HO H

O_2N NO_2

N

O^-

b. DISPLACEMENT OF GROUPS OTHER THAN HYDROGEN

(1) ***Pyridines.*** Although Meisenheimer-type compounds have been known and studied for many years in the benzene series, the pyridine analogs have been prepared only recently (381–383). Spectroscopic evidence has shown the σ-complex nature of these compounds (**IA-59**) which have been prepared by the reaction of methoxide ion in methanol with activated pyridine derivatives (383).

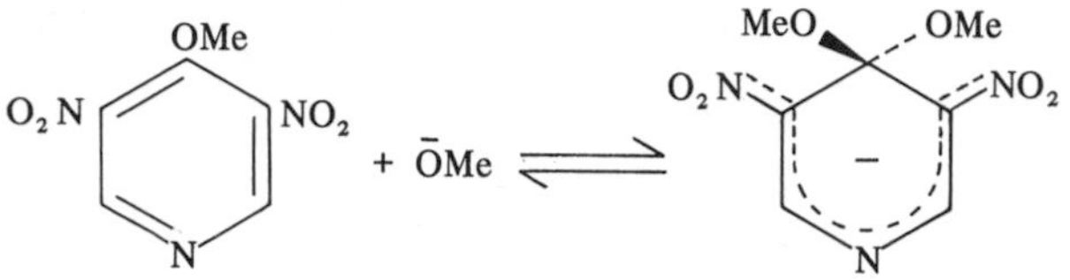

OMe O_2N NO_2 N + $\bar{O}Me$ ⇌ MeO OMe O_2N NO_2 N

IA-59

2-Methoxy-3,5-dinitropyridine reacts with methoxide ion in DMSO-d_6 to give **IA-60** (1,3-adduct), which shows no evidence of rearranging to a 2,2-dimethoxy adduct (1,1-adduct) (384, 385).

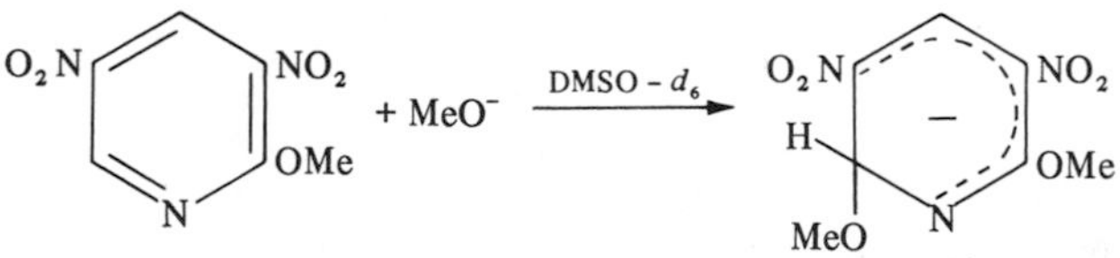

O_2N NO_2 OMe N + MeO^- $\xrightarrow{DMSO-d_6}$ O_2N NO_2 H OMe MeO N

IA-60

The kinetics of the interaction of 3,5-dinitro-4-methoxypyridine with methoxide ion in methanol and in MeOH-DMSO mixtures have recently been studied using a stopped-flow technique (386). In methanol, when the methoxide ion concentration is below 0.01 *M*, the 1,1-adduct **IA-59** is formed directly. At concentrations above this, the 1,3-adduct **IA-61** is formed (kinetic control) which rearranges to the thermodynamically more stable **IA-59**. In DMSO-MeOH solution, formation of **IA-61** was again usually found to precede that of **IA-59**. In agreement with the conclusions drawn by Illuminati and his co-workers (383) it was found that the influence of an aza group *para* to the geminal position on the stability of the 1,1-complex is intermediate between that of the nitro- and cyano-group, consistent with the more effective delocalization of the negative charge by an aza-group. Replacing a nitro or a cyano by an aza-group in the *ortho* position of the sp^3 hybridized carbon causes a 4.3 and 60-fold increase, respectively, in the equilibrium constant for the formation of the adduct.

MeO OMe OMe OMe
O_2N NO_2 OMe^- O_2N NO_2 O_2N NO_2
H
N N MeO N
IA-59 IA-61

In addition to studies on nucleophilic displacements of the various halides (*vide infra*), a number of other leaving groups have been examined. Phosphorus halides readily displace a 2- or 4-nitramino group to give the corresponding halide derivative (387). The methylsulfonyl group is twice as good a leaving group as is the methylsulfinyl group and either group is about 20 times more readily displaced from C-4 than from C-2 (388-390). The kinetics of displacement of the nitro-group from 2- and 4-nitropyridine by MeO^- in MeOH have been studied: $k_2^{35°}$ for 2-nitro is 0.35×10^{-4} l/(mole)(sec) while for 4-nitro it is 13.9×10^{-4} l/(mole)(sec); that is, $k_{4\text{-}NO_2}^{35°}/k_{2\text{-}NO_2}^{35°} = 39.7$ (391). In a 2-halo-4-nitropyridine, the nitro-group is more readily displaced than the 2-halogen by either hot aqueous potassium hydroxide or by ammonia, but methoxide ion and diethylamine displace the halide, albeit only to a small extent (392).

Displacement of halogen substituents has been studied from a number of different points of view (for reviews see Refs. 172, 393, and 394). The ease of displacement of halogens depends on (*a*) the nature of the nucleophile, (*b*) nature of the halogen, and (*c*) the other substituents on the pyridine ring. Under the proper conditions both bromide and chloride at C-2 are displaced by fluoride (395). Displacement of chloride from C-2 is substantially aided by electron withdrawing groups at C-3 and C-5, and even more so if the electron withdrawing groups occupy both C-3 and C-5 (396). The presence of a mesomerically or inductively electron donating group at C-4 retards displacement of a 2-chloride (396, 397), apparently by making $\Delta S^{\ddagger}$ more negative.

As discussed earlier, in the displacement of a 2-bromide or a 2-chloride by methoxide ion, a 3-methyl-group exerts less of a deactivating effect than does a 5-methyl-group because of an activating ion-dipole attractive interaction between the 3-methyl-group and the approaching nucleophile (398). On the other hand, a 3-methyl-group accelerates displacement by thiophenoxide ion of bromide at C-2, while still retarding displacement of a 6-bromo-substituent: ${}^{o\text{-Me}}_{\text{H}}K = 2.46$, ${}^{p\text{-Me}}_{\text{H}}K = 0.63$ (331). With phenoxide or thiomethoxide ions, the polarizability of the nucleophile is not sufficient for the London dispersion attractive interactions between it and the 3-methyl group to overcome the deactivating influence of the methyl group upon C-2, but 2-halo-3-methylpyridine is still more reactive than 2-halo-5-methylpyridine (332). The same reactivity ratios hold for the 2-fluoropyridine derivatives (332); for example, for 2-fluoro-3- and -5-methylpyridine with KSMe and KOMe in methanol at 110°: $k_{o\text{-Me}}/k_{p\text{-Me}} = 6.5$ for thiomethoxide ion displacement and $k_{o\text{-Me}}/k_{p\text{-Me}} = 1.3$ for methoxide attack, but in both cases, $k_{o\text{-Me}} < k_{\text{H}}$ (332).

The reactivity of 2-fluoropyridine permits displacement of the fluorine atom by α-tetralone oximate ion (399).

The displacement of halide ions in polyhalopyridines has been the object of much intensive study recently e.g., 399a. The relative reactivities of the various halogen atoms and the reactions they undergo are discussed extensively in Chapters VI and VII.

3- And 4-bromopyridine form complexes with $CoCl_2$ which, with methylmagnesium bromide, form 3- and 4-pyridyl radicals (400).

Displacement of a halogen can occur by an elimination-addition (EA) mechanism, in addition to the "normal" addition-eliminations (AE_n) discussed above, and the more complex mechanisms found by van der Plas in the diazine series. The EA-mechanism in the halopyridine series leads to the formation of pyridynes (dehydropyridines). The subject has been reviewed recently (401, 402) and is discussed in Chapters IV, VI, and IX.

Product studies on the reactions of 2-bromopyridine derivatives with amide ion have uncovered some interesting rearrangements and substitution patterns. 2,6-Dibromopyridine and amide ion give 6-amino-2-methylpyrimidine (**IA-62**), possibly through initial addition at C-4, tautomerization, ring cleavage, and recyclization (403). On the other hand, 3-amino-2-bromopyridine gave, under similar conditions, 3-cyanopyrrole (**IA-63**), by tautomerization, ring cleavage, and recyclization. The 3-hydroxy analog gives pyrrole-2-carboxamide, however,

Br Br N $\xrightarrow{NH_2^-}$ H₂N H Br Br N $\longrightarrow$ HN H H H Br Br N $\longrightarrow$

HN̄ Br Br N $\longrightarrow$ N CH₃ NH₂ N

IA-62

implying bond cleavage between C-3 and C-4 instead of between C-2 and C-3 as in the other cases (405).

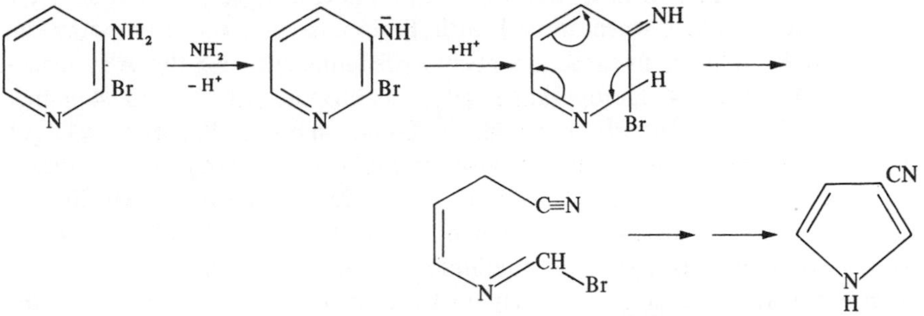

IA-63

In view of the above-mentioned reviews, only a brief account of pyridyne chemistry is given here. The origins of the rapid development of this field are well known. At about the same time, the den Hertog (406) and Kauffmann groups (407) found that treatment of 3-chloropyridine with an amide ion provided mixtures of 3- and 4-substituted aminopyridines, implying initial proton abstraction and subsequent formation of a 3,4-pyridyne intermediate (EA mechanism).

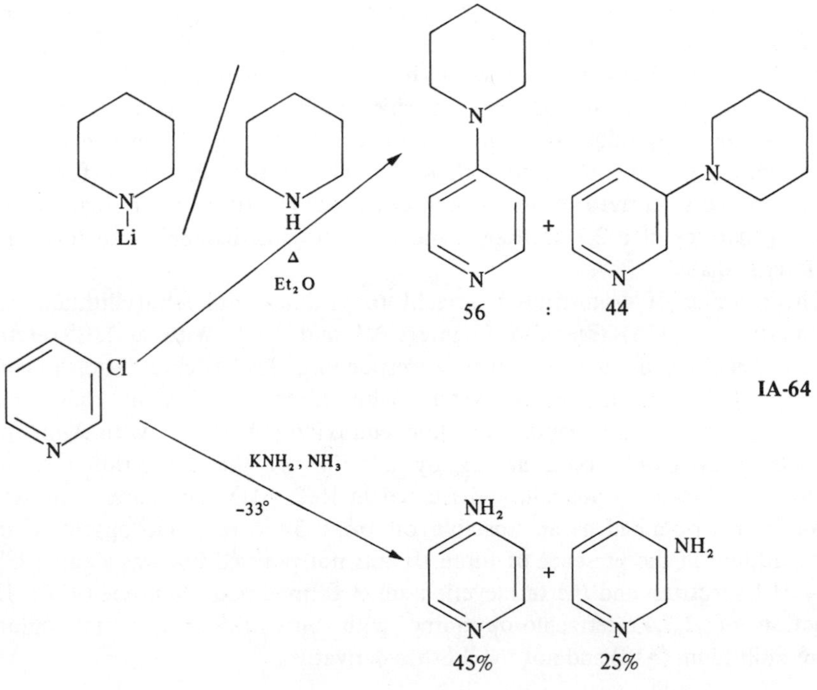

IA-64

Whether or not a pyridine is formed is very dependent on the nature of the halogen and of the base/nucleophile. 3-Bromo- and 3-iodopyridine give results similar to those illustrated in **IA-64**. 3-Fluoropyridine, however, gives 96% of the 3-substituted product when treated with lithiopiperidide (408) or potassium amide (409); that is, the reaction occurs predominantly *via* the AE_n mechanism. Moreover, 4-chloropyridine gives a mixture of 3- and 4-aminopyridine when treated with sodamide in liquid ammonia but only 0.4% of 3-piperidinopyridine on treatment with lithiopiperidide (408). The increased participation of the AE_n mechanism in this case is in the order lithium diisopropylamide < diethylamide < piperidide, and has been explained on the assumption that, with decreasing bulkiness, the amines react more rapidly at C-4 to give the AE_n σ-complex than they abstract *ortho*-hydrogen. The bulk of the leaving halogen is also important (410). 3,4-Pyridynes have been trapped not only with amines [and mixtures of amines (401)] but also with excess thiomethoxide ion to give equal amounts of 3- and 4-methylthiopyridine (411).

Nuclear substituents exert orientation effects analogous to those found in the carbocyclic arynes (412–414), and influence the extent of direct addition or rearrangement (415).

As discussed earlier, 2,3-pyridynes are calculated to be appreciably less stable than 3,4-pyridynes (173) and experimentally, the 2,3-pyridynes are, in fact, much more elusive. Products derived from 2,3-pyridynes can be obtained if other possibilities (e.g., 3,4-pyridyne formation) are blocked (413), or are not available. Ethoxy-2,3-pyridynes have been generated from 2,3-dibromopyridines with lithium amalgam, and trapped with furan. On the other hand, no pyridyne could be generated and trapped in this manner from the reaction of 3-bromo-4-ethoxypyridine with potassium amide in liquid ammonia (416). Treatment of 3-bromopyridine with KNH_2/NH_3 in the presence of furan gave only products derived from 3,4-pyridyne (417). It is clear that in the dehalogenations of a 2,3-dihalopyridine a 2-pyridyl carbanion is not formed as an intermediate.

The reaction of 4-substituted tetrachloropyridines with *n*-butyllithium have been studied (418) (see also Chapters VI and VII). With a 2,3,5,6-tetrachloro-4-dialkylaminopyridine the corresponding 2,5,6-trichloro-4-dialkylamino-3-pyridyllithium is formed which, with furan in ether at room temperature gives the 5,8-dihydroquinoline endoxide (**IA-65**), or with *N*-methylpyrrole gives **IA-66**, both arising by addition of the 2,3-pyridine to the diene (see, however, possibility discussed in Ref. 401). The parent endoxide (**IA-67**) was obtained as an unstable oil from 3-bromo-2-chloropyridine and butyllithium in the presence of furan. It was not purified but was identified by its NMR spectrum and the tetracyclic adduct formed with 2-butene (418). The reaction of 2,3,4,6-tetrachloropyridine with butyllithium gave tetrachloro-4-pyridillithium (418) and not the 3-lithio-derivative.

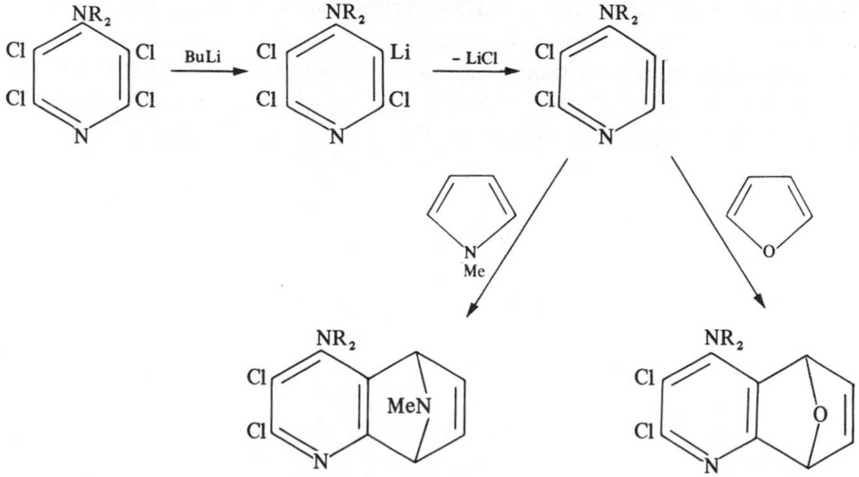

NR$_2$ Cl Cl Cl Cl N BuLi NR$_2$ Cl Li Cl Cl N − LiCl NR$_2$ Cl Cl N

N Me O

NR$_2$ Cl Cl MeN N

IA-66

NR$_2$ Cl Cl O N

IA-65

Br Cl N BuLi Et_2O − 70° Li Cl N O R.T. O N

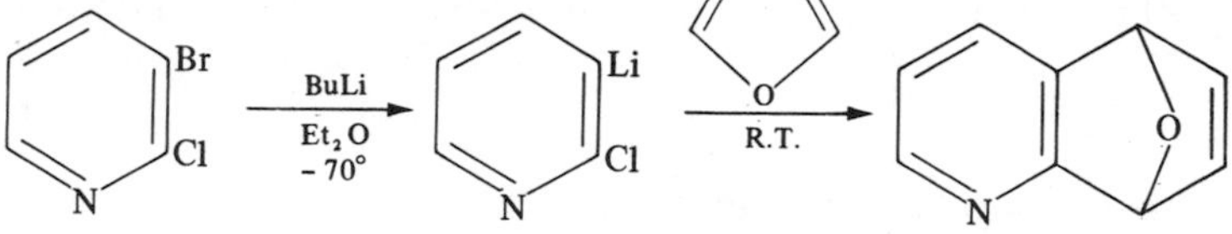

IA-67

When 3-bromo-4-ethoxypyridine is treated with potassium amide in liquid ammonia, 2-amino-4-ethoxypyridine is obtained (some bromine migration occurs also). Since formation of a 2,3-pyridyne has been ruled out in these cases, it has been proposed that this is an example of a "telesubstitution" reaction, proceeding *via* an intermediate such as **IA-68** (a *meta*-dehydropyridine).

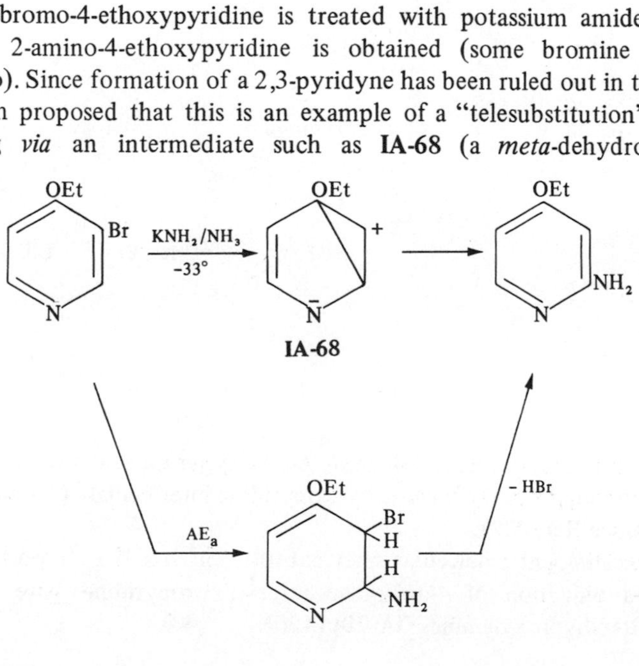

OEt Br N KNH_2/NH_3 −33° OEt + N̄ OEt NH$_2$ N

IA-68

AE$_a$ OEt Br H H NH$_2$ N − HBr

Alternatively, an "abnormal" addition-elimination mechanism (AE_a) may be involved here. Another example of a telesubstitution is that of 2-bromo-6-ethoxypyridine with potassium amide with or without diethyl ketone present (419). The products formed **(IA-69)** have again been rationalized on the basis of two competing reactions: the normal AE_n addition leading to

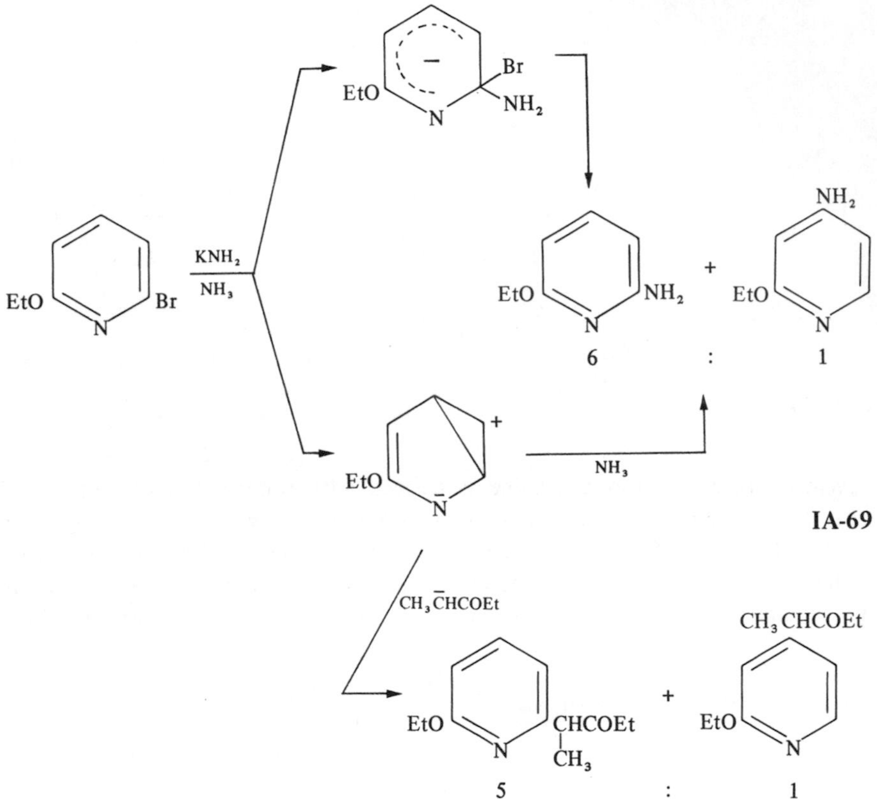

IA-69

2-amino-6-ethoxypyridine, and 4-amino-6-ethoxypyridine, and the two alkylated products coming from the *meta*-dehydropyridine intermediate (for an alternative discussion, see Ref. 402).

An interesting intramolecular *ipso* substitution has been reported wherein peroxyacid oxidation of 4-dialkylaminotetrafluoropyridines gave the *N,N,O*-trisubstituted hydroxylamines **(IA-70)** (420).

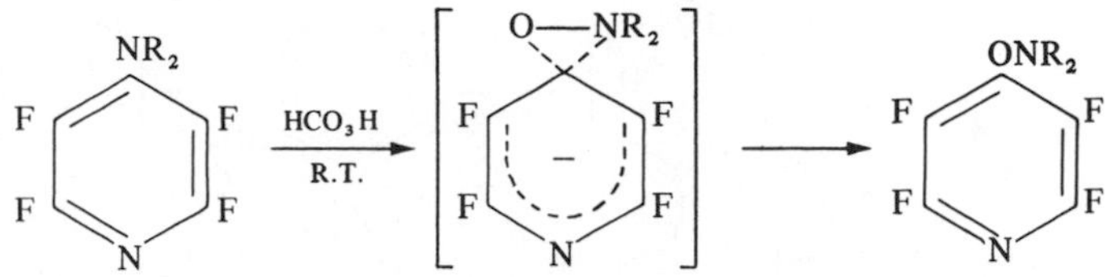

IA-70

(2) Pyridine-1-oxides. Pyridine-1-oxides are much more susceptible to both electrophilic and nucleophilic substitution than are pyridines themselves. In general, a substituent is more easily displaced from a pyridine-1-oxide than from the corresponding pyridine; for example, 2-bromopyridine-1-oxide is 760 times more reactive toward methoxide ion in methanol than is 2-bromopyridine (398), and the order of ring substituent reactivity is pyridinium salt > *N*-oxide > free bases > homocyclic, at all positions (421). The rate or free-energy order of position reactivity is 4 > 2 > 3 for pyridine, 4 ~ 2 > 3 for their *N*-oxides and 2 > 4 > 3 for the pyridinium compounds (for a review, see Ref. 394). The activating power of the heteroaromatic *N*-oxide function in S_NAr displacements is slightly greater than that of a nitro group, but steric inhibition of mesomerism cannot affect this activation, and the steric requirements for hydrogen bond formation are different. The k_2/k_4 ratio for the reactions of piperidine with 2- and 4-nitropyridine-1-oxides is very high: 1700 at 20° [$k_2 = 5400 \times 10^{-6}$ 1/(mole)(sec); $k_4 = 3.2 \times 10^{-6}$ 1/(mole)(sec)], and this has been rationalized in terms of the more powerful inductive effect of the *N*-oxide function at C-2 than at C-4, coupled with the formation of a hydrogen bond between the piperidine *N*-hydrogen atom and the *N*-oxide oxygen and with the relief of strain due to coulombic dipolar repulsion in the ground state of the 2-nitro-compound. At C-2, the NO_2 group is far more labile than Br toward piperidine (422, 423). Under the proper conditions (excess dialkyl amine) a halogen at C-2 is displaced before a nitro group in the same molecule at C-4 (424,425). A fluorine atom at C-3 is displaced by a number of nucleophiles faster than a nitro group at C-4 (426); the 3-substituent probably forces the nitro group out of coplanarity with the ring in the ground state so that attack at C-3 leads to steric acceleration in the transition state since the sp^3 hybridization at C-3 then permits the 4-nitro group to achieve coplanarity with the ring. With the nitro group out of the plane of the ring in the ground state its −M effect at C-3 is considerably weakened.

In contrast to the behavior of pyridines, treatment of 2-chloropyridine-1-oxide with potassium amide gives a low yield of 3-aminopyridine-1-oxide, presumably *via* 2,3-pyridyne-1-oxide, together with the 2-amino-derivative (formed predominantly by an AE_n mechanism) (427–429a). Since the C-3 proton is relatively not acidic, it is not surprising that the AE_n mechanism competes very effectively with the EA pathway. Very recently, the reaction of chloropicoline-1-oxides with potassium amide in liquid ammonia was investigated and found to yield pyridynes in some cases (429b). 2- And 3-bromo-4-ethoxypyridine-1-oxides are also believed to react with sodamide largely *via* 4-ethoxy-2,3-pyridyne-1-oxide to give, in each case, 3-amino-4-ethoxypyridine-1-oxide exclusively. This reflects the strong directive influence of the *N*-oxide function (427). As predicted (183), *N*-oxide formation permits facile proton abstraction from C-2 in 3-halopyridines. Thus while 3-fluoropyridine-1-oxide appears to be unaffected on treatment with sodamide in liquid ammonia (proton-abstraction would remain undetected in a protic solvent), 3-chloro- and 3-bromopyridine-1-oxides are converted to 3-amino-pyridine-1-oxide. Had this been a normal AE_n substitution, the fluoro derivative should have reacted the fastest (430). A small amount of 2-aminopyridine-1-oxide was also isolated, confirming the formation of 2,3-pyridyne-1-oxide.

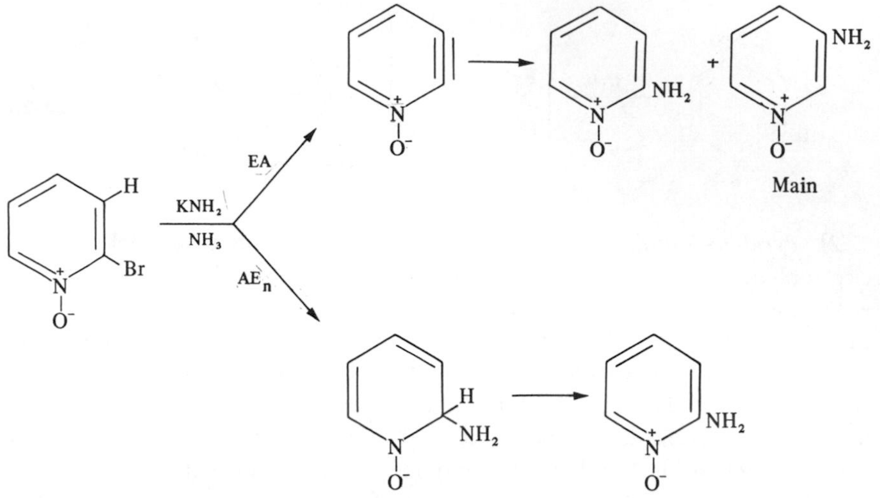

D. *Intramolecular Nucleophilic Substitution*

Many examples of intramolecular cyclization reactions involve a pyridine ring system, and this is probably the most active field of research in pyridine chemistry at the present time. Only a few important examples are discussed here.

The preparation of α- and γ-chloro heterocycles from pyridine-1-oxides and phosphorus oxychloride and pentachloride proceeds, at least in part, by an intramolecular mechanism. The qualitative effect of substituents upon the orientation of the entering chlorine has been reviewed (56, 172). 3-Substituents direct mainly to the 2-position. The effect of a 3-methyl substituent has been studied quantitatively (431) and the product ratios are summarized in Table IA-6. The decrease in the amount of substitution at C-4 when $POCl_3$ was used relative to that with PCl_5 was attributed to a cyclic transition state being

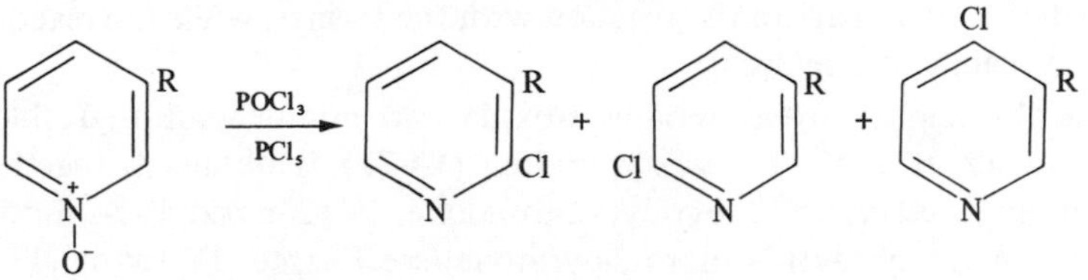

TABLE IA-6. Isomer Ratios for the Chlorination of Some Pyridine-1-oxides (431)

		Ratio of *x*-chloro substituted pyridines		
N-Oxide of	Phosphorus halide	2-	4-	6-
Pyridine	PCl_5	1	1.41	1
	$POCl_3$	1	0.94	1
	PCl_5–$POCl_3$ (1 : 4.5)	1	1.36	1
3-Picoline	PCl_5	1	1.82	0.71
	$POCl_3$	1	1.48	0.83
	PCl_5–$POCl_3$ (1 : 4.5)	1	1.43	0.76
3,4-Lutidine	PCl_5	1	–	0.57
	$POCl_3$	1	–	1.21
	$POCl_3$ + excess NaCl	1	–	1.17

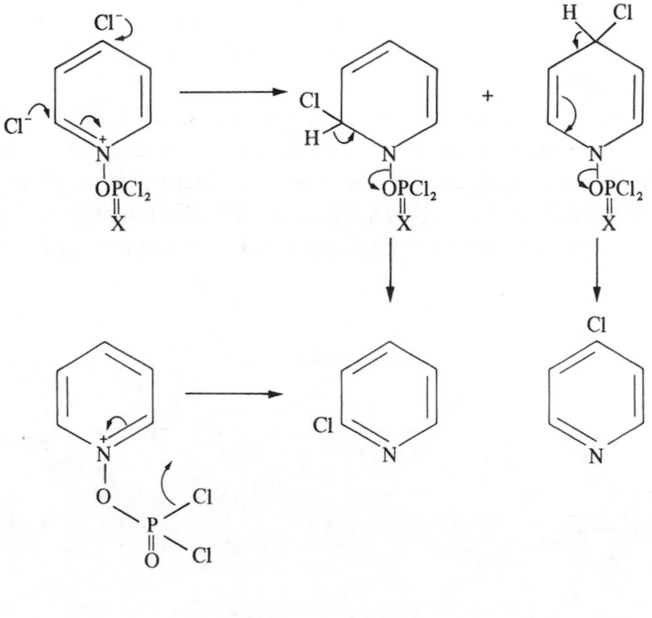

involved at least in part in the reaction with the former, while the reaction with PCl_5 was mainly intermolecular.

When the adduct from pyridine-1-oxide and *p*-toluenesulfonyl chloride is heated to about 205° 3-tosyloxypyridine (**IA-71**) is obtained, together with 2,3′-dipyridyl ether, *N*-(2′-pyridyl)-2-pyridone, *N*-(2′-pyridyl)-5-chloro-2-pyridone, and *N*-(2′-pyridyl)-3-chloro-2-pyridone (see Chapter IV for details) (432). 3-Picoline-1-oxide gives, among other products, 5-tosyloxy-3-picoline (433). On the basis of ^{18}O labeling studies Oae and his co-workers (434) modified an earlier mechanistic proposal (435) and suggested that formation of **IA-71** proceeded *via* an intimate ion pair depicted as **IA-72**. This is essentially

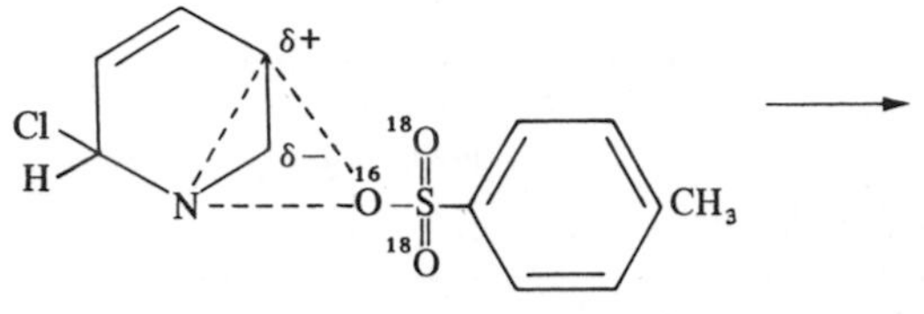

IA-72

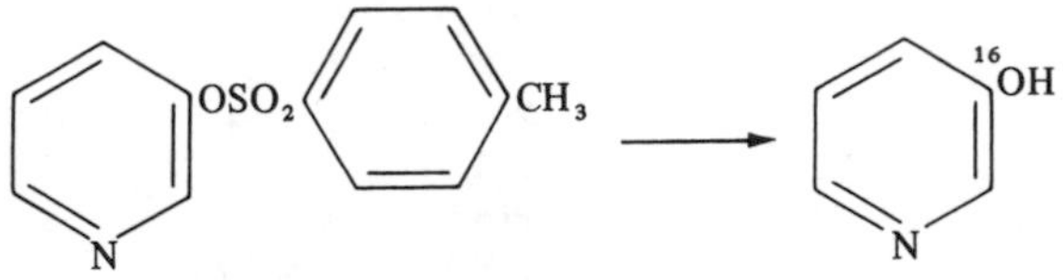

IA-71

equivalent to a 1,5-sigmatropic shift for which there is supporting evidence in 1,2-dihydropyridines (*vide infra*).

The reaction of pyridine-1-oxides with ketenes is discussed in Chapter IV (Section IV.E). A variety of products are obtained, depending on the ketene and the *N*-oxide. For example, one of the products obtained from diphenylketene and pyridine-1-oxide is the ylide **IA-73**, and a 1,3-dipolar addition to pyridine (formed in the reaction) has been postulated (436) to account for its formation:

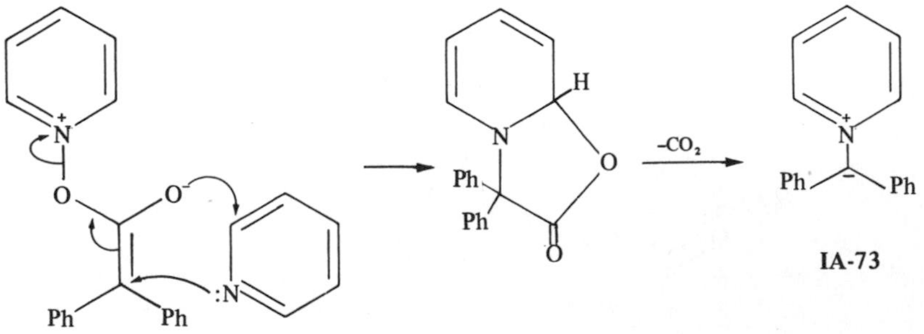

IA-73

The reaction of dimethylketene with pyridine-1-oxide is even more complicated (437) and a number of intramolecular nucleophilic cyclization products have been postulated (see Chapter IV for details).

1,3-Dipolar addition of phenylisocyanate to pyridine-1-oxide leads to an intermediate (**IA-74**) which, at the temperature of the reaction, spontaneously

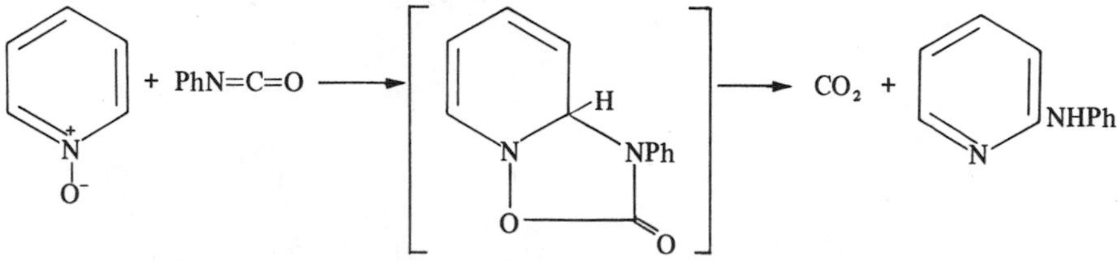

+ PhN=C=O ⟶ [H, N, NPh, O, O] ⟶ CO_2 + NHPh, N

IA-74

loses CO_2 to give 2-anilinopyridine (438a). Such an intermediate has now been isolated from the reaction of phenylisocyanate and 3-picoline-1-oxide and shown to be relatively thermally stable but to be decomposed readily by base (438b). A similar 1,2-dihydro-intermediate is presumably formed in the reaction of pyridine-1-oxide with perfluoropropene to give 2-(1,2,2,2-tetrafluoroethyl)pyridine and COF_2 (439).

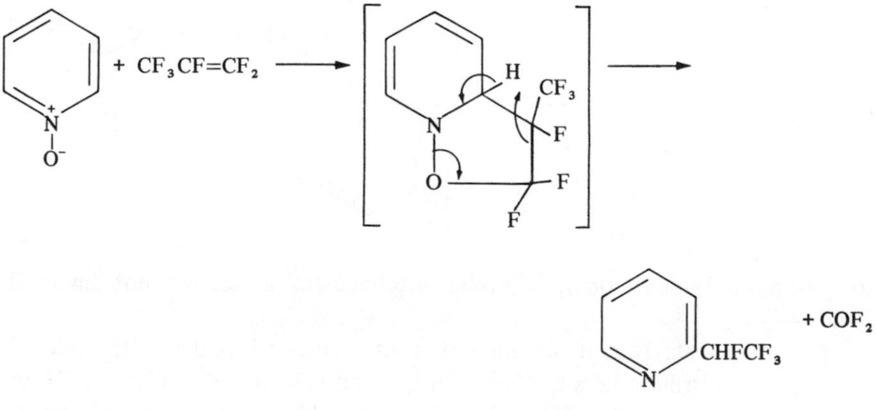

+ $CF_3CF{=}CF_2$ ⟶ [H, CF_3, F, F, F, N, O] ⟶

CHFCF_3 + COF_2

The reaction of pyridine-1-oxide with 2-bromopyridine has been known for some years to give 1-(2-pyridyl)-2-pyridone (**IA-75**) (440, 441). More recently, some of the by-products formed have been identified as a dipyridyl ether **IA-76**, 2-pyridone, and 1-(2-pyridyl)-3-bromopyridine and 2-bromopyridine (442–444). This and related reactions are discussed in detail in Chapter IV, Section IV.5. Taking into account these results (445) and related ones (446), formation of these products may be explained by a modification of the scheme proposed earlier (441). In the reaction of 4-bromopyridines with *N*-oxides, which also lead to 1-(4-pyridyl)-4-pyridones, a cyclic intermediate is clearly not involved (442–444).

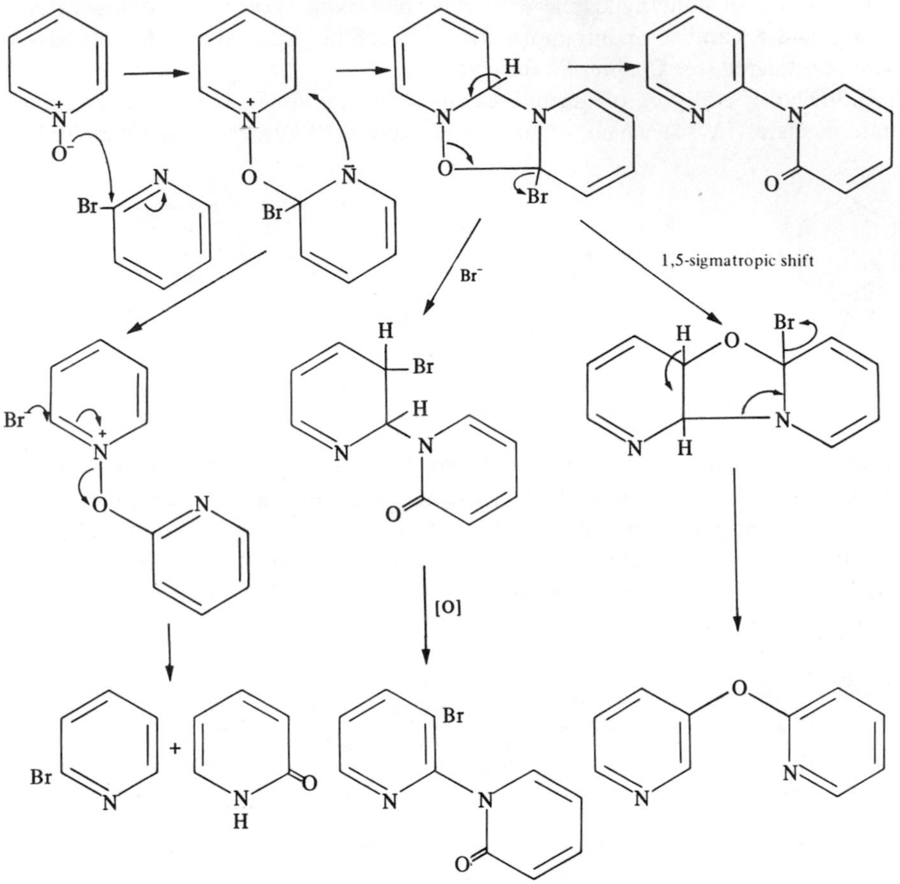

The direct alkyl- and arylamidation of pyridine-1-oxides with imidoyl chlorides and nitrilium salts has been studied and is reviewed in Chapter IV. In addition of the amides **(IA-75)**, benzanilide and 3-chloropyridines were isolated, often in respectable yields, when *N*-phenylbenzimidoyl chloride was used (447, 448). If a thiol was added to the reaction mixture, it could intercept the 1,2-dihydro adduct and lead to a 3-alkyl- or -arylthiopyridine (449). The influence of a 3-substituent upon the orientation of the entering amide function has been studied quantitatively (448), as has the influence of the nature of R and R′ in the imidoyl chloride (see Chapter IV). If the 2- and 6-positions are blocked, the 1,2-dihydro-intermediate can undergo a 1,5-sigmatropic shift to a 2,3-dihydro-intermediate **(IA-76)** which, upon aromatization, gives an *O*-3-

+ RC=NR′ Cl → Intra- or intermolecular → - HCl → RCONHR′ + Cl → −HCl IA-75

pyridyl-*N*-phenylbenzimidate (e.g. **IA-77**) (446a). Similar products have been obtained from 2,6-dihalo- and -diphenylpyridine-1-oxides (446b). A similar 1,5-sigmatropic shift has more recently been observed in the reaction of 3,5-dibromopyridine-1-oxide with phenylisocyanate (438b).

Me Me + PhC=NPh Cl → [Me Me NPh O Ph Cl] → [H O Cl Ph Me Me N Ph] → NPh OCPh Me Me

IA-76 **IA-77**

Attempted intramolecular acylaminations using the *N*-aminopyridinium salts (**IA-78**), instead of the *N*-oxides, *via* the isolable amidine salts **IA-79** have failed to date (450). On the other hand, *N*-ylides **IA-80** underwent intramolecular

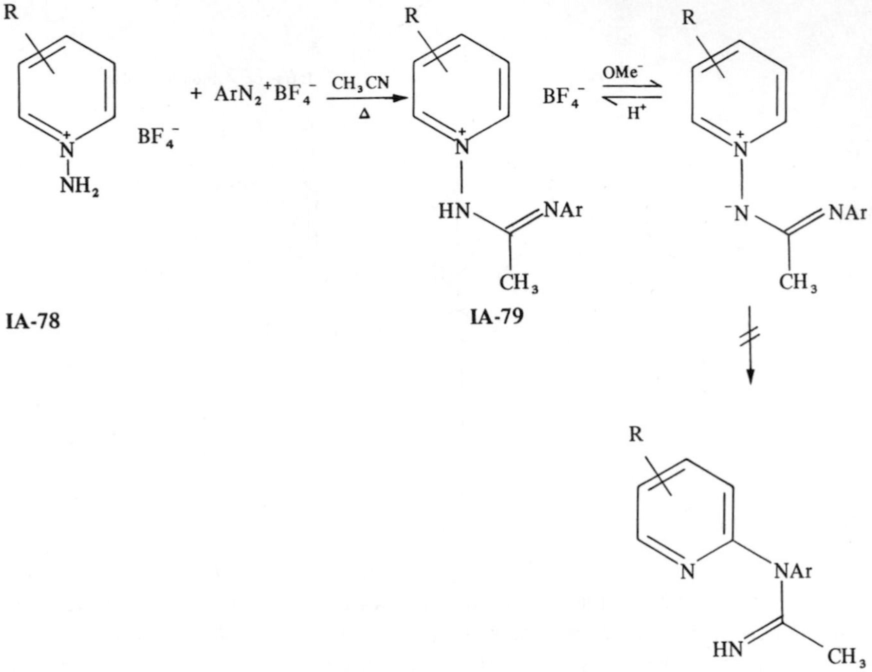

R NH$_2$ BF$_4^-$ + $ArN_2^+BF_4^-$ $\xrightarrow[\Delta]{CH_3CN}$ R HN NAr CH$_3$ BF$_4^-$ $\underset{H^+}{\overset{OMe^-}{\rightleftharpoons}}$ R $^-$N NAr CH$_3$

IA-78 **IA-79**

R N NAr HN CH$_3$

cyclization in cold chloroform solution (451). It is of considerable interest that, if the ylide **IA-80** bore one 3-substituent, cyclization at the sterically more hindered site predominated over cyclization at C-6. The 1,2-dihydro-product (**IA-81**) could be aromatized with palladium on carbon or with tetracyanoethylene, or by photolysis in acetone at 25°. No ring-opening to give the 2-substituted pyridine was reported.

Reaction of phenylpropiolonitrile with pyridine-1-oxide in boiling ethylene chloride gave only a very small amount of the expected intramolecular cyclization produce **(IA-82)**. The main product was the zwitterion **IA-83** together with some 2:1 product **IA-84** (452, 453). It was suggested that a possible mechanism for the production of **IA-83** from the initial adduct **(IA-85)** formed could involve benzoylcyanocarbene. Ylide **IA-86** was synthesized and found to be stable under the reaction conditions (453a).

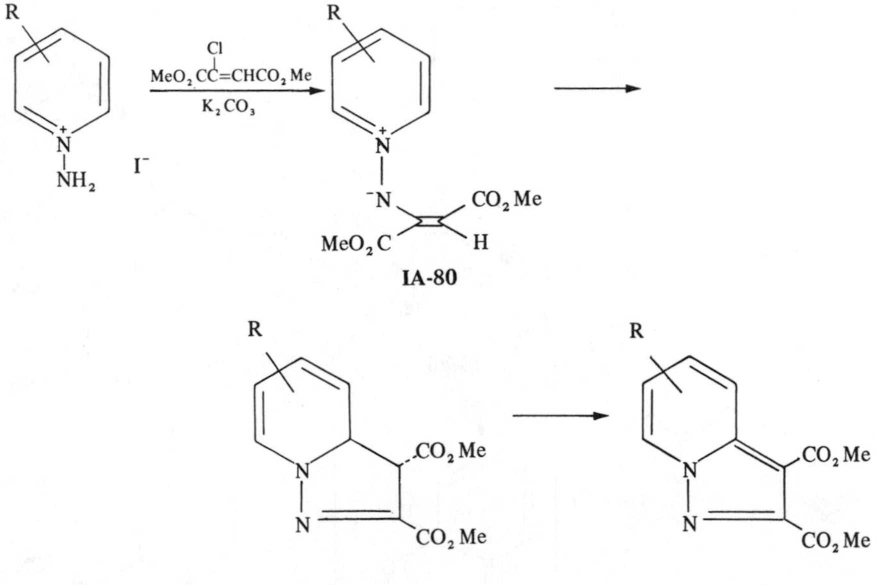

R, $\overset{+}{N}$, NH_2, I^-

Cl, $MeO_2C\overset{|}{C}=CHCO_2Me$, K_2CO_3

R, $\overset{+}{N}$, ^-N, CO_2Me, MeO_2C, H

IA-80

R, N, N, CO_2Me, CO_2Me

IA-81

R, N, N, CO_2Me, CO_2Me

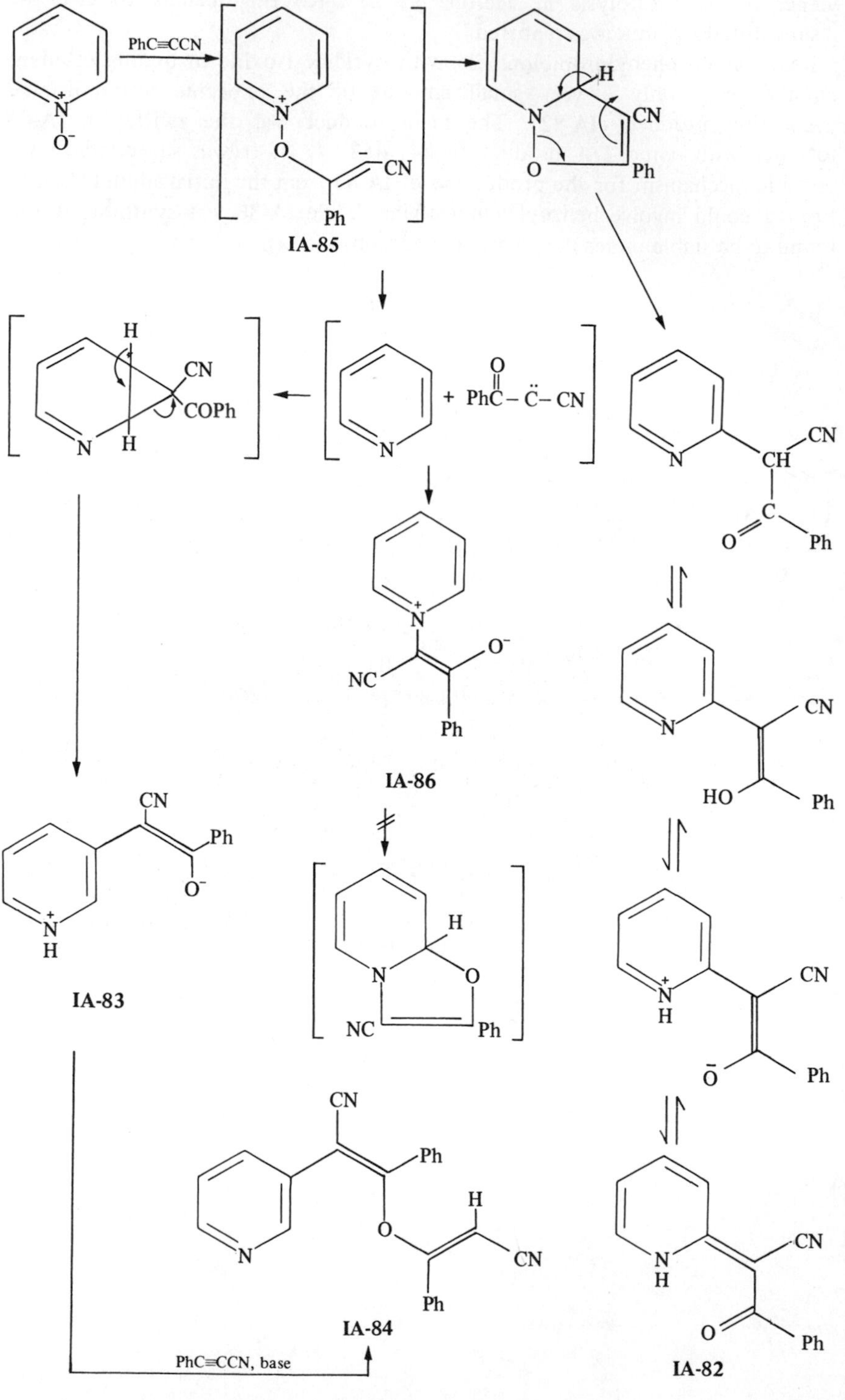

PhC≡CCN

CN

Ph

IA-85

H

CN

Ph

O

N

CN

COPh

H

H

N

+ PhC(=O)–C̈–CN

N

CN

CH

C

O

Ph

NC

O⁻

Ph

IA-86

CN

HO

Ph

CN

Ph

O⁻

IA-83

H

O

NC

Ph

CN

Ph

CN

Ph

H

O

CN

Ph

IA-84

PhC≡CCN, base

N

H

CN

O

Ph

IA-82

Benzoylcyanocarbene was generated in the presence of pyridine and indeed found to yield **IA-83** and **IA-86**, but the ratio of these products was the reverse of what was found in the reaction of pyridine-1-oxide with phenylcyanoacetylene. It appears that while **IA-86** is probably formed from the carbene as suggested, the latter can at best be only a minor route to the β-alkylated products. Two possible alternate pathways are feasible (453b): ring opening of the 1,2-dihydrointermediate to give the resonance stabilized dipolar species, followed by ring closure to the azanorcaradiene and ring opening to **IA-83**.

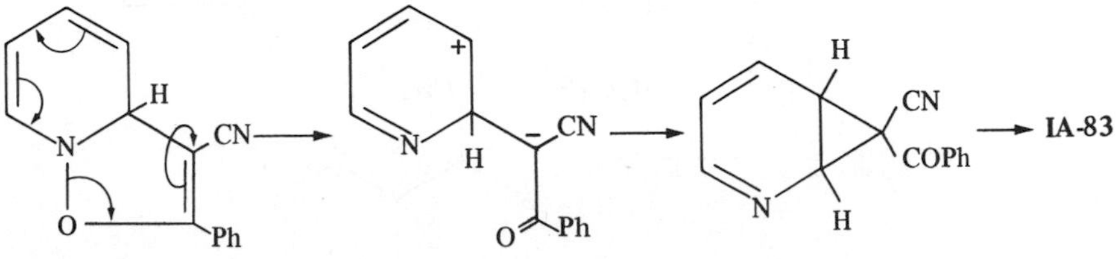

Alternatively and more likely, one can visualize a symmetry allowed concerted rearrangement ($_{\sigma}2_s + _{\pi}2_a + _{\pi}4_s$) of the 1,2-dihydro intermediate directly to the azanorcaradiene (453b). This latter process recieves considerable support from the reaction of pyridine-1-oxide and some of its derivatives with benzyne (453a).

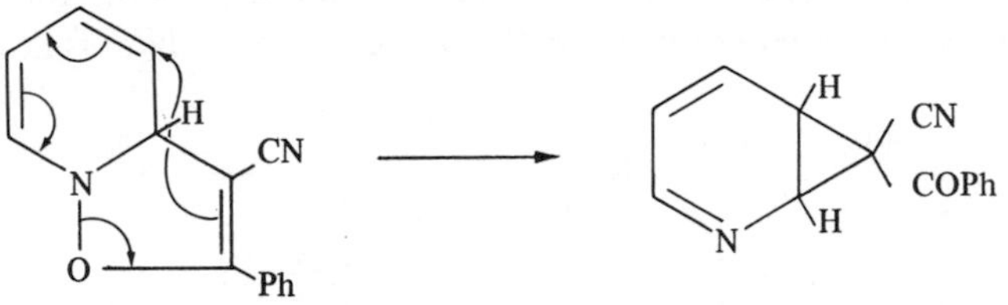

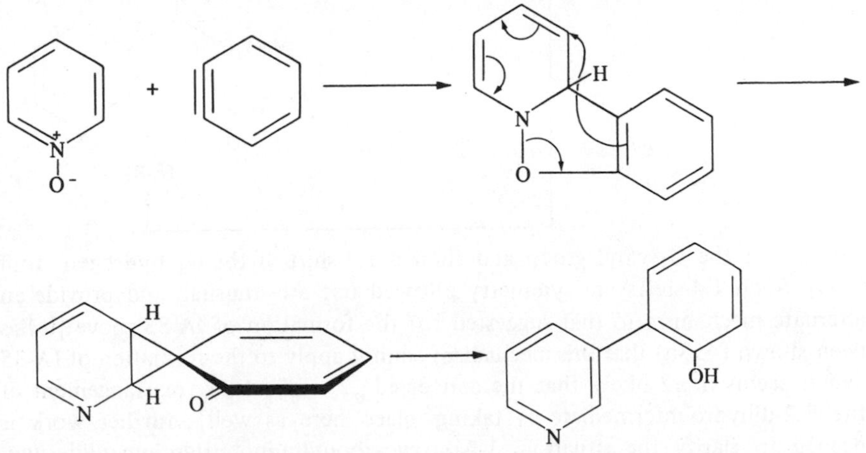

The main product is the 3-*o*-hydroxyphenylpyridine, whose formation may similarly be viewed as involving a concerted rearrangement of the type discussed above, going *via* a spirodienone intermediate. Formation of an intermediate dipolar species (a phenoxide ion in this case) would undoubtedly stop the reaction at that stage and not lead to the rearrangement to C_3. If the 3- and 5-positions are blocked, e.g. 3,5-lutidine-1-oxide, the main product is the 2-*o*-hydroxyphenylpyridine (453a).

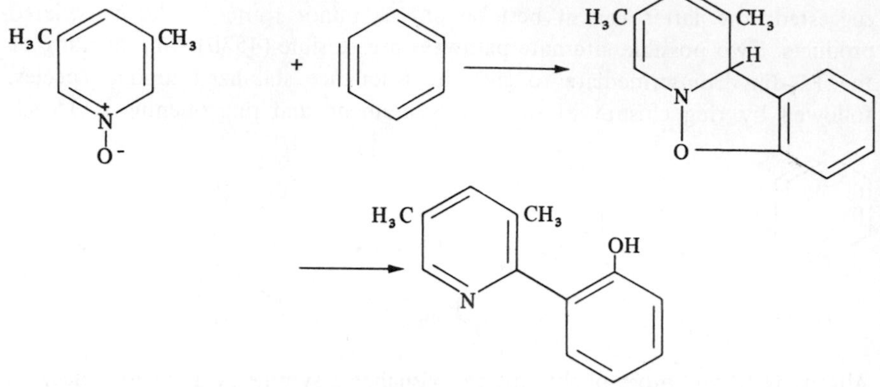

Addition of dimethyl acetylenedicarboxylate to 1-alkoxycarbonylimino-2-methylpyridinium ylide and its derivatives gave, in addition to the expected 1,3-cycloadduct, 3-vinylpyridine derivatives (**IA-87**) (208). A similar reaction occurred with ethyl propiolate. The mechanism proposed for the formation of **IA-87** involves an N–N bond fission in the dihydro-intermediate followed by a

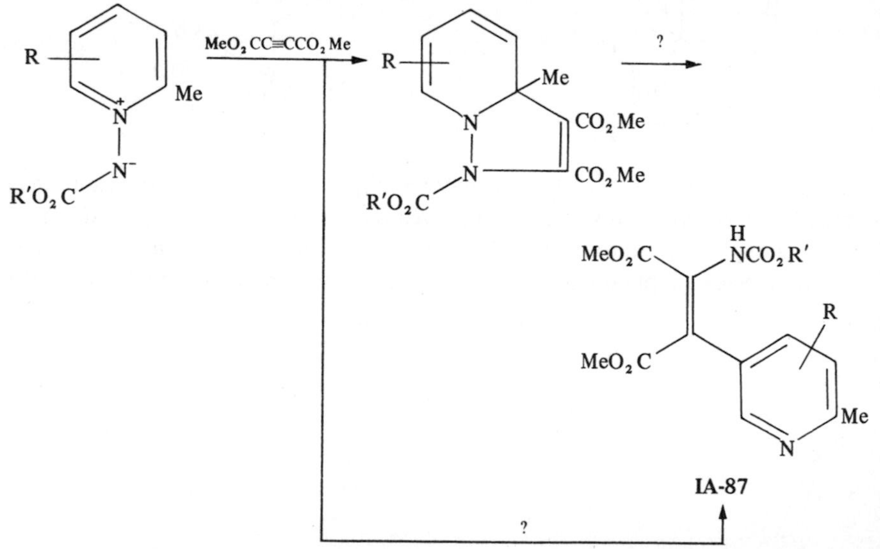

IA-87

1,4-shift of the C_2-vinyl group and then a 1,4-shift of the C_5 hydrogen atom (208). Such 1,4-shifts are symmetry allowed but are unusual, and provide an alternate mechanism to that suggested for the formation of **IA-85** above [it has been shown (453b) that this mechanism cannot apply to the formation of **IA-85** and it seems more likely that the concerted $_{\sigma}2_s + {}_{\pi}2_a + {}_{\pi}4_a$ rearrangement of the 1,2-dihydro-intermediate is taking place here as well]; further work is needed to clarify the situation. 1-Alkoxycarbonyliminopyridinium ylide and benzyne give pyrido [1,2-*b*] indazole (453a).

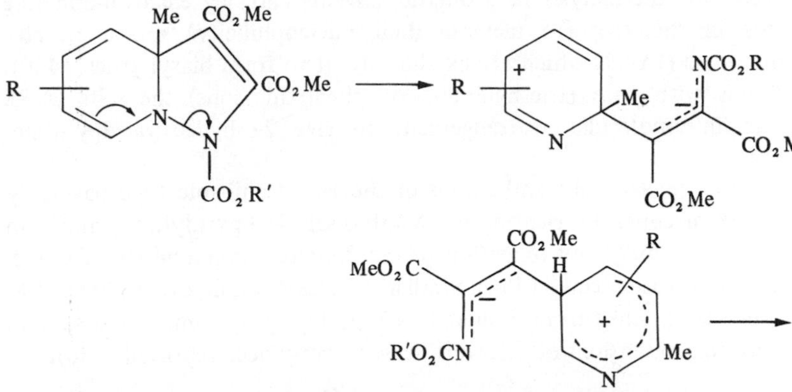

R — Me CO_2Me CO_2Me N N CO_2R' → R + N Me NCO_2R CO_2Me CO_2Me →

MeO_2C CO_2Me H R + $R'O_2CN$ N Me → IA-87

Pyridine-1-oxides react with aryldiazonium salts bearing electron-attracting substituents in the aromatic ring to give *N*-aryloxypyridinium salts **(IA-88)**

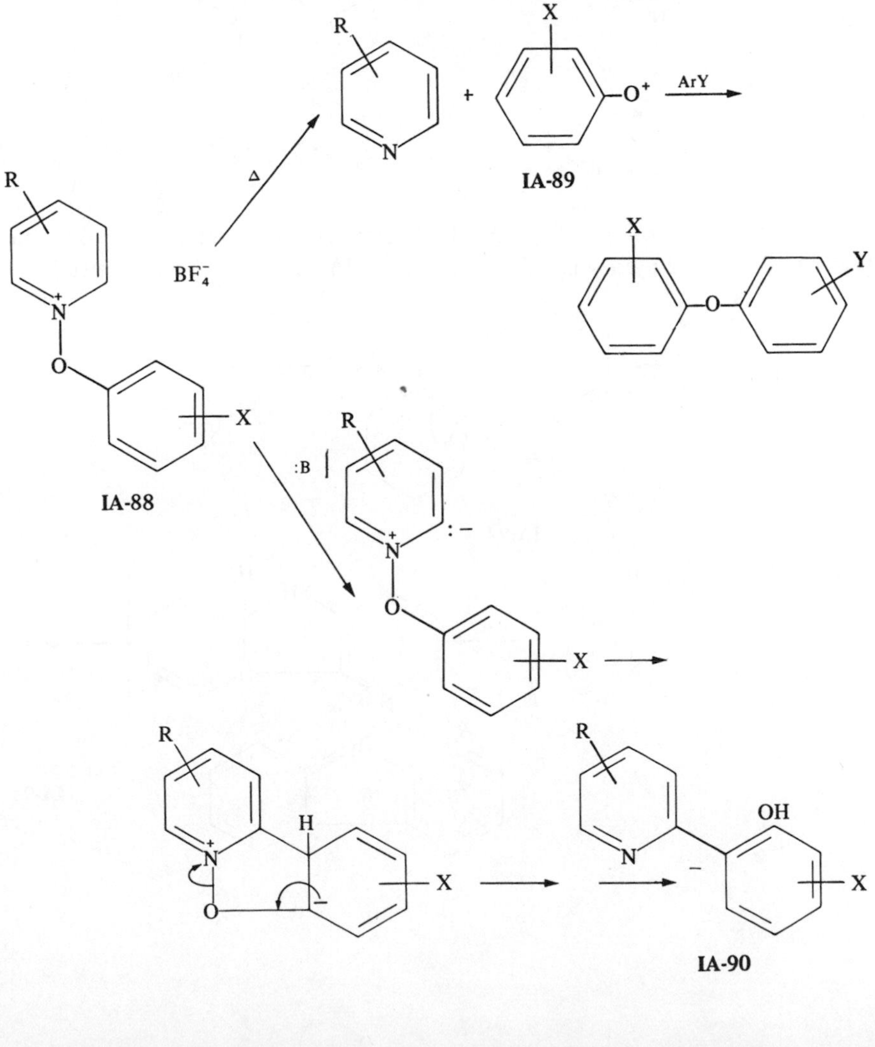

R N + X O^+ ArY → IA-89

R $\overset{+}{N}$ O X BF_4^- Δ

IA-88

X O Y

:B R $\overset{+}{N}$:– O X →

R $\overset{+}{N}$ O H X → → R N OH – X

IA-90

(141). These, on thermolysis in aromatic solvents carrying electron-donating substituents in the ring (to increase their nucleophilicity), appear to give aryloxenium ions (**IA-89**) which attack the solvent to form biaryl ethers (450). On treatment with base (triethylamine or phenoxide ions) the salts **IA-88** undergo an intramolecular rearrangement to give 2-*o*-hydroxyarylpyridines (**IA-90**) (141).

Intramolecular nucleophilic aminations of the Tschitschibabin type have only been reported infrequently. Heating the *N*-lithio salt β-(3-pyridyl)ethylamine in dioxane gives 7-azaindole by oxidation of the dihydro-compound first formed. No attack at C-4 was detected (454). Similarly, 1,2,3,4-tetrahydro-3-phenyl-1,8-naphthyridine is obtained from 2-phenyl-3-(3′-pyridyl)propylamine and sodium (455). Interesting extensions of this work have recently been reported (456).

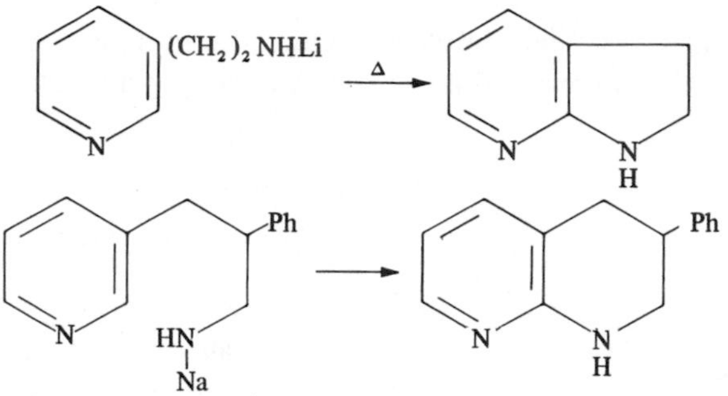

$(CH_2)_2NHLi$ Δ N N H

Ph HN Na N Ph N N H

The formation of pyrido[1′,2′ : 1,2]imidazo[4,5-*b*]pyrazines (**IA-91**) on heating 3-amino-2-pyrazylpyridinium salts (**IA-92**) has been viewed as

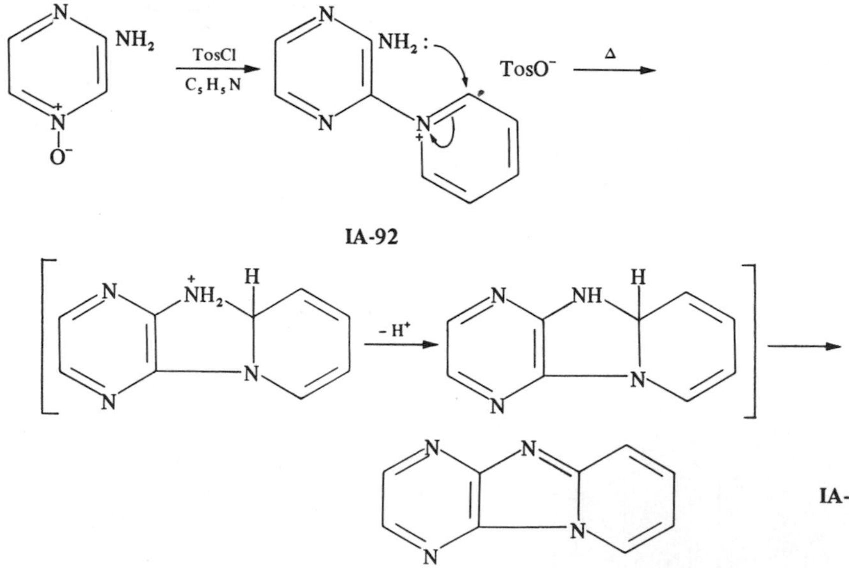

N NH_2 N O^- TosCl C_5H_5N N NH_2: N N TosO⁻ Δ

IA-92

N $\overset{+}{N}H_2$ H N N -H⁺ N NH H N N

N N N N

IA-91

proceeding by intramolecular nucleophilic attack by the amino function upon the electron deficient α-position of the pyridinium ring (457).

E. *Reactions with Acid Anhydrides and Related Compounds*

The reactions between pyridine-1-oxides and acid anhydrides or acid halides have been intensively investigated ever since Katada's first discovery of the reaction (56, 458–461). The reaction and related ones are discussed in detail in Chapter IV, Section IV.2.E, and also in Chapter XII.

A total of seven products **(IA-93)** may be isolated from the reaction of pyridine-1-oxide with acetic anhydride; the main product is either 2-pyridone or 2-acetoxypyridine, depending upon the reaction conditions (462, 463).

+ Ac_2O $\xrightarrow{\Delta}$ 2-pyridone (75%) + 2-acetoxypyridine ($OCOCH_3$) (9.5%)

+ 3-acetoxypyridine ($OCOCH_3$) (3.5%) + 2-acetylpyridine ($COCH_3$) (2%) + 2-aminopyridine (NH_2) (2%) **IA-93**

+ 1-(2-pyridyl)-2-pyridone (7%) + pyridine

The mechanism shown in **IA-94** appears to be the most reasonable explanation of most of the products based upon Oae's findings of complete ^{18}O scrambling (464) and the lack of a hydrogen-deuterium isotope effect (465, 466). The intermolecular attack by acetate ion must be the rate-determining step.

3-Acetoxypyridine arises from the 1,2-dihydro-intermediate. This could occur in one of three ways: intermolecular addition of acetate to the 1,2-dihydro-compound; 1,5-sigmatropic rearrangement of the *N*-acetoxy-group to the 3-position; or intramolecular migration of the 2-acetoxy-group to C-3 *via* a cyclic intermediate **(IA-95)** (467). 2-Aminopyridine is a hydrolysis product of

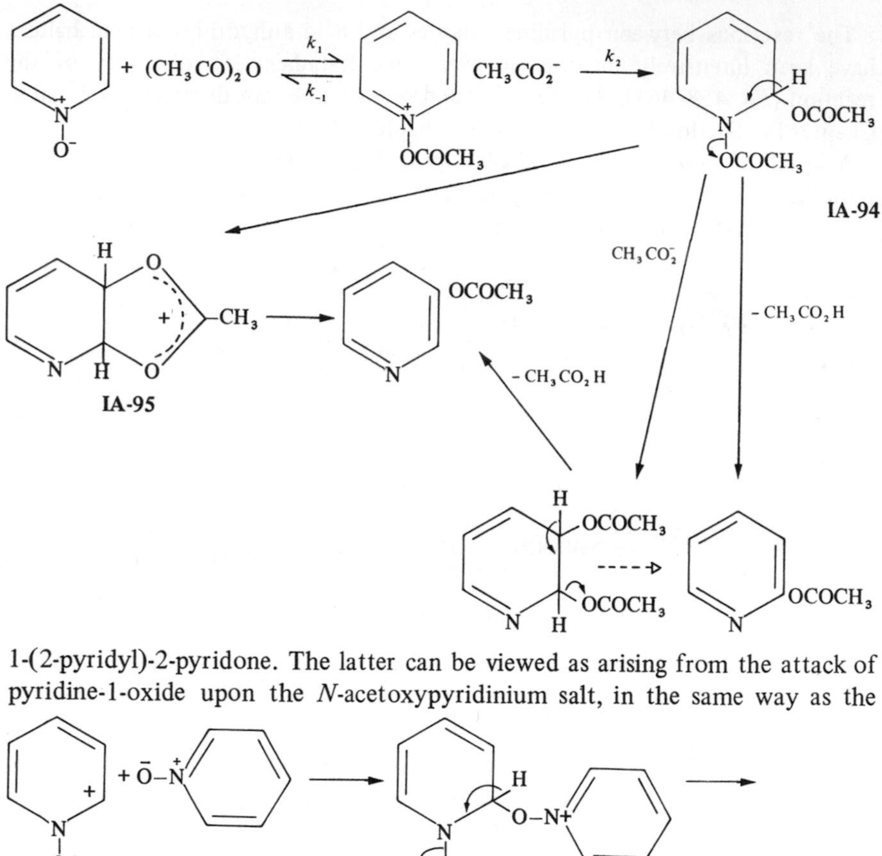

+ $(CH_3CO)_2O$ $\underset{k_{-1}}{\overset{k_1}{\rightleftharpoons}}$ $CH_3CO_2^-$ $\xrightarrow{k_2}$ H OCOCH$_3$ OCOCH$_3$ OCOCH$_3$

IA-94

H O + CH$_3$ OCOCH$_3$ N H O N

IA-95

$CH_3CO_2^-$ $- CH_3CO_2H$ $- CH_3CO_2H$

H OCOCH$_3$ OCOCH$_3$ N H N OCOCH$_3$

1-(2-pyridyl)-2-pyridone. The latter can be viewed as arising from the attack of pyridine-1-oxide upon the *N*-acetoxypyridinium salt, in the same way as the

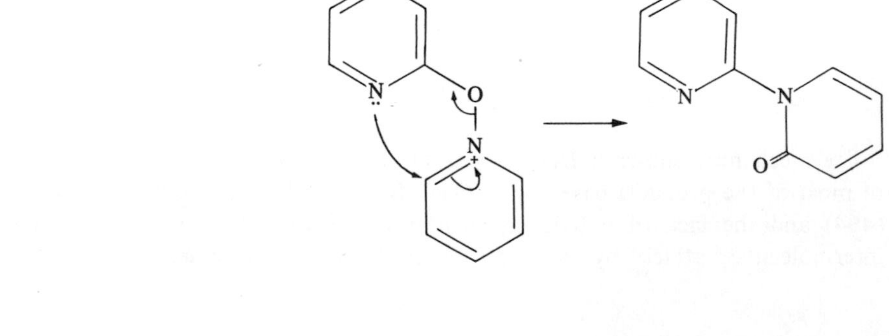

+ N OAc + $\bar{O}$–$\overset{+}{N}$ H O–N+ N OAc

N O N+ N N O

reaction of pyridine-1-oxide with 2-bromopyridine gives the same pyridone. The intermediate *N*-acetoxypyridinium ion has been trapped by its reaction with anisole and with benzonitrile (467) to give a mixture (22% yield) of 2-(*o*-, *m*-, and *p*-methoxyphenyl)pyridine (50.7 : 15.5 : 33.9, respectively), with anisole and 8.5% of *N*-2-pyridylacetamide, and trace quantities of *N*-2-pyridylbenzamide and 2-(*m*- and *p*-cyanophenylpyridine) ($m:p = 71:29$) (with benzonitrile).

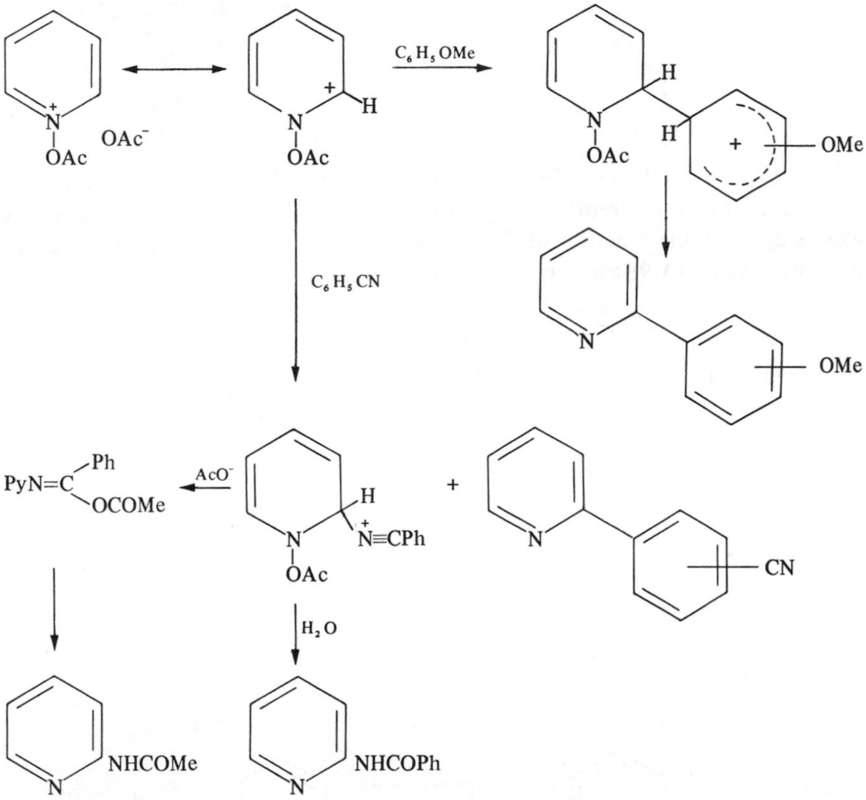

The reactions between 2- or 4-picoline-1-oxides and acid anhydrides are much more complex than those just discussed (458–461). The main product of the reactions is usually a 2- (or 4-)acyloxyalkylpyridine. Mechanisms involving caged radical pairs (e.g., **IA-96**) have been put forward to explain the products formed and ^{18}O-labeling studies (468–470). Alternatively, nucleophilic addition to anhydro-base intermediates have been proposed, again based upon product studies (471), kinetic studies (472, 473), and revised ^{18}O-scrambling experiments (474). The ^{18}O-scrambling results definitely eliminated a concerted rearrangement as in **IA-97** for 2-picoline-1-oxide. Product studies of the reaction with phenylacetic acid anhydride and MO calculations on possible intermediates indicated that ion pairs are more reasonable as intermediates than are radical pairs (471, 475).

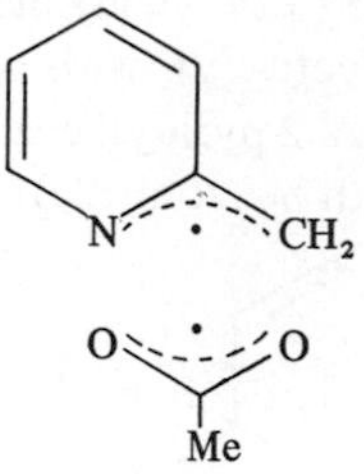

IA-96

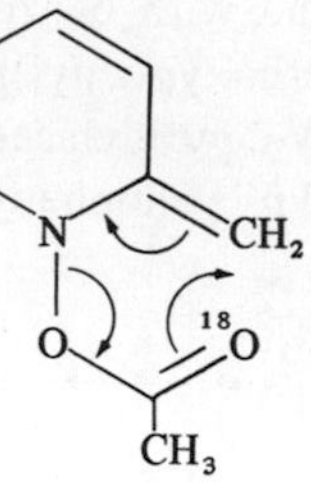

IA-97

Bodalski and Katritzky (476) confirmed Oae's ^{18}O-scrambling results but interpreted them differently based on some of the products they obtained, for example, from the reaction of 2-cyclopentylmethylpyridine-1-oxide and acetic anhydride (e.g., **IA-98**) in terms of a reaction involving ion pairs, such as **IA-99**.

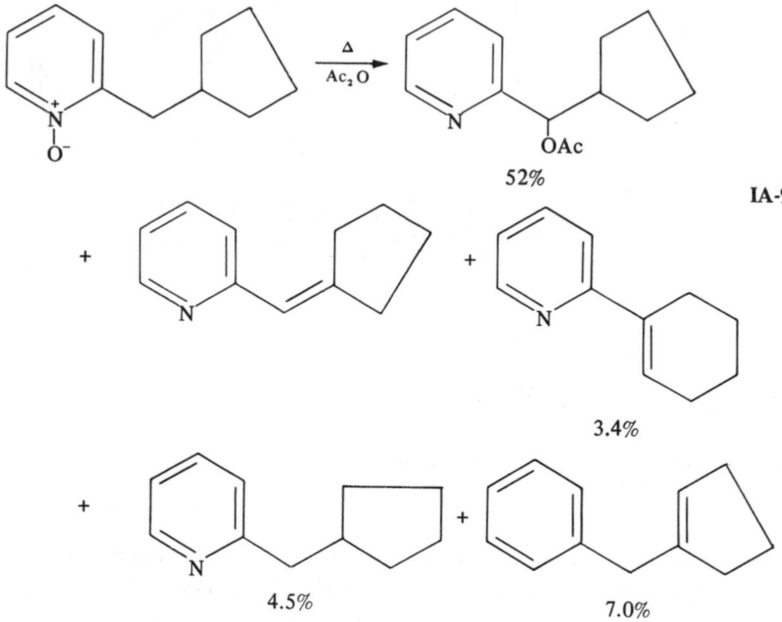

52%

IA-98

3.4%

4.5%

7.0%

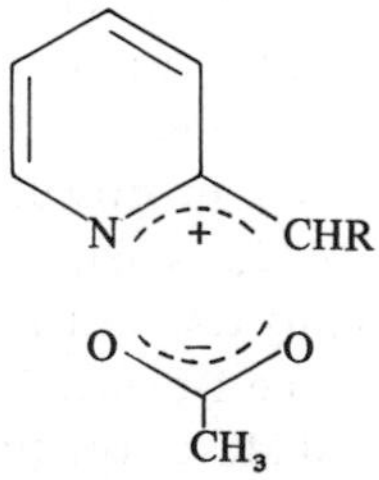

IA-99

Methane and methyl acetate, which are also formed in the reaction of 2-picoline-1-oxide with Ac_2O, are undoubtedly products of a radical reaction, and this has been confirmed by CIDNP measurements. On the other hand, no emission lines due to 2-acetoxymethylpyridine were observed, so that it is not formed by a radical process (477).

As indicated above, ^{18}O-labeling studies of the reaction of 4-picoline-1-oxide with Ac_2O (to give 4-acetoxymethylpyridine and 3-acetoxy-4-picoline) indicated that the reaction proceeded *via* intermolecular nucleophilic addition of acetate ion to the anhydro-base:

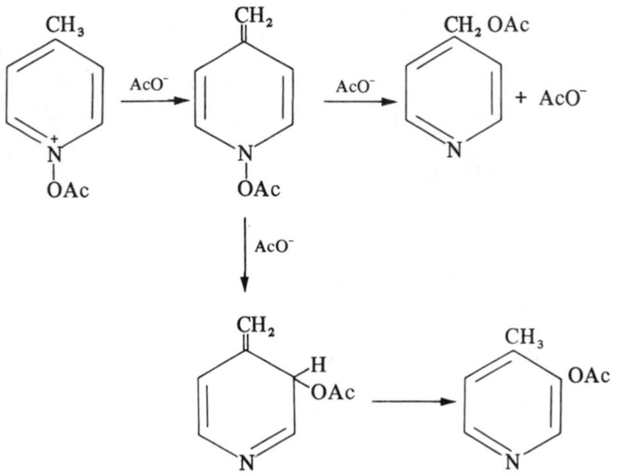

Some 4-ethylpyridine and methane are also formed in this reaction and it seemed likely that these were products of the reaction of methyl radicals (from the decarboxylation of acetoxyl radicals $CH_3CO_2\cdot \rightarrow CO_2 + CH_3\cdot$) and 4-picoline:

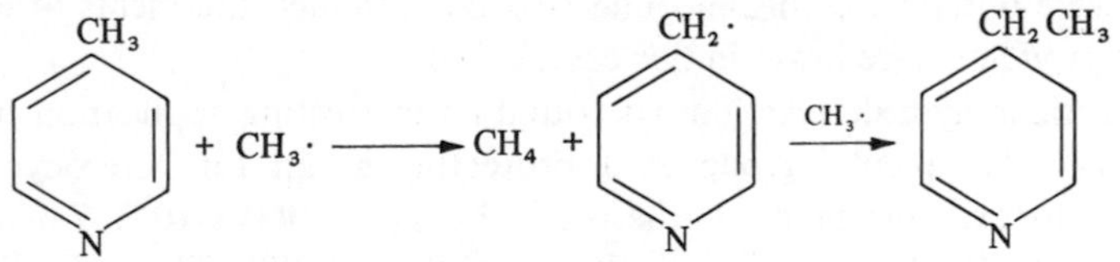

This has now been confirmed by the observation of chemically induced nuclear spin polarization emission spectra (478). Thus the anhydro base cleaves to give both radicals and ion pairs, the former giving 4-ethylpyridine and methane and some 4-acetoxymethylpyridine, and the latter (major pathway) being the main source of the acetoxylated products. The formation of other products and the reaction with trichloroacetic and other anhydrides are discussed in Chapter IV.

The picolyl cations formed as intermediates in the reactions of 2- and 4-picoline-1-oxide with acetic anhydride have been trapped partially by the addition of anisole and benzonitrile (479). Thus the reaction of 4-picoline-1-oxide with acetic anhydride in the presence of anisole gave 4-acetoxymethylpyridine and 3-acetoxy-4-methylpyridine in a ratio of 55:45, together with 20% of a mixture of 4-*o*-, 4-*m*-, and 4-*p*-methoxybenzylpyridines (**IA-100**) (relative yields 65:7:28). The same reaction in benzonitrile gave the esters in the same

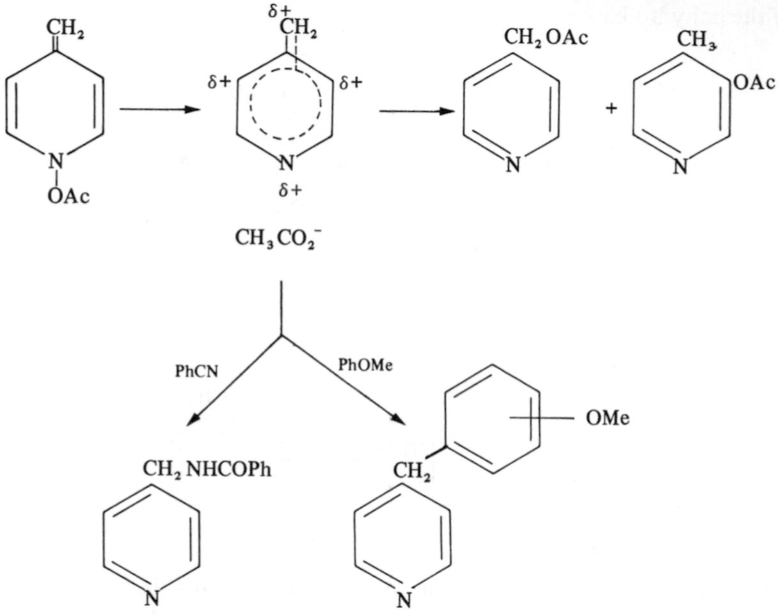

CH_2 N OAc → δ+ CH_2 δ+ δ+ N δ+ → CH_2OAc N + CH_3 OAc N

$CH_3CO_2^-$

PhCN PhOMe

$CH_2NHCOPh$ N **IA-101**

CH_2 OMe N **IA-100**

ratio together with 11% of *N*-4-picolyl-*N*-acetylbenzamide (**IA-101**). Similar results were obtained with 2-picoline-1-oxide, although the yields of externally trapped products were lower in this case.

The acetic anhydride reaction has found an interesting application in the use of the 1-oxido-2-picolyl group as a protecting group for heterocyclic imino functions, for example, in the synthesis of 3-benzyluracil (**IA-102**), 7-benzylhypoxanthine, and 7-α-D-arabinofuranoylhypoxanthine (480). The imine is treated with 1-oxido-2-picolyl chloride or 2-diazomethylpyridine-1-oxide to give the hetero-*N*-methylpyridine-1-oxide. Following whatever manipulation is desired the picolyl group is eliminated by treatment with acetic anhydride at room temperature.

When mercaptans are added to pyridine-1-oxides and acid anhydrides or acid halides the intermediate *O*-acyloxy- or *O*-*p*-toluenesulfonyloxy salts react with the thiols to give complex product mixtures [see Section III.5.C.a.(2)]. The mechanism of these reactions is discussed in detail in Chapter IV (Section IV.2.C).

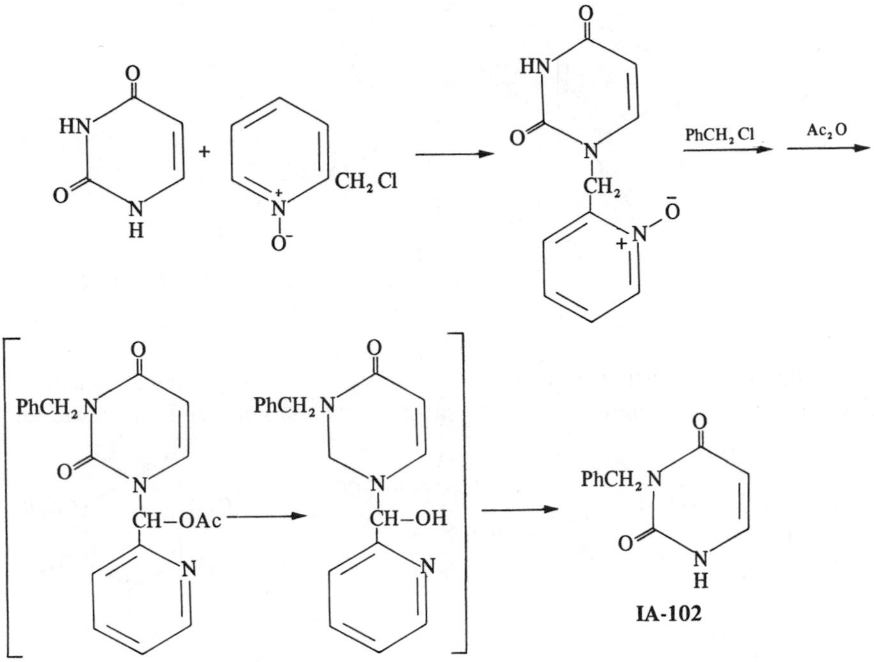

6. Photochemically Induced Transformations

The photochemistry of pyridines, pyridine-1-oxides, pyridinium salts, and pyridinium ylides has been studied at an accelerated rate over the past few years.

A. *Pyridines*

Irradiation of pyridine in acetonitrile at 2537 Å produces unstable "Dewar pyridine" (**IA-103**) ($\tau_{1/2}^{25°} = 2.5$ min) (481). In water, a 2-hydroxy derivative is formed which quickly suffers ring cleavage (481, 482) to give **IA-104**. This involves the $n \rightarrow \pi^*$ singlet excited state of the pyridine (482). In acidic methanol solution pyridine is photoalkylated to yield 2- and 4-methylpyridine, 1-(2-pyridyl)-2-(4-pyridyl)ethane and 1,2-di-(4-pyridyl)ethane (483). Since no photoalkylation was found in the absence of hydrochloric acid, the authors assume that the $\pi \rightarrow \pi^*$ excited state causes the initial hydrogen abstraction (the n electrons presumably being bound to a proton).

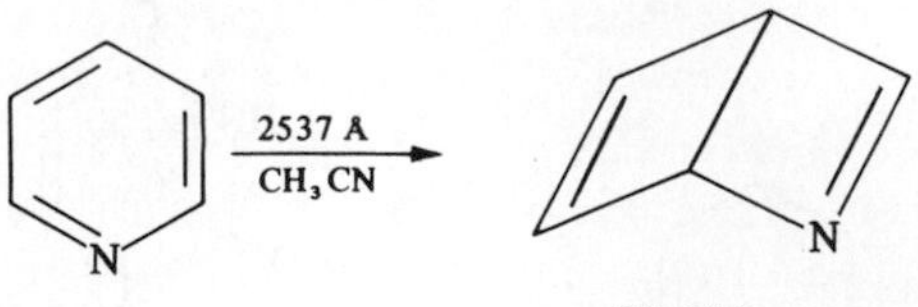

2537 Å
CH_3CN

IA-103

2537 Å

NH

OH

$H_2NCH{=}CH{-}CH{=}CHCHO$

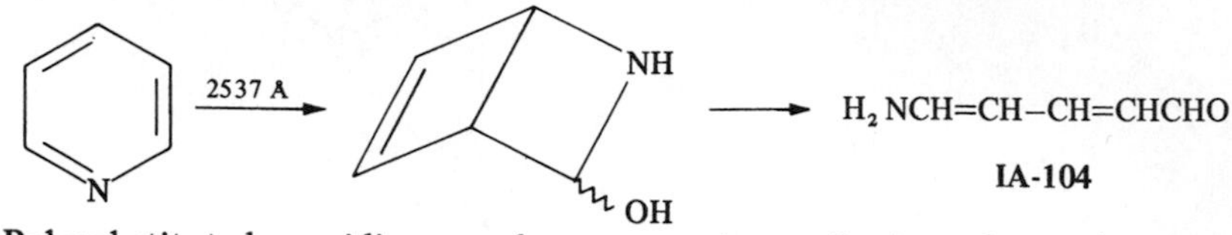

IA-104

Polysubstituted pyridines undergo a variety of photochemical reactions depending on the nature of the substituents present and the reaction

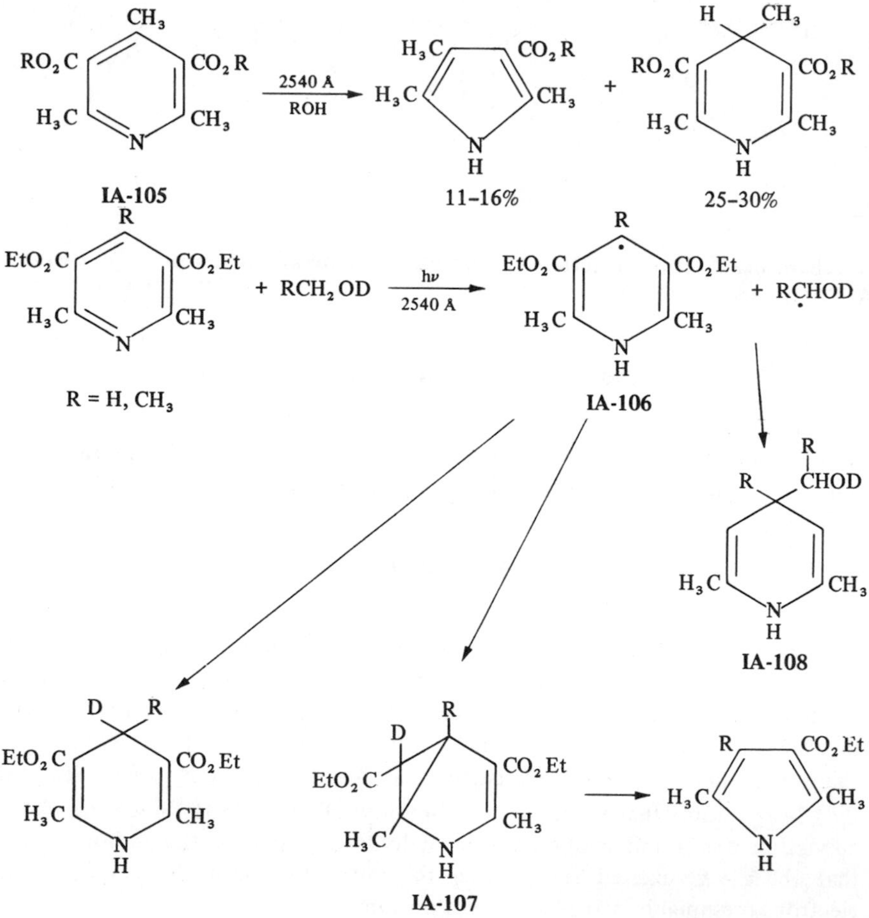

2540 Å
ROH

IA-105

11–16%

25–30%

R = H, CH$_3$

$h\nu$
2540 Å

+ RCH_2OD

+ RCHOD

IA-106

IA-108

IA-107

conditions. Irradiation of the 2,4,6-collidine-3,5-dicarboxylic ester (**IA-105**) leads to both ring contraction and to reduction products (484). It is postulated that the initial step is proton abstraction from the ethanol solvent (by the pyridine nitrogen atom) to give **IA-106** which can then either abstract the oxygen-bound hydrogen (deuterium) or undergo rearrangement to **IA-107** and subsequent ring contraction. A small amount of radical coupling leads to **IA-108**. Nicotinonitrile and ethyl nicotinate also undergo photoaddition of methanol (485).

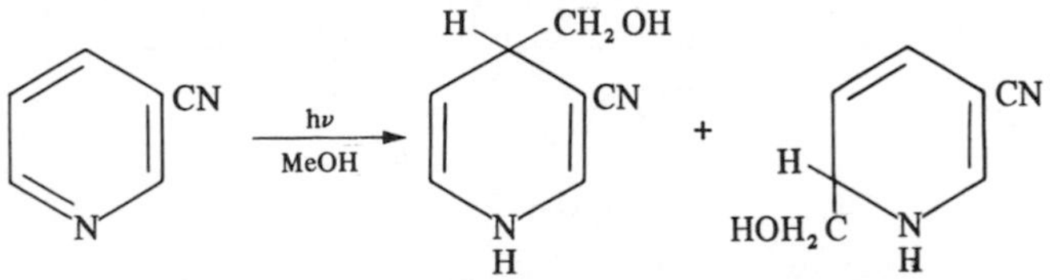

Pyridylacetic acids undergo photochemical decarboxylation at p*H* 4.0–4.2 using 2537 Å light (486).

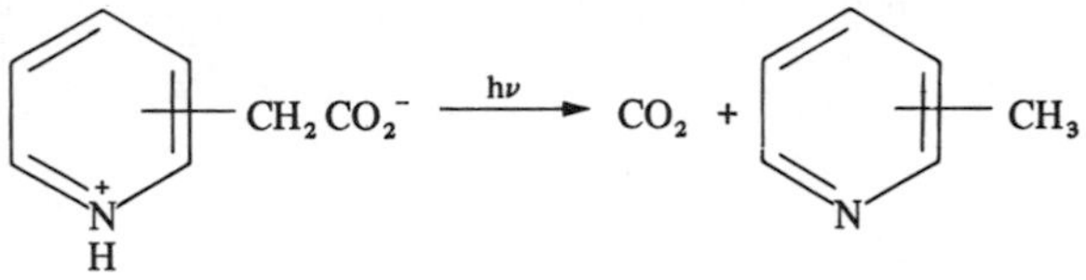

Pyridine-2,3- and -3,4-dicarboxylic acids are specifically decarboxylated photochemically to the 2- and 4-carboxylic acids, respectively, which is the reverse of the trend observed in the thermal decarboxylations (487).

The gas phase photochemistry of picolines and lutidines has been investigated (488). These have been found to photorearrange: for example, irradiation of 2-picoline (low conversions) at a pressure of 10 mm of Hg at 2537 Å gave mainly 4-picoline, together with some 2,4-lutidine and a little pyridine, while 4-picoline rearranged to give 2-picoline and a little pyridine–in both cases, the original excess picoline was recovered. These rearrangements proceed either by a methylation-demethylation process or by the formation of azaprismane intermediates:

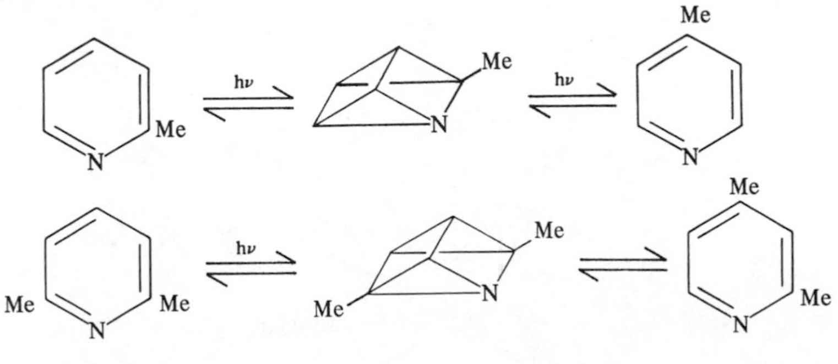

When pyridine and 2- and 4-picoline are irradiated in cyclohexane solution the former gives 2- and 4-cyclohexylpyridine while 2-picoline gives 4- and 6-cyclohexyl-2-methylpyridine (489). Photointerconversion of 2- and 4-picoline also occurs. A radical mechanism has been proposed. E.s.r. studies show that irradiation of pyridine in methylcyclohexane gives radical **IA-109** (490).

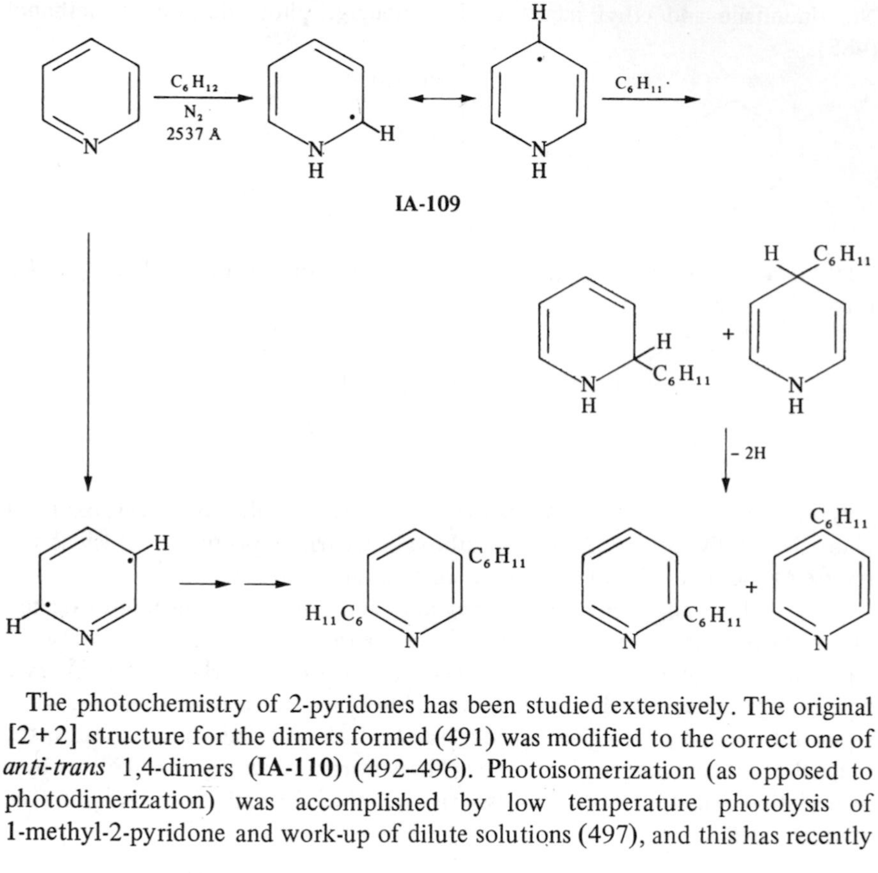

C_6H_{12}, N_2, 2537 Å; $C_6H_{11}\cdot$; $-2H$

IA-109

The photochemistry of 2-pyridones has been studied extensively. The original [2 + 2] structure for the dimers formed (491) was modified to the correct one of *anti-trans* 1,4-dimers **(IA-110)** (492–496). Photoisomerization (as opposed to photodimerization) was accomplished by low temperature photolysis of 1-methyl-2-pyridone and work-up of dilute solutions (497), and this has recently

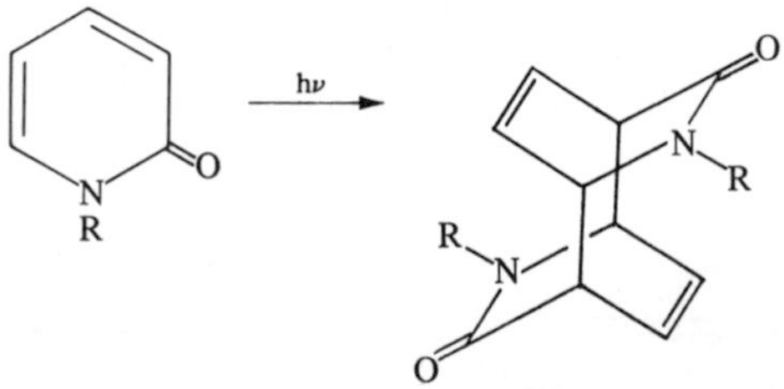

hν

IA-110

been generalized and extended to the synthesis of photo-2-pyridones (2-aza-3-oxobicyclo[2.2.0]hex-5-enes) (**IA-111**), including the parent compound (R=R′=R″=H) (498). It was suggested that the stability of photo-2-pyridones relative to Dewar pyridines was mainly due to the lower ground state energies expected for **IA-111**.

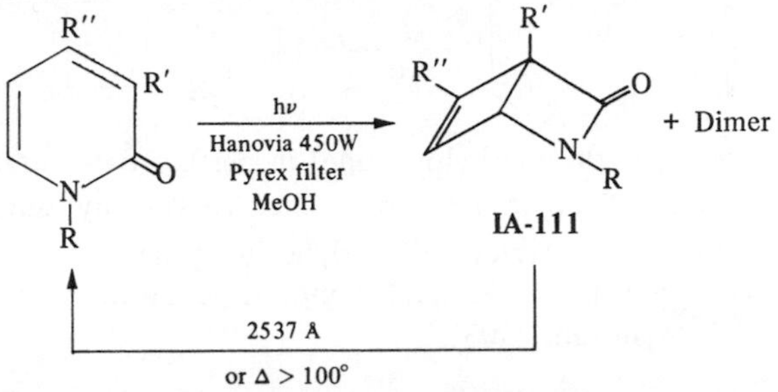

hν, Hanovia 450W, Pyrex filter, MeOH → + Dimer

2537 Å or Δ > 100°

IA-111

2-Aminopyridines photodimerize only in aqueous acidic solution to give **IA-112** (494). The stereochemistry of dimers **IA-110** and **IA-112** has been established unambiguously by X-ray crystallographic studies (496, 499). Photodimerizations are thought to proceed *via* the $\pi \rightarrow \pi^*$ singlet (494).

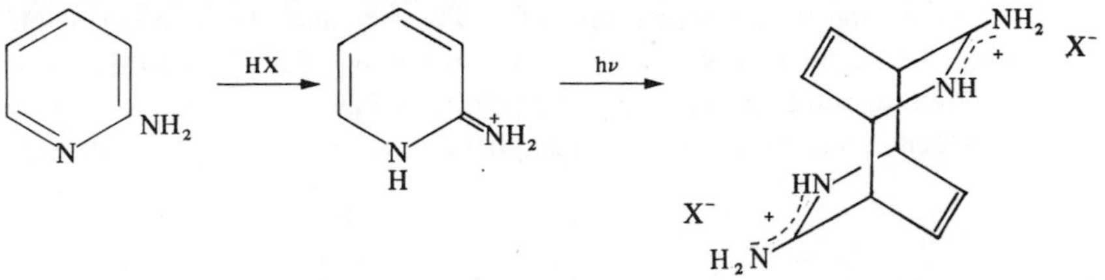

HX → hν →

IA-112

Photolysis of pentachloropyridine in benzene yields 2,4,5,6-tetrachloro-3-phenylpyridine; if the photolysis is carried out in dioxane or ether solution the product is 2,4,5,6-tetrachloropyridine (500). Irradiation of a number of substituted polyhalogenopyridines (e.g., **IA-113**) has led to practicable syntheses of fused polyhalogeno-heterocycles (501).

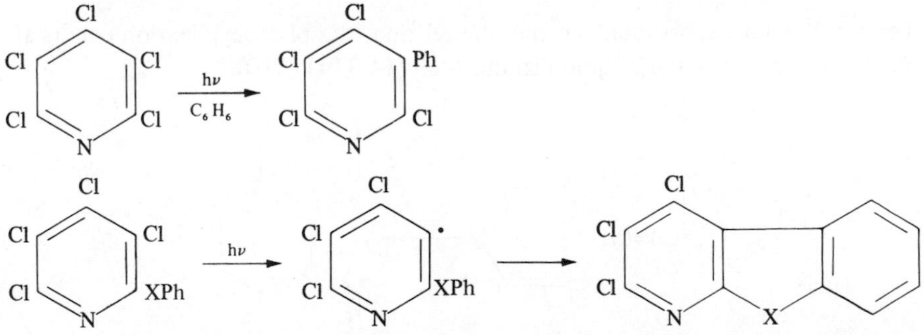

hν, C_6H_6 →

hν → →

IA-113 X = O, S, NH

The photocyclization of di-(2-pyridyl) ketone and 2-benzoylpyridine in aqueous solution has been studied (502) and is thought to involve an n → π* triplet excited state of the ketones which abstracts a hydrogen atom from the solvent.

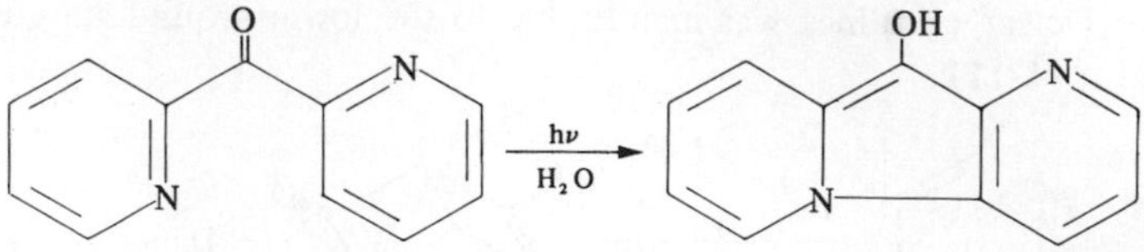

Irradiation of penta(perfluoroethylpyridine) in perfluoro-*n*-pentane with light >270 nm at room temperature gives the Dewar pyridine derivative, pentakis(pentafluroethyl)-1-azabicyclo[2.2.0] hexa-2,5-diene. Light of > 200nm produces the corresponding prismane, pentakis(pentafluoroethyl)-1-azatetracyclo-[$2.2.0.0^{2,6}.0^{3,5}$] hexane (503).

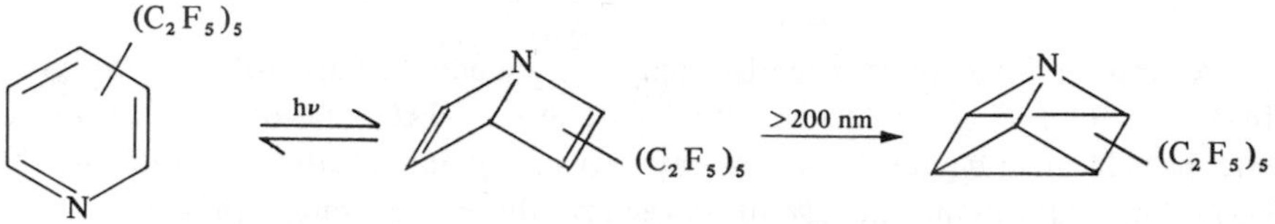

The photochemical isomerization of stilbenes and the photochemical cyclization of *cis*-stilbene is well known. Not surprisingly, 2- (504–506), 3- (504, 506, 507), and 4-stilbazoles (504, 507–509) behave similarly. 3-Stilbazole cyclized preferentially to C-4 to give 2-azaphenanthrene (507). If the 2-stilbazole bears a 2′-chloro substituent on the phenyl ring, the photocyclization occurs at nitrogen to give a benzo[*c*] quinolizinium salt **(IA-114)** (510).

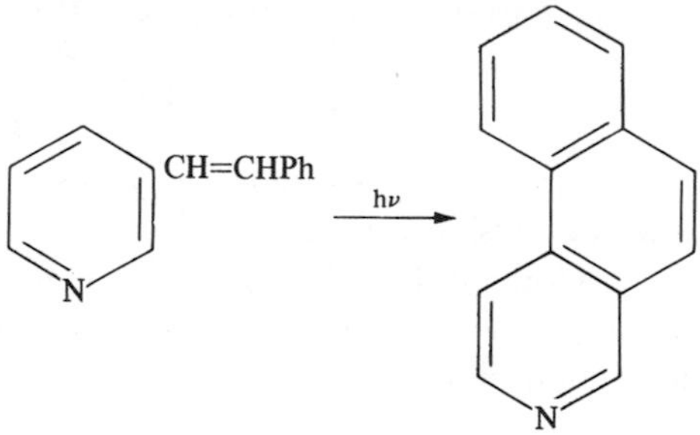

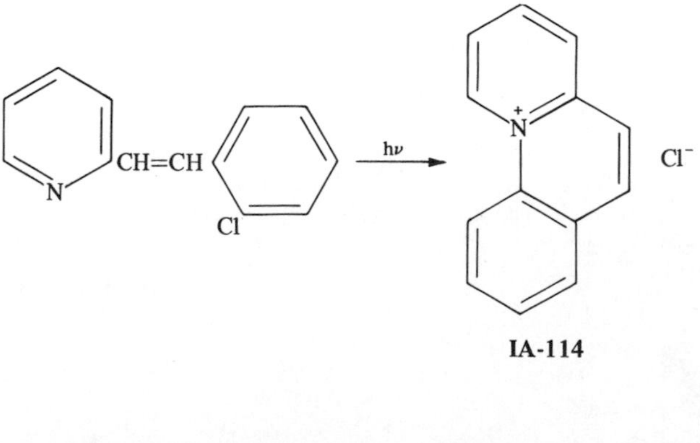

IA-114

B. *Pyridine-1-oxides*

The photolytic behavior of *N*-oxides depends largely on the nature of the substituents present, reaction involving either the substituent, particularly if it is a nitro- or azido-group, or the *N*-oxide oxygen. Some of this work has been reviewed (511) (see also Chapter IV, Section IV). The nitro group is readily displaced photochemically either by piperidine (512), or by oxygen in alcoholic solution (513–515). Under a nitrogen atmosphere irradiation of an alcoholic solution of 4-nitropyridine-1-oxides produces 4-hydroxylaminopyridine-1-oxides (513, 515, 516). While 4-hydroxylaminopyridine-1-oxide can be photooxidized to 4-nitropyridine-1-oxide in the presence of oxygen (514), 4-nitrosopyridine-1-oxide is reduced to 4-hydroxylaminopyridine-1-oxide both in the presence and the absence of oxygen. It has been suggested (515) that the reaction proceeds through an excited singlet and then an undefined short-lived intermediate which is partitioned, depending on conditions, to either 4-hydroxylamino- or 4-hydroxypyridine-1-oxide. The lack of photoreactivity of the *N*-oxide function under these circumstances ($\lambda > 300$ nm used) implies that the light induced reactions take place first at the substituents. Sunlight effects the photoconversion of 4-azidopyridine-1-oxide in acetone to 4,4′-azopyridine-1,1′-dioxide (517).

Molecular orbital theory has been applied to the prediction of the photochemical reactions of aromatic amine oxides (36, 518). The energy of C–O bond formation in a given *N*-oxide was put as $\Delta E \propto p_{ij}^{LVMO}\, \beta_{i,j}$ (where p_{ij}^{LVMO} is the bond order of the lowest vacant molecular orbital between atoms i and j – O and C in this case, and β_{ij} is the exchange integral (518). The ease of bond formation should depend solely upon the magnitude of p_{ij}^{LVMO} (>0.06) and, in the case of pyridine-1-oxide itself, p_{ij}^{LVMO} is calculated by HMO to be +0.1514, while $p_{OC}^{HOMO} = -0.2535$. With 4-nitropyridine-1-oxide p_{OC}^{LVMO} is computed by PPP-SCF MO calculations to be +0.0307 (i.e., <0.06), accounting for the fact that oxaziridine ring formation does not occur in the latter case. The fact that for pyridine-1-oxide (and other *N*-oxides which do form oxaziridines) the bond orders of the LVMO between O and C are positive and >0.06 while those of the highest occupied molecular orbitals (HOMO) are negative indicates that the formation of the oxaziridine (**IA-115**) is preferred from the excited species of the *N*-oxide rather than from its ground state.

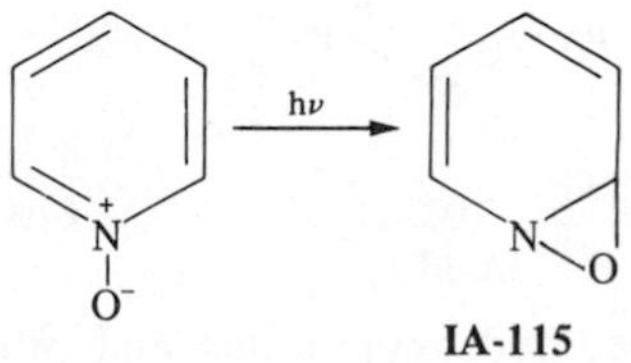

hν

IA-115

Vapor phase photolysis of pyridine-1-oxide (519) or of the picoline-1-oxides (520, 521) results mainly in deoxygenation, perhaps with the release of atomic oxygen (see below, however). The process is temperature independent at 2537 Å (π-π^*) but temperature dependent at 3261 Å (n-π^*), implicating an excited triplet as the reactive intermediate at 2537 Å. A similar process must also occur in the condensed phase, since, among the many photochemical reactions of pyridine-1-oxide, it has been found to transfer an oxygen atom either to its own C-3 position (522) (**IA-116**) or to a suitable acceptor, for example, solvent (522-524) (**IA-117**). It is possible, however, that, in solution, deoxygenation proceeds *via* prior rearrangement to the oxaziridine (**IA-115**), which can either

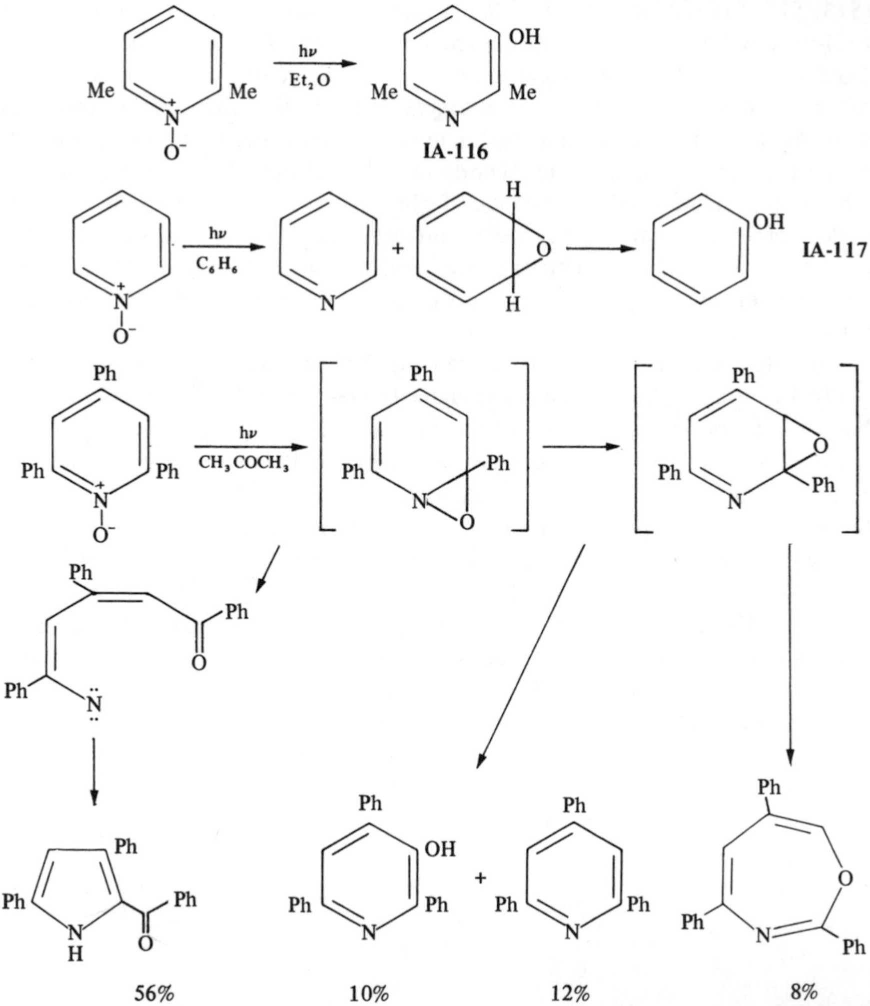

hν, Et_2O — **IA-116**

hν, C_6H_6 — **IA-117**

hν, CH_3COCH_3

56% **IA-118** — 10% **IA-119** — 12% — 8% **IA-120**

rearrange further to give 3-hydroxypyridine and other products, or act as an oxidizing agent for solvent molecules, for example, benzene to give phenol (522), naphthalene oxide and 1-naphthol with naphthalene (523), and alcohols to aldehydes (525). The photolysis of 2,4,6-triphenylpyridine-1-oxide in acetone illustrates the type of products formed in these reactions (524). The main product isolated was 2-benzoyl-3,5-diphenylpyrrole (**IA-118**), together with minor amounts of 3-hydroxy-2,4,6-triphenylpyridine (**IA-119**), the parent pyridine, and what is probably 2,4,6-triphenyl-1,2-oxazepine (**IA-120**).

The 2-formylpyrroles often found (525-527) are believed to be formed from **IA-115** and its derivatives. Irradiation of methanol or ethanol solutions of pyridine-1-oxide gave pyridine, 2-formylpyridine, the acetals of *N*-formylpyrrole, and a trace of *N*-formylpyrrole, together with formaldehyde or acetaldehyde, repectively. The latter are thought to arise by oxidation of the solvent by the oxaziridine intermediate (525).

Photolysis of pentachloropyridine-1-oxide yields (1,2,3,4,4-pentachloro-1,3-butadienyl)isocyanate and traces of pentachloropyridine (500).

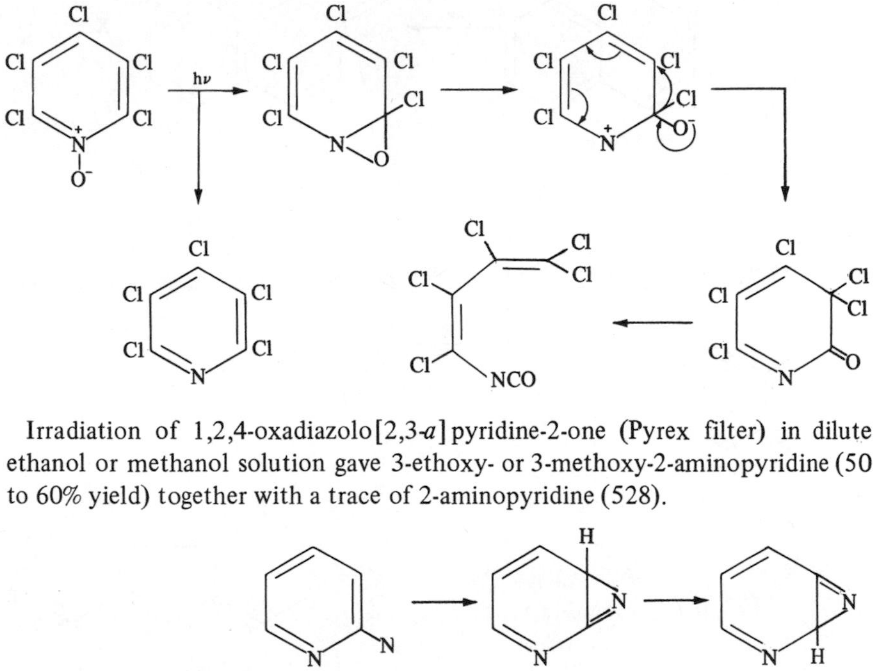

Cl Cl Cl Cl Cl $h\nu$ NCO

Irradiation of 1,2,4-oxadiazolo[2,3-*a*]pyridine-2-one (Pyrex filter) in dilute ethanol or methanol solution gave 3-ethoxy- or 3-methoxy-2-aminopyridine (50 to 60% yield) together with a trace of 2-aminopyridine (528).

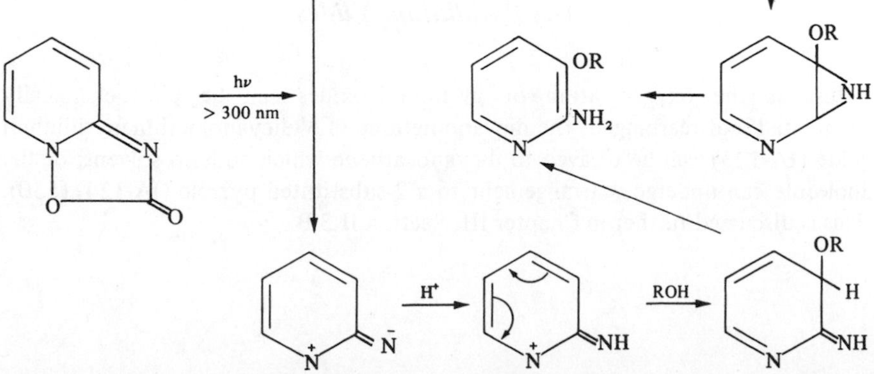

$h\nu$ > 300 nm OR NH_2 NH H H^+ ROH

C. *Pyridinium Salts*

Not surprisingly, the photochemistry of pyridinium salts is similar to that of pyridines. 1-Styrylpyridinium salts **(IA-121)** can be cyclized to phenanthridizinium salts (529) and 1-tolyl-2-halopyridinium salts give benzo[*b*]indolizines **(IA-122)** (505).

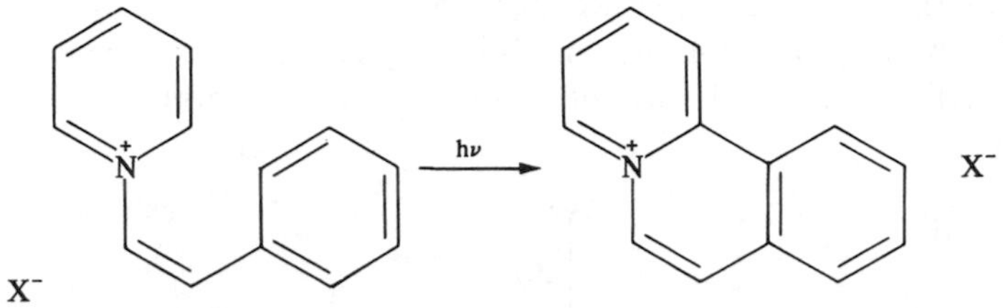

IA-121

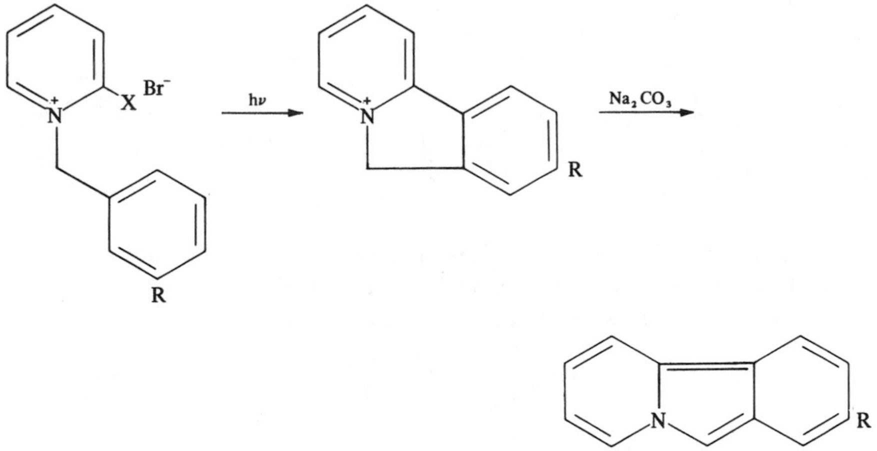

IA-122

D. *Pyridinium Ylides*

Just as the oxygen atom of pyridine-1-oxides can be photochemically eliminated and rearranged, the dicyanomethine of *N*-dicyanomethinepyridinium ylide **(IA-123)** can be cleaved to dicyanocarbene which adds to solvent; or the molecule can undergo rearrangement to a 2-substituted pyrrole **(IA-124)** (530). This is discussed further in Chapter III, Section II.3.B.

IA-123 IA-124

Similarly, carboethoxynitrene is said to be generated from **IA-125** (530, 531), and in a "photochemical Curtius reaction," methylisocyanate is formed from **IA-126** (532), although in both cases the major product is the diazepine resulting from rearrangement, presumably *via* a diaziridine intermediate which has been shown to form in the direction of least steric hindrance (553, 535). No singlet sulfonylnitrenes could be detected by the photolysis of *N*-sulfonyliminopyridinium ylides **(IA-127)** (536a). Instead only hydrogen abstraction product **(IA-128)** and ring-expansion products **(IA-129)** could be isolated (536a, b).

IA-125 95% 2.7%

IA-126

IA-127 IA-128 IA-129

No

Photolysis of *N*-vinyliminopyridinium ylides (**IA-80**) in acetone at 0° gave **IA-81** and its dehydrogenation product, but no diazepine formation was observed (451).

The photolysis of 1-aryliminopyridinium ylides (**IA-130**) leads to N–N bond cleavage with the postulated formation of aryl nitrenes (537, 538). Again, no diazepine formation was reported.

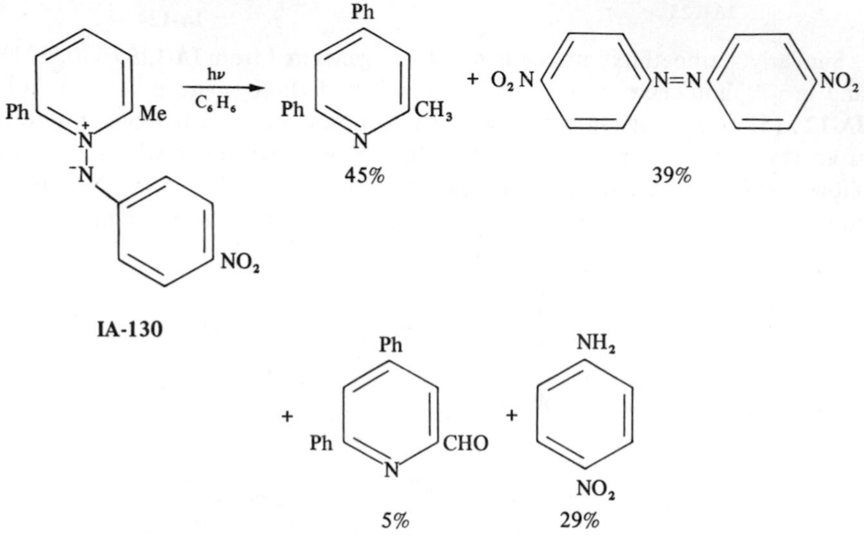

Ph Me $\xrightarrow[C_6H_6]{h\nu}$ Ph Ph CH_3 + O_2N N=N NO_2

NO_2 45% 39%

IA-130

\+ Ph Ph CHO + NH_2 NO_2

5% 29%

7. Oxidation and Reduction

The *N*-oxidation and deoxygenation of pyridines and their *N*-oxides, respectively, are discussed in Chapter IV.

8. Ring-Cleavage Reactions

Ring-opening occurs in a number of reactions. Some of these—for example, the reaction of pyridine-1-oxides with phenylmagnesium bromide (Section III.5.B.b), and the rearrangements of some 2-bromopyridines in the presence of potassium amide in liquid ammonia [Section III.5.C.b.(1)]—have already been mentioned. Ring-openings and recyclizations have also been mentioned in the section on photochemical reactions.

Hydrolysis of *N*-methyl-2- and 4-cyanopyridinium salts gives the corresponding *N*-methyl-2- and 4-pyridones and the *N*-methylpyridinium carboxamide salts (539). Hydrolysis of the 3-cyano-1-methylpyridinium salt (**IA-131**), a much slower reaction than the hydrolysis of the other isomers, gives a small yield of a ring-cleavage product (**IA-132**). Such a ring-cleavage is not observed in the alkaline ferricyanide oxidation of **IA-131**. A kinetic study has implicated

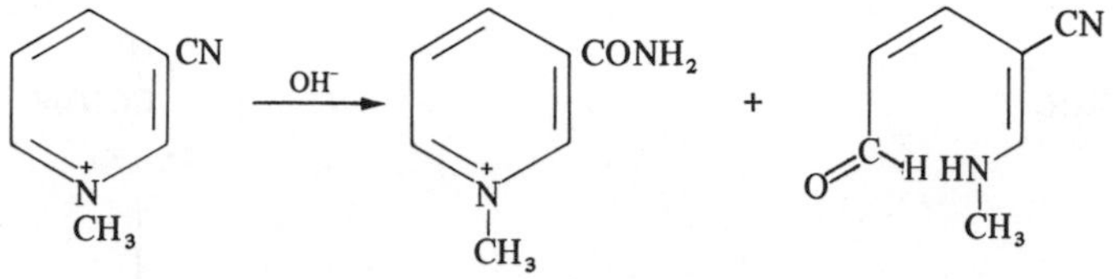

CN; OH⁻; CONH$_2$; $\overset{+}{N}$; CH_3; +; CN; C; H HN; CH_3

IA-131 **IA-132**

2-hydroxy-1-methoxy-1,2-dihydropyridine (**IA-133**) as an intermediate in the reactions of *N*-methoxypyridinium perchlorate with aqueous base to give the *O*-alkyl derivative of glutacondialdehyde monoxime (540). The ring-opening with piperidine has been mentioned earlier (**IA-58**). The base-catalyzed

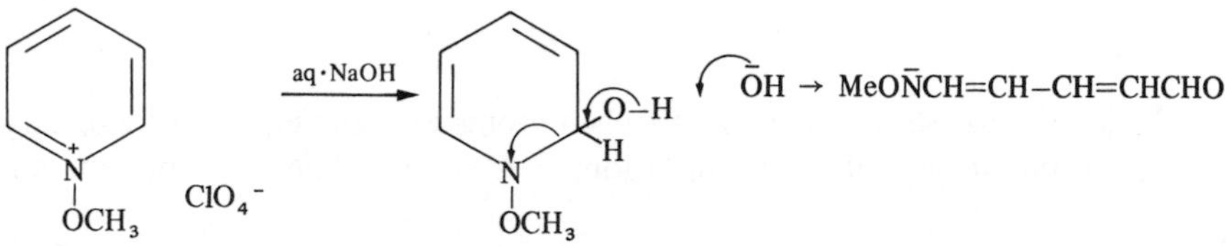

$\overset{+}{N}$; OCH_3; ClO_4^-; aq·NaOH; O–H; H; N; OCH_3; $\overline{O}H$ → $MeO\overline{N}CH{=}CH{-}CH{=}CHCHO$

IA-133

ring-opening of 1-(*N, N*-dimethylcarbamoyl)pyridinium chloride (**IA-134**) leads to **IA-135**. This can either give glutacondialdehyde and 1,1-dimethylurea (at p*H* > 13.5) or yield pyridine and dimethylamine (at p*H* 8–13). The latter pathway does not involve reversion to the pseudo-base but rather closure to **IA-136** which gives the products (541).

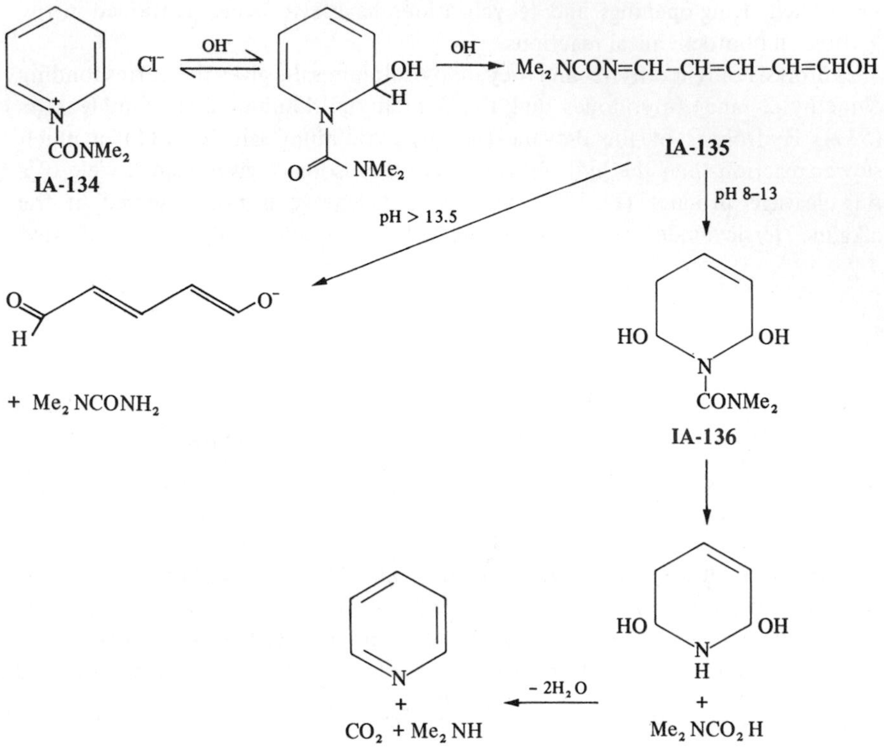

IA-134

$Me_2NCON{=}CH{-}CH{=}CH{-}CH{=}CHOH$

IA-135

pH 8–13

pH > 13.5

$+ Me_2NCONH_2$

IA-136

$- 2H_2O$

$CO_2 + Me_2NH$

Me_2NCO_2H

[15]*N*-labeling has shown that the nitrogen atoms in 2-aminopyridine exchange their positions under hydrolytic conditions. 3-Aminopyridine does not behave in

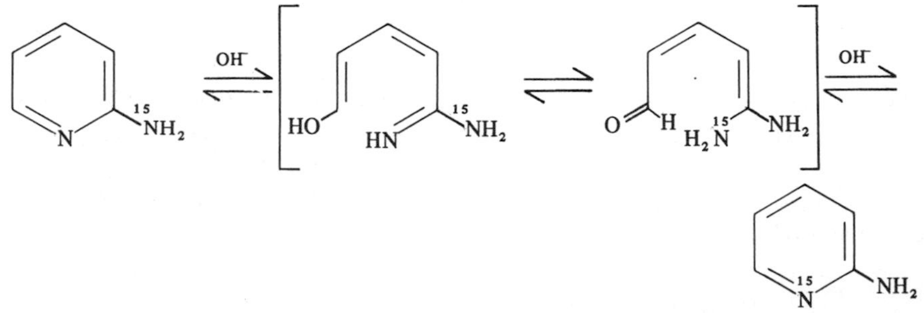

this way (542). Diazotization of 1-alkyl-3-aminopyridinium bromides results in the formation of β-(1-alkyl-1,2,3-triazol-4-yl)acroleins (**IA-137**) (543).

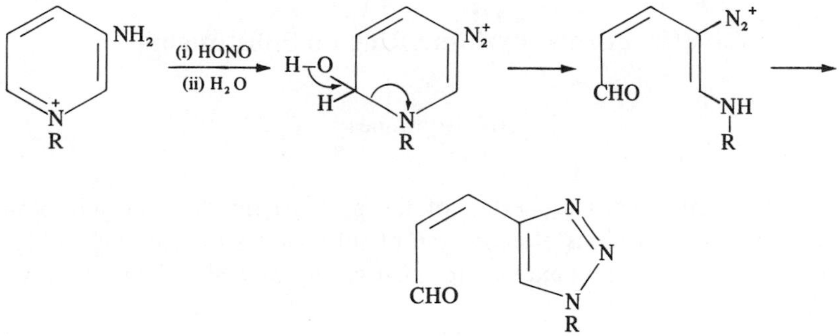

(i) HONO
(ii) H_2O

IA-137

The thermolysis of 2-azidopyridine-1-oxides in nonprotic solvents leads to the formation of 2-cyano-1-hydroxypyrroles in high yield (544). In nucleophilic solvents, 3-substituted 2,3-dihydro-2-pyrrolone derivatives are formed by a Michael-type addition of the solvent to the open-chain unsaturated nitrile (**IA-138**).

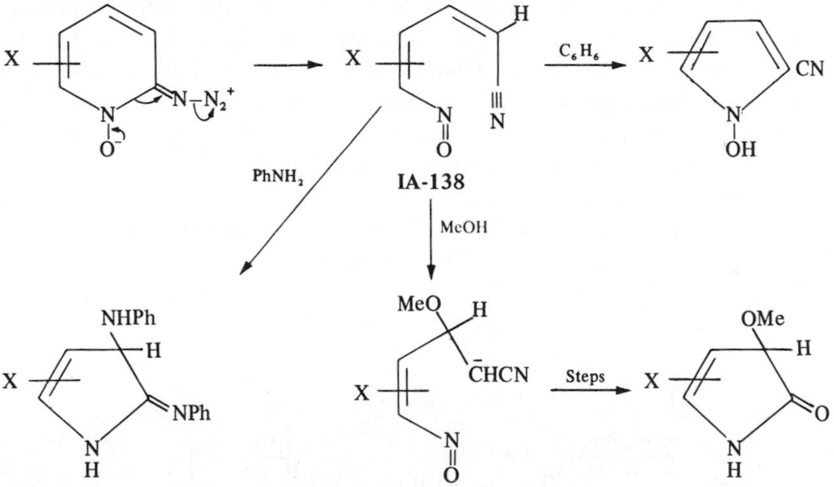

C_6H_6

$PhNH_2$

IA-138

MeOH

Steps

Flash vacuum pyrolysis of **IA-139** at 600°/0.1 mm gives a mixture of 2- and 3-cyanopyrrole (545). Other similar reactions have been discussed (546).

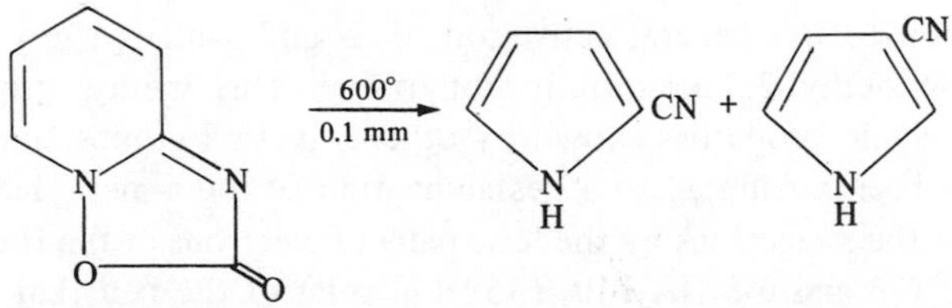

600°
0.1 mm

IA-139

IV. Effect of the Pyridine Ring on Substituents

1. Alkylpyridines

The electron withdrawing properties of the pyridine ring have already been discussed extensively in terms of activation of substituents toward nucleophilic displacement. This activation extends to alkyl groups as well and thus bestows some acidity to these.

These protons are sufficiently acidic to be subject to base-catalyzed hydrogen-deuterium exchange under relatively mild conditions, 4-methyl protons exchanging 50% faster than 2-methyl hydrogens (547). Despite this trend, 2-methyl groups are metallated faster by organolithium compounds than are 4-methyl groups in 2,4-lutidine or *sym*-collidine for example (548). This is probably due to an initial complexation of the organolithium reagent at the pyridine nitrogen atom. A 3-methyl-group which, in general, resembles a benzene methyl group most closely, is also activated sufficiently to undergo metallation, albeit under more stringent conditions (549). The kinetics and mechanism of H-D exchange in the 2-methyl groups in 2-methyl-, 2,6-dimethyl-, and 4-amino-2,6-dimethylpyridine in dilute acid have been investigated. The mechanism involves specific acid-general base catalysis, the conjugate acid of the substrate being deprotonated at carbon by base (550).

The reactions of the carbanions are normal and include carboxylation (548), addition to olefins (551–553) or Schiff bases (554), addition to carbonyls (296, 555), and alkylation (555, 556). These carbanions can also be nitrated with *n*-propyl nitrate (557, 558).

$$\text{Py-CH}_3 \xrightarrow[\text{(ii) } n\text{-PrONO}_2,\ -35^\circ]{\text{(i) NH}_2^-/\text{NH}_3} \text{Py-CH=NO}_2^-\ \text{NO}^+ \xrightarrow[\text{R.T.}]{\text{H}^+} \text{Py-CH}_2\text{NO}_2$$

An exception to this general activation of α- and γ-methyl groups is in the behavior of 4-methyl-2,3,5,6-tetrafluoropyridine. This methyl group exhibits none of the acidic properties shown by other γ-methyl groups, and its lack of reactivity has been attributed to a destabilization of the 4-methyl-carbanion by a repulsion of the π-electrons by the lone pairs of electrons of the fluorine atoms (I_π-effect) at C-3 and C-5 (**IA-140**) (559), in spite of the fact that for fluorine $+M < -I$.

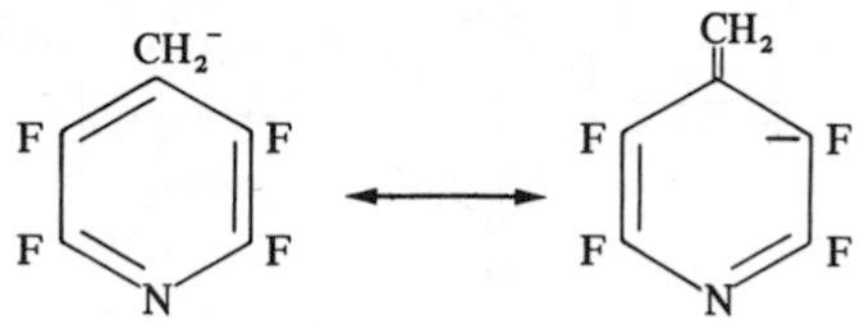

IA-140

2- And 4-vinylpyridines are also activated and can serve as acceptors in Michael-type additions (560–562).

2. Halopyridines and Nitropyridines

The activation of halo- and nitropyridines toward nucleophilic displacement has been discussed thoroughly (Section III.5.C.b.). A 3-halo- or -nitro-group behaves in the same way as the correspondingly substituted benzene.

Halopyridines and nitropyridines are discussed in Chapters VI and VIII, respectively.

3. Tautomerism of Hydroxyl, Thiol, and Amino Groups

In the solid-state and in solution hydroxypyridines exist to a slightly lesser extent as the keto-tautomers than do the mercaptopyridines, while the aminopyridines prefer the amino- to the imino-form (563, 564). These equilibrium constants have been determined by the measurement of pK_a values and a variety of spectral parameters (563–567). The tendency to form the keto-tautomer has been correlated with the difference in the delocalization energies of the two tautomers (564), and with solvent polarity (568) using the equation:

$$\ln\left(\frac{K_{T_1}}{K_{T_2}}\right) = a(P_1 - P_2)$$

where K_{T_i} = equilibrium constant in solvent i
P_i = polarity of solvent i

Prototropic tautomerism of heteroaromatic six-membered compounds has been reviewed (46).

3-Hydroxy- (569) and 3-mercapto-pyridine (563) exist as mixtures of the enol and zwitterionic forms in aqueous solution (see also p. 20). In solvents of low dielectric constant the hydroxy form greatly predominates. While

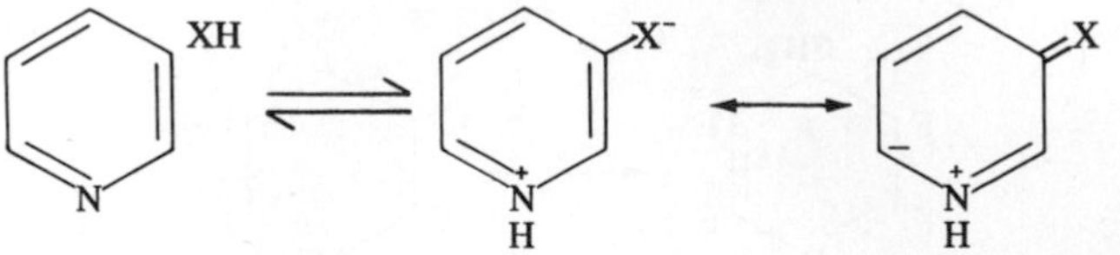

2,6-dihydroxypyridines (glutaconimides) probably exist predominantly in the hydroxylactam form (46), UV and IR data indicate that 4-amino-2,6-dihydroxypyridine (glutazine) exists as an equilibrium mixture of **IA-141a** and **1A-141b** (570). On the other hand, UV and NMR evidence show that

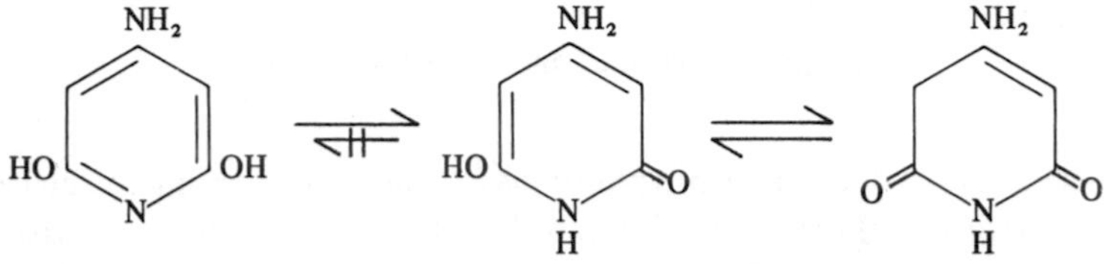

IA-141a **IA-141b**

2,4-dihydroxy-6-methylpyridine exists exclusively in the 4-hydroxy-6-methyl-2-pyridone form (571, 572). Halogens at C-2 and/or C-6 increase the proportion of

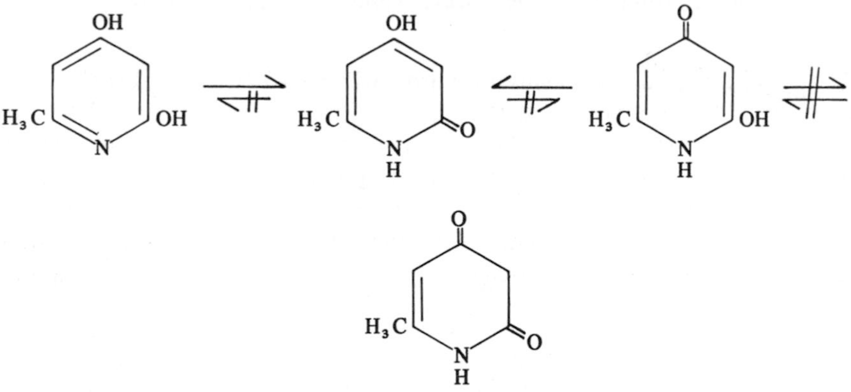

the hydroxy tautomer in a 2- or 4-hydroxypyridine (573). Ring strain due to the fusion of a five-membered ring to a 2-pyridone reduces the proportion of the

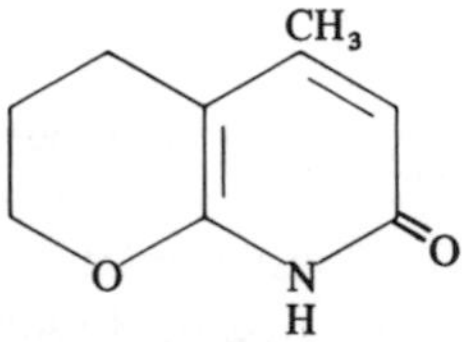

IA-142 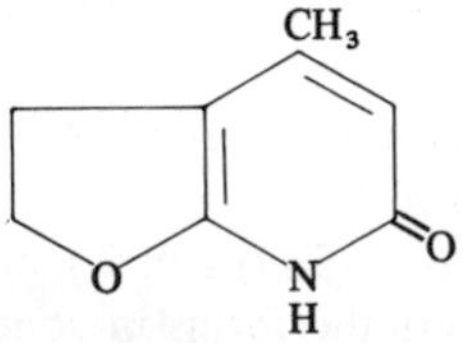 **IA-143**

keto-tautomer (574). Thus **IA-142** exists to the extent of 10% in the enolic form in ethanol solution, while 75% of **IA-143** is enolized in the same solvent.

Consideration of tautomeric equilibria has allowed the estimation of the aromatic stabilization energy of 2-pyridone as 25 kcal/mole [compared with a value of 12 to 15 kcal/mole deduced from the estimated ring current (575)]. The corresponding energy for 2-pyridinethione was 26.2 kcal/mole and for a 2-pyridone imine, 23 kcal/mole (576).

4-Hydroxypyridine-1-oxide exists as a mixture of comparable amounts of both forms in aqueous solution (577, 578). Similarly, 2-hydroxypyridine-1-oxide appears to exist as a mixture of the strongly hydrogen-bonded tautomeric forms, with the 2-pyridone form predominating (577).

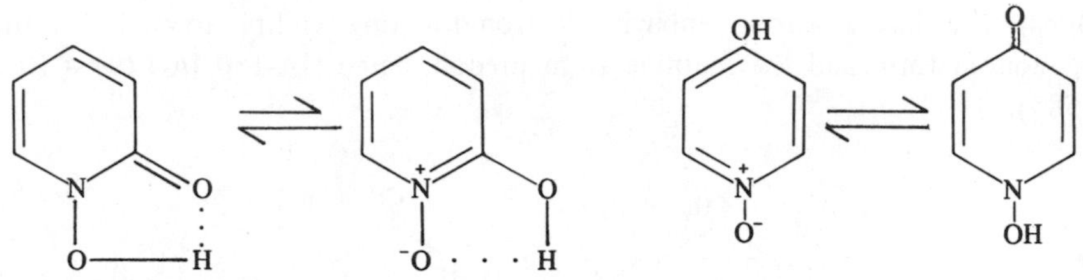

The pK_a and infrared spectral studies have shown that aminopyridines exist predominantly in the amino- rather than the imino-form (579, 580) in the solid-state and in solution.

From a study of the ionization potentials of the potentially tautomeric hydroxy- and mercapto-pyridines, it was concluded that, *in the gaseous phase*, these exist as the hydroxy- and mercapto-tautomers. On the other hand, a comparison of the ionization potentials of 2-, 3-, and 4-aminopyridine, 2-, 3-, and 4-methylaminopyridine and 1-methyl-2-pyridone imine, 1-methyl-4-pyridone imine, and the 1-methyl-3-iminopyridinium zwitterion led to the qualitative conclusion that the aminopyridines exist in the amino-form in the gaseous state (581).

2- And 4-mercaptopyridine-1-oxides exist mainly as the thiopyridone tautomer (**IA-145/IA-144** = $10^{1.72}$; **IA-147/IA-146** = $10^{0.56}$) (582) but, once again, introduction of the *N*-oxide oxygen atom increases the mesomeric stabilization of the enol form and decreases that of the -one form.

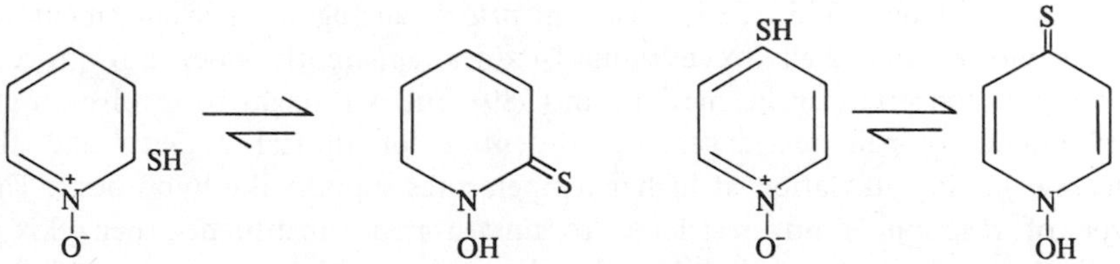

IA-144 **IA-145** **IA-146** **IA-147**

2- And 4-methylsulfonamidopyridines exist predominantly in the imino form (583). On the other hand, benzoyl- and phenylsulfonylmethylpyridines do not exist at all in the methine or quinonoïd form (584). Indeed, 2-pyridine methide **(IA-148)** is calculated to have only about half the resonance stabilization of

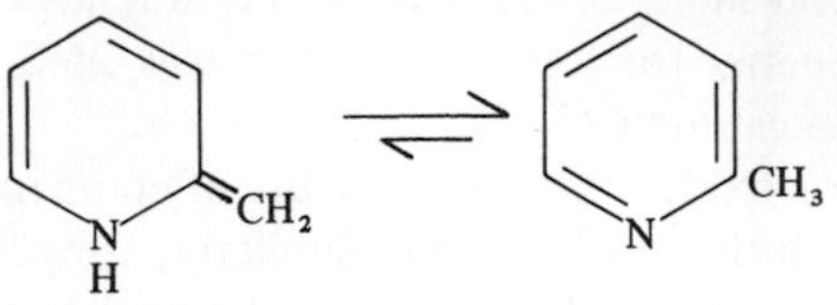

IA-148

pyridine itself (576). On the other hand, the diphenylphosphine oxide group apparently has a strong enough electron-donating ability to stabilize the quinonoïd form and the methine form predominated **(IA-150/IA-149)** = 10^8) (585).

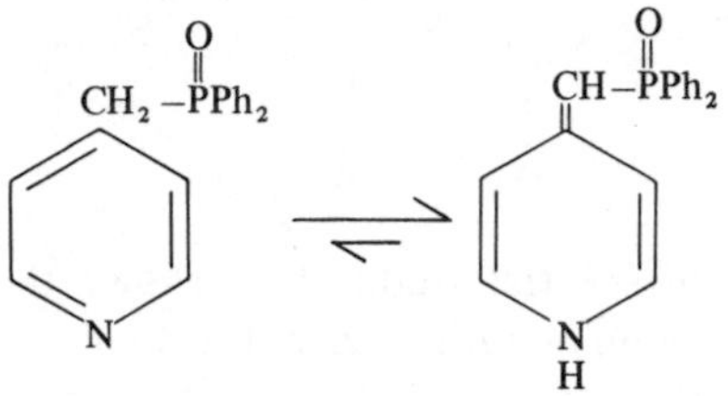

IA-149 **IA-150**

V. Miscellaneous Topics

1. Claisen and Smiles Rearrangements

The thermal Claisen rearrangement of 2-allyloxypyridines occurs either in the absence of solvent (245°) (586), in dimethyl- (245°) (586), or diethylaniline (250°) (587), or it may be catalyzed by Lewis acids (90°) (588). Under the action of acid catalysts, the rearrangement proceeds primarily to the ring nitrogen (588), but in the absence of solvent or in dialkylaniline, formation of 3-allyl-2-pyridone occurs to a slightly greater extent than does that of 1-allyl-2-pyridone (586, 587). The *a priori* analogous rearrangement of 2-alkenyloxy- and 2-allyloxypyridine-1-oxides apparently does not proceed through a concerted cyclic mechanism (589) but is thought to involve a tight carbonium ion-pair rearrangement. Retention of optical activity and the increase in ring alkylation at higher temperatures support the hypothesis. This type of reaction is not restricted to unsaturated substituents: benzyloxy-, methoxy-, and ethoxypyridine-1-oxides also undergo this rearrangement (590a). These rearrangements are discussed in detail in Chapter XII.

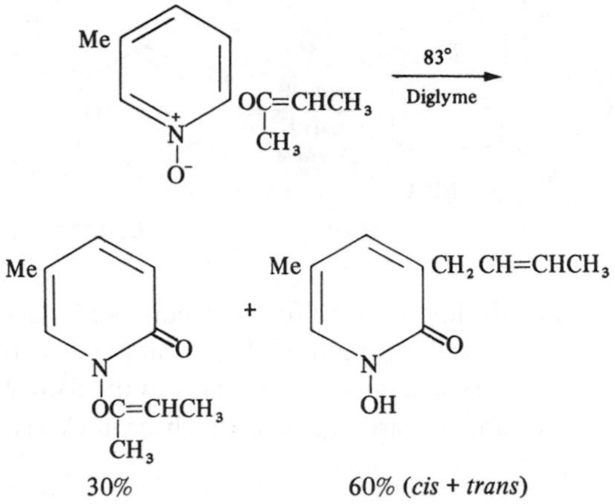

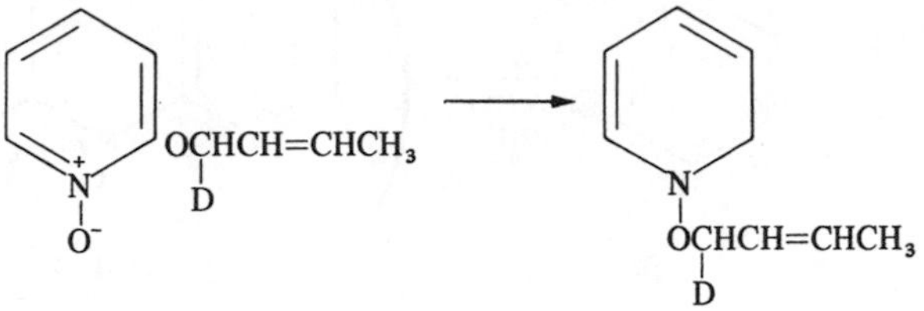

Me OC=CHCH$_3$ CH$_3$ $\overset{+}{N}$ O$^-$ → 83° Diglyme

Me N OC=CHCH$_3$ CH$_3$ 30% + Me CH$_2$CH=CHCH$_3$ N OH 60% (*cis* + *trans*)

$\overset{+}{N}$ O$^-$ OCHCH=CHCH$_3$ D → N OCHCH=CHCH$_3$ D

The equilibrium between methoxypyridines and 1-methylpyridones (the Dimroth rearrangement) catalyzed by *N*-methylmethoxypyridinium salts is almost entirely in favor of the pyridone unless the 2- and 6-positions are blocked (98) (see also Chapter XII). On the other hand the equilibrium between *N*-mesyl-2-pyridones and 2-pyridyl methanesulfonates lies far on the side of the *O*-sulfonates (590b).

The Smiles rearrangement **(IA-151)** of pyridine derivatives occurs under a variety of acidic and basic conditions (591). In at least one case, ring closure to a thiazine derivative **(IA-152)** occurred instead of the usual rearrangement (592).

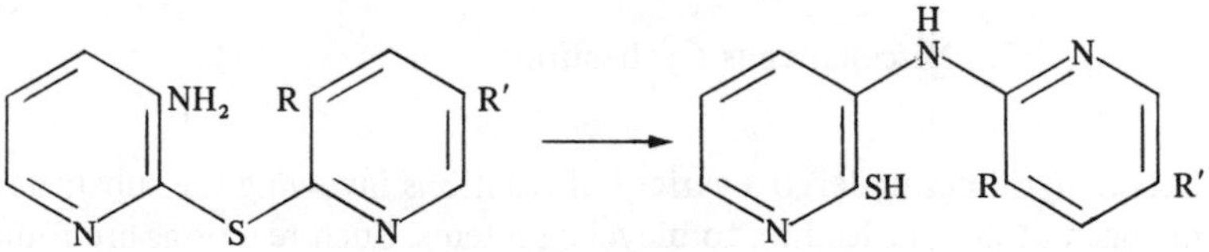

NH$_2$ R R' N S N → H N N SH R R' N **IA-151**

R = H , R' = NO_2
R = NO_2, R' = H
R = CH_3, R' = NO_2
R = NO_2, R' = CH_3

Cl NO$_2$ S NHAc N — Δ, aq.KOH, MeOH, Acetone → IA-152

The rearrangement of dipyridyl sulfones occurs with loss of SO_2 (593), although the sulfur atom is also sometimes lost from pyridyl aryl sulfides (594). In some cases, the use of strong mineral acid leads to the formation of disulfides (**IA-153**) (595a). The Smiles rearrangement has been reviewed in Chapter XV (see also 595b).

Cl Cl N S NH$_2$ — Δ, 20 *N* HCl, EtOH → IA-153

An *o*-aminophenoxypyridine also undergoes a Smiles-type rearrangement to give a high yield of **IA-154** (596).

N O H$_2$N — 60°, 2 *N* HCl, MeOH → IA-154

2. Miscellaneous Cyclizations

2-Substituted pyridines undergo a variety of reactions involving the substituent and the nitrogen atom and leading to bicyclic systems. Such reactions are found throughout these volumes, and only a few illustrative examples are given here. The reaction of 2-aminopyridine with ethyl bromoacetate to give 2-oxoimidazo[1,2-*a*]pyridine hydrobromide (**IA-155**) (597) has been extended: 2-amino- and 2-methoxypyridines and α-bromoketones give imidazo[1,2-*a*]pyridinium salts (**IA-156**) (598) and oxazolo[3,2-*a*]pyridinium salts (**IA-157**) (599), respectively.

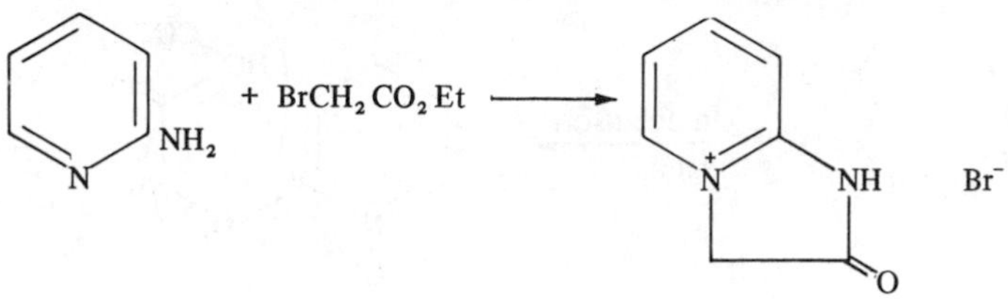

NH$_2$ + $BrCH_2CO_2Et$ ⟶ N$^+$ NH Br$^-$ O

IA-155

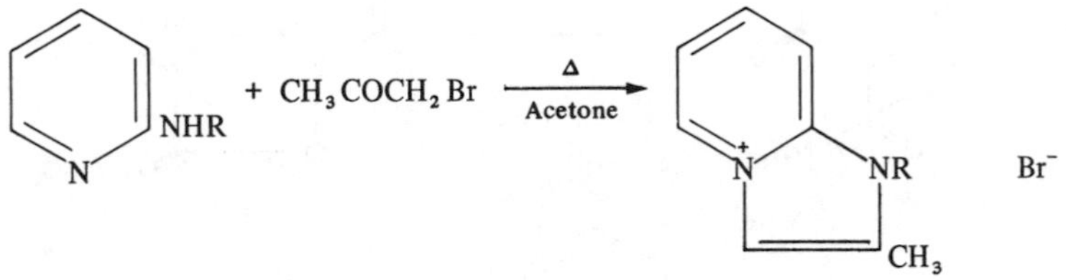

NHR + CH_3COCH_2Br $\xrightarrow[\text{Acetone}]{\Delta}$ N$^+$ NR Br$^-$ CH_3

IA-156

OCH_3 + $RCOCH_2Br$ ⟶ N O CH_2COR $\xrightarrow[\text{(ii) } HClO_4]{\text{(i) } H_2SO_4}$

N$^+$ O ClO_4^- R

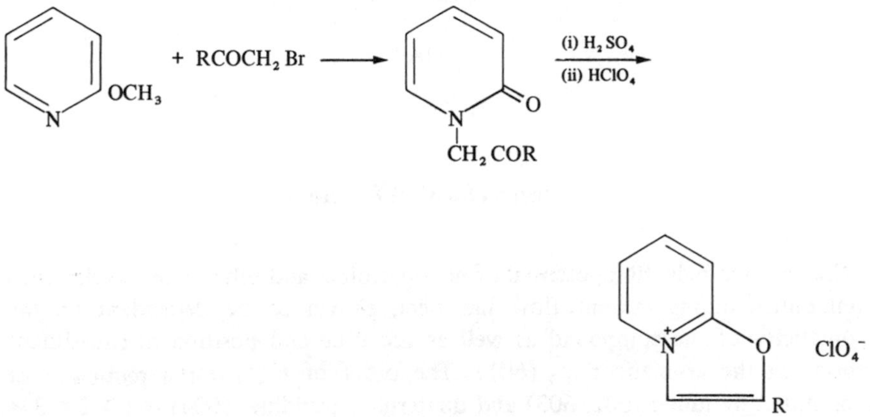

IA-157

Pyridine-1-oxides take part in a wide variety of reactions which result either in the formation of a new heterocyclic system or proceed *via* a cyclic intermediate or transition state. Some of these, for example, the acylamination of *N*-oxides, have already been discussed. Treatment of the sulfonamide **IA-158** with base gives the pyrido[1,2-*b*]isoxazolone **IA-159**, which probably arises as shown (600).

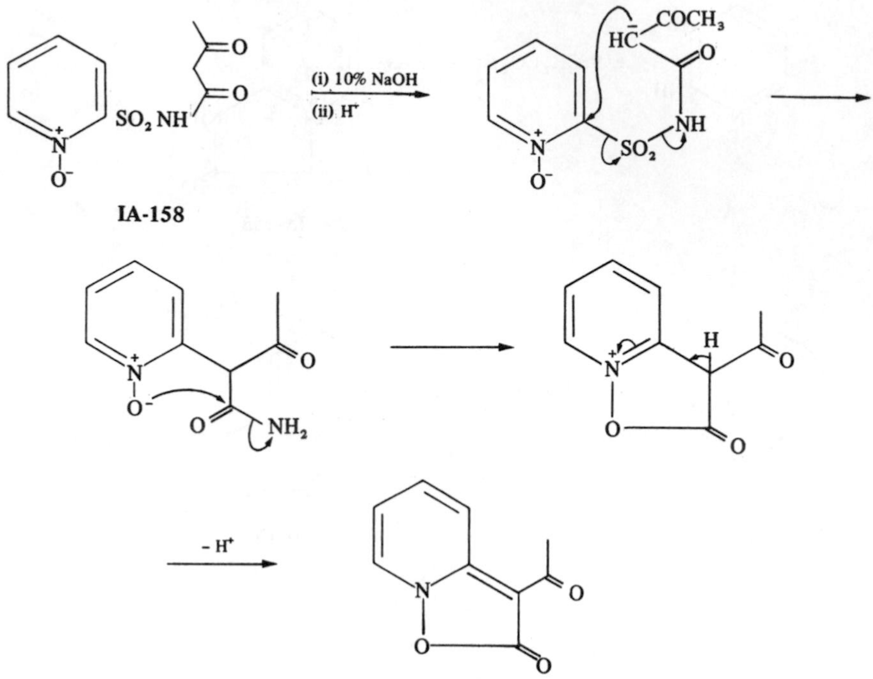

(i) 10% NaOH
(ii) H^+

IA-158

$- H^+$

IA-159

3. Electrochemical Reactions

The cathode reduction potential of nitropyridines and other heterocyclic nitro compounds during current flow has been shown to be dependent on the aromaticity of the compound as well as the type and position of substituent groups in the aromatic rings (601). The order of $E_{1/2}$ for the reduction of substituted pyridines (602, 603) and quaternary pyridines (604) is $4 > 2 \gg 3 \gg$ PhH. Halopyridines are reduced to pyridine and X^- following the order above (603) in a reaction similar to that observed in the reduction of bromobenzene in pyridine, which yields benzene, 4,4′-bipyridyl, and the phenylpyridines (605).

Isonicotinoylanilide is reduced to the corresponding aldehyde, alcohol, or amine at -0.80 V depending on the solvent acidity; in basic solution at -1.40 V, however, it is the ring which is reduced to give 1,4-dihydroisonicotinoylanilide (606). 4,4′-Bipyridyl also undergoes a 2-electron reduction at -1.42 V at pH 10 (607). Isonicotinic acid and its quaternary salt are reduced to the corresponding aldehyde in acidic medium (608).

2,6-Dimethoxypyridine undergoes anodic oxidation in alcohol solution to 2,3,3,6,6-pentamethoxy-3,6-dihydropyridine **(IA-160)** mainly, a reaction which, it has been suggested, involves consecutive two-electron oxidations (609).

The polarographic reduction of pyridinium salts produces free radicals which often dimerize to give bipyridyls (610–612). The ESR spectrum of the free radical intermediates from the reduction of **IA-161** to the Diquat derivative **IA-162** consists of 11 lines at low resolution (612). The spectrum is similar to that obtained (133 lines at high resolution) from the macroelectrolysis of Diquat at −0.75 V (612–614). The intermediacy of such free radicals is of great importance in determining the mode of action of Diquat and Paraquat (615). Herbicidal activity in quaternary salts derived from 2,2′- and 4,4′-bipyridyl was confined to compounds with redox potentials (E_0') between −300 and −500 mV in which the two pyridine rings are in the same plane (Paraquat, $E_0' = -446$ mV; Diquat, $E_0' = -349$ mV). The low activity of benzyl viologen ($E_0' = -350$ mV) has been attributed to its larger molecular size which may result in slower penetration into the plant (616).

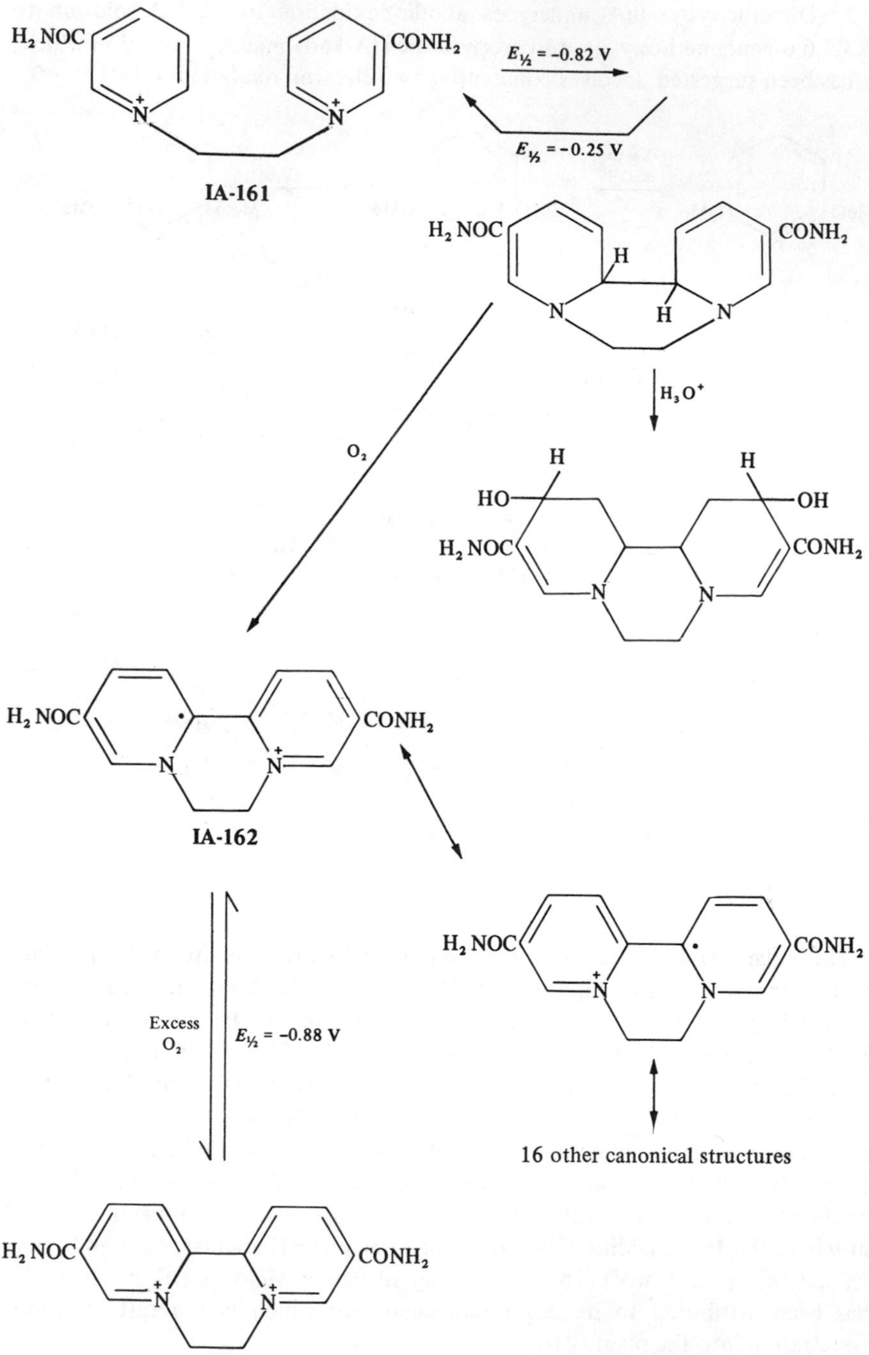

H_2NOC $CONH_2$ $E_{1/2} = -0.82$ V $E_{1/2} = -0.25$ V

IA-161

H_2NOC H H $CONH_2$

H_3O^+

H H HO OH H_2NOC $CONH_2$

O_2

H_2NOC $CONH_2$

IA-162

H_2NOC $CONH_2$

Excess O_2 $E_{1/2} = -0.88$ V

16 other canonical structures

H_2NOC $CONH_2$

4. Linear Free Energy Relationships

Numerous efforts have been made to extend the linear free energy relations (Hammett equation and constants) to nonbenzenoïd systems including pyridines. The paucity of suitable compounds for study has limited efforts in this area as has also the occurrence of tautomerism, quaternization, protonation, and solvation of the ring nitrogen atom by the reagents used or the reaction media.

The $\sigma_{-N=}$ constants for the ring nitrogen atom have been measured for reactions at the α-, β-, and γ-positions. These are obtained from a variety of different measurements, for example, pyrolysis of arylacetates (617), pK_a measurements (618–621), PMR chemical shifts of the α-proton in β-pyridylacrylic acids (622a), of the *ortho* and *para* protons in 2-substituted pyridines (622b), integrated absorption intensities in substituted pyridines (623), alkaline hydrolysis of methyl esters (624), methoxydechlorination of chloropyridines (625), and many other data. Some of these values are summarized in Table IA-7, which is by no means exhaustive. Though the values listed vary over a wide range, they are consistently in the order $\sigma_\gamma > \sigma_\alpha > \sigma_\beta$.

Substituent constants for the pyridinium ion have also been estimated and the mean values of $\sigma_{-\overset{+}{N}H=}$ from several reactions are: $\sigma_\alpha = 3.11$; $\sigma_\beta = 2.10$: $\sigma_\gamma = 2.57$ (620). σ-Constants have also been derived for the $\overset{+}{N}$–$\bar{O}$ group in pyridine-1-oxides and again a series of values have been obtained, depending on the reactions studied, for example, $\sigma_\alpha^{=N(O)-} = 0.65$ from the alkaline hydrolysis of ethyl β-pyridylacrylates (621); $\sigma_\alpha^- = 1.52$; $\sigma_\beta^- = 1.178$; $\sigma_\gamma^- = 1.526$ from the methanolysis of halopyridine-1-oxides (625–627) and $\sigma_\alpha^0 = 1.0$; $\sigma_\beta^0 = 0.7$; $\sigma_\gamma^0 = 0.5$ from the PMR spectra of pyridines (627). It has been suggested that a new

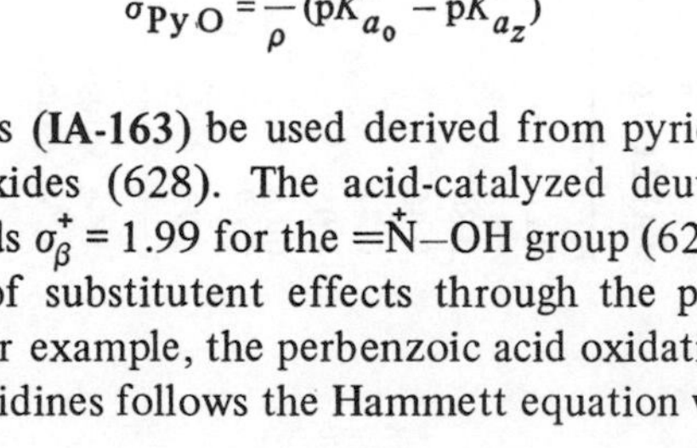

$$\sigma_{\mathrm{Py\,O}} = \frac{1}{\rho}(\mathrm{p}K_{a_0} - \mathrm{p}K_{a_z}) \qquad \ldots \qquad \textbf{IA-163}$$

set of *sigma* constants (**IA-163**) be used derived from pyridine-1-oxide data for use with other *N*-oxides (628). The acid-catalyzed deuteration of 4-aminopyridine-1-oxide yields $\sigma_\beta^+ = 1.99$ for the $=\overset{+}{N}$–OH group (629).

The transmission of substitutent effects through the pyridine ring has also been investigated. For example, the perbenzoic acid oxidation of a number of 2- and 3-substituted pyridines follows the Hammett equation with $\rho = -2.35$ (630).

TABLE 1A-7. $\sigma_{-N=}$ Values for the Pyridine Ring Nitrogen Atom

σ_α	σ_β	σ_γ	Solvent	Reaction	Ref.
0.80[a]	0.30[a]	0.87[a]	Gas phase	Pyrolysis of pyridylacetates	617
0.45	0.62	0.67	H_2O	pK_a	618, 619, 621
0.71	0.65	0.94		Mean values of data from several groups	620
1.0	0.6	0.8	DMSO	PMR of α-proton in β-pyridylacrylic acids	622
0.68	0.33	0.66	CCl_4	Integrated infrared absorption intensities	623
1.0[b]	0.59[b]	1.17[b]	MeOH	Methanolysis of halopyridines	625, 627
0.75	0.65	0.96	85% MeOH	Alkaline hydrolysis of methyl esters	624

[a] σ^+ values.
[b] σ^- values.

The effect of substituents upon the basicity of pyridines has already been mentioned in Section II.1, as has the Hammett correlation between absorption frequencies and σ-values (Section II.3.B). The values of the N–O stretching vibrations in substituted pyridine-1-oxides have been correlated with σ, σ^+ and σ^- constants (631-633). The use of the 1260 cm^{-1} band for purposes of correlation has been questioned because of the possibility of coupling of the $\overset{+}{N}-\bar{O}$ vibration with other vibrations occurring in this range (51, 634). The transmission of substituent effects in systems 2-X-pyr-5-Y and 2-Y-pyr-5-X has been studied (635-639), and also in systems 3-X-pyr-5-Y (640). Numerous other studies of this type have been carried out and these are reviewed critically by Tomasik (641).

As reported in Section II.1, Hammett correlations have also been used in the evaluation of tautomeric equilibria.

References

1. D. G. Anderson, J. R. Chipperfield, and D. E. Webster, *J. Organometal. Chem.*, **12**, 323 (1968).
2. R. H. Linnell, *J. Org. Chem.*, **25**, 290 (1960).
3. C. D. Ritchie and P. D. Heffley, *J. Amer. Chem. Soc.*, **87**, 5402 (1965). B. G. Ramsey and F. A. Walker, *J. Amer. Chem. Soc.*, **96**, 3314 (1974).
4. I. I. Grandberg, G. K. Faizova, and A. N. Kost, *Khim. Geterotsikl. Svedin.*, 561 (1966); *Chem. Abstr.*, **66**, 10453b (1967).
5. J. M. Essery and K. Schofield, *J. Chem. Soc.*, 2225 (1963).
6. G. Favini, *Gazz. Chim. Ital.*, **93**, 635 (1963).
7. J. S. Driscoll, W. Pfleiderer, and E. C. Taylor, *J. Org. Chem.*, **26**, 5230(1961).
8. K. Schofield, "Heterocyclic Nitrogen Compounds–Pyrroles and Pyridines", Butterworths, London, 1967.
9. A. Bryson, *J. Amer. Chem. Soc.*, **82**, 4871 (1960).
10. J. M. Essery and K. Schofield, *J. Chem. Soc.*, 3939 (1961).
11. J. C. Doty, J. T. R. Williams, and P. J. Grisdale, *Can. J. Chem.*, **47**, 2355 (1969).
12. E. Imoto and Y. Otsuji, *Bull. Univ. Osaka Prefect, Ser. A.*, **6**, 115 (1958); *Chem. Abstr.*, **53**, 3027h (1959).
13. K. Clarke and K. Rothwell, *J. Chem. Soc.*, 1885 (1960).
14. H. C. Brown and B. Kanner, *J. Amer. Chem. Soc.*, **88**, 986 (1966).
15. D. H. McDaniel and M. Özcan, *J. Org. Chem.*, **33**, 1922 (1968).
16. C. T. Mortimer and K. J. Laidler, *Trans. Faraday Soc.*, **55**, 1731 (1959).
17. L. Sacconi, P. Paoletti, and M. Ciampolini, *J. Amer. Chem. Soc.*, **82**, 3831 (1960).
18. H. H. Jaffé and G. O. Doak, *J. Amer. Chem. Soc.*, **77**, 4441 (1955).
19. H. H. Jaffé, *J. Org. Chem.*, **23**, 1790 (1958).
20. M. Charton, *J. Amer. Chem. Soc.*, **86**, 2033 (1964).
21. A. Albert, "Physical Methods in Heterocyclic Chemistry," A. R. Katritzky, Ed., Vol. I, Academic Press, New York, 1963, p. 31.
22. G. B. Barlin and W. Pfleiderer, *J. Chem. Soc. B*, 1425 (1971).
23. P. Forsythe, R. Frampton, C. D. Johnson, and A. R. Katritzky, *J. C. S. Perkin II*, 671 (1972).

24. T. A. Mastrukova, Y. N. Sheinker, I. K. Kuznetsova, E. M. Peresleni, T. B. Sakharova, and M. I. Kabachnik, *Tetrahedron,* **19**, 357 (1963).
25. L. D. Hansen, E. A. Lewis, J. J. Christensen, R. M. Izatt, and D. P. Wrathall, *J. Amer. Chem. Soc.*, **93**, 1099 (1971).
26. L. Joris and P. von Ragué Schleyer, *Tetrahedron*, **24**, 5991 (1968).
27. S. Walker, "Physical Methods in Heterocyclic Chemistry," A. R. Katritzky, Ed., Vol. I, Academic Press, New York, 1963, Chap. 5, p. 189.
28. R. A. Barnes, *J. Amer. Chem. Soc.*, **81**, 1935 (1959).
29. C. W. N. Cumper and A. I. Vogel, *J. Chem. Soc.*, 4723 (1960).
30. C. W. N. Cumper, R. F. A. Ginman, and A. I. Vogel, *J. Chem. Soc.*, 4518, 4525 (1962).
31. J. Barassin and H. Lumbroso, *Bull. Soc. Chim. Fr.*, 492 (1961).
32. A. N. Sharpe and S. Walker, *J. Chem. Soc.*, 4522 (1961).
33. A. R. Katritzky and P. Simmons, *J. Chem. Soc.*, 1511 (1960).
34. R. J. W. Lefèvre and D. S. N. Murthy, *Aust. J. Chem.*, **19**, 1321 (1966).
35. L. Sobczyk, *Rocz. Chem.*, **33**, 743 (1959); *Chem. Abstr.*, **54**, 1953c (1960).
36. C. Leibovici and J. Streith, *Tetrahedron Lett.*, 387 (1971).
37. S. F. Mason, "Physical Methods in Heterocyclic Chemistry," A. R. Katritzky, Ed., Vol. II, Academic Press, New York, 1963, Chap. 7, p. 1.
38. S. F. Mason, *J. Chem. Soc.*, 1240, 1247, 1253 (1959).
39. A. R. Katritzky and P. Simmons, *J. Chem. Soc.*, 4901 (1960).
40. L. Sobczyk, *Bull. Acad. Pol. Sci., Ser. Sci. Chim.*, **9**, 237 (1961); *Chem. Abstr.*, **59**, 7344h (1963).
41. T. N. Misra, *Indian J. Phys.*, **34**, 381 (1960); *Chem. Abstr.*, **55**, 13049b (1961).
42. R. D. Brown and M. L. Heffernan, *Aust. J. Chem.*, **12** 554 (1959).
43. G. Favini, I. Vandoni, and M. Simonetta, *Theor. Chim. Acta*, **3**, 45 (1965).
44. Y. Ferré, E.-J. Vincent, H. Larivé, and J. Metzger, *Bull. Soc. Chim. Fr.*, 2570 (1971).
45. L. Goodman and R. W. Harrell, *J. Chem. Phys.*, **30**, 1131 (1959).
46. A. R. Katritzky and J. M. Lagowski, *Adv. Heterocycl. Chem.*, **1**, 328, 339 (1963).
47. K. Nakamoto and A. E. Martell, *J. Amer. Chem. Soc.*, **81**, 5857, 5863 (1959).
48. A. R. Katritzky, A. R. Hands, and R. A. Jones, *J. Chem. Soc.*, 3165 (1958).
49. V. S. Korobkov and A. V. Voropaeva, and I. Kh. Fel'dman, *Zh. Obshch. Khim.*, **31**, 3136 (1961); *Chem. Abstr.*, **56**, 15059b (1962).
50. S. F. Mason, *J. Chem. Soc.*, 2437 (1960).
51. S. Ghersetti, G. Maccagnani, A. Mangini, and F. Montanari, *J. Heterocycl. Chem.*, **6**, 859 (1969).
52. E. M. Kosower, *J. Amer. Chem. Soc.*, **80**, 3253, 3261, 3267 (1958).
53. I. Kubota and T. Yamikawa, *Bull. Chem. Soc. Jap.*, **35**, 1046 (1962).
53a. R. A. Mackay and E. J. Poziomek, *J. Amer. Chem. Soc.*, **94**, 6107 (1972).
54. J. W. Verhoeven, I. P. Dirkx, and Th. J. deBoer, *Tetrahedron*, **25**, 4037 (1969).
55. T. Kubota, M. Yamakawa, Y. Mizuno, and K. Nishikida, "Abstracts of Papers, 3rd International Congress of Heterocyclic Chemistry," Sendai, Japan, 1971, paper B-26-6, p. 223.
56. A. R. Katritzky and J. M. Lagowski, "Chemistry of the Heterocyclic *N*-Oxides," Academic Press, New York, 1971, Chap. 1, p. 14.
57. R. Isaac, F. F. Bentley, H. Sternglanz, W. C. Coburn, Jr., C. V. Stephenson, and W. S. Wilcox, *Appl. Spectrosc.*, **17**, 90 (1963); *Chem. Abstr.*, **59**, 7077f (1963).
58. H. E. Podall, *Anal. Chem.*, **29**, 1423 (1957).
59. R. A. Abramovitch, C. S. Giam, and A. D. Notation, *Can. J. Chem.*, **38**, 624 (1960).
60. E. Janeckova and J. Kuthan, *Z. Chem.*, **5**, 349 (1965).

61. A. R. Katritzky and A. P. Ambler, "Physical Methods in Heterocyclic Chemistry," A. R. Katritzky, Ed., Vol. II, Academic Press, New York, 1963, Chap. 10, pp. 274–291.
62. R. A. Coburn and G. O. Dudek, *J. Phys. Chem.*, **72**, 3681 (1968).
63. A. R. Katritzky and R. A. Jones, *J. Chem. Soc.*, 3674 (1959).
64. R. A. Jones and R. P. Rao, *Aust. J. Chem.*, **18**, 583 (1965).
65. A. P. Rud'ko, I. N. Chernyuk, Yu. S. Rozum, and G. T. Pilyugin, *Ukr. Khim. Zh.*, **34**, 1275 (1968); *Chem. Abstr.*, **70**, 110146e (1969).
66. A. R. Katritzky, *Rec. Trav. Chim. Pays-Bas*, **78**, 995 (1959).
67. J. Susko and M. Szafran, *Bull. Acad. Pol. Sci., Ser. Sci. Chim.*, **10**, 233 (1962); *Chem. Abstr.*, **58**, 7519c (1963).
68. V. E. Blokhin, Z. Yu. Kokoshko, E. P. Darienko, and Z. V. Pushkoreva, *Zh. Obshch. Khim.*, **39**, 1623 (1969); *Chem. Abstr.*, **71**, 101167f (1969).
69. A. R. Katritzky, C. R. Palmer, F. J. Swinbourne, T. T. Tidwell, and R. D. Topsom, *J. Amer. Chem. Soc.*, **91**, 636 (1969).
70. S. Ghersetti, C. Maccagnani, A. Mangini, and F. Montanari, *J. Heterocycl. Chem.*, **6**, 859 (1969).
71. R. A. Abramovitch and T. Takaya, *J. Org. Chem.*, **37**, 2022 (1972).
72. K. Ramaiah and V. R. Srinivasan, *Proc. Indian Acad. Sci.*, **50A**, 213 (1959); *Chem. Abstr.*, **54**, 5247i (1960).
73. K. Ramaiah and V. R. Srinivasan, *Proc. Indian Acad. Sci.*, **55A**, 221 (1962); *Chem. Abstr.*, **57**, 6761e (1962).
74. R. F. M. White, "Physical Methods in Heterocyclic Chemistry," A. R. Katritzky, Ed., Vol. II, Academic Press, New York, 1963, Chap. 9, p. 103.
75. L. M. Jackman, "Nuclear Magnetic Resonance Spectroscopy," 2nd ed., Pergamon Press, Oxford, 1969.
76. W. Brügel, *Z. Electrochem.*, **66**, 159 (1962).
77. R. A. Abramovitch and J. B. Davis, *J. Chem. Soc. B*, 1137 (1966).
78. J. W. Emsley, J. Feeney, and L. H. Sutcliffe, "High Resolution Nuclear Magnetic Resonance Spectroscopy," Pergamon Press, Oxford, 1966.
79. D. G. deKowalewski and E. C. Ferrá, *Mol. Phys.*, **13**, 547 (1967).
80. W. B. Smith and J. L. Roark, *J. Phys. Chem.*, **73**, 1049 (1969).
81. R. H. Cox and A. A. Bothner-By, *J. Phys. Chem.*, **73**, 2465 (1969).
82. T. K. Wu, *J. Phys. Chem.*, **71**, 3089 (1967).
83. M. Katcka and T. Urbánski, *Bull. Acad. Pol. Sci., Ser. Sci. Chim.*, **14**, 347 (1968).
84. B. D. N. Rao and P. Verkatewarlu, *Proc. Indian Acad. Sci.*, **54A**, 305 (1961); *Chem. Abstr.*, **56**, 12450f (1962).
85. V. J. Kowalewski and D. G. deKowalewski, *J. Chem. Phys.*, **36**, 266 (1962).
86. V. J. Kowalewski and D. G. deKowalewski, *J. Chem. Phys.*, **37**, 2603 (1962).
87. J. A. Pople, H. J. Bernstein, and W. G. Schneider, *Ann. N. Y. Acad. Sci.*, **70**, 806 (1958).
88. S. Castellano, C. Sun, and R. Kostelnik, *J. Chem. Phys.*, **46**, 327 (1967).
89. T. Schaefer and W. G. Schneider, *Can. J. Chem.*, **41**, 966 (1963).
90. I. C. Smith and W. G. Schneider, *Can. J. Chem.*, **39**, 1158 (1961).
91. J. Kuthan and V. Skala, *Z. Chem.*, **6**, 422 (1966).
92. M. L. Martin, J. P. Dorie, and F. Peradejordi, *J. Chim. Phys.*, **64**, 1193 (1967).
93. G. G. Dvoryantseva, V. P. Lezina, V. F. Bystrov, T. N. Ul'Yanova, and G. P. Syrova, *Izv. Akad. Nauk SSSR, Ser. Khim.*, 994 (1968); *Chem. Abstr.*, **69**, 76367v(1968).
94. Y. Sasaki, M. Hanataka, I. Shiraishi, M. Suzuki, and K. Nishimoto, *Yakugaku Zasshi*, **89**, 21 (1969); *Chem. Abstr.*, **70**, 92138x (1969).
95. H.-H. Perkampus, and V. Krüger, *Ber. Bunsenges Phys. Chem.*, **71**, 447 (1967).

96. J. A. Elvidge and L. M. Jackman, *J. Chem. Soc.*, 859 (1961).
97. L. M. Jackman, Q. N. Porter, and G. R. Underwood, *Aust. J. Chem.*, **18**, 1221 (1965).
98. P. Beak, J. Bonham, and J. Lee, Jr., *J. Amer. Chem. Soc.*, **90**, 1569 (1968).
99. A. H. Gawer and B. P. Dailey, *J. Chem. Phys.*, **42**, 2658 (1965).
100. T. K. Wu and B. P. Dailey, *J. Chem. Phys.*, **41**, 3307 (1964).
101 B. M. Lynch and H. J. M. Dou, *Tetrahedron Lett.*, 2627 (1966).
102. V. M. S. Gil and J. N. Murrell, *Trans. Faraday Soc.*, **60**, 248 (1964).
103. J. Abblard, C. Decoret, L. Cronenberger, and H. Pacheco, *Bull. Soc. Chim. Fr.*, 2466 (1972).
104. M. Freymann, R. Freymann, and C. Geissner-Prettre, *Arch. Sci.* (Geneva), **13** *Spec. No.*, 506 (1960); *Chem. Abstr.*, **58**, 3289d (1963).
105. A. R. Katritzky, F. J. Swinbourne, and B. Ternai, *J. Chem. Soc. B*, 235 (1966).
106. C. R. Kanekar and H. V. Venkatasetly, *Curr. Sci.*, **34**, 555 (1965); *Chem. Abstr.*, **64**, 2885h (1966).
107. J. N. Murrell and V. M. S. Gil, *Trans. Faraday Soc.*, **61**, 402 (1965).
108. J. A. Ladd and V. I. P. Jones, *Spectrochim. Acta*, **23A**, 2791 (1967).
109. T. M. Spotswood and C. I. Tanzer, *Tetrahedron Lett.*, 911 (1967).
110. P. Laszlo and J. L. Soong, Jr., *J. Chem. Phys.*, **47**, 4772 (1967).
110a. G. Kotowycz, T. Schaefer, and E. Bock, *Can. J. Chem.*, **42**, 2541 (1964).
110b. M. S. Sun, F. Grein, and D. G. Brewer, *Can. J. Chem.*, **50**, 2626 (1972).
111. C. Beauté, Z. W. Wolkowski, and N. Thoai, *Tetrahedron Lett.*, 817 (1971).
112. W. L. F. Armarego, T. J. Batterham, and J. R. Kershaw, *Org. Magn. Resonance*, **3**, 575 (1971). G. Beech and R. J. Morgan, *Tetrahedron Lett.*, 973 (1974).
113. J. Reuben and J. S. Leigh, Jr., *J. Amer. Chem. Soc.*, **94**, 2789 (1972).
114. R. D. Fischer and R. V. Ammon, *Angew. Chem. Int. Ed.*, **11**, 675 (1972).
115. G. Montaudo and P. Finocchiaro, *Tetrahedron Lett.*, 3429 (1972). G. Montaudo, G. Kruk, and J. W. Verhoeven, *Tetrahedron Lett.*, 1841 (1974).
116. J. Lee and K. G. Orrell, *J. Chem. Soc.*, 582 (1965).
117. J. G. Rowbotham and T. Schaefer, *Can. J. Chem.*, **50**, 2344 (1972).
118. J. W. Emsley and L. Phillips, *J. Chem. Soc. B*, 434 (1969).
119. J. D. Baldeschwieler and E. W. Randall, *Proc. Chem. Soc.*, 303 (1961).
120. P. C. Lauterbur, *J. Chem. Phys.*, **43**, 360 (1965).
121. J. B. Stothers, *Quart. Rev.*, **19**, 144 (1965).
122. H. L. Retcofsky and R. A. Friedel, *J. Phys. Chem.*, **71**, 3592 (1967).
123. H. L. Retcofsky and R. A. Friedel, *J. Phys. Chem.*, **72**, 290 (1968).
124. H. L. Retcofsky and R. A. Friedel, *J. Phys. Chem.*, **72**, 2619 (1968).
125. T. Tokuhiro, N. K. Wilson, and G. Fraenkel, *J. Amer. Chem. Soc.*, 90, 3622 (1968). Y. Takeuchi and N. Dennis, *J. Amer. Chem. Soc.*, 96, 3657 (1974). M. Hansen and H. J. Jacobsen, *J. Magn. Resonance*, 10, 74 (1973).
126. F. Vögtle and H. Risler, *Angew. Chem. Int. Ed.*, **11**, 727 (1972).
127. Q. N. Porter and J. Baldas, "Mass Spectrometry of Heterocyclic Compounds," Wiley-Interscience, New York, 1971, Chap. 11, pp. 376–398.
128. H. Budzikiewicz, C. Djerassi, and D. H. Williams, "Mass Spectrometry of Organic Compounds," Holden-Day, San Francisco, 1967.
129. D. H. Williams and J. Ronayne, *Chem. Commun.*, 1129 (1967).
130. R. J. Dickinson and D. H. Williams, *J. C. S. Perkin II*, 1363 (1972).
131. W. G. Cole, D. H. Williams, and A. N. H. Yeo, *J. Chem. Soc. B*, 1284 (1968).
132. G. H. Keller, L. Bauer, and C. L. Bell, *J. Heterocycl. Chem.*, **5**, 647 (1968).
133. M. Ogata, Y. Miyaji, and H. Ichikawa, "Abstracts of Papers, 3rd International Congress of Heterocyclic Chemistry," Sendai, Japan, 1971, paper C-24-7, p. 335.
134. J. Bonham, E. McLeister, and P. Beak, *J. Org. Chem.*, **32**, 639 (1967).
135. R. Lawrence and E. S. Waight, *J. Chem. Soc. B*, 1 (1968).
136. E. M. Kaiser, *J. Heterocycl. Chem.*, **5**, 571 (1968).

137. K. F. King, F. M. Hershenson, L. Bauer, and C. L. Bell, *J. Heterocycl. Chem.*, **6**, 851 (1969).
138. R. Neeter and N. M. M. Nibbering, *Tetrahedron,* **28**, 2575 (1972).
139. R. Grigg and B. G. Odell, *J. Chem. Soc. B,* 218 (1966).
140. N. Bild and M. Hesse, *Helv. Chim. Acta,* **50**, 1885 (1967).
141. R. A. Abramovitch, S. Kato, and G. M. Singer, *J. Amer. Chem. Soc.*, **93**, 3074 (1971).
142. K. Tsuji, H. Yoshida, K. Hayashi, and S. Okamura, *Nippon Hoshasen Kobunshi. Kenkyu Kyokai Nenpo*, **6**, 163 (1964–1965); *Chem. Abstr.*, **65**, 199b (1966).
143. K. Tsuji, H. Yoshida and K. Hayashi, *J. Chem. Phys.*, **45**, 2894 (1966).
144. H. J. Bower, J. A. McRae, and M. C. R. Symons, *J. Chem. Soc. A,* 2696 (1968).
145. K. Kuwata, T. Ogawa, and K. Hirota, *Bull. Chem. Soc. Jap.*, **34**, 291 (1961).
146. H. J. Bower, J. A. McRae, and M. C. R. Symons, *J. Chem. Soc. A,* 1918 (1968).
147. E. M. Kosower and E. J. Poziomek, *J. Amer. Chem. Soc.*, **86**, 5515 (1964).
148. E. M. Kosower and J. L. Cotter, *J. Amer. Chem. Soc.*, **86**, 5524 (1964).
149. P. Atlanti, J. F. Biellman, R. Briere, H. Lamaire, and A. Rassat, *Tetrahedron,* **28**, 2827 (1972).
150. E. G. Janzen and J. W. Happ, *J. Phys. Chem.*, **73**, 2335 (1969).
151. L. Guibé, *C. R. Acad. Sci., Paris, Ser. C,* **250**, 3014 (1960).
152. L. Guibé, *Ann. Phys.* (Paris), **7**, 177 (1962); *Chem. Abstr.*, **58**, 6352c (1963).
153. E. A. C. Lucken, *Trans. Faraday Soc.*, **57**, 729 (1961).
154. E. Schemp and P. J. Bray, *J. Chem. Phys.*, **49**, 3450 (1968).
155. C. Goffart, J. Momigny, and P. Natalis, *Int. J. Mass Spectrom. Ion Phys.*, **3**, 371 (1969); *Chem. Abstr.*, **72**, 89451e (1970). M. A. Weiner and M. Lattman, *Tetrahedron Lett.*, 1709 (1974).
156. A. D. Baker, D. Betteridge, N. R. Kemp, and R. E. Kirby, *Chem. Commun.*, 286 (1970).
157. J. del Bené and H. H. Jaffé, *J. Chem. Phys.*, **48**, 1807 (1968).
158. C. T. Kyte, G. H. Jeffrey, and A. I. Vogel, *J. Chem. Soc.*, 4454 (1960).
159. J. S. Fitzgerald, *Aust. J. Appl. Sci.*, **12**, 51 (1961).
160. A. L. Brown and K. R. Buck, *Chem. Ind.* (London), 714 (1961).
161. R. A. Abramovitch and A. R. Vinutha, *J. Chem. Soc. C,* 2104 (1969).
162. A. Veillard and G. Berthier, *Theor. Chim. Acta,* **4**, 347 (1966).
163. K. Nishimoto and L. S. Forster, *Theor. Chim. Acta*, **4**, 155 (1966).
164. M. J. S. Dewar and G. J. Gleicher, *J. Chem. Phys.*, **44**, 759 (1966).
165. A. K. Chandra and S. Basu, *J. Chem. Soc.*, 1623 (1959).
166. G. Favini and M. Simonetta, *Gazz. Chim. Ital.*, **90**, 363 (1960).
167. J. M. Hernando and M. D. Herrezuelo, *Ann. Quim.*, **65**, 1089 (1969); *Chem. Abstr.*, **72**, 103972j (1970).
168. L. Paoloni, M. Tosato, and M. Cignitti, *Theor. Chim. Acta,* **14**, 221 (1969).
169. G. Berthier, B. Lévy, and L. Paoloni, *Theor. Chim. Acta*, **16**, 316 (1970).
170. J. P. Cartier and C. Sandorfy, *Can. J. Chem.*, **41**, 2759 (1963).
171. W. Adam and A. Grimison, *Tetrahedron*, **21**, 3417 (1965).
172. R. A. Abramovitch and J. G. Saha, *Adv. Heterocycl. Chem.*, **6**, 229 (1966).
173. W. Adam, A. Grimison, and R. Hoffmann, *J. Amer. Chem. Soc.*, **91**, 2590 (1969).
174. V. N. Drozd, V. I. Minkin, and Yu. A. Osthoumov, *Zh. Org. Khim.*, **4**, 1501 (1968); *Chem. Abstr.*, **69**, 105773g (1968).
175. J. S. Kwiatkowski and B. Zurawski, *Acta Phys. Pol.*, **32**, 893 (1967); *Chem. Abstr.*, **70**, 10956a (1969).
176. J. Paleček, V. Skála, and J. Kuthan, *Collect. Czech. Chem. Comm.*, **34**, 1110 (1969).
177. H. J. M. Dou, G. Vernin, and J. Metzger, *Bull. Soc. Chim. Fr.*, 4593 (1971).
178. W. Adam, A. Grimison, and G. Rodríguez, *Tetrahedron,* **23**, 2513 (1967).

179. J. W. Emsley, *J. Chem. Soc. A*, 1387 (1968).
180. J. Kuthan, M. Ferles, and N. V. Koshmina, *Tetrahedron,* **26**, 4361 (1970).
181. J. Kuthan, N. V. Koshmina, J. Paleček, and V. Skála, *Collect. Czech. Chem. Comm.*, **35**, 2787 (1970).
182. E. M. Evleth, *Theor. Chim. Acta,* **11**, 145 (1968).
183. R. A. Abramovitch, G. M. Singer, and A. R. Vinutha, *Chem. Commun.*, **55** (1967).
184. R. M. Acheson and G. A. Taylor, *J. Chem. Soc.*, 1691 (1960) and Refs. cited therein.
185. R. M. Acheson and J. N. Bridson, *J. Chem. Soc. C,* 1143 (1969).
186. E. Pohjala, *Tetrahedron Lett.*, 2585 (1972).
187. J. S. Grossert, *Chem. Commun.*, 305 (1970).
188. G. M. Singer, Ph. D. Dissertation, University of Saskatchewan, 1970.
189. A. de Souza Gomes and M. M. Joullié, *J. Heterocycl. Chem.*, **6**, 729 (1969).
190. A. I. Meyers and P. Singh, *Chem. Commun.*, 576 (1968).
191. H. Tomisawa, H. Hongo, and R. Fujita, "Abstracts of Papers, 3rd International Congress of Heterocyclic Chemistry," Sendai, Japan, 1971, paper C-23-4, p. 290.
192. L. Bauer, C. L. Bell, and G. E. Wright, *J. Heterocycl. Chem.*, **3**, 393 (1966); E. B. Sheinin, G. E. Wright, C. L. Bell, and L. Bauer, *ibid.*, **5** 859 (1968).
193. A. R. Katritzky and Y. Takeuchi, *J. Amer. Chem. Soc.*, **92**, 4134 (1970); *J. Chem. Soc., C,* 874, 878 (1971).
194. Y. Takeuchi, N. Dennis, A. R. Katritzky, and I. Taulov, "Abstracts of Papers, 3rd International Congress of Heterocyclic Chemistry," Sendai, Japan, 1971, paper B-26-9, p. 232. N. Dennis, A. R. Katritzky, T. Matsuo, S. K. Parion, and Y. Takeuchi, *J. C. S. Perkin I*, 746 (1974).
195a. N. Dennis, A. R. Katritzky, S. K. Parion, and Y. Takeuchi, *J. C. S. Chem. Comm.*, 707 (1972).
195b. N. Dennis, B. Ibrahim, A. R. Katritzky, and Y. Takeuchi, *J. C. S. Chem. Comm.*, 292 (1973).
196. J. Markert and E. Fahr, *Tetrahedron Lett.*, 4337 (1967).
197. J. Miyasaka, *Iyo Kizai Kenkyusho Hokoku,* **2**, 67 (1968); *Chem. Abstr.*, **70**, 106324u (1969).
198. T. Sasaki, K. Kanematsu, and M. Murata, *Tetrahedron,* **27**, 5121 (1971).
199. R. Daniels and O. L. Salerni, *Proc. Chem. Soc.*, 286 (1960).
200. G. A. Taylor, *J. Chem. Soc.*, 3332 (1965).
201. T. Kato, H. Yamanaka, T. Niitsuma, K. Wagatsuma, and M. Oizumi, *Chem. Pharm. Bull.* (Tokyo), 12, 910 (1964); *Chem. Abstr.*, 63 2949d (1965).
202. G. Stöckelmann, H. Specker, and W. Riepe, *Chem. Ber.*, **102**, 455 (1969).
203. Z. J. Allan, J. Podstata, and Z. Vrba, *Tetrahedron Lett.*, 4855 (1969).
204. T. Kato, Y. Goto, and Y. Yamamoto, *Yakugaku Zasshi,* **82**, 1649 (1962); *Chem. Abstr.*, **59**, 2765d (1963).
205. V. Boekelheide and N. A. Fedoruk, *J. Amer. Chem. Soc.*, **90**, 3830 (1968).
206. C. Leonte and I. Zugravescu, *Tetrahedron Lett.*, 2029 (1972).
207. V. Boeckelheide and N. A. Fedoruk, *J. Org. Chem.*, **33**, 2062 (1968).
208. T. Sasaki, K. Kanematsu, and A. Kakehi, *J. Org. Chem.*, **36**, 2978 (1971).
209. T. Sasaki, K. Kanematsu, A. Kakehi, and G. Ito, *Bull. Chem. Soc. Jap.*, **45**, 2050 (1972).
210. R. A. Abramovitch and V. Alexanian, unpublished results, 1972.
211. T. Okamoto, M. Hirobe, and Y Tamai, *Chem. Pharm. Bull.* (Tokyo), 11, 1089 (1963).
212. J. Delarge, *Farmaco, Ed. Sci.*, **20**, 629 (1965); *Chem. Abstr.*, **64**, 3467e (1966).
213. A. R. Katritzky and C. D. Johnson, *Angew. Chem. Int. Ed.*, **6**, 608 (1967).

214. M. W. Austin, M. Brickman, J. H. Ridd, and B. V. Smith, *Chem. Ind.* (London), 1057 (1962).
215. C. D. Johnson, A. R. Katritzky, and M. Viney, *J. Chem. Soc. B,* 1211 (1967).
216. C. D. Johnson, A. R. Katritzky, B. J. Ridgewell, and M. Viney, *J. Chem. Soc. B,* 1204 (1967).
217. P. J. Brignell, A. R. Katritzky, and H. O. Tarhan, *J. Chem. Soc. B,* 1477 (1968).
218. A. R. Katritzky, H. O. Tarhan, and S. Tarhan, *J. Chem. Soc. B,* 114 (1970).
219. R. C. De Selms, *J. Org. Chem.,* **33,** 478 (1968).
220. L. D. Smirnov, R. E. Lokhov, V. P. Lezina, B. E. Zaitsev, and K. M. Dyumaev, *Izv. Akad. Nauk. SSSR, Ser. Khim.,* 1567 (1969); *Chem. Abstr.,* **71,** 112765a (1969).
221. P. J. Brignell, P. E. Jones, and A. R. Katritzky, *J. Chem. Soc. B,* 117 (1970).
222. E. E. Garcia, C. V. Greco, and I. M. Hunsberger, *J. Amer. Chem. Soc.,* **82,** 4430 (1960).
223. T. Batkowski, D. Tomasik, and P. Tomasik, *Rocz. Chem.,* **41,** 2101 (1967).
224. C. R. Kolder and H. J. den Hertog, *Rec. Trav. Chim. Pays-Bas,* 79, 474 (1960).
225. K. Schofield, *Quart. Rev.,* **4,** 382 (1950).
226. K. Lewicka and E. Plazek, *Rocz. Chem.,* **40,** 405 (1966); *Chem. Abstr.,* **65,** 7134g (1966).
227. A. R. Katritzky and B. J. Ridgewell, *J. Chem. Soc.,* 3753 (1963).
228. P. Bellingham, C. D. Johnson, and A. R. Katritzky, *J. Chem. Soc. B,* 1226 (1967).
229. A. El-Anani, P. E. Jones, and A. R. Katritzky, *J. Chem. Soc. B,* 2363 (1971).
230. H. von Peckman, *Ber.,* **28,** 1624 (1895).
231. R. A. Barnes, "Pyridine and Its Derivatives," E. Klingsberg, Ed., 1st ed. Wiley, New York, 1959, Chap. 7, p. 8.
232. N. Kornblum and G.-P. Coffey, *J. Org. Chem.,* **31,** 3447 (1966).
233. S. Nesnow and R. Shapiro, *J. Org. Chem.,* **34,** 2011 (1969).
234. H. Bojarska-Dahlig and I. Gruda, *Rocz. Chem.,* **33,** 505 (1959); *Chem. Abstr.,* **54,** 522h (1960).
235. D. G. Anderson and D. E. Webster, *J. Organometal. Chem.,* **13,** 113 (1968).
236. D. G. Anderson and D. E. Webster, *J. Chem. Soc. B,* 765 (1968).
237. D. G. Anderson and D. E. Webster, *J. Chem. Soc. B,* 878 (1968).
238. F. H. Pinkerton and S. F. Thomas, *J. Heterocycl. Chem.,* **6,** 433 (1969).
239. H. C. H. A. van Riel, F. C. Fischer, J. Lugtenburg, and E. Havinga, *Tetrahedron Lett.,* 3085 (1969).
240. D. Heinert and A. E. Martell, *Tetrahedron,* **3,** 49 (1958).
241. L. D. Smirnov, V. P. Lezina, V. F. Bystrov, and K. M. Dyumaev, *Izv. Akad. Nauk SSSR, Ser. Khim.,* 198 (1965); *Chem. Abstr.,* **62** 11774d (1965).
242. G. Klopman, *J. Amer. Chem. Soc.,* **90,** 223 (1968).
243. M. van Ammers and H. J. den Hertog, *Rec. Trav. Chim. Pays-Bas,* **78,** 586 (1959).
244. H. C. van der Plas, H. J. den Hertog, M. van Ammers, and B. Haase, *Tetrahedron Lett.,* 32 (1961).
245. R. B. Moodie, K. Schofield, and M. J. Williams, *Chem. Ind.* (London), 1577 (1964).
246. J. Gleghorn, R. B. Moodie, K. Schofield, and M. J. Williams, *J. Chem. Soc. B,* 870 (1966).
247. C. D. Johnson, A. R. Katritzky, N. Shakir, and M. Viney, *J. Chem. Soc. B,* 1213 (1967).
248. A. R. Katritzky and M. Kingsland, *J. Chem. Soc. B,* 862 (1968).
249. G. P. Bean, P. J. Birgnell, C. D. Johnson, A. R. Katritzky, B. J. Ridgewell, H. O. Tarhan, and A. M. White, *J. Chem. Soc. B,* 122 (1967).
250. M. van Ammers, H. J. den Hertog, and B. Haase, *Tetrahedron,* **18,** 227 (1962).

251. J. Arotsky, H. C. Mishra, and M. C. R. Symons, *J. Chem. Soc.*, 12 (1961).
252. M. Hammana and M. Yamazaki, *Chem. Pharm. Bull.* (Tokyo), 9, 414 (1961).
253. K. M. Dyumaev and R. E. Lokhov, *J. Org. Chem. U.S.S.R.*, **8**, 416 (1972).
254. A. R. Fersht and W. P. Jencks, *J. Amer. Chem. Soc.*, **91**, 2125 (1969).
255. J. Jones and J. Jones, *Tetrahedron Lett.*, 2117 (1964).
256. G. A. Olah, J. A. Olah, and N. A. Overchuck, *J. Org. Chem.*, **30**, 3373 (1965).
257. C. A. Cupas and R. L. Pearson, *J. Amer. Chem. Soc.*, **90**, 4742 (1968).
258. A. R. Katritzky and B. J. Ridgewell, *J. Chem. Soc.*, 3882 (1963).
259. M. M. Boudakian, F. F. Frulla, D. F. Gavin, and J. A. Zaslowsky, *J. Heterocycl. Chem.*, **4**, 375 (1967).
260. M. M. Boudakian, D. F. Gavin, and R. J. Polak, *J. Heterocycl. Chem.*, **4**, 377 (1967).
261. T. Caronna, G. P. Gardini, and F. Minisci, *Chem. Commun.*, 201 (1969).
262a. F. Minisci, G. P. Gardini, R. Galli, and F. Bertini, *Tetrahedron Lett.*, 15 (1970).
262b. F. Minisci, R. Galli, V. Malatesta, and T. Caronna, *Tetrahedron*, **26**, 4083 (1970).
263. Yu. I. Chumakov and V. F. Novikova, *Khim. Prom.*, (2) 47 (1968); *Chem. Abstr.*, **69**, 43743m (1968).
264. P. A. Claret and G. H. Williams, *J. Chem. Soc. C*, 146 (1969).
265. P. Ph. H. L. Otto and J. P. Wibaut, *Rec. Trav. Chim. Pays-Bas*, **78**, 345 (1959).
266. R. A. Abramovitch and K. Kenaschuck, *Can. J. Chem.*, **45**, 509 (1967).
267. J. -M. Bonnier and J. Court, *C. R. Acad. Sci., Paris, Ser. C.*, **265**, 133 (1968).
268. H. J. M. Dou, G. Vernin, and J. Metzger, *Bull. Soc. Chim. Fr.*, 4189 (1971).
269. H. J. M. Dou, G. Vernin, and J. Metzger, *Bull. Soc. Chim. Fr.*, 3553 (1971).
270. K. C. Bass and P. Nababsing, *J. Chem. Soc. C*, 388 (1969).
271. R. A. Abramovitch and J. G. Saha, *J. Chem. Soc.*, 2175 (1964).
272. R. A. Abramovitch and J. G. Saha, *Tetrahedron*, **21**, 3297 (1965).
273. H. J. M. Dou, G. Vernin, and J. Metzger, *Tetrahedron Lett.*, 953 (1968).
274. P. J. Bunyan and D. H. Hey, *J. Chem. Soc.*, 3787 (1960).
275. R. Grashey and R. Huisgen, *Chem. Ber.*, **92**, 2641 (1959).
276. H. J. M. Dou and B. M. Lynch, *Bull. Soc. Chim. Fr.*, 3815 (1966).
277. G. Vernin, H. J. M. Dou, and J. Metzger, *C. R. Acad. Sci., Paris, Ser. C*, **264**, 1762 (1967).
278. J. -M. Bonnier and J. Court, *C. R. Acad. Sci., Paris, Ser. C*, **263**, 262 (1966).
279. J. -M. Bonnier, J. Court, and T. Fay, *Bull. Soc. Chim. Fr.*, 1204 (1967).
280. R. A. Abramovitch and M. Saha, *J. Chem. Soc. B*, 733 (1966).
281. R. A. Abramovitch, *Intra-Sci. Chem. Rep.*, **3** (3) 211 (1969).
282. R. A. Abramovitch and M. Saha, *Can. J. Chem.*, **44**, 1765 (1966).
283. J. Court, S. Vidal, and J. -M. Bonnier, *Bull. Soc. Chim. Fr.*, 3107 (1972).
284. J. -M. Bonnier and J. Court, *Bull. Soc. Chim. Fr.*, 1834 (1972).
285. E. C. Kooyman, in "Advances in Free-Radical Chemistry," G. H. Williams, Ed., Vol. 1, Logos Press, London, 1965, Chap. 4, p. 151.
286. E. C. Kooyman, in "Congress Lectures, XIX International Congress of Pure and Applied Chemistry, London, 1963," Butterworth, London, 1963, p. 193.
287. L. G. Shevchuk, T. A. Boyarskaya, and N. A. Vysotskaya, *Chem. Heterocycl. Compounds*, **5**, 563 (1969).
288. L. K. Dyall and K. H. Pausacker, *J. Chem. Soc.*, 18 (1961).
289. R. A. Abramovitch and O. A. Koleoso, *J. Chem. Soc. B*, 1292 (1968).
290. R. M. Elofson, F. F. Gadallah, and K. F. Schulz, *J. Org. Chem.*, **36**, 1526 (1971).
291. I. F. Tupitsyn and N. K. Semenova, *Tr. Gos. Inst. Prikl. Khim.*, **49**, 120 (1962); *Chem. Abstr.*, **60**, 6721d (1964).
292. J. A. Zoltewicz and G. M. Kauffman, *J. Org. Chem.*, **34**, 1405 (1969).

293. K. W. Ratts, R. K. Howe, and W. G. Phillips, *J. Amer. Chem. Soc.*, **91**, 6115 (1969).
294. J. A. Zoltewicz and J. D. Meyer, *Tetrahedron Lett.*, 421 (1968).
295. J. A. Zoltewicz, C. L. Smith, and J. D. Meyer, *Tetrahedron*, **24**, 2269 (1968).
296. R. A. Abramovitch, E. M. Smith, E. E. Knaus, and M. Saha, *J. Org. Chem.*, **37**, 1690 (1972).
297. I. F. Tupitsyn, N. N. Zatsepina, and A. V. Kirova, *Isotopenpraxis*, **3**, 136 (1967); *Chem. Abstr.*, **71**, 21351w (1969).
298. J. A. Zoltewicz and C. L. Smith, *Tetrahedron*, **25**, 4331 (1969).
299. J. A. Zoltewicz, G. M. Kauffman, and C. L. Smith, *J. Amer. Chem. Soc.*, **90**, 5939 (1968).
300. H. E. Dubb, M. Saunders, and J. H. Wang, *J. Amer. Chem. Soc.*, **80**, 1767 (1958).
301. P. Beak and J. Bonham, *J. Amer. Chem. Soc.*, **87**, 3365 (1965).
302. I. F. Tupitsyn, N. N. Zatsepina, N. S. Kolodina, and A. A. Kane, *Reakts. Sposobnost. Org. Soedin.*, **5**, 931 (1968); *Chem. Abstr.*, **71**, 49041 (1969).
303. H. L. Jones and D. L. Beveridge, *Tetrahedron Lett.*, 1577 (1964).
304. R. A. Abramovitch, M. Saha, E. M. Smith, and R. T. Coutts, *J. Amer. Chem. Soc.*, **89**, 1537 (1967).
305. R. A. Abramovitch, E. M. Smith, and R. T. Coutts, *J. Org. Chem.*, **37**, 3584 (1972).
306. R. A. Abramovitch and E. E. Knaus, *J. Heterocycl. Chem.*, **6**, 989 (1969).
307. R. A. Abramovitch, J. Campbell, E. E. Knaus, and A. Silhankova, *J. Heterocycl. Chem.*, **9**, 1367 (1972).
308. E. E. Knaus, Ph.D. Dissertation, University of Saskatchewan, 1970.
309. F. Kröhnke, *Angew. Chem Int. Ed.*, **2**, 225, 380 (1963).
310. R. K. Howe and K. W. Ratts, *Tetrahedron Lett.*, 4743 (1967).
311. H. Quast and E. Frankenfeld, *Angew. Chem. Int. Ed.*, **4**, 691 (1965).
312. K. Ziegler and H. Zieser, *Ber.*, **63**, 1847 (1930).
313. R. A. Abramovitch, F. Helmer, and J. G. Saha, *Can. J. Chem.*, **43**, 725 (1965).
314. R. A. Abramovitch and G. A. Poulton, *Chem. Commun.*, 275 (1967); R. A. Abramovitch and G. A. Poulton, *J. Chem. Soc. C*, 128 (1970).
315. R. A. Abramovitch, W. C. Marsh, and J. G. Saha, *Can. J. Chem.*, **43**, 2631 (1965).
316. P. Doyle and R. R. J. Yates, *Tetrahedron Lett.*, 3371 (1970).
317. R. A. Abramovitch and B. Vig, *Can. J. Chem.*, **41**, 1961 (1963).
318a. C. S. Giam and J. L. Stout, *Chem. Commun.*, 478 (1970).
318b. C. S. Giam and S. D. Abbott, *J. Amer. Chem. Soc.*, **93**, 1294 (1971).
318c. C. S. Giam and E. E. Knaus, *Tetrahedron Lett.*, 4961 (1971).
319. G. Fraenkel and J. C. Cooper, *Tetrahedron Lett.*, 1825 (1968).
320. R. Foster and C. A. Fyfe, *Tetrahedron*, **25**, 1489 (1969).
321. C. S. Giam and J. L. Stout, *Chem. Commun.*, 142 (1969).
321a. R. F. Francis, W. Davis, and J. T. Wisener, *J. Org. Chem.*, **39**, 59 (1974).
322. J. Kuthan, E. Janeckova, and M. Havel, *Collect. Czech. Chem. Comm.*, **29**, 143 (1964).
323. R. E. Lyle and D. A. Nelson, *J. Org. Chem.*, **28**, 169 (1963).
324. D. Bryce-Smith and A. C. Skinner, *J. Chem. Soc.*, 577 (1963).
325. D. Bryce-Smith, P. J. Morris, and B. J. Wakefield, *Chem. Ind.* (London), 495 (1964).
326. R. A. Abramovitch and C. S. Giam, *Can. J. Chem.*, **41**, 3127 (1963).
327. R. A. Abramovitch and C. S. Giam, *Can. J. Chem.*, **42**, 1627 (1964).
328. R. A. Abramovitch and A. D. Notation, *Can. J. Chem.*, **38**, 1445 (1960).
329. R. A. Abramovitch and G. A. Poulton, *J. Chem. Soc. B*, 267 (1967).
330. R. A. Abramovitch and G. A. Poulton, *J. Chem. Soc. B*, 901 (1969).
331. R. A. Abramovitch, F. Helmer, and M. Liveris, *J. Org. Chem.*, **34**, 1730 (1969).

332. A. J. Newman, Ph.D. Dissertation, University of Alabama, 1972. R. A. Abramovitch and A. J. Newan, Jr., *J. Org. Chem.*, in press.
333. F. Haglid and J. O. Norén, *Acta Chem. Scand.*, **21**, 335 (1967).
334. F. Haglid, *Acta Chem. Scand.*, **21**, 329 (1967).
335. R. A. Abramovitch and C. S. Giam, *Can. J. Chem.*, **40**, 213 (1962).
336. G. Fraenkel and J. W. Cooper, *J. Amer. Chem. Soc.*, **93**, 7228 (1971).
337. G. Fraenkel, C. C. Ho, Y. Liang, and S. Yu, *J. Amer. Chem. Soc.*, **94**, 4732 (1972).
338. J. Kato and H. Yamanaka, *J. Org. Chem.*, **30**, 910 (1965).
339. T. J. van Bergen and R. M. Kellogg, *J. Org. Chem.*, **36**, 1705 (1971).
340. O. S. Otroshchenko, A. S. Sadykov, M. U. Utebaev, and A. I. Isametova, *Zh. Obshch. Khim.*, **33**, 1038 (1963); *Chem. Abstr.*, **59**, 10142 (1963).
341. R. A. Abramovitch, G. A. Poulton, and G. M. Singer, unpublished results (1967).
342. O. Červinka, *Collect. Czech. Chem. Comm.*, **27**, 567 (1962).
343. O. Červinka, *Chem. Ind.* (London), 1482 (1960).
344. A. D. Miller, C. Osuch, N. N. Goldberg, and R. Levine, *J. Amer. Chem. Soc.*, **78**, 674 (1956).
345. D. N. Kursanov and N. K. Baranetskaya, *Izv. Akad. Nauk SSSR, Otd. Khim. Nauk*, 1703 (1961); *Chem. Abstr.*, **56**, 3447i (1962).
346. J. H. Ager and E. L. May, *J. Org. Chem.*, **25**, 984 (1960).
347. J. H. Ager and E. L. May, *J. Org. Chem.*, **27**, 245 (1962).
348. K. Kovács and T. Vajda, *Acta Chim. Acad. Sci. Hung.*, **29**, 245 (1961); *Chem. Abstr.*, **57**, 5892g (1962).
349. Y. Ban and T. Wakamatsu, *Chem. Ind.* (London), 710 (1964).
350. L. D. Smirnov, V. P. Lezina, V. F. Bystrov, and K. M. Dyumaev, *Izv. Akad. Nauk. SSSR, Ser. Khim.*, 198 (1965); *Chem. Abstr.*, **62**, 11774d (1964).
351. Th. Kauffman and W. Schoeneck, *Angew. Chem.*, **71**, 285 (1959).
352. Th. Kauffman, H. Hacker, and H. Müller, *Chem. Ber.*, **95**, 2485 (1962).
353. A. F. Pozharskii and A. M. Simonov, "Amination of Heterocyclics by the Tschitschibabin Reaction," Rostov University Press, Rostov-on-Don, U.S.S.R., 1971.
354. R. A. Abramovitch, F. Helmer, and J. G. Saha, *Chem. Ind.* (London), 659 (1964); *Can. J. Chem.*, **43**, 725 (1965).
355. L. S. Levitt and B. W. Levitt, *Chem. Ind.* (London), 1621 (1963).
356. F. W. Bergstrom, H. Sturz, and H. Tracy, *J. Org. Chem.*, **11**, 239 (1946).
357. T. Vajda and K. Kovács, *Rec. Trav. Chim. Pays-Bas*, **80**, 47 (1961).
358. R. A. Abramovitch and A. R. Vinutha, *J. Chem. Soc. B*, 131 (1971).
359. A. R. Katritzky and E. Lunt, *Tetrahedron*, **25**, 4291 (1969).
360. E. Ochiai and I. Nakayama, *Yakugaku Zasshi*, **65B**, 582 (1945).
361. W. E. Feeley and E. M. Beavers, *J. Amer. Chem. Soc.*, **81**, 4004 (1959).
362. M. Ferles and M. Jankovský, *Collect. Czech. Chem. Comm.*, **33**, 3848 (1968).
363. H. Tani, *Yakugaku Zasshi*, **80**, 1418 (1960); *Chem. Abstr.*, **55**, 6477i (1961).
364. N. Nishimoto and T. Nakashima, *Yakugaku Zasshi*, **82**, 1267 (1962); *Chem. Abstr.*, **59**, 3886a (1963).
365. K. Wallenfels and H. Dieckmann, *Ann. Chem.*, **621**, 166 (1959).
366. H. Tani, *Yakugaku Zasshi*, **81**, 141 (1961); *Chem. Abstr.*, **55**, 14449i (1961).
367. H. Tani, *Yakugaku Zasshi*, **81**, 182 (1961); *Chem. Abstr.*, **55**, 14450b (1961).
368. R. N. Lindquist and E. H. Cordes, *J. Amer. Chem. Soc.*, **90**, 1269 (1968).
369. L. J. Winters, N. G. Smith, and M. I. Cohen, *Chem. Commun.*, 642 (1970).
370. T. Severin, H. Lerche, and D. Bätz, *Chem. Ber.*, **102**, 2163 (1969).
371. P. Blumbergs, A. B. Ash, F. A. Daniher, C. L. Stevens, H. O. Michael, B. E. Hackley, Jr., and J. Epstein, *J. Org. Chem.*, **34**, 4065 (1969).
372. R. E. Lyle and G. Gauthier, *Tetrahedron Lett.*, 4615 (1965).

373. R. E. Lyle, *Chem. Eng. News.*, (Jan. 10, 1966), p. 72.
374. D. Redmore, *J. Org. Chem.*, **35**, 4114 (1970).
375. R. Eisenthal, A. R. Katritzky, and E. Lunt, *Tetrahedron*, **23**, 2775 (1967).
376. E. M. Kosower, *Prog. Phys. Org. Chem.*, **3**, 81 (1965).
377. M. Hamana and K. Funakoshi, *J. Pharm. Soc. Jap.*, **84**, 28 (1964).
377a M. Nakanishi and M. Yatabe, *Heterocycles*, **2**, 259 (1974).
378. E. J. Moriconi, F. J. Creegan, C. K. Donovan, and F. A. Spano, *J. Org. Chem.*, **28**, 2215 (1963).
379. E. J. Moriconi and F. A. Spano, *J. Amer. Chem. Soc.*, **86**, 38 (1964).
380. E. Ochiai and C. Kaneko, *Chem. Pharm. Bull.* (Tokyo), 8, 28 (1960); *Chem. Abstr.*, **55**, 5491g (1961).
381. C. A. Fyfe, *Tetrahedron Lett.*, 659 (1968).
382. G. Illuminati and F. Stegel, *Tetrahedron Lett.*, 4169 (1968).
383. P. Bemporad, G. Illuminati, and F. Stegel, *J. Amer. Chem. Soc.*, **91**, 6742 (1969).
384. C. Abbolito, C. Iavarone, G. Illuminati, F. Stegel, and A. Vazzoler, *J. Amer. Chem. Soc.*, **91**, 6746 (1969).
385. M. E. C. Biffin, J. Miller, A. G. Moritz, and D. B. Paul, *Aust. J. Chem.*, **22**, 2561 (1969).
386. F. Terrier, A. P. Chatrousse, and R. Schaal, *J. Org. Chem.*, **37**, 3011 (1972).
387. T. Talik and Z. Talik, *Rocz. Chem.*, **42**, 2061 (1968); *Chem. Abstr.*, **70**, 114970s (1969).
388. G. B. Barlin and W. V. Brown, *J. Chem. Soc. B*, 648 (1967).
389. G. B. Barlin and W. V. Brown, *J. Chem. Soc. B*, 1435 (1968).
390. G. B. Barlin and W. V. Brown, *J. Chem. Soc. C*, 921 (1969).
391. A. Dondoni, A. Mangini, and G. Mossa, *J. Heterocycl. Chem.*, **6**, 143 (1969).
392. Z. Talik, *Bull. Acad. Polon. Sci., Ser. Sci. Chim.*, **9**, 567 (1961); *Chem. Abstr.*, **60**, 2884h (1964).
393. R. G. Shepherd and J. L. Fedrick, *Adv. Heterocycl. Chem.*, **4**, 145 (1965).
394. J. Miller, "Aromatic Nucleophic Substitution," Elsevier, Amsterdam, 1968, Chap. 7, p. 234.
395. G. C. Finger, L. D. Starr, D. R. Dickerson, H. S. Gutowsky, and J. Hamer, *J. Org. Chem.*, **28**, 1666 (1963).
396. M. Forchiassin, G. Illuminati, and G. Sleiter, *J. Heterocycl. Chem.*, **6**, 879 (1969).
397. N. B. Chapman, D. K. Chaudhury, and J. Shorter, *J. Chem. Soc.*, 1975 (1962).
398. R. A. Abramovitch, F. Helmer, and M. Liveris, *J. Chem. Soc. B*, 492 (1968).
399. T. Sheradsky and G. Salemnick, *J. Org. Chem.*, **36**, 1061 (1971).
399a. R. D. Chambers, W. K. R. Musgrave, J. S. Waterhouse, D. L. H. Williams, J. Burdon W. B. Hollyhead, and J. C. Tatlow, *J. C. S. Chem. Comm.*, 239 (1974).
400. D. I. Davies, J. N. Done, and D. H. Hey, *J. Chem. Soc. C*, 2019 (1969).
401. Th. Kauffmann and R. Wirthweim, *Angew. Chem. Int. Ed.*, **10**, 20 (1971).
402. R. W. Hoffman, "Dehydrobenzene and Cycloalkynes," Academic Press, New York, 1967, Chap. 6, p. 275.
403. H. J. den Hertog, H. C. van der Plas, M. J. Pieterse, and J. W. Streef, *Rec. Trav. Chim. Pays-Bas*, **84**, 1569 (1965).
404. H. J. den Hertog, R. J. Martens, H. C. van der Plas, and J. Bon, *Tetrahedron Lett.*, 4325 (1966).
405. W. A. Roelfsema and H. J. den Hertog, *Tetrahedron Lett.*, 5089 (1967).
406. M. J. Pieterse and H. J. den Hertog, *Rec. Trav. Chim. Pays-Bas*, **80**, 1376 (1961)
407. Th. Kauffmann and F. P. Boettcher, *Angew. Chem.*, **73**, 65 (1961).
408. Th. Kauffmann and F. P. Boettcher, *Chem. Ber.*, **95**, 1528 (1962).
409. R. J. Martens, H. J. den Hertog, and M. van Ammers, *Tetrahedron Lett.*, 3207 (1964).
410. Th. Kauffmann and R. Nürnberg, *Chem. Ber.*, **100**, 3427 (1967).
411. J. A. Zoltewicz and C. Nisi, *J. Org. Chem.*, **34**, 765 (1969).

412. M. J. Pieterse and H. J. den Hertog, *Rec. Trav. Chim. Pays-Bas,* **81**, 855 (1962).
413. H. J. den Hertog, M. J. Pieterse, and D. J. Buurman, *Rec. Trav. Chim. Pays-Bas,* **82**, 1173 (1963).
414. J. W. Streef and H. J. den Hertog, *Rec. Trav. Chim. Pays-Bas,* **85**, 803 (1966).
415. J. W. Streef and H. J. den Hertog, *Tetrahedron Lett.*, 5945 (1968).
416. H. N. M. van der Lans and H. J. den Hertog, *Rec. Trav. Chim. Pays-Bas,* **87**, 549 (1968).
417. W. J. van Zoest and H. J. den Hertog, footnote in Ref. 416.
418. J. D. Cook and B. J. Wakefield, *J. Chem. Soc. C,* 1973 (1969).
419. H. Boer and H. J. den Hertog, *Tetrahedron Lett.*, 1943 (1969).
420. S. M. Roberts and H. Suschitzky, *J. Chem. Soc. C,* 1485 (1969).
421. M. Liveris and J. Miller, *J. Chem. Soc.*, 3486 (1963).
422. R. M. Johnson, *J. Chem. Soc. B,* 1058 (1966).
423. R. M. Johnson, *J. Chem. Soc. B,* 1062 (1966).
424. Z. Talik, *Rocz. Chem.*, **35**, 475 (1961); *Chem. Abstr.*, **57**, 15065h (1962).
425. Z. Talik, *Bull. Acad. Polon. Sci., Ser. Sci. Chim.*, **9**, 561 (1961); *Chem. Abstr.*, **60**, 2884d (1964).
426. T. Talik and Z. Talik, *Rocz. Chem.*, **38**, 777 (1964); *Chem. Abstr.*, **61**, 10653c (1964).
427. R. J. Martens and H. J. den Hertog, *Rec. Trav. Chim. Pays-Bas,* **83**, 621 (1964).
428. Th. Kauffmann and R. Wirthwein, *Angew. Chem. Int. Ed.*, **3**, 806 (1964).
429a. T. Kato, T. Niitsuma, and N. Kusaka, *Yakugaku Zasshi,* **84**, 432 (1964); *Chem. Abstr.*, **61**, 4171d (1964).
429b. T. Kato and T. Niitsuma, *Heterocycles*, **1**, 233 (1973).
430. R. J. Martens and H. J. den Hertog, *Rec. Trav. Chim. Pays-Bas,* **86**, 655 (1967).
431. R. A. Abramovitch and E. M. Smith, unpublished results, 1969; E. M. Smith, Ph. D. Dissertation, University of Saskatchewan, 1969.
432. H. J. den Hertog, D. J. Buurman, and P. A. de Villiers, *Rec. Trav. Chim. Pays-Bas,* **80**, 325 (1961).
433. E. Matsumura, *J. Chem. Soc. Jap.*, **74**, 363, 446 (1953); *Chem. Abstr.*, **48**, 6442b (1954).
434. S. Oae, T. Kitao, and Y. Kitaoka, *Tetrahedron,* **19**, 827 (1963).
435. E. Ochiai and T. Yokokawa, *J. Pharm. Soc. Jap.*, **75**, 213 (1955); *Chem. Abstr.*, **50**, 1819b (1956).
436. T. Koenig and T. Barklow, *Tetrahedron,* **25**, 4875 (1969).
437. R. N. Pratt, D. P. Stokes, G. A. Taylor, and S. A. Procter, *J. Chem. Soc. C,* 1472 (1971).
438a. H. Seidl, R. Huisgen, and R. Grashey, *Chem. Ber.*, **102**, 926 (1969).
438b. T. Hisano, S. Yoshikawa, and K. Muraoka, *Org. Prep. and Proc. Intern*., **5**, 95 (1973). T. Hisano, T. Matsuoka, and M. Ichikawa, *Heterocycles,* **2**, 163 (1974).
439. E. A. Mailey and L. R. Ocone, *J. Org. Chem.*, **33**, 3343 (1968).
440. K. Takeda, K. Hamamoto, and H. Tone, *Yakugaku Zasshi,* **72**, 1427 (1952; cited in Ref. 461, p. 325.
441. F. Ramirez and P. W. von Ostwalden, *J. Amer. Chem. Soc.* **81**, 156 (1959).
442. S. Kajihara, *Nippon Kagaku Zasshi,* **85**, 672 (1964); *Chem. Abstr.*, **62**, 14624e (1965).
443. S. Kajihara, *Nippon Kagaku Zasshi,* **86**, 839 (1965); *Chem. Abstr.*, **65**, 16935c (1966).
444. S. Kajihara, *Nippon Kagaku Zasshi,* **86**, 1060 (1965); *Chem. Abstr.*, **65**, 16936f (1966).
445. A. Deegan and F. L. Rose, *J. Chem. Soc. C,* 2756 (1971).
446a. R. A. Abramovitch and R. B. Rogers, *Tetrahedron Lett.*, 195 (1971).
446b. R. A. Abramovitch and T. D. Bailey, unpublished results.
447. R. A. Abramovitch and G. M. Singer, *J. Amer. Chem. Soc.*, **91**, 5672 (1969).
448. R. A. Abramovitch and R. B. Rogers, "Abstracts of Papers, 3rd International Congress of Heterocyclic Chemistry," Sendai, Japan, 1971, paper B-26-11, p. 240. *J. Chem.*, in press.
449. R. A. Abramovitch, E. M. Smith, and P. Tomasik, unpublished results (1970).
450. R. A. Abramovitch and S. Kato, unpublished results (1972).

451. T. Sasaki, K. Kanematsu, and A. Kakehi, *J. Org. Chem.*, **37**, 3106 (1972).
452. R. A. Abramovitch, G. Grins, R. B. Rogers, J. L. Atwood, M. D. Williams, and S. Crider, *J. Org. Chem.*, **37**, 3383 (1972).
453a. R. A. Abramovitch and I. Shinkai, unpublished results (1972).
453b. R. A. Abramovitch and I. Shinkai, *J. C. S. Chem. Comm.*, 569 (1973).
454. R. A. Abramovitch and J. B. Davis, unpublished results (1965); reported in Ref. 172, p. 345.
455. E. M. Hawes and D. G. Wibberley, *J. Chem. Soc. C*, 315 (1966).
456. E. M. Hawes and H. L. Davis, *J. Heterocycl. Chem.*, **10**, 39 (1973).
457. F. Uchimaru, S. Okada, A. Kosasayama, and T. Konno, "Abstracts of Papers, 3rd International Congress of Heterocyclic Chemistry," Sendai, Japan, 1971, paper B-27-2, p. 248.
458. M. Katada, *J. Pharm. Soc. Jap.*, **67**, 51 (1947); *Chem. Abstr.*, **45**, 9536d (1951).
459. G. M. Badger and J. W. Clark-Lewis, in "Molecular Rearrangements," P. de Mayo, Ed., Part I, Interscience, New York, 1963, p. 639.
460. V. J. Traynelis, "Mechanisms of Molecular Migrations," B. S. Thyagarajan, Ed., Interscience, New York, 1969, p. 1.
461. E. Ochiai, "Aromatic Amine Oxides," Elsevier Publishing, Amsterdam, 1967.
462. D. M. Pretorius and P. A. de Villiers, *J. S. Afr. Chem. Inst.*, **28**, 48 (1965).
463. A. Klaebe and A. Lattes, *J. Chromatogr.*, **27**, 502 (1967).
464. S. Oae and S. Kozuka, *Tetrahedron*, **20**, 2691 (1964).
465. S. Oae and S. Kozuka, *Tetrahedron*, **21**, 1971 (1965).
466. J. H. Markgraf, H. B. Brown, Jr., S. C. Mohr, and R. G. Peterson, *J. Amer. Chem. Soc.*, **85**, 958 (1963).
467. T. Cohen and G. L. Deets, *J. Org. Chem.*, **37**, 55 (1972).
468. V. J. Traynelis and Sr. A. I. Gallagher, *J. Amer. Chem. Soc.*, **87**, 5710 (1965).
469. P. W. Ford and J. M. Swan, *Aust. J. Chem.*, **18**, 867 (1965).
470. S. Oae, T. Kitao, and Y. Kitaoka, *J. Amer. Chem. Soc.*, **84**, 3359, 3362 (1962).
471. T. Cohen and J. H. Fager, *J. Amer. Chem. Soc.*, **87**, 5701 (1965).
472. S. Oae, S. Tamagaki, T. Negoro, K. Ogino, and S. Kazuka, *Tetrahedron Lett.*, 917 (1968).
473. S. Furukawa, *Yakugaku Zasshi*, **59**, 487 (1959); *Chem. Abstr.*, **53**, 18028f (1959).
474. S. Oae, S. Tamagaki, T. Negoro, and S. Kozuka, *Tetrahedron*, **26**, 4051 (1970).
475. T. Koenig, *J. Amer. Chem. Soc.*, **88**, 4045 (1966).
476. R. Bodalski and A. R. Katritzky, *J. Chem. Soc. C*, 831 (1968).
477. H. Iwamura, M. Iwamura, T. Nishida, and I. Miura, *Tetrahedron Lett.*, 3117 (1970).
478. H. Iwamura, M. Iwamura, T. Nishida, and S. Sato, *J. Amer. Chem. Soc.*, **92**, 7474 (1970).
479. T. Cohen and G. L. Deets, *J. Amer. Chem. Soc.*, **94**, 932 (1972).
480. Y. Mizuno, W. Limn, K. Tsuchida, and K. Ikeda, *J. Org. Chem.*, **37**, 39 (1972). Y. Mizuno, T. Endo, T. Miyaoka, and K. Ikeda, *J. Org. Chem.*, **39**, 1250 (1974).
481. K. E. Wilzbach and D. J. Rausch, *J. Amer. Chem. Soc.*, **92**, 2178 (1970).
482. J. Joussot-Dubien and J. Houdard, *C. R. Acad. Sci., Paris, Ser. C*, **267**, 866 (1968).
483. E. F. Travecedo and V. I. Stenberg, *Chem. Commun.*, 609 (1970).
484. R. M. Kellogg, T. J. van Bergen, and H. Wynberg, *Tetrahedron Lett.*, 5211 (1969).
485. M. Natsume and M. Wada, "Abstracts of Papers, 3rd International Congress of Heterocyclic Chemistry," Sendai, Japan, 1971, paper B-26-2, p. 207.
486. F. R. Stermitz and W. H. Huang, *J. Amer. Chem. Soc.*, **92**, 1446 (1970).
487. C. Azuma and A. Sugimori, *Kogyo Kagaku Zasshi*, **72**, 239 (1969); *Chem. Abstr.*, **70**, 96575k (1969).
488. S. Caplain and A. Lablache-Combier, *Chem. Commun.*, 1247 (1970).
489. S. Caplain, J. P. Cattean, and A. Lablache-Combier, *Chem. Commun.*, 1475 (1970).
490. C. Chachatz and A. Forchioni, *C. R. Acad. Sci., Paris, Ser. C*, **264**, 1421 (1967).

491. E. C. Taylor and W. W. Paudler, *Tetrahedron Lett.*, 1 (1960).
492. W. A. Ayer, R. Hayatsu, P. de Mayo, S. T. Reid, and J. B. Stothers, *Tetrahedron Lett.*, 648 (1961).
493. E. C. Taylor, R. O. Kan, and W. W. Paudler, *J. Amer. Chem. Soc.*, **83**, 4484 (1961).
494. E. C. Taylor and R. O. Kan, *J. Amer. Chem. Soc.*, **85**, 776 (1963).
495. L. A. Paquette and G. Slomp, *J. Amer. Chem. Soc.*, **85**, 765 (1963).
496. M. Laing, *Proc. Chem. Soc.*, 343 (1964).
497. E. J. Corey and J. Streith, *J. Amer. Chem. Soc.*, **86**, 950 (1964).
498. R. C. de Selms and W. R. Schleigh, *Tetrahedron Lett.*, 3563 (1972).
499. R. O. Kan, "Organic Photochemistry," McGraw Hill, New York, 1966, p. 180, quoting unpublished work by R. A. Jackson and B. Gorres.
500. E. Ager, G. E. Chivers, and H. Suschitzky, *J. C. S. Chem. Comm.*, 504 (1972).
501. J. Bratt and H. Suschitzky, *J. C. S. Chem. Comm.*, 949 (1972).
502. C. R. Hurt and N. Filipesco, *J. Amer. Chem. Soc.*, **94**, 3649 (1972).
503. M. G. Barlow, J. G. Dingwall, and R. N. Haszeldine, *Chem. Commun.*, 1580 (1970).
504. G. E. Loader and C. J. Timmons, *J. Chem. Soc. C*, 1078 (1966).
505. H. H. Perkampus, G. Kassebeer, and P. Müller, *Ber. Bunsenges. Phys. Chem.*, **71**, 40 (1967).
506. G. Galiazzo, P. Bortolus, and G. Cauzzo, *Tetrahedron Lett.*, 3717 (1966).
507. G. Galiazzo, P. Bortolus, G. Cauzzo, and U. Mazzucato, *J. Heterocycl. Chem.*, **6**, 465 (1969).
508. C. E. Loader and C. J. Timmons, *J. Chem. Soc. C*, 1457 (1967).
509. D. G. Whitten and M. T. McCall, *Tetrahedron Lett.*, 2755 (1968).
510. A. Fozard and C. K. Bradsher, *J. Org. Chem.*, **31**, 2346 (1966).
511. G. G. Spence, E. C. Taylor, and O. Buchardt, *Chem. Rev.*, **70**, 231 (1970).
512. R. M. Johnson and C. W. Rees, *J. Chem. Soc. B*, 15 (1967).
513. C. Kaneko, I. Yokoe, and S. Yamada, *Tetrahedron Lett.*, 775 (1967).
514. C. Kaneko, Sachiko, and I. Yokoe, *Chem. Pharm. Bull.* (Tokyo). **15**, 356 (1967); *Chem. Abstr.*, **67**, 64212 (1967).
515. N. Hata, E. Okutsu, and I. Tanaka, *Bull. Chem. Soc. Jap.*, **41**, 1769 (1968); *Chem. Abstr.*, **70**, 3032t (1969).
516. C. Kaneko, S. Yamada, and I. Yokoe, *Tetrahedron Lett.*, 4729 (1966).
517. S. Kamiya, *Chem. Pharm. Bull.* (Tokyo), **10**, 471 (1962); *Chem. Abstr.*, **58**, 4545c (1963).
518. C. Kaneko, S. Yamada, H. Ichikawa, and T. Kubota, "Abstracts of Papers, 3rd International Congress of Heterocyclic Chemistry," Sendai, Japan, 1971, paper B-26-4, p. 215.
519. N. Hata and I. Tanaka, *J. Chem. Phys.*, **36**, 2072 (1962).
520. N. Hata, *Bull. Chem. Soc. Jap.*, **34**, 1440 (1961); *Chem. Abstr.*, **56**, 4286b (1962).
521. N. Hata, *Bull. Chem. Soc. Jap.*, **34**, 1444 (1961); *Chem. Abstr.*, **56**, 4286e (1962).
522. J. Streith, B. Danner, and C. Sigwalt, *Chem. Commun.*, 979 (1967).
523. D. M. Jerina, D. R. Boyd, and J. W. Daly, *Tetrahedron Lett.*, 457 (1970).
524. P. L. Kumler and O. Buchardt, *Chem. Commun.*, 1321 (1968).
525. A. Alkaitis and M. Calvin, *Chem. Commun.*, 292 (1968).
526. J. Streith and C. Sigwalt, *Tetrahedron Lett.*, 1347 (1966).
527. M. Ishikawa, C. Kaneko, I. Yokoe, and S. Yamada, *Tetrahedron*, **25**, 295 (1969).
528. H. Takayama, and T. Okamoto, "Abstracts of Papers, 3rd International Congress of Heterocyclic Chemistry," Sendai, Japan, 1971, paper B-24-7, p. 199.
529. R. E. Doolittle and C. K. Bradsher, *J. Org. Chem.*, **31**, 2616 (1969).
530. J. Streith, A. Blind, J.-M. Cassal, and C. Sigwalt, *Bull. Soc. Chim. Fr.*, 948 (1969).
531. J. Streith and J. M. Cassal, *Angew Chem. Int. Ed.*, **7**, 129 (1968).
532. V. Snieckus, *Chem. Commun.*, 831 (1969).

533. T. Sasaki, K. Kanematsu, A. Kakehi, I. Ichikawa, and K. Hayakawa, *J. Org. Chem.*, **35**, 426 (1970).
534. A. Balasubramanian, J. M. McIntosh, and V. Snieckus, *J. Org. Chem.*, **35**, 433 (1970).
535. J. Streith, J. P. Luttringer, and M. Nastasi, *J. Org. Chem.*, **36**, 2962 (1971).
536a. R. A. Abramovitch and T. Takaya, *J. Org. Chem.*, **38**, 3311 (1973).
536b. J. Streith and J. M. Cassal, *Tetrahedron Lett.*, 4541 (1968).
537. V. Snieckus and G. Kan, *Chem. Commun.*, 172 (1970).
538. C. W. Bird, I. Partridge, and D. Y. Wong, *J. C. S., Perkin I*, 1020 (1972).
539. E. M. Kosower and J. W. Patton, *Tetrahedron*, **22**, 2081 (1966).
540. R. Eisenthal and A. R. Katritzky, *Tetrahedron*, **21**, 2205 (1965).
541. S. L. Johnson and K. A. Rumon, *Tetrahedron Lett.*, 1721 (1966).
542. M. Wahren, *Z. Chem.*, **6**, 181 (1966).
543. W. König, M. Coenen, F. Bahr, B. May, and A. Bassl, *J. Prakt. Chem.*, **33**, 54 (1966).
544. R. A. Abramovitch and B. W. Cue, Jr., *J. Org. Chem.*, **38**, 173 (1973); *Heterocycles*, **2**, 297 (1974).
545. R. F. C. Brown and R. J. Smith, *Chem. Commun.*, 795 (1969).
546. W. D. Crow, A. R. Lea, and M. N. Paddon-Row, *Tetrahedron Lett.*, 2235 (1972).
547. T. I. Abramovitch, I. P. Gragerov, and V. V. Perekalin, *Zh. Obshch. Khim.*, **31**, 1962 (1961); *Chem. Abstr.*, **55**, 27373i (1961).
548. A. D. Cale, Jr., R. W. McGinnis, Jr., and P. C. Teague, *J. Org. Chem.*, **25**, 1507 (1960).
549. H. Pines and D. Wunderlich, *J. Amer. Chem. Soc.*, **81**, 2568 (1959).
550. J. A. Zoltewicz, and P. E. Kandetzki, *J. Amer. Chem. Soc.*, **93**, 6562 (1971).
551. B. Notari and H. Pines, *J. Amer. Chem. Soc.*, **82**, 2945 (1960).
552a. Yu. I. Chumakov and V. M. Ledovskikh, *Tetrahedron*, **21**, 937 (1965).
552b. T. Kato, Y. Goto, M. Hikichi, and T. Kawamata, *Yakugaku Zasshi*, **84**, 869 (1964); *Chem. Abstr.*, **62**, 1629f (1965).
553. H. Pines and S. V. Kannan, *Chem. Commun.*, 1360 (1969).
554. M. E. Derieg, I. Douvan, and R. I. Fryer, *J. Org. Chem.*, **33**, 1290 (1968).
555. R. E. Smith, S. Boatman, and C. R. Hauser, *J. Org. Chem.*, **33**, 2083 (1968).
556. S. Boatman, T. M. Harris, and C. R. Hauser, *J. Org. Chem.*, **30**, 3593 (1965).
557. H. Feuer and J. P. Lawrence, *J. Amer. Chem. Soc.*, **91**, 1856 (1969).
558. H. Feuer, J. P. Lawrence, and J. K. Doty, "Abstracts of Papers, 3rd International Congress of Heterocyclic Chemistry," Sendai, Japan, 1971, paper C-23-3, p. 287.
559. R. D. Chambers, B. Iddon, W. K. R. Musgrave, and L. Chadwick, *Tetrahedron*, **24**, 877 (1968).
560. E. Profft and W. Steinke, *Chem. Ber.*, **94**, 2267 (1961).
561. A. N. Kost, P. B. Terentev, and M. A. Chernova, *Vestn. Mosk. Univ. Ser. II, Khim.*, **19**, 59 (1964); *Chem. Abstr.*, **61**, 3064e (1964).
562. V. Boekelheide and R. Scharrer, *J. Org. Chem.*, **26**, 3802 (1961).
563. A. Albert and G. B. Barlin, *J. Chem. Soc.*, 2384 (1959).
564. J. Kuthan, V. Skála, and J. Paleček, *Z. Chem.*, **8**, 305 (1968).
565. E. Spinner and J. C. B. White, *J. Chem. Soc. B*, 991 (1966).
566. A. Gordon, A. R. Katritzky, and S. K. Roy, *J. Chem. Soc. B*, 556 (1968).
567a. R. A. Y. Jones, A. R. Katritzky, and J. M. Lagowski, *Chem. Ind.* (London), 870 (1960).
567b. Yu. L. Frolov, V. K. Voronov, N. M. Deriglazov, L. V. Belousova, N. M. Vitkovzkaya, S. M. Tyrina, and G. G. Skvortsova, *Chem. Heterocycl. Compounds*, **8**, 89 (1972).
568. A. Gordon and A. R. Katritzky, *Tetrahedron Lett.*, 2767 (1968).
569. S. F. Mason, *J. Chem. Soc.*, 674 (1960).
570. H. Sterk and H. Junek, *Monatsh. Chem.*, **98**, 1763 (1967).
571. C. S. Wang, *J. Heterocycl. Chem.*, **7**, 389 (1970).
572. K. Yamada, H. Nakata, and Y. Hirata, *Bull. Chem. Soc. Jap.*, **33**, 1298 (1960).
573. A. R. Katritzky, J. D. Rowe, and S. K. Roy, *J. Chem. Soc., B*, 758 (1967).
574. E. Spinner and G. B. Yeon, *Tetrahedron Lett.*, 5691 (1968).

575. J. A. Elvidge and L. M. Jackman, *J. Chem. Soc.*, 859 (1961).
576. M. J. Cook, A. R. Katritzky, P. Linda, and R. D. Tack, *Chem. Commun.*, **510**, (1971); *J. C. S. Perkin II*, 1295 (1972).
577. J. N. Gardner and A. R. Katritzky, *J. Chem. Soc.*, 4375 (1957).
578. R. A. Y. Jones, A. R. Katritzky, and J. M. Lagowski, *Chem. Ind.* (London), 870 (1960).
579. S. J. Angyal and C. L. Angyal, *J. Chem. Soc.*, 1461 (1952).
580. S. F. Mason, *J. Chem. Soc.*, 3619 (1958).
581. T. Grønneberg and K. Undheim, *Tetrahedron Lett.*, 3193 (1972).
582. R. A. Jones and A. R. Katritzky, *J. Chem. Soc.*, 2937 (1960).
583. R. A. Jones and A. R. Katritzky, *J. Chem. Soc.*, 378 (1961).
584. S. Golding, A. R. Katritzky, and H. Z. Kucharska, *J. Chem. Soc.*, 3090, 3093 (1965).
585. A. R. Katritzky and B. Ternai, *J. Chem. Soc. B*, 631 (1966).
586. F. J. Dinan and H. Tieckelmann, *J. Org. Chem.*, **29**, 892 (1964).
587. R. B. Moffett, *J. Org. Chem.*, **28**, 2885 (1963).
588. H. F. Stewart and R. P. Seibert, *J. Org. Chem.*, **33**, 4560 (1968).
589. J. E. Litster and H. Tieckelmann, *J. Amer. Chem. Soc.*, **90** 4361 (1968).
590a. F. J. Dinan and H. Tieckelmann, *J. Org. Chem.*, **29**, 1650 (1964).
590b. R. A. Abramovitch and G. N. Knaus, *J. C. S. Chem. Comm.*, 238 (1974).
591. O. R. Rodig, R. E. Collier, and R. K. Schlatzer, *J. Org. Chem.*, **29**, 2652 (1964).
592. Y. Maki, M. Sato, and K. Yamane, *Yakugaku Zasshi*, **85**, 429 (1965); *Chem. Abstr.*, **63**, 5477f (1965).
593. Y. Maki, K. Yamane, and T. Masugi, *Gifu Yakka Daigaku Kiyo*, **15**, 31 (1965); *Chem. Abstr.*, **68**, 21796h (1968).
594. Y. Maki, K. Kawasaki, and K. Sato, *Gifu Yakka Daigaku Kiyo*, **13**, 34 (1963); *Chem. Abstr.*, **60**, 13220e (1964).
595a. Y. Maki, K. Yamane, and M. Sato, *Yakugaku Zasshi*, **86**, 50 (1966); *Chem. Abstr.*, **64**, 11165a (1966).
595b. J. Skarzewski and Z. Skrowaczewska, *Wiadom Chem.*, 155 (1974).
596. D. Jerchel and L. Jakob, *Chem. Ber.*, **92**, 724 (1959).
597. A. Van Dormael, *Bull. Soc. Chim. Belg.*, **58**, 167 (1949).
598. C. K. Bradsher, E. F. Litzinger, Jr., and M. F. Zinn, *J. Heterocycl. Chem.*, **2**, 331 (1965).
599. C. K. Bradsher and M. F. Zinn, *J. Heterocycl. Chem.*, **4**, 66 (1967).
600. R. Dohmori, *Chem. Pharm. Bull.* (Tokyo), **12**, 588, 595 (1964); *Chem. Abstr.*, **61**, 4305 (1964).
601. R. Glicksman and C. K. Morehouse, *J. Electrochem. Soc.*, **107**, 717 (1960).
602. J. Tirouflet and M. Person, *Ric. Sci.*, **30**, Suppl. 5, 269 (1960); *Chem. Abstr.*, **55**, 20898i (1961).
603. J. Volke, R. Kubicek, and F. Santovy, *Collect. Czech. Chem. Comm.*, **27**, 680 (1962).
604. N. G. Lordi and E. M. Cohen, *Anal. Chim. Acta*, **25**, 281 (1961).
605. T. T. Tsai, W. E. McEwen, and J. Kleinberg, *J. Org. Chem.*, **26**, 318 (1961).
606. H. Lund, *Acta Chem. Scand.*, **17**, 2325 (1963).
607. M. T. Falqui and M. Secci, *Ann. Chim.* (Rome), **48**, 1168 (1958); *Chem. Abstr.*, **53**, 12062c (1959).
608. H. Lund, *Acta Chem. Scand.*, **17**, 972 (1963).
609. N. L. Weinberg and E. A. Brown, *J. Org. Chem.*, **31**, 4054 (1966).
610. W. M. Schwarz, E. M. Kosower, and I. Shain, *J. Amer. Chem. Soc.*, **83**, 3164 (1961).
611. S. G. Mairanovskii, *Trudy Chetvertogo Soveshchaniya Elektrokhim., Moscow*, 223 (1956) (publ. 1959); *Chem. Abstr.*, **54**, 9558h (1960).
612. D. J. McClemens, A. K. Garrison, and A. L. Underwood, *J. Org. Chem.*, **34**, 1867 (1969).
613. J. Engelhardt and W. P. McKinney, *J. Agr. Food Chem.*, **14**, 377 (1966).

614. W. R. Boon, *Chem. Ind.* (London), 752 (1965).
615. A. D. Dodge, *Endeavour*, **111**, 130 (1970).
616. R. F. Homer, G. C. Mees, and T. E. Tomlinson, *J. Sci. Food Agr.*, **11**, 309 (1960).
617. R. Taylor, *J. Chem. Soc.*, 4881 (1962).
618. T. Nakajima and A. Pullman, *J. Chim. Phys.*, **55**, 793 (1958).
619. A. Shin-Chuen Chia and R. F. Trimble, Jr., *J. Phys. Chem.*, **65**, 863 (1961).
620. J. H. Blanch, *J. Chem. Soc. B*, 937 (1966).
621. P. R. Flakner and D. Harrison, *J. Chem. Soc.*, 2148 (1962).
622a. A. R. Katritzky and F. J. Swinbourne, *J. Chem Soc.*, 6707 (1965).
622b. G.P. Syrova and Yu. N. Sheinker, *Chem. Heterocycl. Compounds*, **8**, 313 (1972).
623. R. Joeckle, E. D. Schmid, and R. Mecke, *Z. Naturforsch., A*, **21**, 1906 (1966).
624. A. D. Campbell, S. Y. Chooi, L. W. Deady, and R. A. Shanks, *Aust. J. Chem.*, **23**, 203 (1970).
625. M. Liveris and J. Miller, *J. Chem. Soc.*, 3486 (1963).
626. J. Miller and Wan Kai-Yan, *J. Chem. Soc.*, 3492 (1963).
627. I. F. Tupitsyn, N. N. Zatsepina, Yu. M. Kapustin, and A. V. Kirova, *Reakts. Sposobnost. Org. Soedin.*, **5**, 613 (1968); *Chem. Abstr.*, **70**, 76941y (1969).
628. J. H. Nelson, R. G. Garvey, and R. O. Ragsdale, *J. Heterocycl. Chem.*, **4**, 591 (1967).
629. G. P. Bean and A. R. Katritzky, *J. Chem. Soc. B*, 864 (1968).
630. A. Dondoni, G. Modena, and P. E. Todesco, *Gazz. Chim. Ital.*, **91**, 613 (1961); *Chem. Abstr.*, **57**, 3359i (1962).
631. G. Costa and D. Blasina, *J. Phys. Chem.* (Frankfurt), **4**, 24 (1955).
632. H. Shindo, *Chem. Pharm. Bull.* (Tokyo), **8**, 845 (1960); *Chem. Abstr.*, **55**, 19480 (1961).
633. H. H. Jaffé, *J. Org. Chem.*, **23**, 1790 (1958).
634. A. R. Katritzky, J. A. T. Beard, and N. A. Coats, *J. Chem. Soc.*, 3680 (1959).
635. E. Imoto and Y. Otsuji, *Bull. Univ. Osaka Prefect., Ser. A*, **6**, 115 (1958); *Chem. Abstr.*, **53**, 3027h (1959).
636. E. Imoto, *Rev. Polarography* (Japan), **9**, 185 (1961).
637. Y. Otsuji, Y. Koda, M. Kubo, M. Furukawa, and E. Imoto, *Nippon Kagaku Zasshi*, **80**, 1293 (1959); *Chem. Abstr.*, **55**, 6476c (1961).
638. P. Tomasik, *Rocz. Chem.*, **44**, 341 (1970).
639. P. Tomasik, *Rocz. Chem.*, **44**, 1211 (1970).
640. Y. Ueno and E. Imoto, *Nippon Kagaku Zasshi*, **88**, 1210 (1967); *Chem. Abstr.*, **69**, 66782n (1968).
641. P. Tomasik, "Application of the Hammett Equation to Heteroaromatic Molecules," to be published, personal communication to R. A. Abramovitch, 1972.

CHAPTER IB

Partially Reduced Pyridines

ROBERT E. LYLE

Department of Chemistry
University of New Hampshire
Durham, New Hampshire

The reduction of the aromatic heterocycle pyridine can lead to several isomeric dihydro- and tetrahydropyridines which are the subject of this section. These compounds are of interest as reaction intermediates and intermediates in organic synthesis as well as representing ring systems of theoretical and biological importance. This discussion provides a comprehensive view of the types of syntheses and properties of partially reduced pyridines, for which direct structural evidence is presented, without attempting to provide an exhaustive coverage of the literature.

I. Dihydropyridines

The dihydropyridines can exist in three isomeric enamine forms (**IB-1**, **IB-2**, or **IB-3**) or six immonium salt isomers (**IB-4**, **IB-5**, **IB-6**, and **IB-7**) which could be considered to be the possible protonated forms of **IB-1**, **IB-2**, or **IB-3**. The relative stabilities of these isomers has been the subject of some discussion mentioned below; however, as is evident from enamine chemistry, the isomers **IB-4** through **IB-7** would not be stable in a basic medium if a hydrogen is located on the 3- or 5-sp^3 carbons and **IB-1** through **IB-3** would not exist under acidic conditions.

A review of the syntheses and properties of dihydropyridines provides a survey of the literature through early 1971 (1).

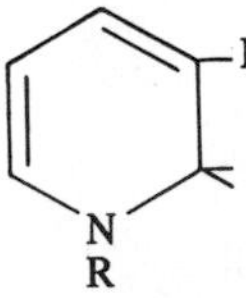

IB-1 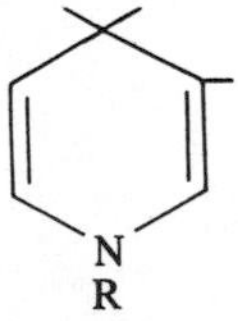 IB-2 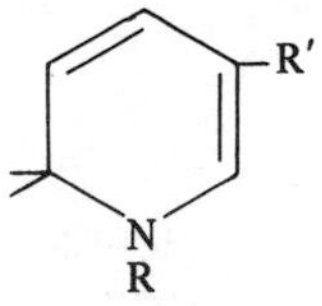 IB-3

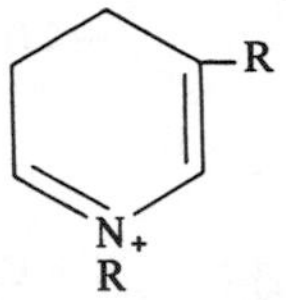

IB-4 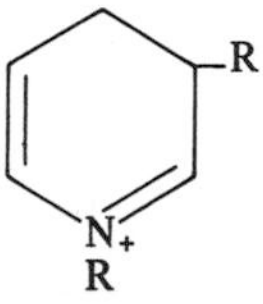 IB-4a

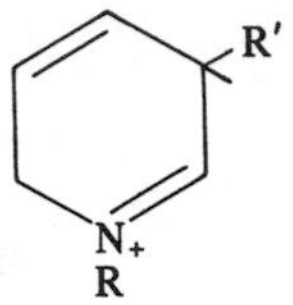

IB-5a 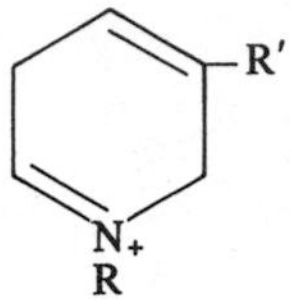 IB-5 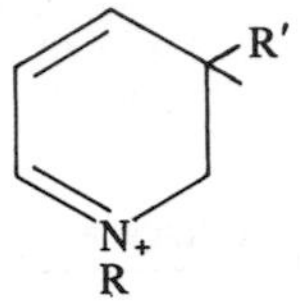 IB-6 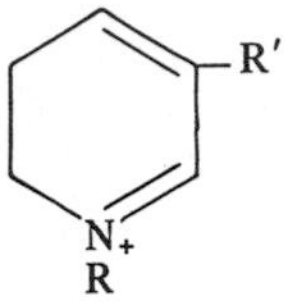 IB-7

1. Syntheses by Cyclization

The formation of dihydropyridines by (i) cyclization of acyclic starting materials and (ii) reduction of pyridines, pyridinium ions, or their derivatives provide the general synthetic routes. The versatility of the Hantzsch synthesis (eq. **IB-1–IB-3**) (2–6) for the preparation of substituted pyridines has led to the synthesis of many dihydropyridines as intermediates. This reaction of an aldehyde, a β-keto ester, or diketone, and ammonia usually gives a 1,4-dihydropyridine (7, 8). The initial reaction of the ammonia or ammonium chloride with the β-dicarbonyl compound gives a vinyl amine (enamine). The aldehyde alkylates two molecules of the enamine. Cyclization with elimination of ammonia then gives the dihydropyridine. The β-aminovinylcarbonyl may be formed *in situ*, be preformed, or be derived from an isoxazole ring (eq. **IB-4**) (9).

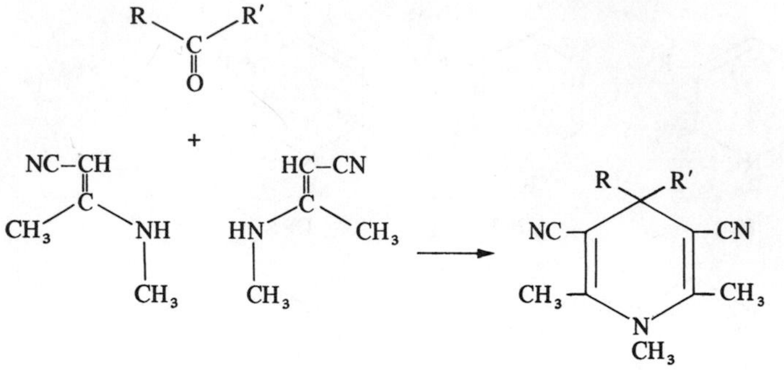

$R = CH_2CN$; $R' = CH_3$
$R - R' = -CH_2CH_2CH_2CH_2CH(Cl)-$

eq. IB-1

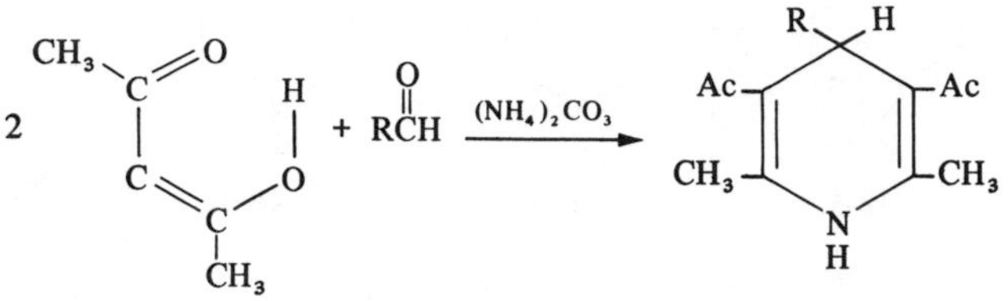

$(NH_4)_2CO_3$

eq. IB-2

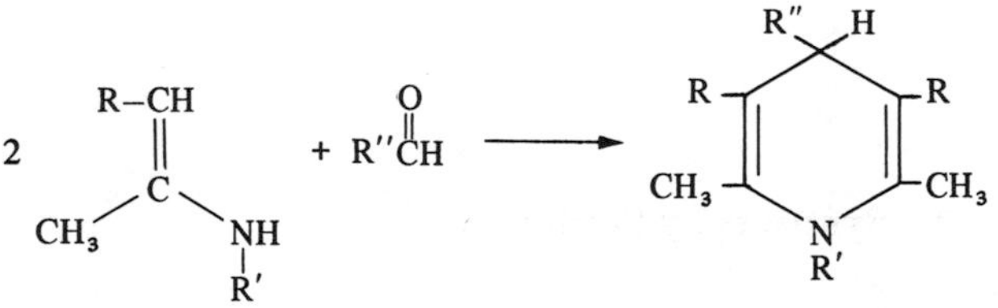

$R = CN, COOC_2H_5, COCH_3$
$R' = H, CH_3$
$R'' = H, COOCH_2Ph, CH_2Cl$

eq. IB-3

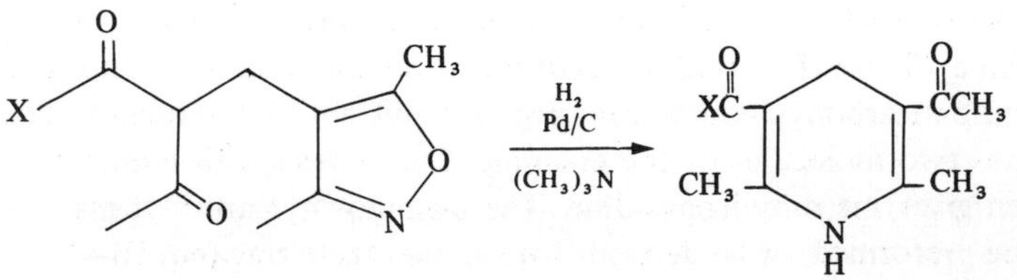

H_2
Pd/C
$(CH_3)_3N$

eq. IB-4

Although the Hantzsch synthesis usually leads to 1,4-dihydropyridines, similar reactions have been shown to give 1,2-dihydropyridines. When the vinylamine is prepared from a β-ketonitrile and a ketone rather than an aldehyde as the electrophile, 1,2-dihydropyridines **(eq. IB-5A)** are reported to be formed (10). Similar results to form **B** in eq. IB-5 were obtained when a ketone was used with acetoacetaldehyde as the source of the vinyl amine (11).

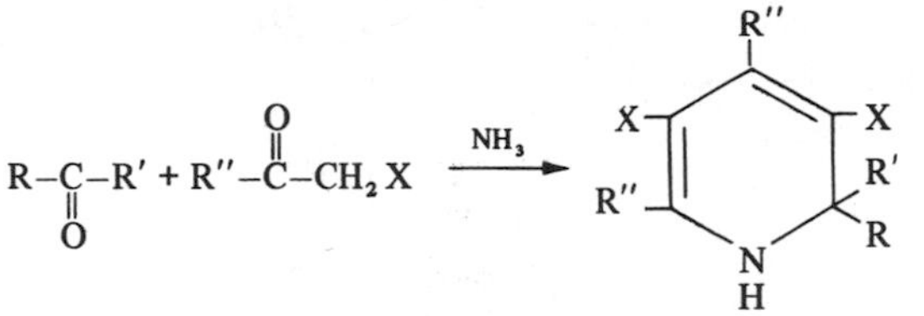

$$R{-}\underset{\underset{O}{\|}}{C}{-}R' + R''{-}\overset{\overset{O}{\|}}{C}{-}CH_2X \xrightarrow{NH_3}$$

A: R–R′ = –$(CH_2)_4$–; R″ = Ph; X = CN

B: R = R′ = CH_3; R″ = H; X = Ac

eq. IB-5

Cycloadditions have been shown to lead to dihydropyridines. The reaction of an anil with the diester of acetylene dicarboxylic acid gives a 1,2-dihydropyridine (12) **(eq. IB-6)**, while the product of a 1,3-dipolar addition of diethoxymethyleneaminodiphenylphosphine with acrylic acid esters underwent a further Diels-Alder reaction with the acrylic acid ester to form a 3,4-dihydropyridine derivative (13) **(eq. IB-7)**. The assignments of structures to these unusual heterocycles were based on NMR and mass spectral data. The 3,4-dihydropyridines are also formed by the ammonia-catalyzed condensation of malononitrile, salicylaldehyde and a ketone (13a).

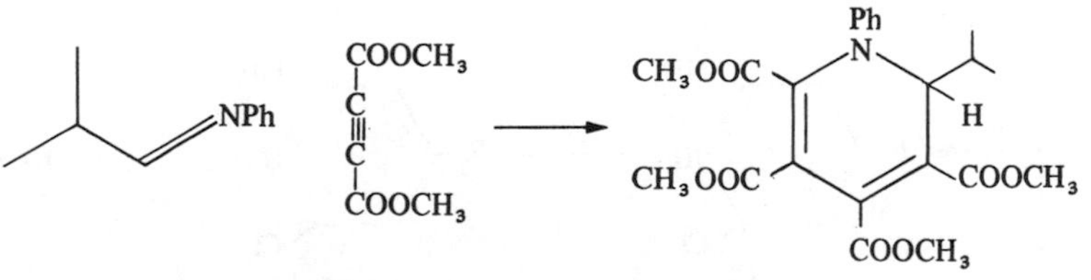

NPh; $COOCH_3$–C≡C–$COOCH_3$ → Ph, N, CH_3OOC, CH_3OOC, H, $COOCH_3$, $COOCH_3$

eq. IB-6

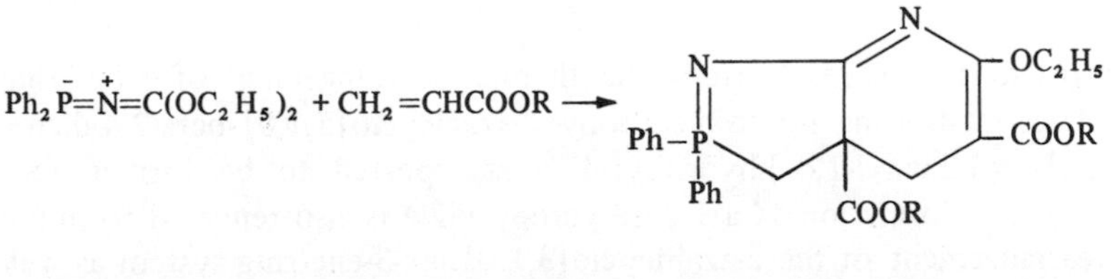

$$Ph_2\overset{-}{P}{=}\overset{+}{N}{=}C(OC_2H_5)_2 + CH_2{=}CHCOOR \longrightarrow$$

N, N, OC_2H_5, Ph–P, Ph, COOR, COOR

eq. IB-7

The reaction of butanal with aniline, a reaction probably occurring *via* an anil intermediate, was reported to give a 1,4-dihydropyridine(14). This product has more recently been shown to be the 1,2-dihydropyridine **IB-8** (R = Ph) (eq.

IB-8) (15). Butanal in acetic acid with ammonium acetate is reported to give the 2,3-dihydropyridine **IB-9** (16), which may be considered to be a protonated form of **IB-8** (R = H). The 2,5-, 3,4-, and 2,3-dihydropyridines have been proposed as the protonated forms of 1,2-, 1,4-, or 1,6-dihydropyridines (17–19).

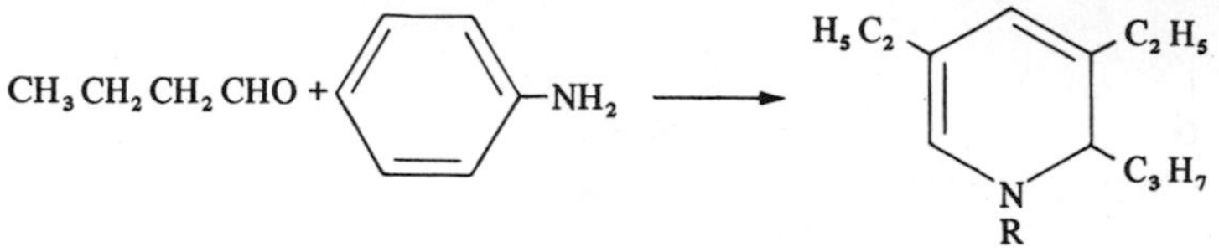

$CH_3CH_2CH_2CHO$ + C_6H_5–NH_2 ⟶

IB-8
R = Ph

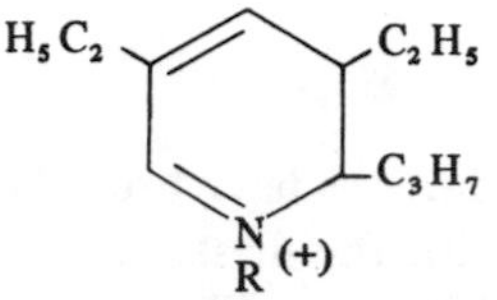

IB-9
R = H

eq. **IB-8**

The cyclization of glutaronitriles with anhydrous hydrogen chloride gives a tetrahydropyridine which, on treatment with base, gives a 3,4-dihydropyridine (**IB-10**) (eq. **IB-9**) (20).

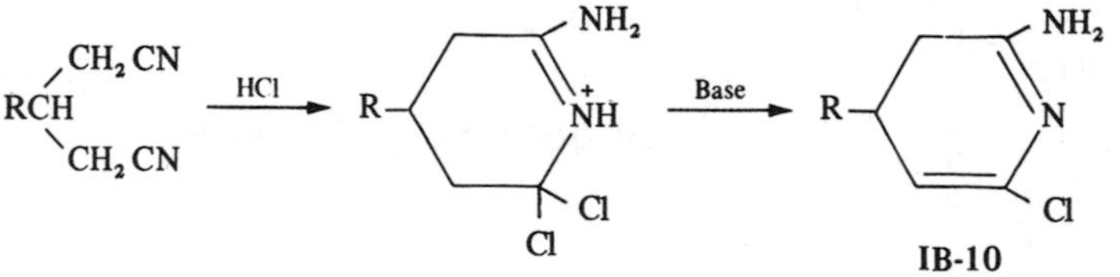

$RCH(CH_2CN)_2$ —HCl→ —Base→

IB-10

eq. **IB-9**

Dihydropyridines can result from the thermal rearrangement of other ring systems. Thus on heating 1-ethoxycarbonyl-1-azabicyclo[5,1,0]-octa-2,4-diene, 1-ethoxycarbonyl-2-vinyl-1,4-dihydropyridine is reported to be formed (eq. **IB-10**) (21a). The formation of a 1,2-dihydropyridine is also reported from the thermal rearrangement of the 2-azabicyclo[3.1.0]hex-3-ene ring system as well (eq. **IB-10**) (21b) and the comparable tricycle **IB-11** gives 1-benzenesulfonyl-1,4-dihydropyridine (**IB-12**) on heating (21c). The 2,3-dihydropyridine **IB-13** is suggested to be a valence tautomer of the 3,4-dihydroazocene **IB-14** (22).

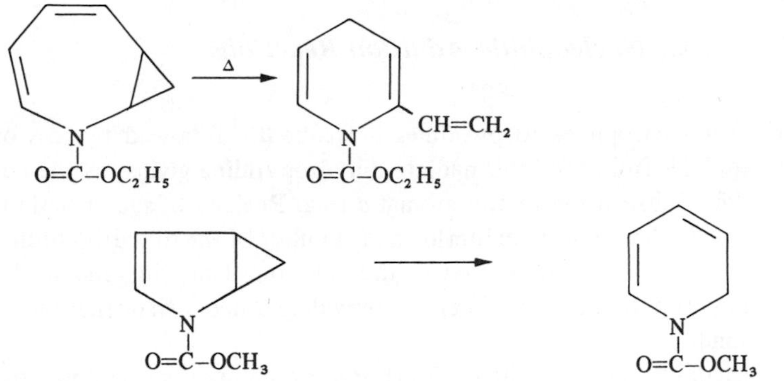

N
O=C–OC$_2$H$_5$
Δ
N
CH=CH$_2$
O=C–OC$_2$H$_5$

eq. IB-10

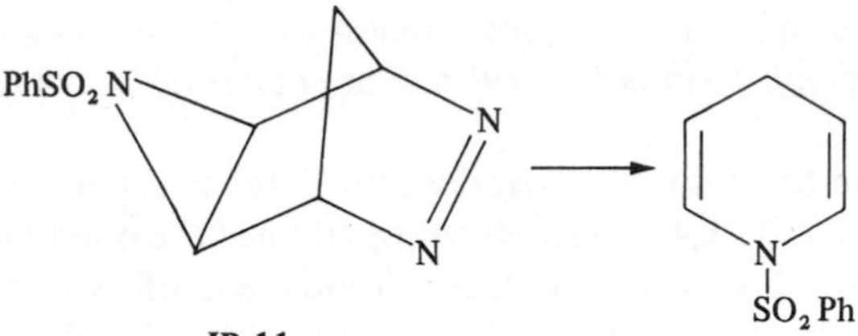

N
O=C–OCH$_3$
N
O=C–OCH$_3$

PhSO$_2$N
N
N
N
SO$_2$Ph

IB-11

IB-12

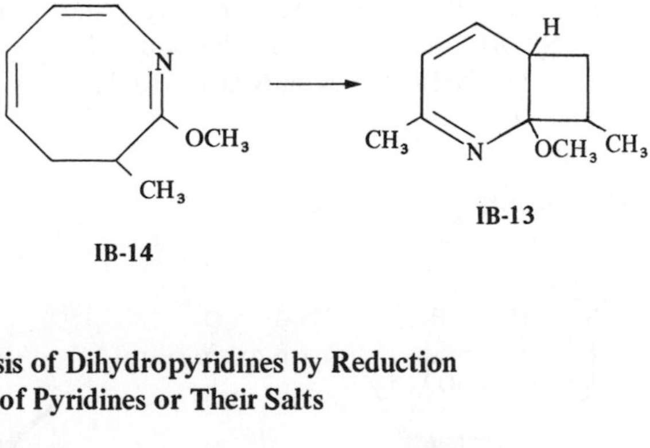

N
OCH$_3$
CH$_3$
H
CH$_3$
N
OCH$_3$ CH$_3$

IB-13

IB-14

2. Synthesis of Dihydropyridines by Reduction of Pyridines or Their Salts

The addition of two electrons, in one or two steps, to the pyridine ring leads to the corresponding substituted dihydropyridine. The high electronegativity of the pyridine aromatic system provides an easily reduced π-system, the reduction of which is further facilitated by salt formation. The discussion of these reactions is organized in terms of a single, two-electron reduction, nucleophilic addition, and two, one-electron reactions, possible pyridinyl radical intermediates.

A. *Nucleophilic Addition Reactions*

The addition of nucleophiles to pyridines is frequently followed by loss of different nucleophiles from the intermediate dihydropyridine giving, as the net result, nucleophilic substitution of the aromatic ring. Reviews of such reactions are available (23–25). The ease of reduction and stability of the dihydropyridine is increased if a pyridinium salt is used in the reaction. Thus, in general, the 1-substituted-dihydropyridines have been more widely studied than the unsubstituted compounds.

When the nucleophile added to the pyridine or pyridinium salts is hydride, the reaction may lead to dihydro-derivatives or, if a source of Lewis acid is available, further reduction to tetrahydropyridines or piperidines can occur. These reactions have been the subject of reviews (17–19) and are discussed in detail later.

The addition of a nucleophile to an unsymmetrically substituted pyridine ring would be expected to lead to a 1,2-, 1,4-, or 1,6-dihydropyridine. The question of identity and regiospecificity of such nucleophilic addition was of greatest significance in considering the structure of the reduced form of the ubiquitous coenzyme dihydronicotinamide adenine dinucleotide (NADH) **(IB-15)** (26). The reduced ring was shown to be a 1,4-dihydropyridine by labeling studies which showed that the carbon that received and transferred deuterium in the NAD⇌NADH reaction was at the 4-position (27–29). This assignment was confirmed by NMR studies on NADH (30–31).

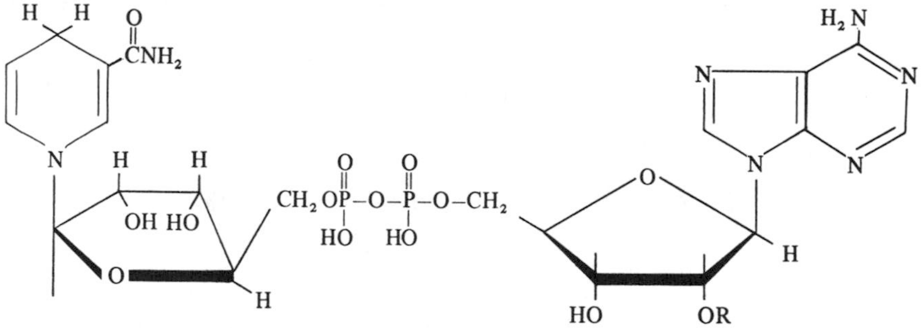

IB-15

NADH, R = H
NADP, R = PO_3H_2

The study of the reaction of nucleophiles with NAD and its analogs has provided most of the information concerning the position of attack by a nucleophile on a pyridine ring and the ideas about the factors which affect the course of the reaction. The orientation of the nucleophilic addition, and accordingly, the substitution prior to elimination, is usually on the carbon atom adjacent to nitrogen at the 2- or 6-positions; however, certain reactions may lead to 1,4-dihydropyridines (see Tables IB-1 and IB-2). The factors which govern the position of reaction have been the subject of considerable controversy (32–37). Both electronic and steric factors seem to be involved; however, the primary directive influence seems to be the electron-attraction of the nitrogen which results in the 1,2-dihydropyridines being the kinetically favored product. If the initial reaction is reversible or if a large group is attached at the 1-position, the 1,4-dihydropyridine results (33, 33a). The greater stability of the 1,4-dihydropyridine system at equilibrium has been shown by several methods (33, 38).

In general, the addition of most nucleophiles to pyridines, 1-oxides, or salts lead to 1,2- or 1,6-dihydropyridines or the 2- or 6-substitution product. The exceptions to this are the addition of cyanide ion and reduction with dithionite ion of pyridinium ions having electron attracting groups at the 3- and/or 5-positions (1, 32–37, 39–42). Some pyridines give 1,4-dihydro-derivatives with other types of nucleophiles under specific conditions. Thus 3-cyanopyridine **(IB-16)** with sodium borohydride in aprotic solvent (43), 3,5-dialkoxycarbonyl- and -keto-pyridines **(IB-17)** with sodium cyanoborohydride (2), 3-benzoylpyridine **(IB-18)** with aryl Grignard reagents (44), and 1-acylpyridinium salts **(IB-19)** with carbanions (45-49) (see Table IB-2) give 1,4-dihydropyridines as the primary or sole product.

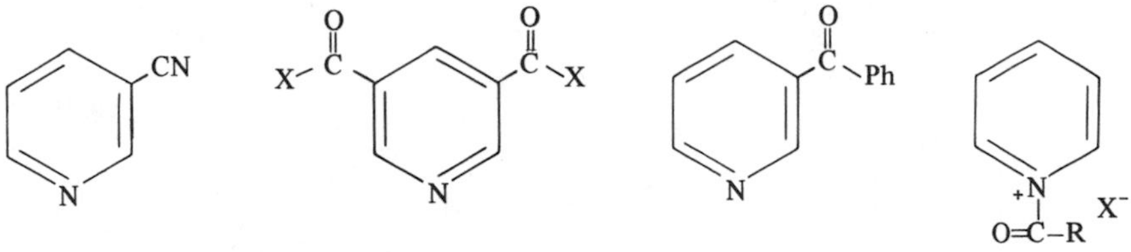

IB-16

X = OR
$X = CH_3$
IB-17

IB-18

IB-19

Sugar residues attached at the 1-position seem to favor hydride addition at the 4-position, for 1-maltosyl- and 1-lactosylpyridinium salts give 1,4-dihydropyridines with sodium borohydride (49a). A polyene bridge of the 3- and 5-positions of pyridine directs hydride addition with lithium aluminum hydride to the 4-position if the resulting π-system contains $4n + 2$ electrons and thus has aromatic stability (49b).

1,2-Dihydropyridines can be prepared readily by the addition of organometallic reagents to pyridines (50–52) or pyridinium salts (53–56) (see Table IB-2 for examples). Pyridinium salts having a 1-substituent with an acidic hydrogen at the proper distance from the ring can form an anion in base, which leads to cyclization forming a fused bicycle with a nitrogen common to both rings (eq. **IB-11**) (57, 58).

The reactions of pyridine-1-oxides and their salts with nucleophiles undoubtedly proceed *via* 1,2- or 1,4-dihydropyridine intermediates to form 2- or 4-substituted pyridines. The reaction of aryl Grignard reagents, for example, was reported to form the 1-hydroxy-2-aryl-1,2-dihydropyridine **(IB-20)** (59); however, the product has now been shown to be an acyclic compound **(IB-21)** (60) (see also Chap. IV).

The reactions of substituted pyridine-1-oxides with mercaptans in acetic anhydride were shown to lead to a mixture of 2- and 3-pyridyl sulfides (60a). This unusual orientation of the entering nucleophile was suggested to result from the formation of a 2,3-episulfonium-2,3-dihydropyridine by decomposition of the intermediate 1-acetoxy-2-thioalkoxy-1,2-dihydropyridine. This reaction pathway was supported by the isolation of 1-acetyl-2-acetoxy-4-*t*-butyl-3,6-di-*t*-butylmercapto-1,2,3,6-tetrahydropyridine from the reaction of 4-*t*-butylpyridine-1-oxide with *t*-butylmercaptan and acetic anhydride in the presence of triethylamine (see also Chaps. IA and IV).

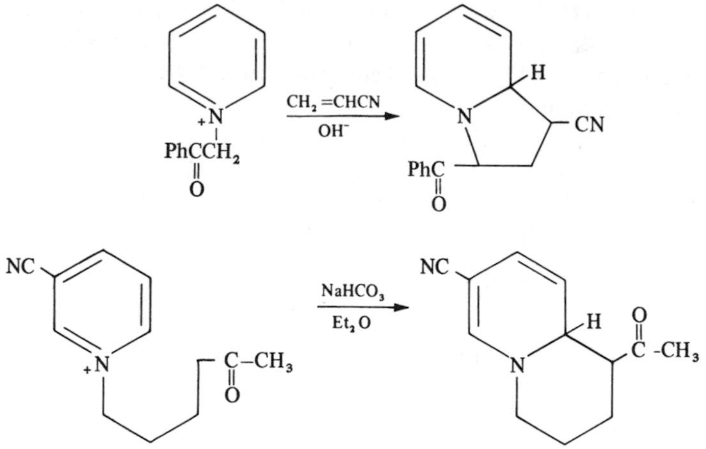

$CH_2{=}CHCN$, OH^-

eq. **IB-11**

$NaHCO_3$, Et_2O

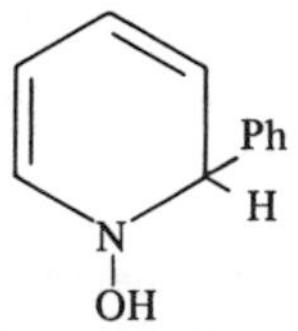

IB-20

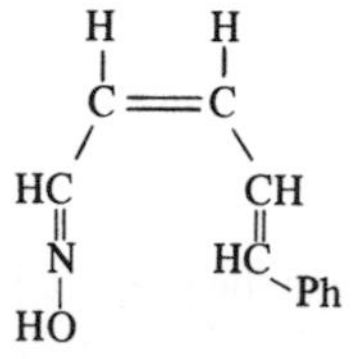

IB-21

TABLE IB-1. Reactions of Pyridines with Nucleophiles and the Properties of the Products

X	R″	Reagent	Type	u.v.[λ_{max} nm(log ϵ)]	i.r.(cm^{-1})	Ref.
Ph	H	PhLi	1[a]			50, 51
$PhCH_2$	H	$PhCH_2MgCl$	1	Not given		230
			2			230
Ph	3-PhCO–	PhMgBr	2	234(4.1), 360(4.0)	1675, 1618	44
H	3-CN	$NaBH_4$	2	330		43
H	3,5-$(CN)_2$	$NaBH_4$	1	213, 254, 382	2192, 1642, 1560, 1540, 1505	231
			2	206, 352	2192, 1685, 1620, 1500	231
H	3,5-$(CN)_2$	$NaAlH_2(OCH_2CH_2OCH_3)_2$	2	See above		232
H	H	$LiAlH_4$	1			
			2			154

[a]The NMR was reported as 4.78(2); 4.43(3); 6.01(4); 4.68(5); 6.79 (6) ppm (position).

TABLE IB-2. Reaction of Pyridinium Salts with Nucleophiles and the Spectroscopic Properties of the Product

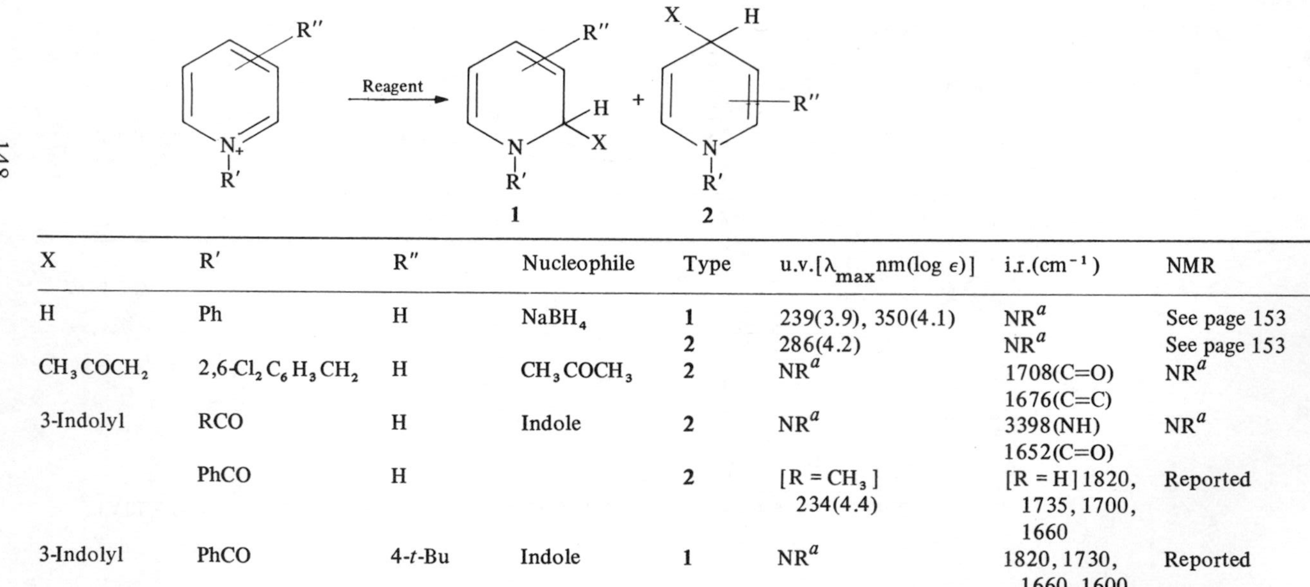

X	R′	R″	Nucleophile	Type	u.v.[λ_{max} nm(log ϵ)]	i.r.(cm^{-1})	NMR	Ref.
H	Ph	H	$NaBH_4$	1	239(3.9), 350(4.1)	NR[a]	See page 153	97
				2	286(4.2)	NR[a]	See page 153	97
CH_3COCH_2	2,6-$Cl_2C_6H_3CH_2$	H	CH_3COCH_3	2	NR[a]	1708(C=O) 1676(C=C)	NR[a]	145
3-Indolyl	RCO	H	Indole	2	NR[a]	3398(NH) 1652(C=O)	NR[a]	47–49
	PhCO	H		2	[R = CH_3] 234(4.4)	[R = H] 1820, 1735, 1700, 1660	Reported	233
3-Indolyl	PhCO	4-*t*-Bu	Indole	1	NR[a]	1820, 1730, 1660, 1600	Reported	233

CN	CH_3	3-CN	CN^-	**2**	330	2220, 1690, 1600	6.9, 6.1, 4.8, 4.5, 3.1	42
CN	CH_3	3-Br-5-$COOC_2H_5$	CN^-	**1**	262, 320	1685, 1630, 1600, 1560	7.3(2), 6.95(4) 5.3(6)	33
				2	343	1680, 1600	7.2(6), 6.4(4) 4.75(2)	33
Ar	CH_3	5-CN	ArMgX	**1**	[Ar = Ph] 222 (4.2), 352(3.6)	2185, 1645, 1575	Reported	54
R	EtOCO	4-R″	RMgX	**1**	292.1	3000, 1700, 1365	5.42(2), 5.70(3) 4.70(5), 3.27(6)	53
H	CH_3OCO	H	$NaBH_4$	**2**	224(4.1)	1726(C=O), 1634(C=C)	6.58(2), 4.60–4.95(3), 2.67–2.95(4)	234
				1	302(3.6)	1718(C=O), 1647, 1585 (C=C)	6.53(6), 4.82–5.95(2,4,5) 4.15(2)	
CN	CH_3	3,5-(CN)	CN^-	**2**	246(3.9), 350(3.9)	1680, 1635	NR[a]	235
				1	359(3.9)	1650, 1640		

[a]NR - not reported.

B. *Two One-Electron Reductions*

The controversy over the mechanism of reduction of NAD to NADH has centered about the question of whether the formation of the 1,4-dihydropyridine proceeds *via* a pyridinyl radical as proposed by Kosower (37, 61, 62) or by the direct addition of a nucleophile. The intermediacy of a charge transfer complex (63) and perhaps a pyridinyl radical has received support by the detection of radical ions from the reaction of cyanide with pyridinium ions (64, 65) to produce viologens **(eq. IB-12)**. The reaction of pyridinyl radical with benzyl halides gives 1,2- and 1,4-dihydropyridines with the benzyl group attached to the sp^3 carbon (63b).

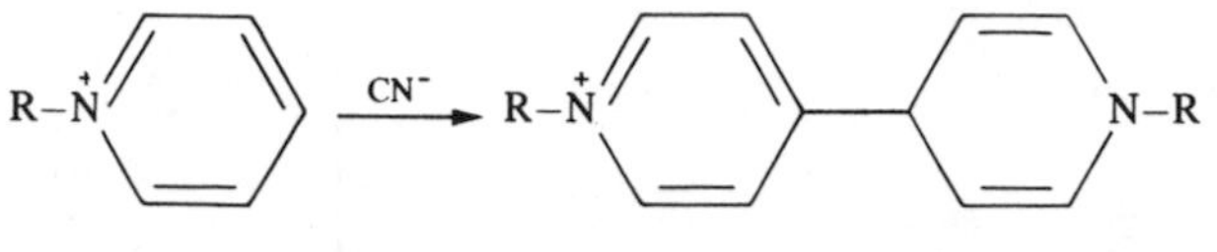

$$\text{R}-\overset{+}{\text{N}}\text{C}_5\text{H}_5 \xrightarrow{\text{CN}^-} \text{R}-\overset{+}{\text{N}}\text{C}_5\text{H}_4-\text{C}_5\text{H}_4\text{N}-\text{R} \qquad \textbf{eq. IB-12}$$

Electrolytic reduction of quaternary salts of 3-substituted pyridines has been reported to give pyridinyl radicals (61, 62), dipyridyls (65, 66), dihydropyridines (66, 67) or tetrahydropyridines (67a). Similar conversion of pyridines to tetrahydro-4,4′-dipyridyls **(IB-22)** can be accomplished by zinc and acetic anhydride reductions (68, 69). Reduction of pyridines or quaternary salts with sodium amalgam can give partially reduced pyridines, piperidines, or tetrahydro-4,4′-bipyridyls, useful as intermediates in the synthesis of the herbicide Paraquat (70). Lithium dispersion and trimethylsilyl chloride converts pyridine to the 1,4-bistrimethylsilyl-1,4-dihydropyridine **(IB-23)** (71). A mixture of 1-trimethylsilyl-1,2- **(IB-24)** and 1,4-dihydropyridines **(IB-25)** was obtained from the reaction of pyridine with trimethylsilyl hydride over palladium (72) (see also Ch. IX for more details).

Photochemical addition of methanol to pyridines gave mixtures of 4-hydroxymethyl-1,4-dihydro- and 6-hydroxymethyl-1,6-dihydropyridine derivatives **(eq. IB-13)** (73, 74). These compounds proved of interest as intermediates in the rearrangement reactions to be discussed later.

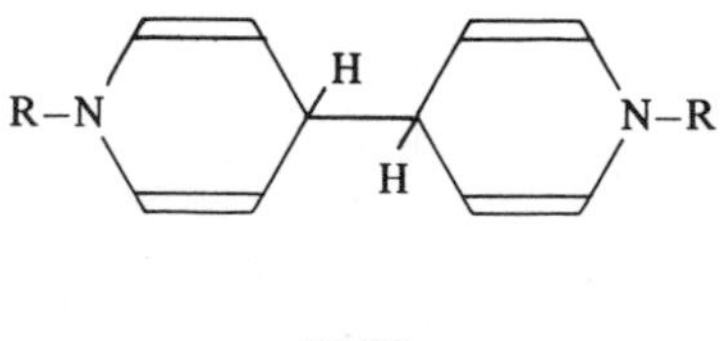

R–N(C₅H₄H)–(HC₅H₄)N–R

IB-22

H, SiMe₃ / N / SiMe₃

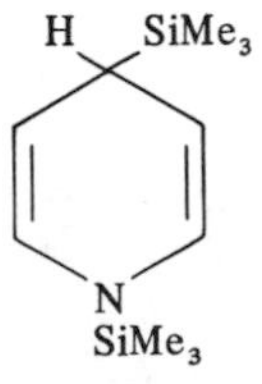

IB-23

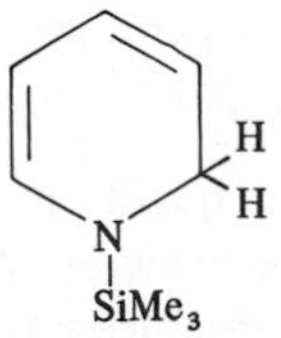

IB-24

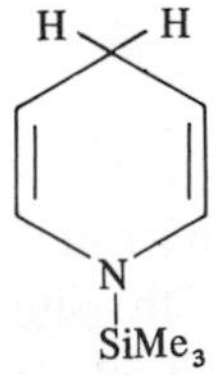

IB-25

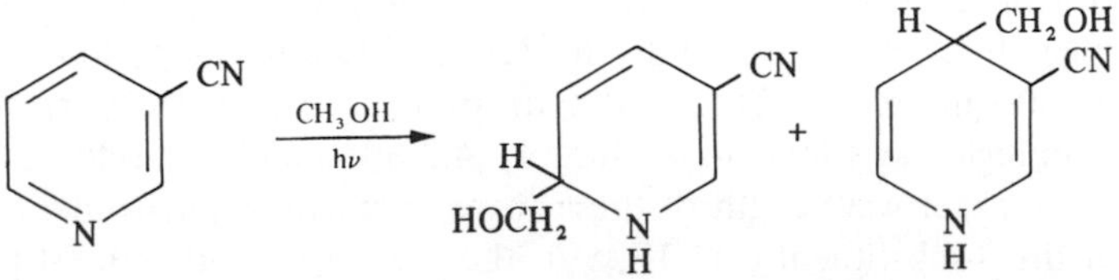

eq. IB-13

C. *Catalytic Hydrogenation*

The catalytic hydrogenation of pyridine and pyridinium salts usually yields the corresponding piperidine (75); however, electron attracting substituents cause the intermediate dihydro- and tetrahydropyridines to be stable and stop the reduction at the partially saturated stage (76). The intermediate 1,2- and 1,4-dihydropyridines can be detected by ultraviolet absorption spectroscopy, and these intermediates can be reduced to the tetrahydropyridine or piperidine (77–80). The intermediate partially reduced pyridines formed by catalytic hydrogenation have been shown to provide convenient routes to the synthesis of alkaloids (81).

3. Properties of Dihydropyridines

The physical and chemical properties of the 1,2- and 1,4-dihydropyridines are determined by the enamine or dienamine systems present in these compounds. The electronic distribution of the π-electrons of these compounds has been estimated by HMO or SCF methods of LCAO-MO calculations (82–87). The delocalization of the electron pair on the heterocyclic nitrogen throughout the enamine system was slight unless there was an unsaturated, electron-attracting substituent on the 3- and 5-positions in which case delocalization is considerable (84, 86). The ultraviolet and infrared spectra and some chemical properties have been related qualitatively to these calculations.

A. *Physical Properties*

a. ULTRAVIOLET ABSORPTION SPECTRA. The qualitative determination of the structures of the dihydropyridines can best be made by spectroscopic means. The ultraviolet absorption spectra of dihydropyridines having a 3-substituent which is electron-withdrawing are definitive for the 1,2-, 1,4-, and 1,6-dihydropyridine chromophores (see Tables IB-1 and IB-2 for data). The 1,2-dihydropyridines, with the extended dienamine chromophore, shows the absorption at longest wavelength (> 350 nm), while the 1,4- and 1,6-dihydropyridines both absorb in the 300 nm region. The 1,6-dihydropyridines show an absorption in the high 200-nm region as do the 3-substituted 1,4,5,6-tetrahydropyridine (1, 17, 76, 88, 89). The exact wavelength of these absorption bands is dependent upon the nature of the 3-substituent (1b, 17, 89), the presence of other substituents (89), the solvent (83), and the nature of the 1-substituent (90).

b. INFRARED SPECTRA. The structures of dihydro and tetrahydropyridines can be deduced from their i.r. spectra; however, this method has received less attention than that based on u.v. spectroscopy. The C=C and C–N ring stretching vibrations in the cyclic enamines lead to absorptions in the 1500 to 1700 cm^{-1} region. These bands have been confused with the C=O stretching vibration of a conjugated 3-substituent on occasion, for the C=O of the vinylogous amide is shifted to very low frequency (about 1600 cm^{-1}) in most cases (44, 54, 91, 92). Other unsaturated functional groups conjugated through a double bond with the ring nitrogen also show stretching vibrations at lower frequencies than expected. Thus the C≡N bond leads to absorption at about 2180 to 2200 cm^{-1} in such systems (93–95).

c. NUCLEAR MAGNETIC RESONANCE SPECTROSCOPY. Proton magnetic resonance spectroscopy was first used to determine the structures of the dithionite reduction products of 3-aminocarbonylpyridinium salts. Reduction of the salts in deuterium oxide and reoxidation of the dihydropyridine gave deuterium labeling at the site of reduction. The NMR spectrum of the pyridinium salt showed the label to be at the 4-position (31), supporting the chemical evidence that the coenzyme NADH also formed by dithionite reduction was a 1,4-dihydropyridine (27–29, 96). The first NMR spectrum of dihydropyridine was reported by Hutton and Westheimer (30) for 1-methyl-1,4-dihydronicotinamide and the 2-, 4-, or 6-deuterio derivative, demonstrating the importance of this method for structural studies of dihydropyridines.

The features of the NMR spectra of the parent dihydropyridines can be illustrated by considering the 1-phenyl-1,4- (**IB-25**) and 1,2-dihydropyridines (**IB-26**) prepared by the sodium borohydride reduction of

1-phenylpyridinium chloride (97). The protons at the sp^3 carbon are furthest upfield with the 4-methylene of the 1,4-dihydropyridine at higher field than the 2-methylene of the 1,2-dihydropyridine due to the deshielding effect of the adjacent nitrogen. This effect of the nitrogen also leads to the protons on the sp^2 carbons at the 2- and/or 6- positions giving signals at the lowest field of the

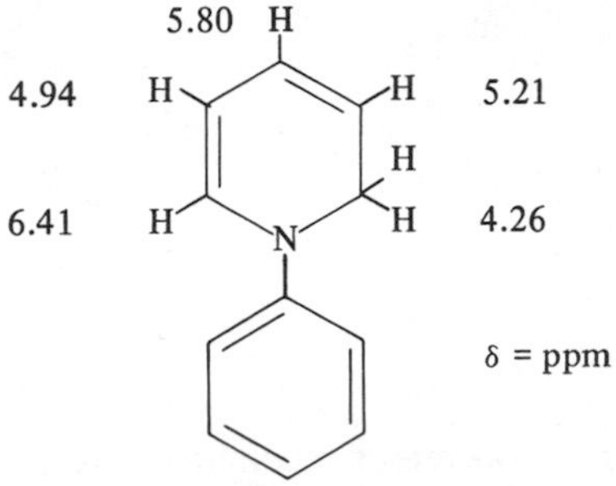

5.80 H; 4.94 H; 5.21 H; H; 6.41 H; 4.26 H; N; δ = ppm

J_{Hz}
$J_{2,3} = 3.6$ $J_{4,5} = 4.5$ $J_{2,4} = 1.5$
$J_{3,4} = 7.7$ $J_{5,6} = 6.9$ $J_{3,5} = J_{3,6} = J_{4,6} = 0.9$

IB-26

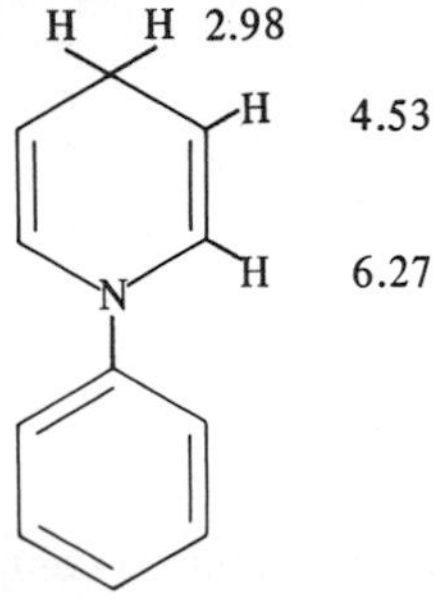

H H 2.98; H 4.53; H 6.27; N

J_{Hz}
$J_{2,3} = 9.0$ $J_{3,4} = 3.9$
$J_{2,4} = 1.6$

IB-25

ring protons. The ring protons are all coupled leading to a 6-spin system which cannot be analyzed by first-order interpretation. Estimates by inspection give important structural information, however, for it is evident that the protons attached to the same double bond (**IB-26**, $J_{3,4}$ and $J_{5,6}$; **IB-25**, $J_{2,3}$) are coupled more strongly than those on carbons connected by a single bond. The allylic couplings are probably negative but no signs are reported except for **IB-27**.

The presence of unsaturated substituents on the heterocyclic ring cause downfield shifts of the signals of the protons nearby as illustrated by the dihydropyridines, **IB-27**, **IB-28**, and **IB-29** (54, 78, 98). Complete analysis of the

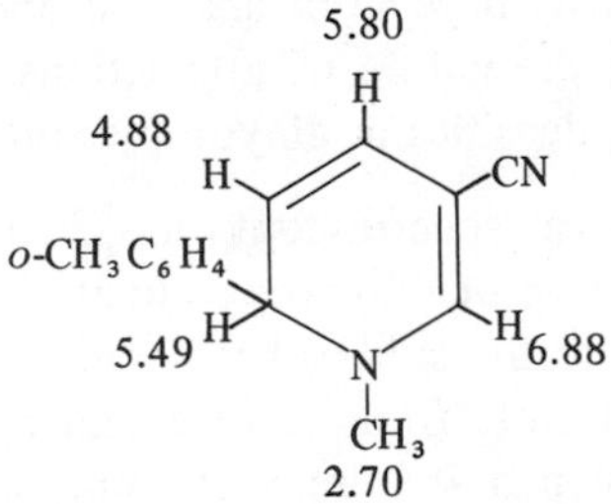

5.80 H; 4.88 H; CN; o-$CH_3C_6H_4$; 5.49 H; H 6.88; N; CH_3 2.70

$J_{4,5} = 10$ $J_{4,6} = -1.5$
$J_{5,6} = 3.8$

IB-27

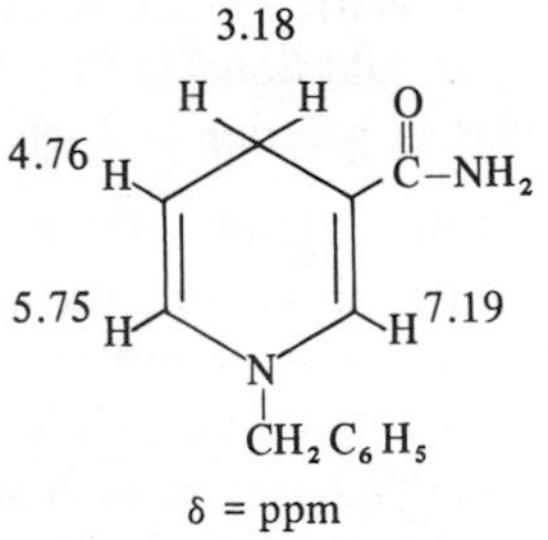

3.18 H H; O ‖ C–NH_2; 4.76 H; 5.75 H; H 7.19; N; $CH_2C_6H_5$; δ = ppm

$J_{4,5} = 3.4$ $J_{2,4} = 0.5$
$J_{5,6} = 8.2$ $J_{4,6} = 1.7$
$J_{2,6} = 1.5$

IB-28

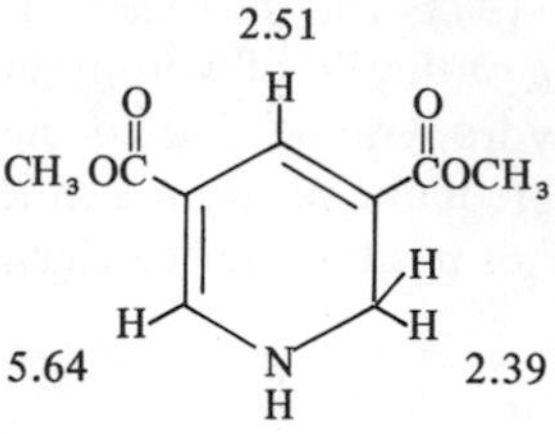

$$\text{IB-29: dimethyl 1,6-dihydropyridine-3,5-dicarboxylate; } \delta\ 2.51\ (\text{H-4}),\ 5.64\ (\text{H-2}),\ 2.39\ (\text{CH}_2\text{-6})$$

$J_{2,4} = 0.75$
$J_{4,6} = 1.5$
$J_{1,6} = 7.0$
$J_{1,2} = 1.7$

IB-29

spectrum requires computer simulation; however, a qualitative interpretation provides sufficient information for structural assignment.

The conformational study of dihydropyridines has been the object of several NMR studies (99–102) for it was assumed that nonequivalence of the methylene protons attached to the sp^3 ring carbon of the dihydropyridine would indicate a nonplanar ring. Because of the chiral centers in the sugar residue of NADH, the protons are inherently diastereotopic (103) and thus should be magnetically nonequivalent. This nonequivalence has now been demonstrated (104, 105). The apparent equivalence of the methylene protons in achiral models of NADH probably indicate a planar ring as indicated by x-ray studies of dihydropyridines (106, 107); however, the possibility of rapid interconversion of two nonplanar conformations cannot be eliminated. Similarly, the observed equivalence of the diastereotopic methylene protons of 1-benzyldihydropyridines having a chiral sp^3 carbon in the dihydropyridine ring indicates the lack of sensitivity of this probe to subtle magnetic differences.

The ^{13}C NMR spectra of NAD and NADH have been analyzed and should provide an additional probe for studying the nature of interactions between molecules in solution and relative electron densities in dihydropyridines (108).

d. FLUORESCENCE SPECTRA. Fluorescence spectroscopy has had limited value as a method for structure determination of dihydropyridines (109–112). This technique has, however, been applied to the study of the conformation of the NADH molecule. The energy transfer from the adenine to the dihydropyridine ring as evidenced by emission at 465 nm with excitation at 340 nm (dihydropyridine) or 260 nm (adenine) indicates a close proximity of the two rings in solution (113). Similar energy transfer was observed with NADH analogs in which the amide was replaced by acetyl; however, conversion of the adenine NH_2 to OH or a change in the stereochemistry of the glycosidic linkage of the sugar caused decrease in the fluorescence on 260-nm irradiation (114).

e. MASS SPECTROSCOPY. The fragmentation of dihydropyridines under electron bombardment has been shown to be of value in determining the nature of substituents attached at the sp^3 carbon of dihydropyridines (115–117). This method has not proven useful in distinguishing among the isomeric dihydropyridine systems (118, for example).

B. *Chemical Properties*

The importance of the 1,4-dihydropyridine system of NADH has resulted in an extensive study of the reactions of this dihydropyridine isomer. The early confusion over the structure of NADH, that is, 1,2- versus 1,4-dihydropyridine ring, has created some confusion over the reactivity of these systems. The reactivity of silver ion, 2,4-dinitrophenylhydrazine reagent, and dienophiles are notable examples (119, 120). The qualitative distinctions between these isomeric forms is most conveniently made by physical methods. In addition to providing valuable information about the important biological oxidation-reduction process the study of the reactions of dihydropyridines has offered a valuable synthetic intermediate in heterocyclic synthesis.

a. OXIDATION OR DEHYDROGENATION REACTIONS. The conversion of NADH to NAD by the enzymatic reduction of a ketone has not been duplicated in generality in a nonenzymatic process. The aldehyde function of 1,10-phenanthroline-2-carboxaldehyde was reduced by 1-propyl-1,4-dihydronicotinamide on catalysis with zinc ion to give a model system (121). Electron-poor ketones such as hexachloroacetone (40), pyruvic acid (96, 122), thiobenzophenone (123), benzoylformic acid (124), 3-arylideneindolinenes (125, 126), and various quinones (96, 120, 127-129) have also been used as model systems to convert 1-benzyl-1,4-dihydronicotinamide to the 1-benzyl-3-aminocarbonylpyridinium salts; however, the reduction is not general.

The oxidation may occur by a loss of a hydride ion from the dihydropyridine or by two one-electron transfers, the reverse of the reduction mechanism discussed in Section I.2.B. Removal of hydride ion from dihydropyridines by the cycloheptatrienyl cation (130) or triphenylmethyl cation (131) certainly must occur by a single two electron oxidation, while the reaction of dihydropyridines with hexachloroacetone in nonpolar solvents (121) to form tetra- and pentachloroacetone, with bromotrichloromethane to form chloroform (132), and with chloroform (133) and 2-sulfhydrylbenzophenone (134) to form pyridinium salts all must occur by pyridinyl radical intermediates *via* two one-electron transfers. The photochemical aromatizations of 3-benzoyl-4-phenyl-1,4-dihydropyridine (135), 2,6-dimethyl-4-*o*-nitrophenyl-3,5-disubstituted-1,4-dihydropyridines (136), and 3,5-diacetyl-2,6-dimethyl-1,4-dihydropyridine (137) also probably occur by a nonionic route. The aromatization reactions of dihydropyridines have been reviewed (1). Representative reagents which cause aromatization are the oxides of nitrogen (11, 138), potassium ferricyanide (139), pyridine-1-oxide (140), chromic acid (141), sulfur (142), palladium catalyst (143, 144), *p*-nitroso-*N,N*-dimethylaniline (145), hydrogen peroxide (146), silver nitrate (119, 120), mercuric acetate (147), iodine (97), *N*-bromosuccinimide (131), disulfides (148), cerric ammonium nitrate (131a), nitrobenzene (149), and cupric ion (149). The variety of methods for preparing pyridines from dihydropyridines in conjunction with the reaction of nucleophiles with pyridines to form dihydropyridines gives a synthetic sequence to substituted pyridines not easily accessible otherwise (131).

Although evidence can be presented to support ionic and radical mechanisms for the chemical conversion of dihydropyridines to pyridines, the question of the mechanism of conversion of NADH to NAD is still uncertain.

b. REACTION OF THE UNSATURATED SYSTEM. The unsaturated systems of the 1,2- and 1,6-dihydropyridines are dienamines and conjugated dienes, and the compounds give reactions of both classes. The 1,4-dihydropyridines can function only as enamines and cannot give diene reactions in a concerted process (15).

(1) ***Enamine Reactions.*** The enamine systems of all the dihydropyridines have been shown to undergo reaction with electrophiles. The most reactive system results from the addition of an organometallic reagent to a pyridine giving the anion of the 1,2-dihydropyridine. These reagents undergo facile reaction with electrophiles to give 2,5-dialkyl- or 2-alkyl-5-acylpyridines **(eq. IB-14)** (50, 150–153). The anion of the 1,2-dihydropyridine formed by the reduction of pyridine with lithium aluminum hydride (154) or sodium in liquid ammonia undergoes reaction with electrophiles to form 3-substituted pyridines (51, 155). The intermediate dihydropyridines that are formed must undergo oxidation by air or disproportionation (50, 51), for the substituted pyridine is isolated unless precautions are observed to obtain the dihydropyridine (50). The anions of 1,4-dihydropyridines are alkylated on the nitrogen rather than at carbon (155a).

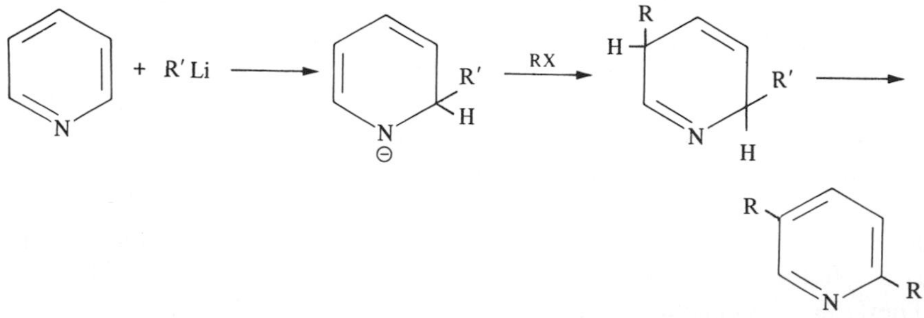

+ R′Li → RX →

eq. IB-14

The reactivity of the dienamine systems at the center of the conjugated function, the 5-position of a 1,2-dihydropyridine, was also observed when the electrophile was a proton. The reduction of pyridinium salts with sodium borohydride in the presence of deuterium oxide gave 1,2,3,6-tetrahydropyridine derivatives labeled at the 3-position as the only product (156).

The enamine function of the 1-substituted-1,4-dihydropyridines is much less reactive toward electrophiles (97); however, characteristic reactions have been reported. The reaction of 1-trimethylsilyl-1,4-dihydropyridine with *p*-toluenesulfonyl isocyanate gave the substituted 1,4-dihydronicotinamide **(eq. IB-15)** (143). The more universal enamine reactivity is observed on reaction with proton acids. The iminium salt formed by *C*-protonation at the 5-position of 3-substituted-1,4-dihydropyridines then acts as an electrophile on reaction with a second molecule of 1,4-dihydropyridine, forming dimeric structures **(eq. IB-16)** (44, 100, 157). Some pyridinium salts behave as electrophiles and form partially reduced bipyridyls on reaction with dihydropyridines (100). If an electron-rich functional group, such as an oxygenated aromatic ring, is present as a substituent on the 1,4-dihydropyridine or 1,2-dihydropyridine, protonation of the enamine at the 5-position can give rise to an intramolecular cyclization (41) **(eq. IB-17)**.

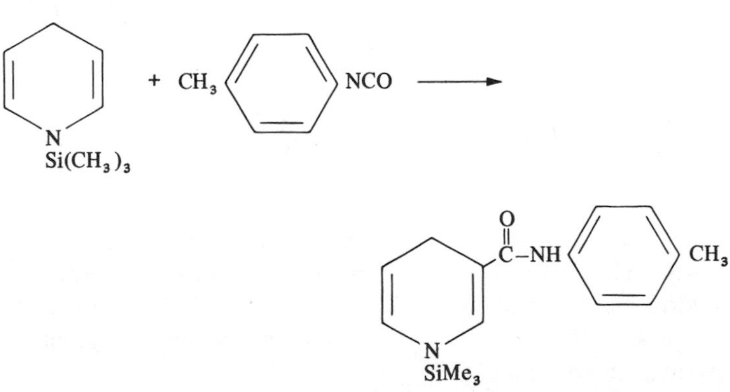

+ CH_3 NCO

N
$Si(CH_3)_3$

O
C–NH CH_3

N
$SiMe_3$

eq. IB-15

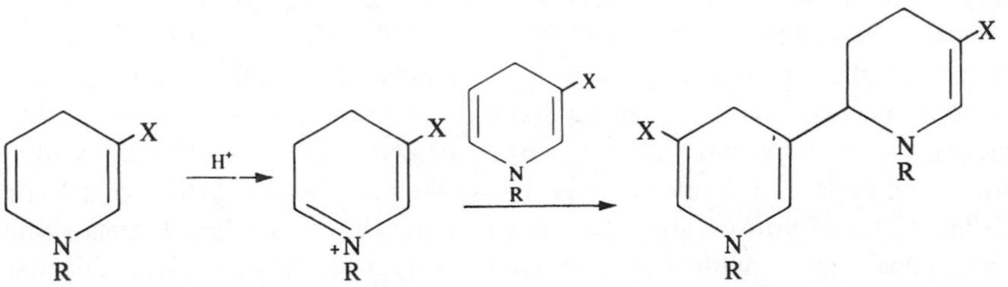

X H^+ X X X X

N R +N R N R N R N R

eq. IB-16

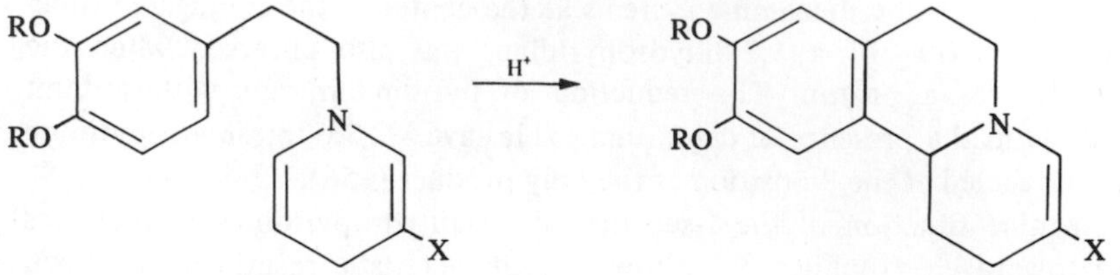

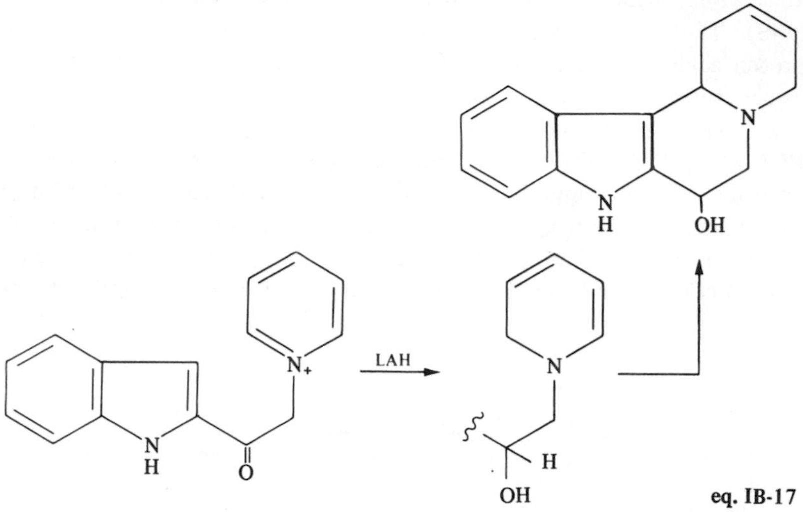

eq. IB-17

The enamine systems of the 1,4- and 1,2-dihydropyridines can react with a proton (17) or Lewis acid (158) to form an immonium salt, which is converted by sodium borohydride to a tetrahydropyridine (see below). The reduction of pyridinium salts with sodium borohydride thus produced tetrahydropyridines through dihydropyridine intermediates (17).

(2) ***Diene Reactions.*** The diene system of the 1,2-dihydropyridine undergoes the Diels-Alder reaction (93–95, 97, 159–161). The ease of reaction depends on the substitution pattern of the 1,2-dihydropyridine; substituents at the terminal carbons of the diene inhibit the reaction (14). The Diels-Alder reaction of methyl vinyl ketone with 1-benzyl-3-cyano-1,6-dihydropyridine provides an important intermediate in the synthesis of ibogaine (162) **(eq. IB-18)**.

1-Phenyl-1,2-dihydropyridine has been reported to act as a diene with *N*-phenylmaleimide. With the acetylenic dienophile, however, the enamine nature of the system prevails and a [4.2.0]-bicyclic intermediate forms and rearranges to a dihydroazocine (162a).

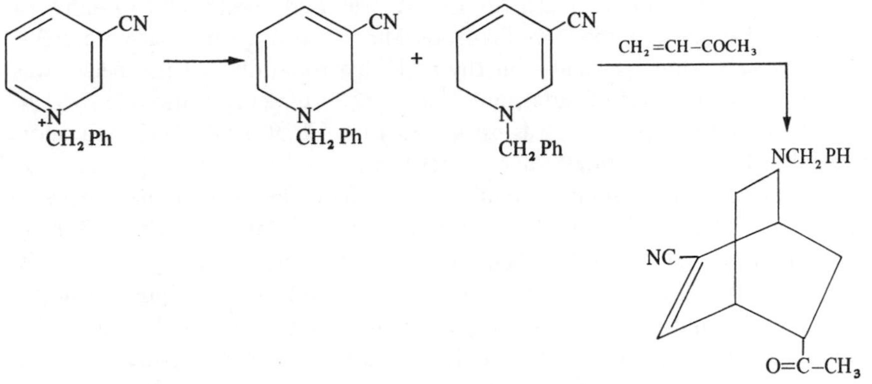

CN, CN, +, CN, $CH_2{=}CH{-}COCH_3$, N+ CH_2Ph, N CH_2Ph, N CH_2Ph, NCH_2PH, NC, $O{=}C{-}CH_3$

eq. IB-18

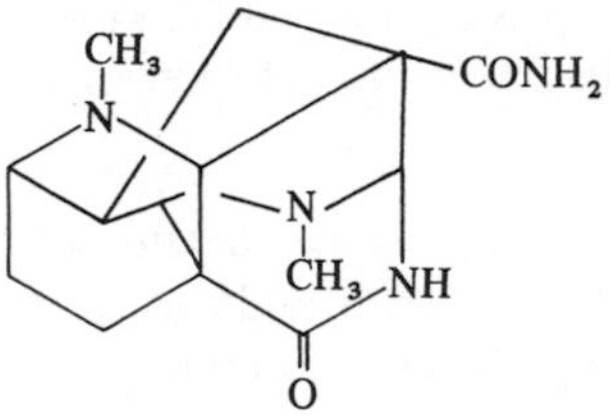

CH_3, N, CONH$_2$, N, CH_3, NH, O

IB-30

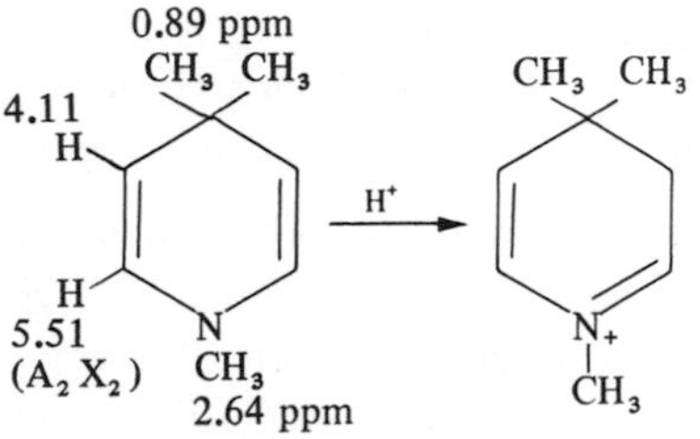

0.89 ppm, CH_3 CH_3, 4.11 H, H 5.51 (A_2X_2), N CH_3 2.64 ppm, H^+, CH_3 CH_3, N+ CH_3

IB-31 IB-32

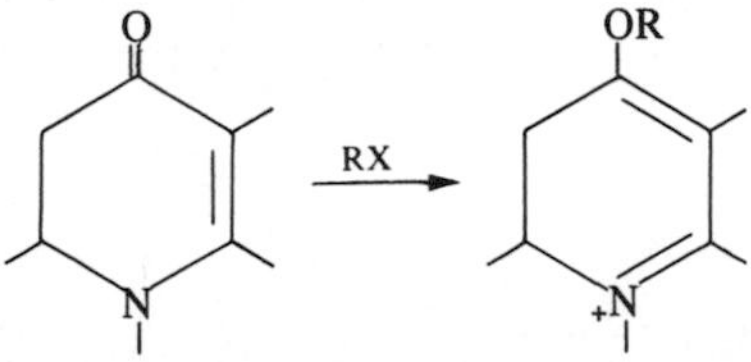

O, RX, OR, N, +N

IB-33a IB-33b

(3) ***The Relative Stabilities of Dihydropyridines.*** The enamine systems of the isomeric dihydropyridines do not appear to be in equilibrium with each other under ordinary conditions (17, 97, 156). The relative stabilities have been indicated by establishing equilibrium by indirect means, however. The rearrangement of cyanide addition compounds in nonpolar solvents suggested that the 1,4-dihydropyridine derivative was the thermodynamically favored (33, 34,

235). The equilibration of 1-trimethylsilyl-1,2- and 1-trimethylsilyl-1,4-dihydropyridines also showed the 1,4-dihydropyridine isomer to be the more stable (72). The 1-methyl-1,2- and 1-methyl-1,4-dihydropyridines have been equilibrated in the presence of potassium *t*-butoxide in dimethyl sulfoxide at 91.6° and the 1,4-dihydropyridine was estimated to be 2.29 ±0.01 kcal/mole more stable than the 1,2-dihydropyridine isomer (38).

Protonation of the enamine system of the isomeric dihydropyridines occurs on carbon and, of course, causes rearrangement of the double bonds (163–166). Protonation of the 5,6-double bond of a 1,4-dihydropyridine gives a highly electrophilic species which can undergo reaction with such nucleophiles as water (166) or with unprotonated 1,4-dihydropyridines to give dimers (44, 100, 157). The dimer of 1-methyl-1,4-dihydronicotinamide undergoes further enamine reactions to form the tetracyclic structure **(IB-30)** (157). Assignments of the structures of the dimers are based on x-ray and spectroscopic data. A series of 1,4-dihydropyridines which could not undergo tautomeric interconversion since the sp^3 carbons were dialkylated **(IB-31)** was prepared by Kosower for study. The structure of the protonated derivative **(IB-32)** was established (167).

Alkylation of the vinylogous amide systems **IB-33a** occurs at oxygen to give 2,3-dihydropyridines of the immonium salt type **(IB-33b)** (168).

c. REARRANGEMENTS OF THE RING. 1,4-Dihydropyridines having a CH_2X substituent, where X is electronegative, on the 4-sp^3 carbon undergo rearrangement to the seven-membered, dihydroazepine, ring on treatment with bases and this azepine then undergoes rearrangement to a cyclopentadiene (3, 142, 169–173) **(eq. IB-19)**. The analogous compounds having a carboxylic group at the 4-position give a complex series of rearrangement products on decarboxylation. These products include pyrroles (174). The analogous 1,2-dihydropyridines

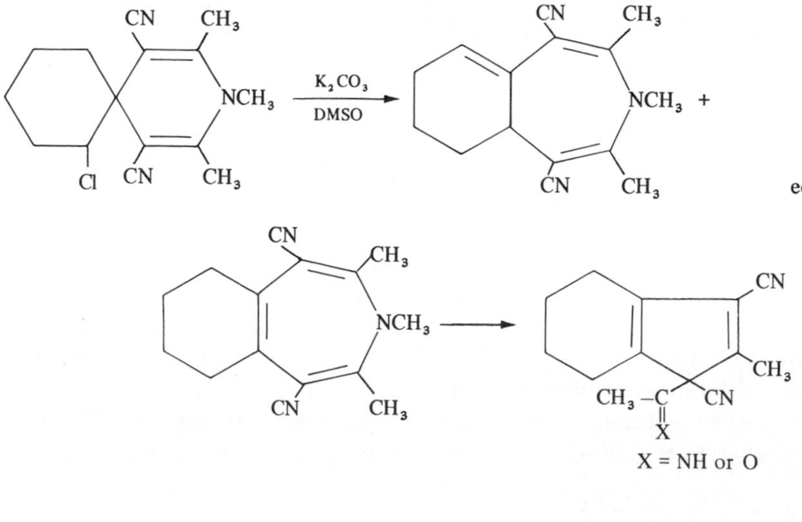

CN CH$_3$ / NCH$_3$ / Cl CN CH$_3$ $\xrightarrow[\text{DMSO}]{K_2CO_3}$ CN CH$_3$ / NCH$_3$ + / CN CH$_3$

eq. IB-19

CN CH$_3$ / NCH$_3$ / CN CH$_3$ ⟶ CN / CH$_3$ / $CH_3-\underset{\underset{X}{\|}}{C}$ CN

X = NH or O

also give this type of rearrangement to dihydroazepines (73). The course of the rearrangement appears to be very sensitive to the nature of the substituent on the 1-position and to reaction conditions (1b, 174).

The addition reactions to the dihydropyridines frequently produce tetrahydropyridines and thus will be discussed in the next section. Typical examples of the spectral data for dihydropyridines are given in Table IB-1.

II. Tetrahydropyridines

The saturation of two of the double bonds of pyridine or the dehydrogenation of piperidine can give one of three isomers, the Δ^1, Δ^2, or Δ^3-piperideines. The reactions of the Δ^1 **(IB-34)** and Δ^2 isomers **(IB-35)** are predictable on the basis of enamine-imine chemistry and are discussed together, since the reaction of the Δ^2-isomer with an electrophile gives the Δ^1-isomer, and removal of a proton from the Δ^1 form can produce a Δ^2-piperideine. The double bond of the Δ^3-piperideines **(IB-36)** behaves much the same as does an isolated double bond and thus the chemistry of these compounds will receive limited consideration. Excellent reviews of the first two systems describe the literature for 1,2,3,4- (175) and 1,2,3,6-tetrahydropyridines (176) through the mid-1960s.

1. Syntheses of Δ^1- (IB-34) and Δ^2-Piperideines (IB-35)

A. *Reduction of Pyridines*

In general, the reduction of the aromatic ring system of pyridine or its salts leads to other reduction products than the 1,2,3,4- or 2,3,4,5-tetrahydropyridines. If an unsaturated, electron-withdrawing group is attached to the 3-position of the parent pyridine, catalytic hydrogenation does lead to the 1,2,3,4-tetrahydro-5-substituted pyridine (76, 81, 177, 178); if, however, the 3-substituent is an acid derivative, facile decarboxylation occurs to give the

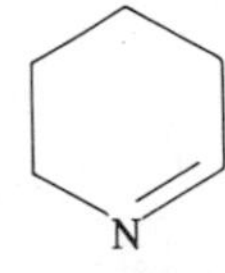

2,3,4,5-Tetrahydro
IB-34

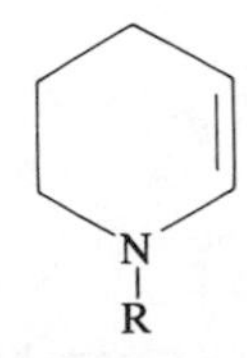

1,2,3,4-Tetrahydro
IB-35

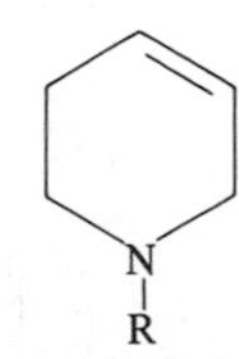

1,2,3,6-Tetrahydro
IB-36

parent Δ^2-piperideine (eq. IB-20) (179). The hydrogenation of pyridinium quaternary salts in basic medium give Δ^1- or Δ^2-piperideines which have been used as reaction intermediates (180, 181).

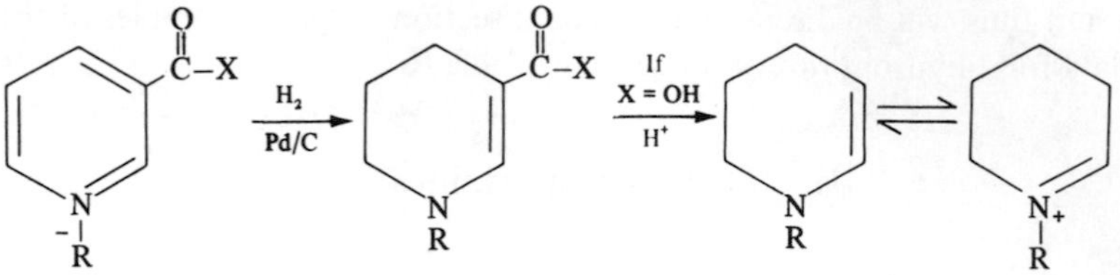

eq. IB-20

Metal hydride reductions do not generally give Δ^2-piperideines however. Reduction of 1-alkyl-2-piperidones or *N*-methyl glutarimides with sodium and alcohol and an electrolytic method respectively gave Δ^2-piperideines (eq. IB-21) (182).

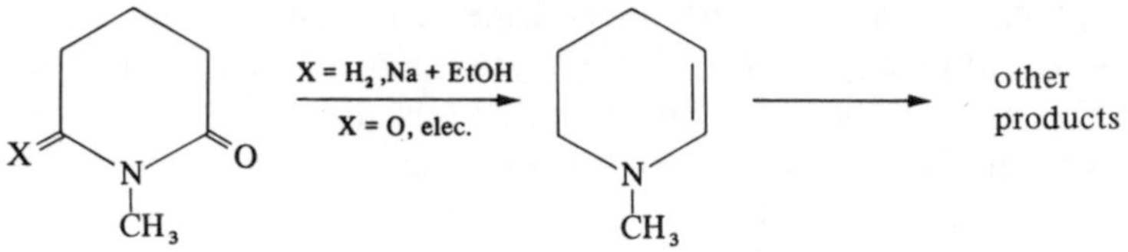

eq. IB-21

B. *Cyclization*

All of the isomeric piperideines can be synthesized by cyclization reactions; however, the reactions of δ-aminoketones (eq. IB-22) (183) or compounds that are converted to δ-aminoketones give Δ^1- or Δ^2-piperideines (175). The alkylation of an anionic enamine system with a 1,3-dihalopropane also gives Δ^2-piperideines (eq. IB-23) (184), and the Diels-Alder reaction of the type shown in eq. IB-24 is reported to give Δ^1-piperideines (185).

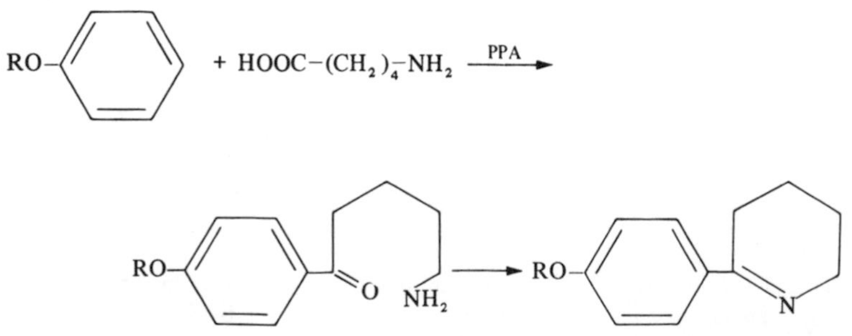

eq. IB-22

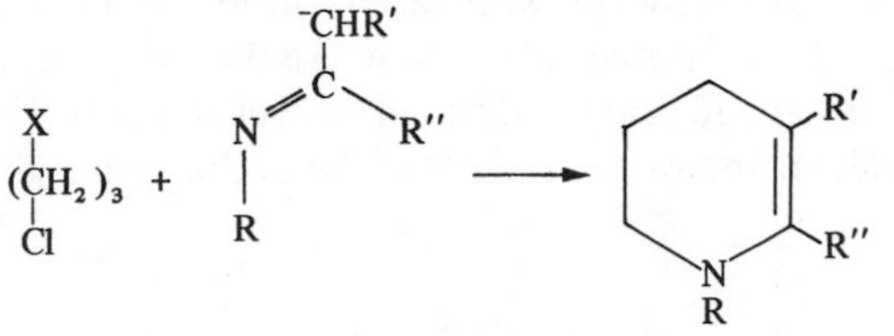

eq. IB-23

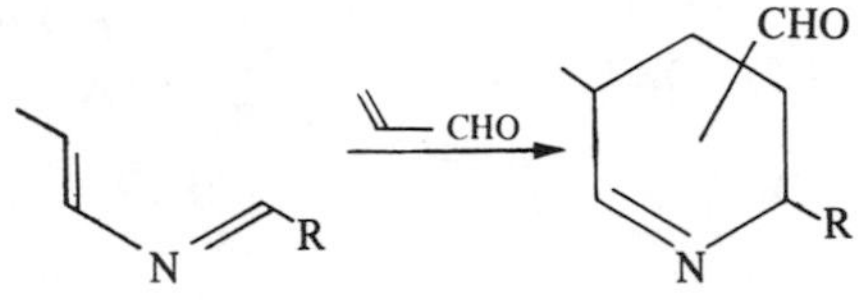

eq. IB-24

C. *Elimination Reaction*

The elimination of a simple molecule between the 1,2-positions of a substituted piperidine provides a method of obtaining Δ^1-(**IB-34**) and Δ^2-piperideines (**IB-35**). The elimination may be caused by loss, as an anion, of an electronegative group with the bonding electrons from the nitrogen or from the 2-position. The means of obtaining the electronegative substituent ranges from *N*-chlorination with *N*-chlorosuccinimide (186) (**eq. IB-25**), oxidative acetoxylation with mercuric acetate (187-189), or addition of a Grignard reagent to a 2-piperidone (190, 191) (**eq. IB-26**). The enamine of 1-substituted-3-piperidones was reported to give unstable Δ^2-piperideines (192).

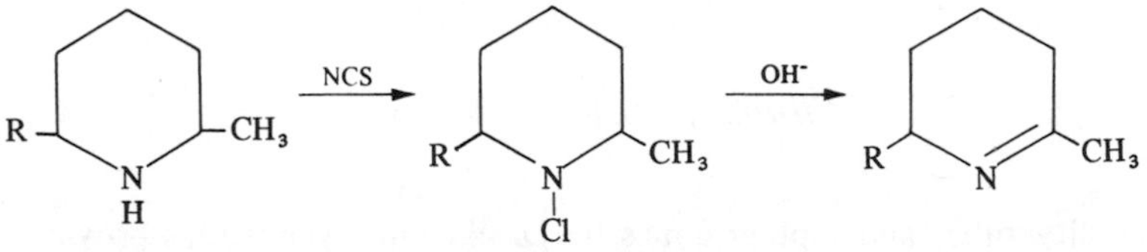

eq. IB-25

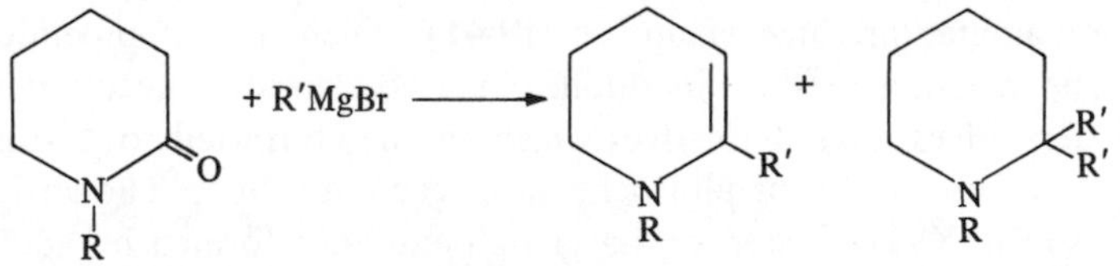

eq. IB-26

The elimination of hydrogen by oxidation and dehydrogenation have produced the Δ^1- or Δ^2-piperideine or a reaction product. Thus 6-acetyl-1,2,3,4,-tetrahydropyridine (**IB-38**), the aroma of bread, was produced by the oxidation of 1-(2-piperidyl)ethanol **(IB-37)** with silver carbonate on Celite (193). The treatment of the 1-substituted piperidine **(IB-39)** with hydrogen peroxide followed by ferrous ion and acid catalyst led to cyclization to **IB-40** indicative of a Δ^1- or Δ^2-piperideine intermediate (194). The cyclization is discussed later.

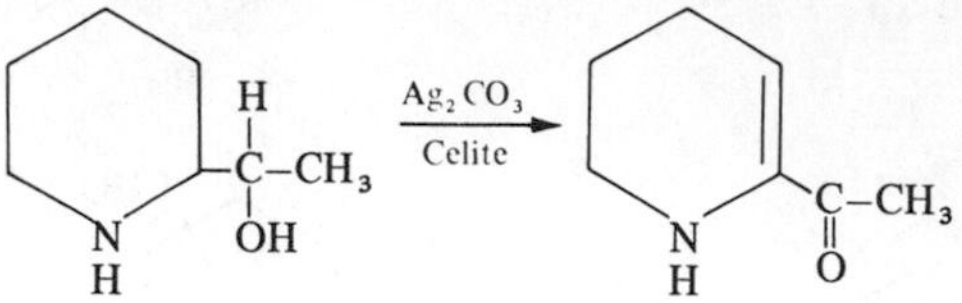

Ag_2CO_3, Celite

IB-37 **IB-38**

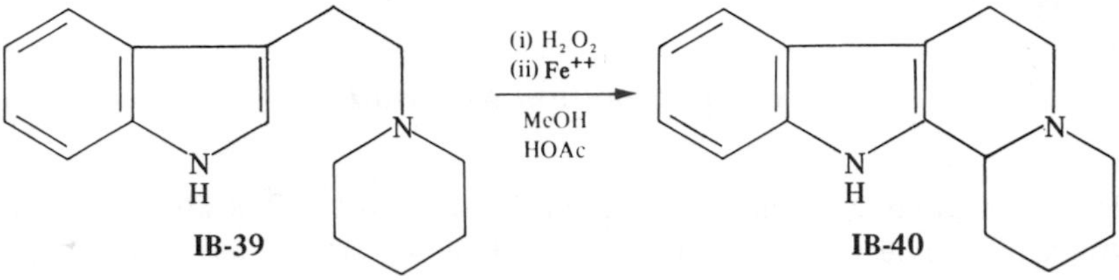

(i) H_2O_2 (ii) Fe^{++}, MeOH, HOAc

IB-39 **IB-40**

2. Syntheses of Δ^3-Piperideines

Δ^3-Piperideines, 1,2,3,6-tetrahydropyridines, can be formed by elimination reactions from substituted piperidines, partial reduction of pyridines, or cyclization reactions.

A. *Elimination Reactions*

The availability of 3- and 4-piperidones by Dieckmann cyclizations provides several routes to Δ^3-piperideines. The reactions of 1-substituted-4-piperidones with secondary amines produce enamines **(IB-41)** which are Δ^3-piperideines (192, 195). The reaction of the piperidones in a condensation reaction with malonic acid derivatives gives derivatives which can be converted to Δ^3-piperideines **(IB-42)** by bases (196) or photochemical reaction (196c). The analgesic activity (see also Ch. XVI) of esters of 4-aryl-4-piperidinols, formed by addition of aryl Grignard or aryllithium reagents to the piperidone, has given a series of piperidinols which can be dehydrated (197) readily to give 4-aryl-Δ^3-piperideines **(IB-43)**. These are also prepared by a Prins reaction from substituted α-methylstyrenes, formaldehyde and primary amines **(eq. IB-27)** (198). The reaction of the ketone with phenol in the presence of an acidic catalyst gives an electrophilic substitution of the phenol to give a 4-aryl-Δ^3-piperideine **(eq. IB-28)** (199). Heating the tosylate of 1-acetyl-3-piperidinol gave 96% of the Δ^3-piperideine by elimination (197e).

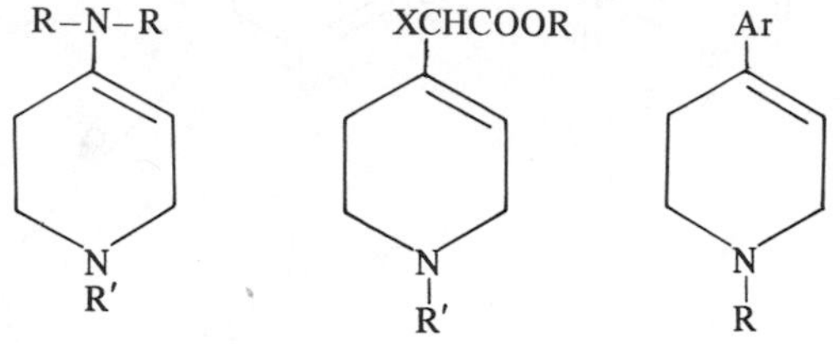

R–N–R XCHCOOR Ar

N N N
R′ R′ R

IB-41 **IB-42** **IB-43**

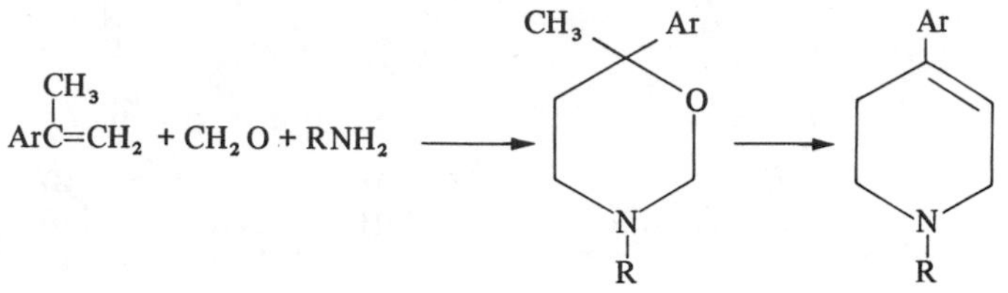

CH_3 Ar Ar
CH_3 O
ArC=CH_2 + CH_2O + RNH_2 ⟶ ⟶ **eq. IB-27** 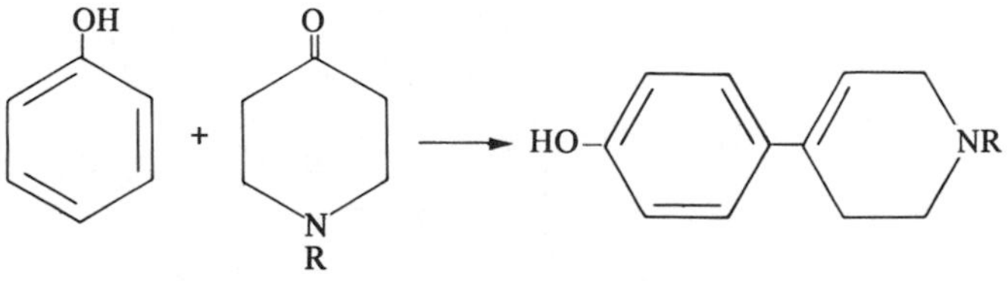
N N
R R

OH O
\+ ⟶ HO– NR **eq. IB-28**
N
R

B. *Reduction Reactions*

The reductions of pyridines and their salts with hydrides was reviewed (17) and the formation of Δ^3-piperideines by other reductions was included in a more recent review (176). These results, in general, show that 1,2-dihydropyridines formed by the methods described under Section I.2.A are protonated by aqueous or alcoholic solvents and further reduced to the 1,2,3,6-tetrahydropyridine (eq. **IB-29**). Thus reduction of pyridines with sodium and alcohol (1) gives a mixture of piperidine and Δ^3-piperideine. The reduction of pyridines with aluminum hydride from $LiAlH_4$ + $AlCl_3$ is a better method for forming Δ^3-piperideines (200). The reaction of 1-alkylpyridinium salts with sodium borohydride usually gives the Δ^3-piperideine as the sole or major product by a reaction sequence shown in eq. **IB-29** (17). Only with large substituents on nitrogen, which cause the initial reaction of the hydride ion to occur at the 4-position, does the reaction lead to piperidines (eq. **IB-30**) (201).

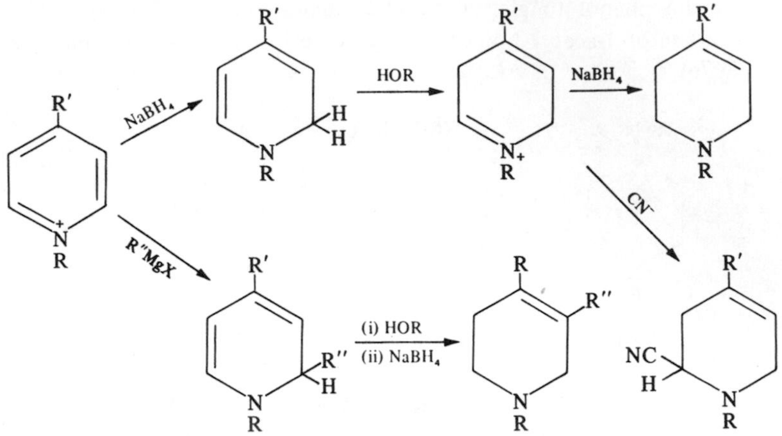

eq. IB-29

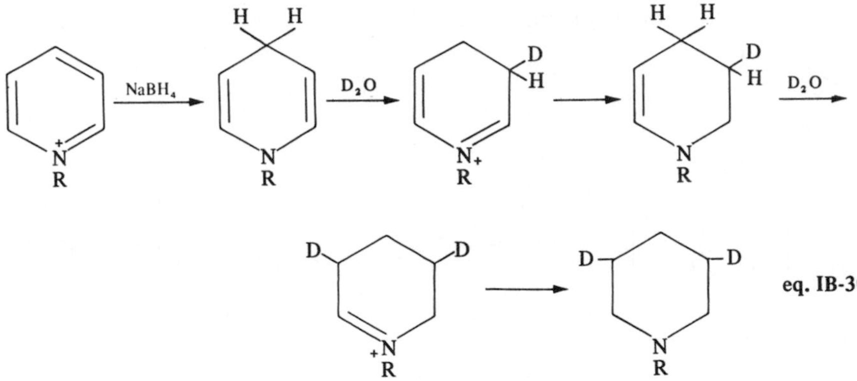

eq. IB-30

Formic acid causes the reduction of pyridinium salts to Δ^3-piperideines by a mechanism quite similar to that of the borohydride reduction (176); however, the yields are usually lower (202, 203). The reduction of 1-substituted-2-pyridones with lithium aluminum hydride also gives Δ^3-piperideines in low yields (203), as did reduction of pyridine-1-oxide with aluminum hydride (203a).

C. *Cyclization*

The Diels-Alder reaction of butadiene with acylated aldehyde aminals, methylenediethylimmonium chloride, or ketimines gives 1,2,3,6-tetrahydropyridines (eq. **IB-31**) (204–206). The quaternary salt of 3-hydroxypyridine has been shown to react as a 1,3-dipole with dipolarophiles such as acrylonitrile to give 3-tropen-2-ones which are bicyclic Δ^3-piperideines (eq. **IB-32**) (207) (see also Chap. IA).

$ClCH_2NEt_2 \longrightarrow CH_2{=}\overset{+}{N}Et_2 \longrightarrow$ [1,1-diethyl-1,2,3,6-tetrahydropyridinium, $\overset{+}{N}Et_2$] eq. IB-31

[3-oxidopyridinium, N-R] $\longleftrightarrow$ [N-R pyridinone zwitterion] $\xrightarrow{CH_2{=}CHCN}$ [RN-bridged bicyclic adduct with NC, H, H and C=O] eq. IB-32

3. Properties of Δ^1- and Δ^2-Piperideines

A. *Chemical Properties*

The reactions of these tetrahydropyridines are correlated by the anticipated reactions of enamines. The reaction of electrophiles at the 3-position creates an electrophilic site at the 2-position in the immonium salt. These intermediates readily add nucleophiles to give the saturated, substituted piperidine.

a. PROTONATION REACTIONS. Deuterium exchange at the 3-position indicates the reactivity of this site towards electrophiles (208). The isolated perchlorate salts, however, may have either the ammonium or immonium salt structure, depending on the substitution pattern of Δ^2-piperideine. A 2-substituent stabilizes the immonium salt form, although in solution an equilibrium exists (209). Sodium borohydride reductions of pyridinium ions occur to some extent by this route and give piperidines (201).

b. CYCLIZATION REACTIONS. The cyclic enamine in the Δ^1- or Δ^2-piperideine form is highly reactive. Trimerization occurs rapidly according to **eq. IB-33** (210). The most ingenious use of the enamine character of the Δ^2-piperidines in the synthesis of alkaloids has been made by Wenkert and his co-workers (81, 177, 179). In general, the 1-(β-arylethyl)-Δ^2-piperideine is prepared by one of the methods described under Section II.a. The enamine then undergoes protonation by the solvent or catalyst, and the resulting immonium salt gives electrophilic substitution of an electron-rich aromatic ring. The preparation of an intermediate in the synthesis of eburnamonine illustrates this method (**eq. IB-34a**) (177). Cyclization may occur by reaction at the 3-position of the indole ring followed by rearrangement (211). The cyclization can occur with a Δ^2-piperideine attached by a chain to the 3-position of the indole, as is most common, or the 2-position **(eq. IB-34b)** (212). The rearrangement of a Δ^3-piperideine, formed by sodium borohydride reduction of the 4-acetylpyridinium ion, to the Δ^2-isomer followed by cyclization has been reported to give dasycarpidone as well as various 2-acyloctahydroindole [2,3-*a*] quinolizine derivatives **(eq. IB-35)** (213). The nucleophile which adds to complete the cyclization does not have to be an indole or benzenoid ring. The cyclization of **IB-44** to the ketal of cyanoquinalizadinone **(IB-45)** probably involves an enolic double bond as the nucleophile (214). Similarly, the annelation of 1-benzyl-5-ethyl-1,2,3,4-tetrahydropyridine **(IB-46)** with methyl vinyl ketone to form the decahydro-7-quinolone **(IB-47)** must involve a carbanion-like character for the methyl group to achieve cyclization (215).

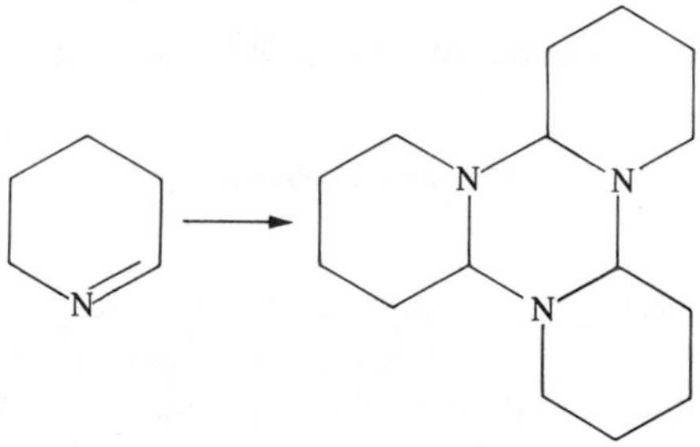

N → N N N

eq. IB-33

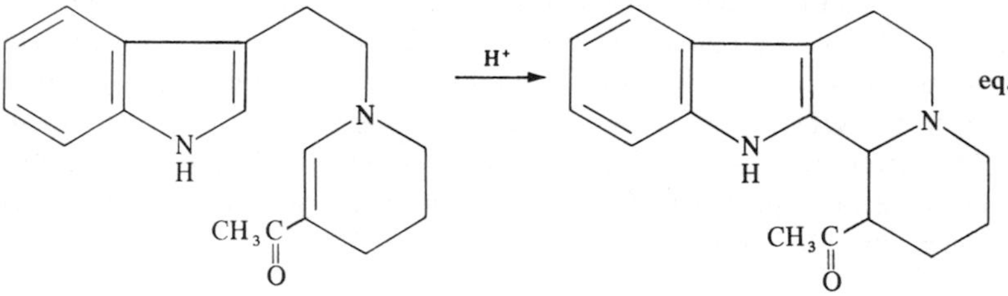

N H, N, CH_3C O → (H^+) N H, N, CH_3C O

eq. IB-34a

N, N H, $CONH_2$ → (H^+,H_2 + Cat.) $H_2\overset{+}{N}$, N H, $CONH_2$

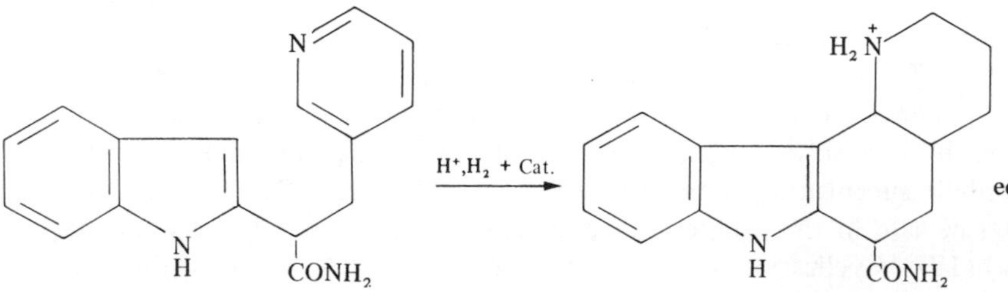

eq. IB-34b

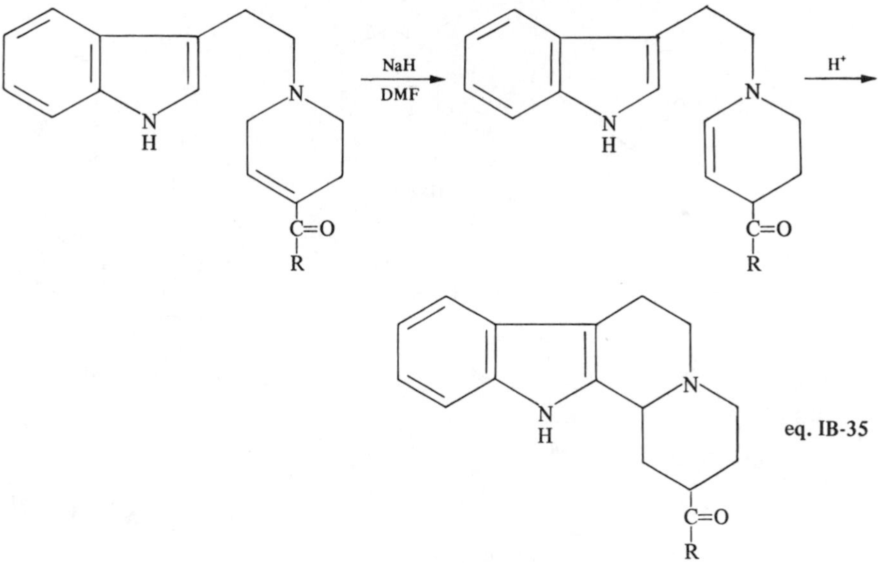

NaH / DMF ; H^+

eq. IB-35

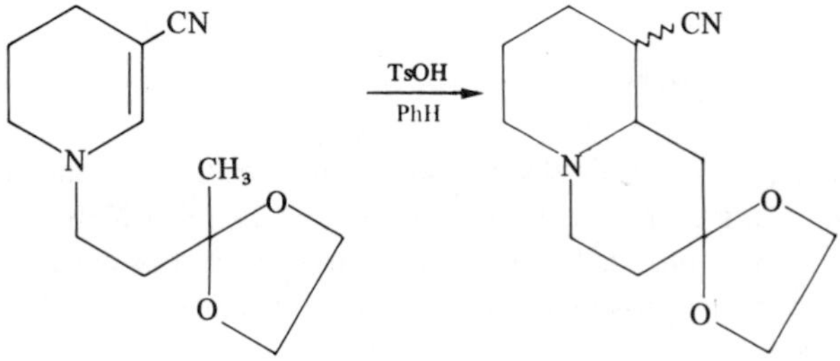

TsOH / PhH

IB-44 IB-45

The Δ^2-piperideine **(IB-46)** undergoes cyclization in a Diels-Alder-like reaction in which the heterocyclic double bond functions as a dieneophile **(eq. IB-36)**, although the mechanism of the cyclization may well be a two-step reaction of an enamine type (216). It is significant to note that the cyclization does not occur with the indole nitrogen of **IB-48** unprotected.

c. ALKYLATIONS AND ACYLATIONS. The enamine system of 1-benzyl-5-ethyl-1,2,3,4-tetrahydropyridine **(IB-46)** was shown to undergo alkylation even though the nucleophilic 5-position was already substituted. The intermediate immonium salt was reduced with sodium borohydride to the piperidine **IB-49** before isolation (215).

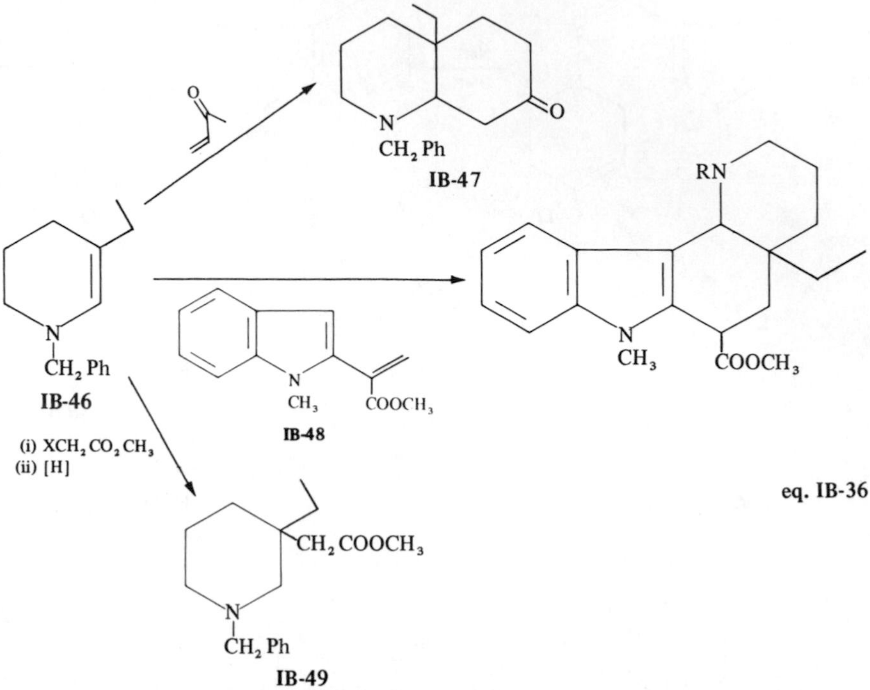

CH$_2$Ph

IB-47

RN

N
CH$_3$ COOCH$_3$

N
CH$_2$Ph
IB-46

N
CH$_3$ COOCH$_3$

IB-48

(i) XCH$_2$CO$_2$CH$_3$
(ii) [H]

eq. IB-36

CH$_2$COOCH$_3$

N
CH$_2$Ph

IB-49

The possibility that tetrahydropyridines can function as nucleophiles in the Δ^2-piperideine form, or as electrophiles in the Δ^1-piperideine form so easily formed by protonation, leads to self-condensation in the presence of a proton source. This approach provided a convenient synthesis of an anabasine derivative **(eq. IB-37)** (180).

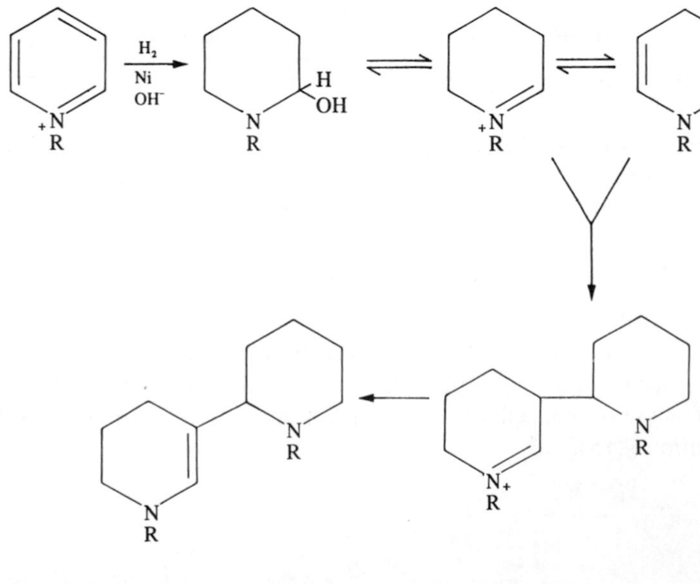

H_2 Ni OH^-

H OH

eq. IB-37

The reaction of the Δ^2-piperideine with acetyl chloride leads to *C*-acylation at the β-carbon if a 1-substituent is present (eq. IB-38) (217). The ring is cleaved if a nitrogen substituent is not present.

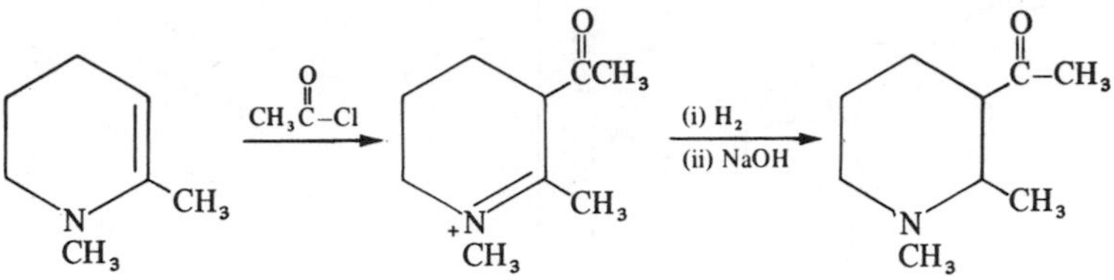

$CH_3\overset{O}{\overset{\|}{C}}{-}Cl$; (i) H_2 (ii) NaOH

eq. IB-38

B. *Physical Properties*

The enamine character of the 1,2,3,4-tetrahydropyridines (Δ^2-piperideines) has prompted investigation of the spectral properties of this system. These properties are very sensitive to the nature of substituents located at the terminus of the enamine, the 5-position. Electron-attracting unsaturated groups cause considerable delocalization of the nonbonding electrons of nitrogen into the π-system.

a. BASICITY. The delocalization of the nitrogen electrons and partial carbanion character of the carbon of an enamine leads to an increase in basicity. Thus the pK_a's of 1-substituted Δ^2-piperideines are significantly larger than those of the saturated derivatives (Table IB-3).

b. ULTRAVIOLET ABSORPTION SPECTRA. The extended π-system of the enamine group of the Δ^2-piperideines leads to lower energy, longer wavelength, absorption bands than those observed with saturated piperidines or Δ^3-piperideines. The presence of an unsaturated substituent at the end of the enamine function causes an absorption at even longer wavelength (Table IB-3).

c. INFRARED SPECTRA. The polarity of the enamine system causes the C=C stretching vibration to give rise to an intense absorption in the 1600 to 1680 cm^{-1} region (Table IB-3). The frequency of this absorption seems to be sensitive to substituents on the enamine system and can be used for structural studies.

d. NUCLEAR MAGNETIC RESONANCE SPECTRA. The chemical shifts of protons attached directly to the enamine system or to substituents on this system reflect the electron density to be predicted on the basis of resonance forms that can be drawn. Clearly this spectral technique can distinguish unequivocally among the various isomeric structures (Table IB-4).

TABLE IB-3. A Comparison of Basicities, Ultraviolet Spectra, and Infrared Spectra of Tetrahydropyridines

	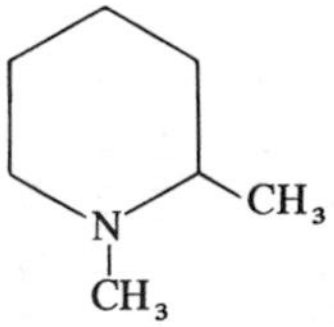 N, CH_3, CH_3	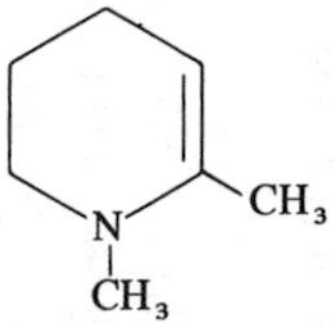 N, CH_3, CH_3	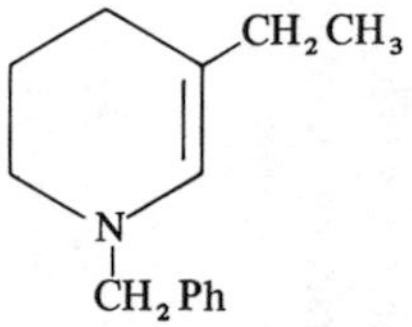 N, CH_2Ph, CH_2CH_3	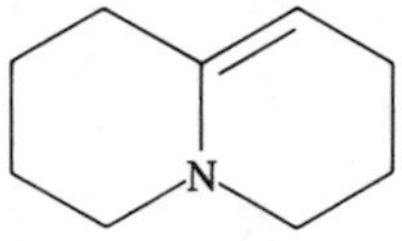 N	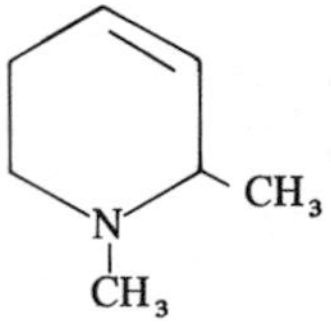 N, CH_3, CH_3	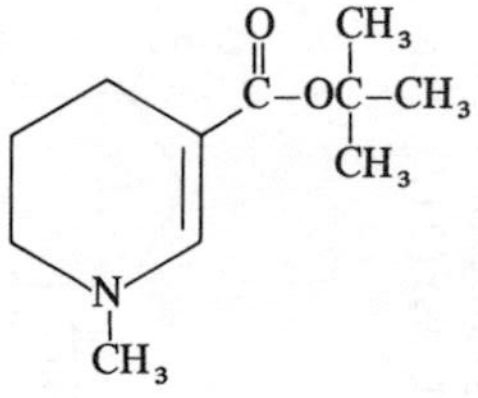 N, CH_3, $C(=O)-OC(CH_3)_3$
pK_a	10.26[a](10.17)[f]	11.43[a](11.12)[f]			9.43[f]	
u.v. (nm)	214	231[c]		228	213.5	295
(log ϵ)	(3.45)[g]	(3.68)[e]		(3.75)[c]	(3.30)[e]	(4.25)[h]
i.r. (cm^{-1})		1650[b], 1653[f]	1675[d]	1652[g]		1685, 1632[h]

[a] Ref. 237.
[b] Ref. 238.
[c] Ref. 239.
[d] Ref. 240.
[e] Ref. 241.
[f] Ref. 242.
[g] Ref. 175.
[h] Ref. 179.

TABLE IB-4. NMR Spectra of Partially Reduced Pyridines

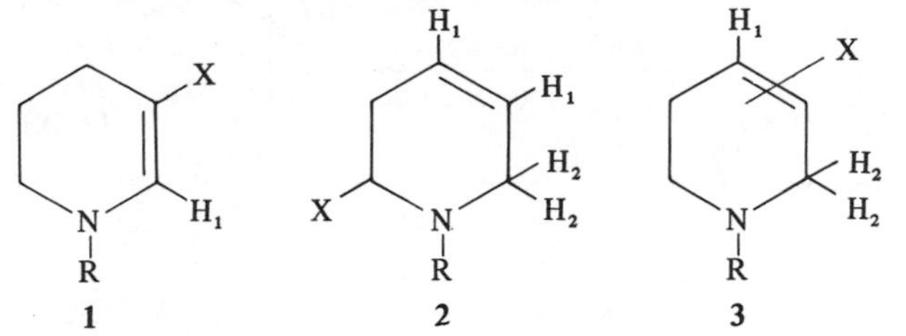

Type	R	X	H_1 (ppm)	X(ppm)	H_2 (ppm)	Ref.
1	$PhCH_2$	C_2H_5	5.60	1.80(2*H*)		215
1	CH_3	CO_2-*t*-Bu	7.21	1.48(9*H*)		179
1	H	CO_2-*t*-Bu	7.37	1.47(9*H*)		179
1	CH_3	CH_3	5.42	1.51(3*H*)		236
2	CH_3	CH_3	5.54	1.01(3*H*)	2.85	236
3	CH_3	4-CH_3	5.25	1.64	2.72	236
3	CH_3	3-CH_3	5.27	1.61	2.64	236

4. Properties of Δ^3-Piperideines

A. *Chemical Properties*

The reactions of 1,2,3,6-tetrahydropyridines (Δ^3-piperideines) can be characterized as those to be expected of an isolated double bond and as secondary or tertiary amine. A few specific reactions are discussed to illustrate this reactivity.

a. ADDITION OF HX TO THE DOUBLE BOND. Hydration of the double bond with aqueous acid of 1,2,3,6-tetrahydropyridines has been reported (198). Hydroboration has been widely investigated (218). The simple 1-alkyl Δ^3-piperideine gave mixtures of 3- and 4-piperidinols, with the former predominating to the extent of about 73% (219, 220). A similar product distribution was observed on hydroboration of methyl 1-benzyloxycarbonyl-1,2,3,6-tetrahydropicolinate (**eq. IB-39**) (221). Hydroboration of 4- or 5-substituted-1,2,3,6-tetrahydropyridines follow the usual orientation of *cis*-addition, with boron preferring the less substituted position (222, 223).

Protonation of the double bond of the Δ^3-piperideines gives carbonium ions capable of electrophilic substitution of an electron rich aromatic ring. The intramolecular model of this reaction has been used to synthesize the morphine antagonists of the benzomorphan system (**IB-56**) (224).

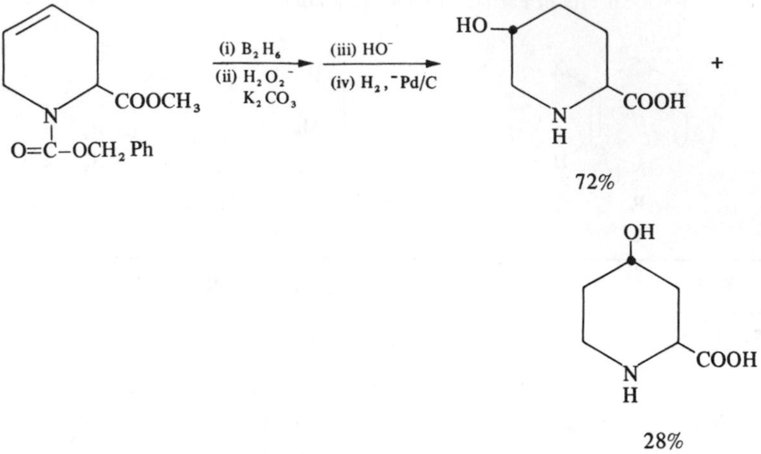

HO, COOCH$_3$, N, O=C–OCH$_2$Ph, (i) B_2H_6, (ii) $H_2O_2^-$ K_2CO_3, (iii) HO^-, (iv) H_2, $^-$Pd/C, N H COOH 72% +, OH, N H COOH 28%

eq. IB-39

b. ADDITION OF XY TO THE DOUBLE BOND. The presence of the basic nitrogen as part of the ring leads to complications in some addition reactions. Epoxidation of the unsaturation with a peracid may give an *N*-oxide during epoxidation, but the protection of the heterocyclic nitrogen as an amide allows epoxidation to the 3,4-epoxide **(IB-51)** (225). The opening of the oxirane ring of **IB-51** gives the *trans*-1-benzoyl-3,4-dihydroxypiperidine **(IB-51a)**. The Δ^3-piperideines undergo addition of carbenes to give [4.1.0] bicyclic products (225a).

The addition of bromine under acidic conditions occurs to give the 3,4-dibromopiperidine in a normal manner (226). A 4-phenyl substituent provides stability to the 4-carbonium ion and thus bromine addition in aqueous medium gives 3-bromo-1-methyl-4-phenyl-4-piperidinol hydrobromide **(IB-52)** and, in acetic acid, gives elimination of hydrogen bromide to 3-bromo-1-methyl-4-phenyl-1,2,3,6-tetrahydropyridine hydrobromide **(IB-53)** on reaction with 1-methyl-4-phenyl-1,2,3,6-tetrahydropyridine **(IB-54)**. The reaction of **IB-52** with base gives the 3,4-epoxide **IB-55** (227).

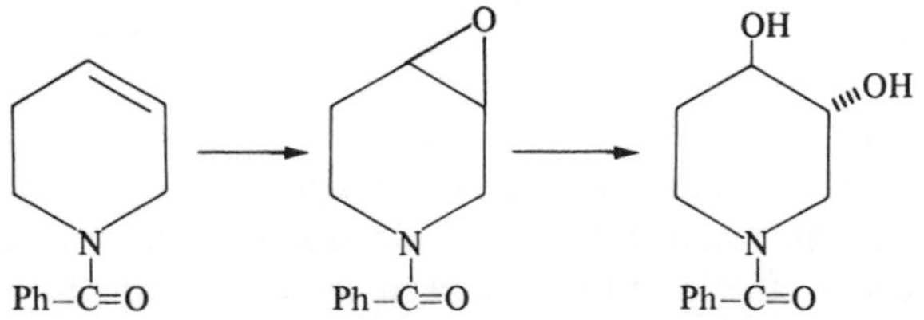

N Ph–C=O **IB-50** → O N Ph–C=O **IB-51** → OH OH N Ph–C=O **IB-51a**

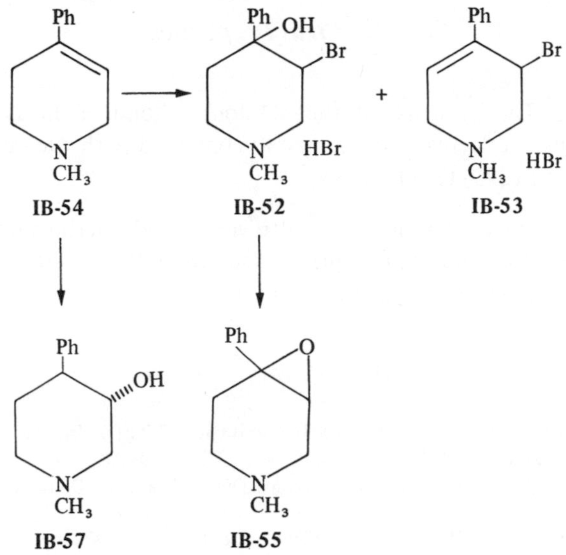

IB-54 IB-52 IB-53

IB-57 IB-55

c. ALLYLIC REACTIVITY AT THE 6-POSITION OF THE 1,2,3,6-TETRAHYDROPYRIDINES. The von Braun reaction of 1-methyl-1,2,3,6-tetrahydropyridine with cyanogen bromide might be expected to give cleavage of the ring since the 6-position is allylic. The loss of the 1-methyl group seems to be the major course of the reaction, however (228).

The quaternary salts of Δ^3-piperideines having a benzylic group on the nitrogen undergo a Stevens rearrangement with a base such as phenyllithium to give a 2-benzyl-Δ^3-piperideine (**eq. IB-40**) which can be cyclized to benzomorphans (**IB-56**) (224).

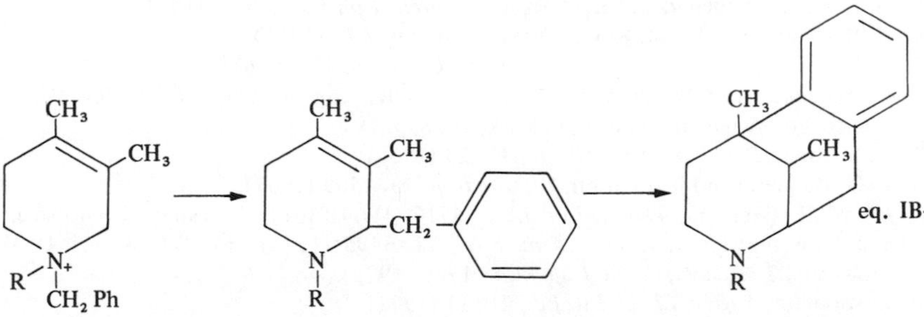

eq. IB-40

IB-56

B. *Physical Properties*

a. BASICITY. The presence of the 3,4-double bond in the Δ^3-piperideines (pK_a ~9.5) causes a decrease in basicity as compared with the saturated piperidine (pK_a ~10.5) (229) (Table IB-3).

b. SPECTRAL DATA. The infrared, ultraviolet, and nuclear magnetic spectra of the Δ^3-piperideines are not exceptional (see Table IB-4). The mass spectra are affected by the allylic amine system, however.

III. References

1. (a) E. Klingsberg, Ed., "Pyridine and Its Derivatives," Part 1, Interscience, New York, 1960 (b) U. Eisner and J. Kuthan, *Chem. Rev.*, **72**, 1 (1972).
2. G. Schroll, S. Nygaard, S. Lawesson, A. M. Duffield, and C. Djerassi, *Ark. Kemi,* **29,** 525 (1968).
3. M. Mahendran and A. W. Johnson, *J. Chem. Soc. C,* 1237 (1971).
4. B. Loev and K. Snader, *J. Org. Chem.*, **30**, 1914 (1965).
5. J. Palecek, K. Vondra, and J. Kuthan, *Collect. Czech. Chem. Commun.*, **34**, 2991 (1969).
6. J. F. Biellmann and H. J. Callot, *Tetrahedron,* **26**, 4799 (1970).
7. A. Sims and P. Smith, *Proc. Chem. Soc.*, 282 (1958).
8. W. Traber and P. Karrer, *Helv. Chim. Acta,* **41**, 2066 (1958).
8a. N. Sugiyana, G. Inouye, and K. Ito, *Bull. Soc. Chem. Jap.*, **35**, 927 (1962).
9. (a) M. Ohashi, H. Kamachi, H. Kakisawa, and G. Stork, *J. Amer. Chem. Soc.*, **89**, 5460 (1967); (b) G. Stork, M. Ohashi, H. Kamachi, and H. Kakisawa, *J. Org. Chem.*, **36**, 2784 (1971).
10. A. Sakurai and H. Midorikawa, *Bull. Chem. Soc. Jap.*, **42**, 220 (1969).
11. J. Palecek and J. Kuthan, *Collect. Czech. Chem. Commun.*, **34**, 3336 (1969).
12. A. de Savignac and A. Lattes, *Bull. Soc. Chim. Fr.*, 4476 (1970).
13. W. Zeiss and A. Schmidpeter, *Tetrahedron Lett.*, 4229 (1972).
13a. A. Sakurai, Y. Motomura, and H. Midorikawa, *Bull. Chem. Soc. Jap.*, **46**, 973 (1973).
14. (a) D. Craig, L. Schaefgen, and W. P. Tyler, *J. Amer. Chem. Soc.*, **70**, 1624 (1948); (b) D. Craig, A. Kuder, and J. Efroymson, *ibid.*, **72**, 5236 (1950).
15. G. Krow, E. Michener, and K. C. Ramey, *Tetrahedron Lett.*, 3653 (1971).
16. H. B. Charman and J. M. Rowe, *Chem. Commun.*, 476 (1971).
17. R. Lyle and P. Anderson, *Adv. Heterocycl. Chem.*, **6**, 45 (1966).
18. E. Schenker, in "Newer Methods of Preparative Organic Chemistry," Vol. IV, M. Foerst, Ed., Academic Press, New York, 1968, p. 197.
19. R. F. Evans, *Rev. Pure Appl. Chem.*, **15**, 23 (1965).
20. L. G. Duquette and F. Johnson, *Tetrahedron,* **23**, 4539 (1967).
21. (a) W. H. Okamura, *Tetrahedron Lett.*, 4717 (1969); (b) S. R. Tanny, J. Grossman, and F. W. Fowler, *J. Amer. Chem. Soc.*, **94**, 6495 (1972); (c) A. I. Meyers, D. M. Stout, and T. Takaya, *Chem. Commun.*, 1260 (1972).
22. L. Paquette, *Angew. Chem. Int. Ed.*, **10**, 11 (1971).
23. G. Illuminati, *Adv. Heterocycl. Chem.*, **3**, 285 (1964).

24. R. A. Abramovitch and J. G. Saha, *Adv. Heterocycl. Chem.*, **6**, 229 (1966).
25. K. Thomas and D. Jerchel, in "Newer Methods of Preparative Organic Chemistry," Vol. III, W. Foerst, Ed., Academic Press, New York, 1964, p. 53.
26. F. H. Westheimer, in "The Enzymes," Vol. 1, 2nd ed., P. D. Boyer, H. Lardy and K. Myrback, Eds., Academic Press, New York, 1959, p. 278.
27. M. E. Pullman, A. San Pietro, and S. P. Colowick, *J. Biol. Chem.*, **206**, 129 (1954).
28. M. E. Pullman and S. P. Colowick, *J. Biol. Chem.*, **206**, 121 (1954).
29. F. A. Loewus, B. Vennesland, and D. L. Harris, *J. Amer. Chem. Soc.*, **77**, 3391 (1955).
30. R. F. Hutton and F. H. Westheimer, *Tetrahedron*, **3**, 73 (1958).
31. H. E. Dubb, M. Saunders, and J. H. Wang, *J. Amer. Chem. Soc.*, **80**, 1767 (1958).
32. R. N. Lindquist and E. H. Cordes, *J. Amer. Chem. Soc.*, **90**, 1269 (1968).
33. R. E. Lyle and G. Gauthier, *Tetrahedron Lett.*, 4615 (1965).
33a. V. Mann, G. Schneider, and F. Kröhnke, *Tetrahedron Lett.*, 683 (1973).
34. R. Foster and C. A. Fyfe, *Tetrahedron*, **25**, 1489 (1969).
35. K. Wallenfels and W. Hanstein, *Ann. Chem.*, **732**, 139 (1970).
36. J. Biellmann and H. Callot, *Bull. Soc. Chim. Fr.*, 1154 (1968).
37. E. Kosower, "Molecular Biochemistry," McGraw-Hill, New York, 1962.
38. F. W. Fowler, *J. Amer. Chem. Soc.*, **94**, 5926 (1972).
39. (a) K. Wallenfels and H. Schuly, *Angew. Chem.*, **70**, 471 (1958); (b) *idem.*, *Ann. Chem.*, **621**, 178 (1959).
40. D. C. Dittmer and R. A. Fouty, *J. Amer. Chem. Soc.*, **86**, 91 (1964).
41. (a) J. H. Supple, D. A. Nelson, and R. E. Lyle, *Tetrahedron Lett.*, 1645 (1963); (b) K. T. Potts and H. G. Shin, *Chem. Commun.*, 857 (1966).
42. K. Wallenfels and H. Schuly, *Ann. Chem.*, **621**, 178 (1959).
43. S. Yamada, M. Kuramoto, and Y. Kikugawa, *Tetrahedron Lett.*, 3101 (1969).
44. R. E. Lyle and D. A. Nelson, *J. Org. Chem.*, **28**, 169 (1963).
45. W. E. McEwen, R. H. Terss, and I. W. Elliott, *J. Amer. Chem. Soc.*, **74**, 3605 (1952).
46. A. S. Bailey, M. C. Chum, and J. J. Wedgwood, *Tetrahedron Lett.*, 5953 (1968).
47. H. Deubel, D. Wolkenstein, H. Jokisch, T. Messerschmitt, S. Brodka, and H. von Dobeneck, *Chem. Ber.*, **104**, 705 (1971).
48. H. von Dobeneck and W. Goltzsche, *Chem. Ber.*, **95**, 1484 (1962).
49. J. Bergman, *J. Heterocycl. Chem.*, **7**, 1071 (1970).
49a. A. Piskorska-Chlebowska, *Rocz. Chem.*, **46**, 2341 (1971); *Chem. Abstr.*, **79**, 32196 (1973).
49b. P. J. Beeby and F. Sondheimer, *Angew. Chem.*, **85**, 404, 406 (1973); P. J. Beeby, J. M. Brown, P. J. Garratt, and F. Sondheimer, *Tetrahedron Lett.*, 599 (1974).
50. C. S. Giam and J. L. Stout, *Chem. Commun.*, 142 (1969); 478 (1970); *Tetrahedron Lett.*, 4961 (1971).
51. C. S. Giam and S. D. Abbott, *J. Amer. Chem. Soc.*, **93**, 1294 (1971).
52. R. A. Abramovitch and G. A. Poulton, *Chem. Commun.*, 274 (1967).
52a. G. Fraenkel and J. C. Cooper, *Tetrahedron Lett.*, 1825 (1968).
53. G. Fraenkel, J. W. Cooper, and C. M. Fink, *Angew. Chem. Int. Ed.*, **9**, 523 (1970).
54. R. E. Lyle and E. White V, *J. Org. Chem.*, **36**, 772 (1971).
55. E. L. May and E. Fry, *J. Org. Chem.*, **22**, 1366 (1957).
56. E. Fry, *J. Org. Chem.*, **28**, 1869 (1963).
56a. L. M. Thiessen, J. A. Lepoivre, and F. C. Alderweireldt, *Tetrahedron Lett.*, 59 (1974).
57. J. Frohlich and F. Kröhnke, *Chem. Ber.*, **104**, 1621 (1971).
58. (a) R. M. Wilson and F. DiNinno, *Tetrahedron Lett.*, 289 (1970); (b) F. DiNinno and W. Heckle, *Tetrahedron Lett.*, 2639 (1972).
59. T. Kato, H. Yamanaka, T. Adachi, and H. Hiranuma, *J. Org. Chem.*, **32**, 3788 (1967).

60. T. J. van Bergen and R. M. Kellogg, *J. Org. Chem.*, **36**, 1705 (1971).
60a. B. A. Mikrut, F. H. Hershenson, K. F. King, L. Bauer, and R. S. Egan, *J. Org. Chem.*, **36,** 3749 (1971) and references cited therein.
61. E. M. Kosower and I. Schwager, *J. Amer. Chem. Soc.*, **86** 4493 (1964).
62. E. M. Kosower and L. Lindquist, *Tetrahedron, Lett.*, 4481 (1965).
63. (a) E. M. Kosower and P. E. Klinedinst, *J. Amer. Chem. Soc.*, **78**, 3493, 3497 (1956); (b) M. Mohammad and E. Kosower, *J. Amer. Chem. Soc.*, **93**, 2709 (1971).
64. (a) L. J. Winters, A. L. Borror, and N. Smith, *Tetrahedron Lett.*, 2313 (1967); (b) L. J. Winters, N. G. Smith, and M. I. Cohen, *Chem. Commun.*, 642 (1970).
65. B. Emmert, *Ber.*, **42**, 1997 (1909).
66. J. N. Burnett and A. L. Underwood, *J. Org. Chem.*, **30**, 1154 (1965).
67. S. J. Leach, J. H. Baxendale, and M. G. Evans, *Aust. J. Chem.*, **6**, 395 (1953).
67a. M. Ferles and A. Attia, *Collect. Czech. Chem. Commun.*, **38,** 2747 (1973).
68. P. Atlani, J. F. Biellmann, R. Briere, H. Lemaire, and A. Rassat, *Tetrahedron,* **28,** 2827 (1972).
69. J. P. Wibaut and S. Vromen, *Rec. Trav. Chim. Pays-Bas,* **67**, 545 (1948).
70. J. E. Colchester and J. H. Entwisle, U. S. Patent 3,478,042 (1969); *Chem. Abstr.*, **72,** 31627a (1970).
71. R. A. Sulzbach, *J. Organometal. Chem.*, **24**, 307 (1970).
71a. R. A. Sulzbach and A. F. M. Iqbal, *Angew. Chem. Int. Ed.*, **10,** 733 (1971).
72. N. C. Cook and J. E. Lyons, *J. Amer. Chem. Soc.*, **87**, 3283 (1965); **88**, 3396 (1966).
73. T. J. van Bergen and R. M. Kellogg, *J. Org. Chem.*, **36**, 978 (1971).
74. M. Natsume and M. Wada, *Tetrahedron Lett.*, 4503 (1971).
75. M. Freifelder, *Adv. Catal.*, **14**, 203 (1963).
76. R. E. Lyle and E. Mallett, *Ann. N. Y. Acad. Sci.*, **145**, 83 (1967).
77. O. Mumm and J. Diederichsen, *Ann. Chem.*, **538**, 195 (1939).
78. U. Eisner, *Chem. Commun.*, 1348 (1969).
79. F. Bohlmann and M. Bohlmann, *Chem. Ber.*, **86**, 1419 (1953).
80. R. Lukes and J. Kuthan, *Collect. Czech. Chem. Commun.*, **25**, 2173 (1960).
81. E. Wenkert, K. Dave, F. Haglid, R. Lewis, T. Oishi, R. Stevens, and M. Terashima, *J. Org. Chem.*, **33**, 747 (1968).
82. B. Pullman and A. Pullman, "Quantum Biochemistry," Wiley, New York, 1963.
83. E. M. Evleth, *J. Amer. Chem. Soc.*, **89**, 6445 (1967).
84. J. Kuthan, *Collect. Czech. Chem. Commun.*, **32**, 1280 (1967).
85. J. Kuthan and L. Musil, *Collect. Czech. Chem. Commun.*, **34**, 3173 (1969).
86. J. Kuthan and J. Palecek, *Collect. Czech. Chem. Commun.*, **34**, 1339 (1969).
87. J. Kuthan, A. Kohoutova, and L. Helesic, *Collect. Czech. Chem. Commun.*, **35**, 2776 (1970).
88. W. Hanstein and K. Wallenfels, *Tetrahedron,* **23**, 585 (1967).
89. P. J. Brignell, U. Eisner, and P. G. Farrell, *J. Chem. Soc. B,* 1083 (1966).
90. N. O. Kaplan and M. M. Ciotti, *J. Biol. Chem.*, **221**, 823 (1956).
91. D. A. Nelson, Ph.D. Dissertation, University of New Hampshire, 1960.
92. E. White, V, Ph.D., Dissertation, University of New Hampshire, 1968.
93. K. Schenker and J. Druey, *Helv. Chim. Acta,* **42**, 1960 (1959).
94. K. Schenker and J. Druey, *Helv. Chim. Acta,* **42**, 1971 (1959); **45**, 1344 (1962).
95. J. Kuthan and R. Bartonickova, *Collect. Czech. Chem. Commun.*, **30**, 2609 (1965).
96. D. Mauzerall and F. H. Westheimer, *J. Amer. Chem. Soc.*, **77**, 2261 (1955).
97. M. Saunders and E. H. Gold, *J. Org. Chem.*, **27**, 1439 (1962).
98. A. C. Lovesey and W. C. J. Ross, *J. Chem. Soc. B,* 192 (1969).
99. W. L. Meyer, H. R. Mahler, and R. H. Baker, *Biochim. Biophys. Acta,* **64**, 353 (1962).

100. D. L. Coffen, *J. Org. Chem.*, **33**, 137 (1968).
101. H. Diekmann, G. Englert, and K. Wallenfels, *Tetrahedron*, **20**, 281 (1964).
102. W. S. Caughey and K. A. Schellenberg, *J. Org. Chem.*, **31**, 1978 (1966).
103. K. Mislow and N. Raban, in "Topics in Stereochemistry," Vol. 1, N. Allinger and E. Eliel, Eds., Wiley, New York, 1967, pp. 22-37.
104. D. P. Hollis, *Org. Magn. Resonance*, **1**, 305 (1969).
105. R. H. Sarma and N. O. Kaplan, *Biochem. Biophys. Res. Commun.*, **36**, 780 (1969).
106. I. L. Karle, *Acta Cryst.*, **14**, 497 (1961).
107. H. Koyama, *Z. Kryst.*, **118**, 51 (1963).
108. B. Birdsall and J. Feeney, *J. Chem. Soc. Perkin II*, 1643 (1972).
109. N. Sugiyama, K. Kubota, G. Inouye, and T. Kubota, *Bull. Chem. Soc. Jap.* **37**, 637 (1964).
110. D. C. Dittmer and J. M. Kolyer, *J. Org. Chem.*, **28**, 2288 (1963).
111. J. Kuthan and E. Janeckova, *Collect. Czech. Chem. Commun.*, **29**, 1654 (1964).
112. J. Ludowieg, N. Bhacca and A. Levy, *Biochem. Biophys. Res. Commun.*, **14**, 431 (1964).
113. G. Weber, *Nature*, **180**, 1407 (1957).
114. S. Shifrin and N. O. Kaplan, *Nature*, **183**, 1529 (1959).
115. R. E. Lyle and E. White V, *Tetrahedron Lett.*, 1871 (1970).
116. G. Hristendahl and K. Undheim, *J.C.S. Perkin II*, 2030 (1972).
117. B. J. S. Wang and E. R. Thornton, *J. Amer. Chem. Soc.*, **90**, 1216 (1968).
118. I. Fleming and J. B. Mason, *J. Chem. Soc. C*, 2509 (1969).
119. P. Karrer, G. Schwarzenbach, F. Benz, and U. Solmssen, *Helv. Chim. Acta*, **19**, 811 (1936).
120. H. Kuhnis, W. Traber, and P. Karrer, *Helv. Chim. Acta*, **40**, 751 (1957).
121. D. J. Creighton and D. S. Sigman, *J. Amer. Chem. Soc.*, **93**, 6314 (1971).
122. K. Wallenfels and D. Hofmann, *Tetrahedron Lett.*, 10 (1959).
123. R. H. Abeles, R. F. Hutton, and F. Westheimer, *J. Amer. Chem. Soc.*, **79**, 712 (1957).
124. R. Abeles and F. H. Westheimer, *J. Amer. Chem. Soc.*, **80**, 5459 (1958).
125. R. W. Huffman and T. C. Bruice, *J. Amer. Chem. Soc.*, **89**, 6243 (1967).
126. K. A. Schellenberg and G. W. McLean, *J. Amer. Chem. Soc.*, **88**, 1077 (1966).
127. E. A. Braude, J. Hannah, and R. Linstead, *J. Chem. Soc.*, 3249 (1960).
128. J. Uldrikis, A. O. Kumerova, and G. Duburs, *Khim. Geterotsikl. Soedin.*, 691 (1973); *Chem. Abstr.*, **79**, 65498 (1973).
129. L. G. Duquette and F. Johnson, *Tetrahedron*, **23**, 4517 (1967).
130. O. M. Grishin and A. Yasnikov, *Ukr. Khim. Zh.*, **34**, 70 (1968); *Chem. Abstr.*, 69, 43241w (1968).
131. R. E. Lyle, C. Ferrando, and E. White, V, unpublished results, University of New Hampshire, 1970.
131a. S. Matsumoto, H. Masuda, K. Iwata, and O. Mitsunobu, *Tetrahedron Lett.*, 1733 (1973).
132. J. L. Kurz, R. Hutton, and F. H. Westheimer, *J. Amer. Chem. Soc.*, **83**, 584 (1961).
133. P. S. Anderson, Ph.D. Dissertation, University of New Hampshire, (1963).
134. K. A. Schellenberg and F. H. Westheimer, *J. Org. Chem.*, **30**, 1859 (1965).
135. D. A. Nelson and J. F. McKay, "Abstracts, 154th National Meeting of the American Chemical Society," Chicago, Ill., Sept. 1967, No. 523.
136. J. A. Berson and E. Brown, *J. Amer. Chem. Soc.*, **77**, 447 (1955).
137. U. Eisner, J. R. Williams, B. W. Matthews, and H. Ziffer, *Tetrahedron*, **26**, 899 (1970).
137a. J. F. Biellman, H. J. Callot, and W. R. Pilgrim, *Tetrahedron* **28**, 5911 (1972).
138. A. Singer and S. M. McElvain, *Org. Syn.*, **14**, 30 (1934).

139. O. M. Grishin and A. Yasnikov, *Zh. Obshch. Khim.*, **43,** 1342 (1973); *Chem. Abstr.*, **79,** 77779 (1973).
140. A. S. Kurbatova, Y. Kurbatov, O. Otroshchenko, and A. Sadykov, *Tr. Samarkand Gos. Univ.*, **167**, 26 (1969); *Chem. Abstr.*, **74**, 141474x (1971).
141. A. Courts and V. Petrow, *J. Chem. Soc.*, 1 (1952).
142. P. J. Brignell, E. Bullock, U. Eisner, B. Gregory, A. W. Johnson, and H. Williams, *J. Chem. Soc.*, 4819 (1963).
143. E. J. Moriconi and R. E. Misner, *J. Org. Chem.*, **34**, 3672 (1969).
144. A. Kamal, M. Ahmad, M. Mohd, and A. M. Hamid, *Bull Soc. Chem. Jap.*, **37**, 610 (1964).
145. H. Albrecht and F. Kröhnke, *Ann. Chem.*, **704,** 133 (1967).
146. G. Vanags and E. I. Stankevich, *Zh. Obshch. Khim*, **30**, 3287 (1960); *Chem. Abstr.*, **55**, 21119 (1961).
147. T. Kametani, K. Ogasawara, and A. Kozaka, *J. Pharm. Soc. Jap.*, **86**, 815 (1966).
148. C. H. Wang, S. Linnell, R. Rosenblum, and N. Wang, *Experientia*, **27**, 243 (1971).
149. D. C. Dittmer and J. Kolyer, *J. Org. Chem.*, **27**, 56 (1962), especially references therein.
150. R. A. Abramovitch and B. Vig, *Can. J. Chem.*, **41**, 1961 (1963).
151. P. Doyle and R. R. J. Yates, *Tetrahedron Lett.*, 3371 (1970).
152. L. D. Quin and D. O. Pinion, *J. Org. Chem.*, **35**, 3130, 3134 (1970).
153. R. Grashey and R. Huisgen, *Chem. Ber.*, **92**, 2641 (1959).
154. P. T. Lansbury and J. O. Peterson, *J. Amer. Chem. Soc.*, **83**, 3537 (1961); **84**, 1756 (1962); **85**, 2236 (1963).
155. F. J. Villani, C. A. Ellis, M. D. Yudis, and J. B. Morton, *J. Org. Chem.*, **36**, 1709 (1971).
155a. R. A. Dommisse and F. C. Alderweireldt, *Bull. Soc. Chim. Belges*, **82**, 441 (1973).
156. P. S. Anderson and R. E. Lyle, *Tetrahedron Lett.*, 153 (1964).
157. (a) H. Ammon and L. Jensen, *J. Amer. Chem. Soc.*, **88**, 613 (1966); (b) J. J. Steffens, J. P. Cross, and D. M. Chipman, *Tetrahedron Lett.* 4883 (1972).
158. F. Liberatore, V. Carelli, and M. Cardellini, *Tetrahedron Lett.*, 4735 (1968).
159. P. Baum, B. S. Thesis, University of New Hampshire, 1968.
160. T. Agawa and S. I. Miller, *J. Amer. Chem. Soc.*, **83**, 449 (1961).
161. R. A. Wiley, B. A. Faraj, A. Jantz, and M. M. Hava, *J. Med. Chem.*, **15**, 374 (1972).
162. G. Büchi, D. Coffen, K. Kocsis, P. Sonnet, and F. Ziegler, *J. Amer. Chem. Soc.*, **88**, 3099 (1966).
162a. R. M. Acheson and G. Paglietti, *Chem. Commun.*, 665 (1973).
163. M. Takeda, A. E. Jacobson, and E. L. May, *J. Org. Chem.*, **34**, 4161 (1969).
164. A. E. Jacobson and E. L. May, *J. Med. Chem.*, **7**, 409 (1964).
165. K. Wallenfels, D. Hofmann, and H. Schulz, *Ann. Chem.*, **621**, 188 (1959).
166. A. C. Anderson and G. Berkelhammer, *J. Amer. Chem. Soc.*, **80**, 992 (1958).
167. E. M. Kosower and T. S. Sorensen, *J. Org. Chem.*, **27**, 3764 (1962).
168. (a) A. I. Meyers, A. Reine, and R. Gault, *J. Org. Chem.*, **34**, 698 (1969); (b) A. I. Meyers and S. Singh, *Tetrahedron*, **25**, 4161 (1969).
169. R. C. Allgrove, L. A. Cort, J. A. Elvidge, and U. Eisner, *J. Chem. Soc. C*, 434 (1971).
170. G. B. Gill, D. J. Harper, and A. W. Johnson, *J. Chem. Soc. C*, 1675 (1968).
171. J. Ashby and U. Eisner, *J. Chem. Soc. C*, 1706 (1967).
172. R. C. Allgrove and U. Eisner, *Tetrahedron Lett.*, 499 (1967).
173. P. J. Brignell, U. Eisner, and H. Williams, *J. Chem. Soc.*, 4226 (1965).
174. J. F. Biellmann and H. J. Callot, *Tetrahedron*, **26**, 4809 (1970).
175. K. Blaha and O. Cervinka, *Adv. Heterocycl. Chem.*, **6**, 147 (1966).

176. M. Ferles and J. Pliml, *Adv. Heterocycl. Chem.*, **12**, 43 (1970).
177. E. Wenkert, *Accounts Chem. Res.*, **1**, 78 (1968).
178. M. Freifelder, *J. Org. Chem.*, **29**, 2895 (1964).
178a. H. B. Renfroe, *Ger. Offen.* 2,257,310 (1973); *Chem. Abstr.*, **79**, 78631g (1973).
179. E. Wenkert, K. G. Dave, and F. Haglid, *J. Amer. Chem. Soc.*, **87**, 5461 (1965).
180. C. Schopf, G. Herbert, R. Rausch, and G. Schröder, *Angew. Chem.*, **69**, 391 (1957).
181. J. Thesing and W. Festag, *Experientia*, **15**, 127 (1959).
182. R. Lukes and J. Kovar, *Collect. Czech. Chem. Commun.*, **19**, 1215, 1227 (1954).
183. W. Koller and P. Schlack, *Chem. Ber.*, **96**, 93 (1963).
184. D. A. Evans, *J. Amer. Chem. Soc.*, **92**, 7593 (1970).
185. G. Y. Kondrateva and Y. S. Dolskaya, *Zh. Org. Khim.*, **6**, 2200 (1970).
186. J. Parello, M. Rivierre, E. Desherces, and A. Lattes, *C. R. Acad. Sci., Ser. C, Paris*, **273**, 1097 (1971).
187. N. J. Leonard, A. S. Hay, R. W. Fulmer, and V. W. Gash, *J. Amer. Chem. Soc.*, **77**, 439 (1955).
188. N. J. Leonard and D. F. Morrow, *J. Amer. Chem. Soc.*, **80**, 371 (1958).
189. J. Gutzwiller, G. Pizzolato, and M. Uskokovic, *J. Amer. Chem. Soc.*, **93**, 5907 (1971).
190. R. Lukes and F. Sorm, *Collect. Czech. Chem. Commun.*, **12**, 356 (1947).
191. R. Lukes and Z. Vesely, *Collect. Czech. Chem. Commun.*, **25**, 2318 (1959).
192. S. Danishefsky and R. Cavanaugh, *J. Org. Chem.*, **33**, 2959 (1968).
192a. P. Stütz and P. A. Stadler, *Tetrahedron Lett.*, 5095 (1973).
193. G. Büchi and H. Wuest, *J. Org. Chem.*, **36**, 609 (1971).
194. C. A. Scherer, C. A. Dorschel, J. M. Cook, and P. W. LeQuesne, *J. Org. Chem.*, **37** 1083 (1972).
195. (a) H. Hirsch, *Chem. Ber.*, **100**, 1289 (1967); (b) A. Z. Britten and J. O'Sullivan, *Tetrahedron*, **29**, 1331 (1973).
196. (a) S. M. McElvain and R. E. Lyle, *J. Amer. Chem. Soc.*, **72**, 384 (1950); (b) L. D. Quin, J. W. Russell, R. D. Prince, and H. Shook, *J. Org. Chem.*, **36**, 1495 (1971); (c) R. J. Sundberg, L. S. Lin, and F. X. Smith, *J. Org. Chem.*, **38**, 2558 (1973); (d) E. Nikitskaya, E. Levkoeva, M. Rubtsov, and V. Usorskaya, USSR Patent 238,545 (1970); *Chem. Abstr.*, **74**, 13020j (1971).
197. (a) A. Ziering, L. Berger, S. D. Heineman, and J. Lee, *J. Org. Chem.*, **12**, 894 (1947); (b) S. M. McElvain and J. C. Safranski, *J. Amer. Chem. Soc.*, **72**, 3134 (1950); (c) S. Oshiro, *J. Pharm. Soc. Jap.*, **75**, 658 (1955); (d) A. F. Casy, A. H. Beckett, and M. A. Iorio, *Tetrahedron*, **23**, 1405 (1967); (e) J. L. Wong and D. O. Helton, *Chem. Commun.*, 352 (1973).
198. C. J. Schmidle and R. Mansfield, *J. Amer. Chem. Soc.*, **77**, 5698 (1955); **78**, 425 1702 (1956).
199. S. M. McElvain and R. S. Berger, *J. Amer. Chem. Soc.*, **77**, 2848 (1955).
200. A. Silhankova, M. Holik, and M. Ferles, *Collect. Czech. Chem. Commun.*, **33**, 2494 (1968).
201. P. S. Anderson, W. E. Krueger and R. E. Lyle, *Tetrahedron Lett.*, 4011 (1965).
202. R. Lukes and J. Pliml, *Collect. Czech. Chem. Commun.*, **24**, 2560 (1959).
203. M. Ferles and M. Holik, *Collect. Czech. Chem. Commun.*, **31**, 2416 (1966).
203a. M. Jankovsky and M. Ferles, *Collect. Czech. Chem. Commun.*, **35**, 2802 (1970).
204. R. Merten and G. Muller, *Angew. Chem.*, **74**, 866 (1962).
205. H. Bohme, K. Hartke, and A. Muller, *Chem. Ber.*, **96**, 607 (1963).
206. W. J. Middleton and C. G. Krespan, *J. Org. Chem.*, **33**, 3625 (1968).
207. A. R. Katritzky and Y. Takeuchi, *J. Amer. Chem. Soc.*, **92**, 4134 (1970).
208. W. H. Daly, J. G. Underwood, and S. C. Kuo, *Tetrahedron Lett.*, 4375 (1971).

209. N. J. Leonard and F. P. Hauck, *J. Amer. Chem. Soc.*, **79**, 5279 (1957).
210. C. Schopf, A. Komzak, F. Braun, and E. Jacobi, *Ann. Chem.*, **559**, 1 (1948).
211. E. E. van Temelen and G. C. Knapp, *J. Amer. Chem. Soc.*, **77**, 1860 (1955).
212. H. P. Husson, C. Thal, P. Potier, and E. Wenkert, *Chem. Commun.*, 480 (1970).
213. M. S. Allen, A. J. Gaskell, and J. A. Joule, *J. Chem. Soc. C*, 736 (1971).
214. E. Wenkert and A. R. Jeffcoat, *J. Org. Chem.*, **35**, 515 (1970).
215. F. E. Ziegler, J. A. Kloek, and P. A. Zoretic, *J. Amer. Chem. Soc.*, **91**, 2342 (1969).
216. F. E. Ziegler and E. B. Spitzner, *J. Amer. Chem. Soc.*, **92**, 3492 (1970).
217. O. Cervinka, *Collect. Czech. Chem. Commun.*, **25** 1174 (1960).
218. H. Kugita and M. Takeda, *Chem. Ind.* (London), 2099 (1964).
219. R. E. Lyle, K. R. Carle, C. R. Ellefson, and C. K. Spicer, *J. Org. Chem.*, **35**, 802 (1970).
220. R. E. Lyle and C. K. Spicer, *Tetrahedron Lett.*, 1133 (1970).
221. Y. Fujita, F. Irreverre, and B. Witkop, *J. Amer. Chem. Soc.*, **86**, 1844 (1964).
222. R. E. Lyle, D. H. McMahon, W. E. Krueger, and C. K. Spicer, *J. Org. Chem.*, **31**, 4164 (1966).
223. (a) Z. Polivak and M. Ferles, *Collect. Czech. Chem. Commun.*, **35**, 2392, (1970), **36**, 4099 (1971); (b) M. Ferles, J. Hauer, J. Kolář, Z. Polivka, and P. Štern, *ibid.*, **37**, 2464 (1972); (c) M. Ferles, P. Štern, P. Trška, and F. Vyšāta, *ibid.*, **37**, 1206 (1973); (d) M. A. Iorio, P. Ciuffa, and G. Damia, *Tetrahedron*, **26**, 5519 (1970).
224. "Synthetic Analgesics," J. Hellerbach, O. Schnider, H. Besendorf, and B. Pellmont, Part IIA Morphinans; N. B. Eddy and E. L. May, Part IIB 6,7-Benzomorphans, Permagon Press, Oxford, England, 1966, pp. 117–137.
225. V. Petrow and O. Stephenson, *J. Pharm. Pharmacol.*, **14**, 306 (1962).
225a. M. Ferles and A. H. Attia, *Z. Chem.*, **11**, 18 (1971).
226. R. Lukes, *Collect. Czech. Chem. Commun.*, **12**, 71 (1947).
227. R. E. Lyle and W. E. Krueger, *J. Org. Chem.*, **30**, 394 (1965); **32**, 3613 (1967).
228. M. Ferles and M. Prystas, *Collect. Czech. Chem. Commun.*, **24**, 3326 (1959).
229. J. Lakomy, A. Silhankova, M. Ferles, and O. Exner, *Collect. Czech. Chem. Commun.*, **33**, 1700 (1968).
230. The NMR was reported as 4.78(2); 4.43(3); 6.01(4); 4.68(5); 6.79(6). G. J. Dubsky and J. Guillarmod, *Helv. Chim. Acta*, **52** 1735 (1969).
231. J. Kuthan and E. Janeckova, *Collect. Czech. Chem. Commun.*, **30**, 3711 (1965).
232. J. Kuthan, J. Prochazkova, and E. Janeckova, *Collect. Czech. Chem. Commun.*, **30**, 3558 (1968).
233. W. Steglich and G. Hofte, *Chem. Ber.*, **102**, 1129 (1969).
234. F. W. Fowler, *J. Org. Chem.*, **37**, 1321 (1972).
235. K. Wallenfels and W. Hanstein, *Angew. Chem.*, **77**, 861 (1965). *Ann. Chem.*, **709**, 151 (1967).
236. Z. Ksandr, Z. Samek, V. Spirko, and M. Ferles, *Collect. Czech. Chem. Commun.*, **31**, 3003 (1966).
237. R. Adams and J. E. Mahon, *J. Amer. Chem. Soc.*, **64**, 2588 (1942).
238. V. Prelog and O. Haflizer, *Helv. Chim. Acta*, **32**, 1851 (1949).
239. N. J. Leonard and D. M. Locke, *J. Amer. Chem. Soc.*, 77, 437 (1955).
240. F. E. Ziegler, *J. Amer. Chem. Soc.*, **91**, 2342 (1969).
241. M. Holik, A. Tesarova, and M. Ferles, *Collect. Czech. Chem. Commun.*, **32**, 1730 (1967).
242. J. Lakomy, A. Silkanova, M. Ferles, and O. Exner, *Collect. Czech. Chem. Commun.*, **33** 1700 (1968).

CHAPTER II

Synthetic and Natural Sources of the Pyridine Ring

N. S. BOODMAN, J. O. HAWTHORNE, P. X. MASCIANTONIO, and A. W. SIMON

U. S. Steel Corporation, Research Laboratory, Monroeville, Pennsylvania

This chapter deals with the preparation of pyridine compounds from nonpyridine sources and is essentially divided into two sections. The first is concerned with naturally occurring materials from which pyridines are obtained either directly, by such methods as degradation, or by other treatment. The second section covers chemical synthesis from cyclic and acyclic nonpyridinoid compounds. The preparation of pyridine derivatives from pyridine compounds is not discussed. However, the biosynthesis of pyridine nucleotides has been included. Literature coverage includes references from 1956 through 1968 and supplements the previously published comprehensive discussion by Brody and Ruby (see 1st ed.).

I. Pyridines from Natural Sources

No significant new source of pyridines from natural sources has been reported. Although considerable research has been reported on sources of alkaloids and enzymes, the only natural source of pyridines results from the carbonization of coal. The quantity of pyridines produced in the United States during 1968 is reported by the United States Tariff Commission to have amounted to more than 10 million pounds. Potential new sources of pyridines from natural products may result from development work on new coal conversion processes to produce gasoline, liquefied coal, or coal gasification products. Continued development of oil shale might also give rise to new natural sources of pyridines.

1. Pyridines in Nature

The subject of pyridines in nature has been treated in many investigations. Recent studies have been particularly related to increased interest in the nature of tobacco and the effects of cigarette smoking on health. Several reviews on the naturally occurring pyridines have been published (123–125).

A. *Pyridine Enzymes and Pyridines in Living Organisms*

The pyridine enzymes have been identified in animal and plant tissues. Many reports continue to be published on the pyridine enzymes and the presence of pyridines in various living substances.

It has been shown (1) that the pyridine ring of ricinine is biosynthesized from succinate, propionate, β-alanine, and acetate derivatives of glycerol by injection of the respective ^{14}C–labeled compounds into the internodular space of young *Ricinus communis* plants.

Schenk and Shütte (2) have reported the presence of 4-aminopipecolinic acid in the alcoholic extract of *Strophanthus scandens*.

Pooled urine from dogs after administration of (–)-cotinine was separated on Dowex 50 resin, and the eluate was shown to contain a 3-pyridyl metabolite believed to be γ-(3-pyridyl)-β-oxo-*N*-methylbutyramide (3).

The pattern of urinary excretion of ^{14}C in the dog after intravenous injection of labeled (–)-nicotine was shown (4) by chromatographic procedures to include the metabolites cotinine, γ-(3-pyridyl)-γ-(methylamino)butyric acid, demethylcotinine, hydroxycotinine, and γ-(3-pyridyl)-β-oxo-*N*-methylbutyramide. Administration of (−)-cotinine to the dog gave 3-pyridine acetic acid in the urine.

Pretreatment with cyclophosphamide enhanced the content of abnormally structured pyridine nucleotides in various tumors by more than twofold but had no effect on normal tissue (5). Thus an increase in the lethal synthesis in tumor cells should be possible in principle without influencing normal tissue.

It has been shown (6) that the metapyrocatechase-type reaction involving oxidative ring fission of catechol compounds results in derivatives of α-hydroxymuconic acid which can undergo four types of reversible isomerizations. On this basis, biosynthesis of the nuclei of pyridine, α-pyrone, and α-tetronic acid are described.

A mutant of *B. megaterium* was obtained that lacks the enzyme to form dehydropicolinate (7) and cannot sporulate. However, dipicolinate added to cultures of the mutant restores the ability to produce heat-stable spores. The enzyme that produced dehydropicolinate during vegetative growth also functions as the source of this compound for dipicolinate synthesis for sporulation.

It was found (8) that 23 varieties of cheese and eight types of processed cheese contained 9 to 55% of the vitamin content of milk including niacin, B_6, pantothenic acid, biotin, and folic acid.

The enzyme responsible for transformation of 3-hydroxyanthranilic acid into quinolinic acid in rat liver (9) exhibits a 50% increase in activity by addition of adenosine triphosphate.

A strain of penicillium (*P. citreoviride*) accumulated dipicolinic acid from a culture medium (10).

It was observed (11) that the amount of dipicolinic acid in a culture of *Bacillus megaterium* in a medium is proportional to the amount of sporulation.

Intraperitoneally administered 3-hydroxyanthranilic acid-carboxyl-^{14}C was metabolized by the rat to give both nicotinamide methochloride showing 60% of the specific activity injected and quinolinic acid with 91% of specific activity administered (12). It was also shown that metabolism of DL-tryptophan-3a,7a,7-^{14}C in the rat gave quinolinic acid with a specific activity 65 to 76% of that injected (13). Studies using enzyme prepared from rat liver also converted 3-hydroxyanthranilic acid to quinolinic acid (33).

A yield of 3.8 g of dipicolinic acid per 670 ml of culture filtrate was obtained from *penicillium spp*. newly isolated from the soil cultured on a medium of glucose, urea, sodium nitrate, potassium monophosphate, magnesium sulfate, and corn steep liquor (14).

Although it was previously known that isonicotinic acid was produced from isoniazid by *Mycobacterium tuberculosis* BCG, evidence has recently shown (15) that 4-pyridinemethanol is a breakdown product also, thereby suggesting that 4-pyridinealdehyde may be an intermediate product in this process.

Pyridine nucleotides, coenzymes, DPN, and TPN were secreted from yeasts (16) during culture by adjusting the pH to 3.0 to 2.0 with acid or by adding NH_4^+ salts. Bakers yeast also yielded DPN which was absorbed on charcoal (107).

A study of the synthesis of dipicolinic acid by *B. megaterium* sporulating cells (17) in common metabolites such as amino and carboxylic acids indicates evidence for two possible biosynthetic mechanisms. These involve either a condensation between aspartic acid and pyruvic acid or a condensation between alanine and oxalacetic acid.

Photosynthetic bacteria were shown (18) to produce a fast increase in bound pyridine nucleotide (PNH) with fluorescence change upon illumination. Under aerobic conditions, both anacystis and chlorella gave a net increase in PNH on illumination. It has also been shown that stearate dehydrogenase need the Codehydrogenase I for its effect and must be classed as a pyridine protein (19). Numerous studies have been completed on the systems involving diphosphopyridine and triphosphopyridine nucleotides (20–22, 24, 27, 28, 41, 43).

The biosynthesis of adenine nucleotide from ingestion of nicotinamide has also been reported (23), and the effect of nicotinic acid ^{14}C has also been studied on nucleotide biosynthesis (25).

The concentration of Codehydrogenase enzyme in the blood has been the subject of investigation related to various ailments (26). Various amino acids from proteins have been isolated and identified from extracts of living substances (42).

The biosynthesis of the pyridine ring from labeled acetate and succinate fed to tobacco plants indicates the 2- and 3-carbons of the pyridine ring arise from the methylene carbons of succinate (117).

The metabolism of alkaloids in various animals and especially the rat has been investigated including studies with radioactive tracer techniques (33, 53, 65, 121). The bacterial decomposition of nicotine and other alkaloids has been examined (67).

Enzymatic action on putrescine and cadaverine has given rise to pyridine derivatives (98).

Biosynthesis of anabasine from sodium acetate-^{14}C in the stems of intact *N. glauca* plants indicates all radioactivity is located at C-2 and C-3 (109). Other studies on the biosynthesis of the pyridine ring of anabasine have also been reported (114).

The serum and blood levels of water-soluble vitamins, including nicotinic acid, have been determined (110).

A study on the effect of B-complex vitamin supplementation of pig rations has been reported with respect to barley grown in the Peace River district of Canada (111), and the nutritional balance of corn diets (112) and protein-depleted chickens (113) has been reported.

B. *The Tobacco Alkaloids*

Identification of tobacco alkaloids in various plant substances has been thoroughly investigated. Recent studies have concentrated on the identification of alkaloids in tobacco smoke with regard to possible relationships to human health and smoking habits. A summary of work on the isolation and identification of tobacco alkaloids appears in Table II-1. The status of low-nicotine tobacco and the physiological role of the alkaloid in the plant have been studied extensively (29).

3-Pyridinol was shown to be present in cigar smoke to the extent of 130 γ per cigar by a GLC analysis of an ether extract (30).

The relationship between tobacco quality, smoking strength, and nicotine content has been established (48). The aging of tobacco and its alkaloid content has also been investigated as well as the importance of hybrid characteristics (49–51).

There has been considerable analytical investigation of tobacco alkaloids with regard to detection and composition (38, 52, 58, 69, 122). Among the techniques used are gas chromatography (31, 61–63, 93), thin layer chromatography (34, 46, 75, 80), and specific test methods (47, 64, 66, 68, 74), including colorimetric spot tests (39, 87, 88). The determination of small quantities of alkaloids in various natural products has been the subject of many published reports (76–79, 81–86, 105, 115–119). Nicotine can be selectively removed and recovered from tobacco by ammonia extraction (94).

The pyrolysis of tobacco has been shown to destroy about 50% of the nicotine content and to give rise to a wide variety of pyridine derivatives (91, 92).

TABLE II-1. Identification of Tobacco Alkaloids

Source	Method	Compounds identified	Ref.
Cigar smoke	GLC	3-Pyridinol	30
Tobacco smoke	Colorimetric	Nicotine	38
Tobacco leaf	Extraction–fractionation	Nicotine	52
Cigarette smoke	Base fractionation	Nicotine, pyridines, lutidines, collidines	69
Tobacco	Review of various methods	Nicotine, nornicotine	122
Tobacco smoke	GLC	Nicotine, nornicotine	31
Nicotine pyrolysis	GLC	Myosmine, vinylpyridine, 3-cyanopyridine	61, 62
Tobacco leaf oils	Steam distillation, chromatography	Nicotine, myosmine, 2,3′-bipyridine	63
Cigar smoke	GLC	Pyridines, picolines, lutidines, nicotine	93
Cigarette smoke	Column chromatography	Pyridine, nicotines	34
Tobacco powder	Extraction, paper chromatography	Nicotine, anabasine, nornicotine	46
Cigarette smoke	Acid absorption-fractionation	Pyridine, nicotine	75
Tobacco powder	Paper chromatography	Nicotine, nornicotine, anabasine	80
Tobacco, tobacco smoke	GLC-UV	Nornicotine	47
Tobacco leaf	Extraction-distillation	Nicotine	64
Cigarette	Steam distillation	Pyridine	66
Tobacco smoke	Alkaline extraction	Pyridine	68
Tobacco smoke	Pfyl technique	Nicotine, pyridine	74
Tobacco leaf	Extraction–paper chromatography	Nornicotine	88
Tobacco leaf	Colorimetric	Nicotine	39
Green tobacco leaf	Extraction-colorimetric	Nicotine	87

C. *Alkaloids in Various Plant Materials*

The detection of alkaloids in different plant species other than tobacco has received considerable attention. These investigations have included studies on various parts of the plants and the effect of various nutritional substances on the concentration of alkaloids. A summary of what alkaloids are found in various plant species is presented in Table II-2. The isolation and structure of ambine and 4-methoxyparacotoin, extracted from rosewood with benzene, has been reported (95). A variety of alkaloids, including γ-coniceine, have been isolated from hemlock fruit and studied as a function of ripening (99, 108).

TABLE II-2. Alkaloids in Various Plant Species

Specimen	Plant section	Alkaloids	Ref.
Rosewood	Total	4-Methoxyparacotoin	95
Hemlock	Fruit	α-Coniceine	99, 108
Alfalfa	Total	Trigonelline	120, 100
Siberian black currant	Fruit	Nicotinic acid	104
Nuphar japonicum	Not specified	Deoxynupharidine	96
Nuphar japonicum	Not specified	Nupharidine	97
Tobacco grafts on tomato	Grafts	Nicotine	90
Nightshade and tomato grafts on tobacco	Graft sections	Nicotine	55
Tobacco grafts on tomato roots	Graft sections	Nicotine	59
Soybean	Leaves	None detected	70

Alfalfa has been shown to contain trigonelline (100), and the stereochemistry of hemlock alkaloids has been the subject of an intensive study (120). Ninety-six species of Hawaiian plants were examined for alkaloid content (101), and thirty-two species were shown to give positive tests.

Plants from the Canary Islands were also shown to contain from 1.3 to 1.6% alkaloids such as *l*-nornicotine and nicotine (102).

Various seeds have been examined by demethylation techniques and nicotinic acid isolated therefrom (103). Nicotinic acid was also isolated in Siberian black currant hybrid bushes (104).

The alkaloid content of 52 interspecific combinations of *nicotiana* were examined to determine the transmittal effect of hybrids on alkaloids (106), and no simple basis for inheritance was evident.

Studies on the synthesis of alkaloids in various sections of plants have shown interesting results with regard to factors affecting the rate of production as well as the importance of feed nutrients, roots, trimming, plant age, and grafting (32, 35–37, 40, 45, 54–57, 59, 60, 70, 89, 90).

Transformation of anabasine, cotinine, and other alkaloids to pyridine compounds with various reagents has been studied (71–73). Nuparidine has also been subjected to structural studies after isolation from "Kawahone" (96, 97).

The structure of piericidin A has been reported to be (**II-A**) based on spectral data, decomposition studies, and comparison with model compounds (118).

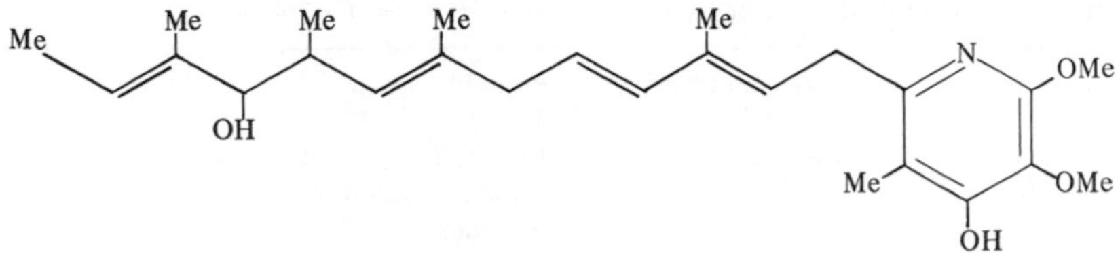

II-A

2. Degradation of Natural Products

A. *Coal*

Most of the lower pyridine bases have been identified in products from both low-temperature and high-temperature pyrolysis, or carbonization, of coals. For purposes of differentiation and categorization, low-temperature carbonization in this context is defined as pyrolysis below about 900°C whereas high-temperature carbonization is effected above this temperature. Bases are also formed by special reactions, including gasification and alkaline hydrolysis of coal (133, 139, 152, 154).

Low-temperature coal tar, in general, contains 1.9 to 2.5% total bases (128, 134, 145, 147). A Bureau of Mines study on a typical low-temperature coal tar, boiling to about 355°C, resulted in identification of 51 individual basic compounds in this boiling range. These compounds are characterized by eight basic structures, including pyridine, cyclopentenopyridine, and phenylpyridine. The alkyl groups present consist primarily of methyl and some ethyl groups; no groups with chain lengths greater than two carbon atoms were detected (148).

Narrow- and wide-range fractions of low-temperature tar were resolved by various treatments to facilitate identification of individual tar-base components (Table II-3). The treatments consisted generally of isolation of the bases by sulfuric acid extraction either preceding or following fractional distillation. Identification was made on the basis of the crystallizing point of a derivative, generally the picrate. In one instance, the tar fraction was cracked thermally prior to work-up to recover the bases, and it was found that the higher methylpyridines had been substantially demethylated to form additional pyridine and 2-picoline (127).

Tars from the high-temperature carbonization of bituminous coal contain less total tar bases than do the corresponding low-temperature tars. A breakdown of products from the high-temperature carbonization of Russian (Cheremkhovo) coal shows the following contents of organic nitrogen bases (tons per 1000 tons of coal): tar, 0.088; tar liquor, 0.053; coke-oven gas, 0.100; total, 0.241 (132). Recoveries and identifications were made by the same techniques as are employed with low-temperature carbonization products (Table II-4).

TABLE II-3. Identification of Bases in Low-Temperature Tar Fractions

Bases	Processing method	Bases identified	Ref.
Tar bases, b.p. 150–190°	Infrared analysis	All lutidines except 2,6-isomer and all collidines except 3,4,5-isomer	131
Tar fraction, b.p. <200°/10 mm Hg	(1) Base extraction (2) Fractional distillation (3) Separation by fractional crystallization of picrate and oxalate derivatives	Pyridine; all picolines and lutidines; 2,6- and 2,4-methylethylpyridines; 3- and 4-ethylpyridine; all collidines except 2,3,4- and 3,4,5-isomers; 2,4-ethylmethylpyridine; 4,2,6-ethyldimethylpyridine	140
Middle oil fraction	(1) Fractional distillation (2) Base extraction	2.5% contained bases of which ca. 40% are pyridines	142
Tar bases, b.p. 165–195°	(1) Fractional distillation (2) Formation of picrates (3) Separation of picrates by fractional crystallization	2,4- and 3,4-Lutidine; 4-ethylpyridine; all collidines except 2,4,5- and 3,4,5-isomers; 2,4- and 2,5-methylethylpyridines; 2,3,5,6- and 2,3,4,6-tetramethylpyridines; 3,4-ethylmethylpyridine	146
Tar bases, b.p. 119–120°	(1) Fractional distillation	2-Picoline, 14% of total bases 3-Picoline, 4% 4-Picoline, 4% 2,6-Lutidine, 12% 2,4-Lutidine, 4% 2,5-Lutidine, 2% 2,4,6-Collidine, 3.2%	147
Tar bases from Cresol No. 1 (49 vol. % b.p. <150°; 70 vol. % b.p. <175°)	(1) Fractional distillation (2) Formation of $HgCl_2$ salts and picrates (3) Separation of derivatives by fractional crystallization	All picolines; 2,3-, 2,4-, and 2,6-lutidines; 4-ethylpyridine; 2,4,6-collidine	151
Tar fraction, b.p. 220–300°	(1) Thermal hydrogenolysis at 650°/70 atm (2) Base extraction (3) Fractional distillation to isolate 110–150° fraction	Pyridine, 16% of base fraction 2-Picoline, 15% 3- and 4-Picoline, 6% Lutidines, 11% Collidines, 3%	127

TABLE II-4. Identification of Bases in High-Temperature Tar Fractions

Bases	Processing method	Bases identified	Ref.
Light-medium tar from Cheremkhovo coals		All picolines; all lutidines except 3,4 isomer; 3-ethylpyridine; 2,6-ethylmethylpyridine	12
Tars bases, b.p. (a) 157.0–178.0° (b) 167.5–181.0° (c) 167.0–181.5° (d) 169.0–183.0°	Analyzed by infrared spectrophotometry	%: (a) (b) (c) (d) 2,3-Lutidine: 10.3, 8.6, 6.6, 6.8 2,4-Lutidine: 19.5, 16.9, 15.6, 16.7 2,5-Lutidine: 7.8, 8.5, 6.0, 7.3 2,4,6-Collidine: 30.5, 43.4, 46.9, 44.8	13
Tar bases, b.p. 178–200°	Analyzed by GLC/UV	3,4-Lutidine; 2,3,5- and 2,4,5-collidine; 2,4- and 2,5-ethylmethylpyridine; 2,3-cyclopentenopyridine; 2,3-, 2,4- and 2,5-dimethyl-6-ethylpyridine; 3- and 4-ethyl-2,6-dimethylpyridine	13
Tar bases, b.p. 209–212°	(1) Fractional distillation (2) Picrate derivatives formed	3,4-Cyclopentenopyridine (b.p. 211–218°/759 mm, d_{20}^{20} 1.045, n_D^{20} 1.5439); 6-methyl-2,3-cyclopentenopyridine (m.p. 31.9°, b.p. 211.9°/761mm, d_{20}^{33} 0.9910, $n_D^{37.5}$ 1.5297); 3,4,5-collidine (m.p. 36.8°, b.p. 211.4°/759 mm, d_{20}^{40} 0.9471, n_D^{37} 1.5132)	13
Light tar bases, b.p. 145–155° and 150–180°	(1) Fractional distillation (2) Picrate derivatives formed and crystallized	3-Ethylpyridine	13
Light tar bases	(1) Fractional distillation (2) Redistillation of 90–170° fraction	Pyridine, 18 pbw, b.p. 112–118° 2-Picoline, 6 pbw, b.p. 126–132° 3-Picoline, 8 pbw, b.p. 140–147° 2,4-Lutidine, 4 pbw, b.p. 152–158° Residue, 8 pbw, b.p. > 160°	14

TABLE II-4. Identification of Bases in High-Temperature Tar Fractions

Bases	Processing method	Bases identified	Ref.
High-boiling tar bases	(1) Recovered from tar oil with aq. H_2SO_4 (2) Crude bases dried (3) Fractional distillation	Lutidines (predom. 2,4-isomer), b.p. 155–165°; collidines (predom. 2,4,6-isomer), b.p. 165–175°	14
Intermediate tar bases	Analysis by i.r.	2,3,4,6-Tetramethylpyridine; 2,3-cyclopentenopyridine; 2,3-cyclopenteno-6-methylpyridine (5% of xylidine fraction of coal-tar bases)	14
Tar pitch fraction	(1) Fractional distillation (2) Acid extraction (3) Fractional distillation of recovered bases	Pitch yielded 12.3% light-middle fraction, b.p. 196–230°, contg. 3 wt. % bases (25% alkylpyridines found after acid extraction); and 10.6% heavy fraction, contg. 6.8% bases	14
Coke-oven gas from Hungarian coal		Pyridine, 150 g per 1000 nm^3 of gas	15
Tar bases, b.p. 180–230°	(1) Acetylation to remove anilines (2) Fractional distillation (3) Analysis by GLC/IR	3,4- and 3,5-Lutidine; 2,3,5-, 2,4,5-, and 3,4,5-collidines; 5-ethyl-3-methylpyridine; 3-ethyl-4-methylpyridine	15
Tar bases, b.p. 200.5–228.0°	(1) Acid extraction (2) Acetylation to remove anilines (3) Fractional distillation (4) Picrate derivatives formed and crystallized	2,3-Cyclopenteno-5-methylpyridine	15

Pyridine bases recovered from coal gasification processes are characterized by a significantly higher concentration of di- and trimethylpyridines than are found in products from conventional coal-carbonization operations (139) (Table II-5). The apparent absence of pyridine from the products of alkaline hydrolysis of coal is also significant (133).

TABLE II-5. Identification of Bases in Coal Gasification and Hydrolysis Products

Source	Processing method	Bases identified	Ref.
Gasoline fraction from Lurgi gasification of Polish coal (700–800°)	(1) Dephenolization (2) Extraction with MeOH	Pyridine, 2.0% of total bases 2-Picoline, 7.0% 3-Picoline, 6.6% 4-Picoline, 16.6% 2,6-Lutidine, 10.5% Higher lutidines, 14.5% Collidines, 6.5%	139
Oil fraction from gasification, b.p. 60–220°	(1) Extraction of bases (2) Analysis by paper chromatography	Pyridine, methylpyridines and dimethylpyridines	152
Gas from underground gasification of coal	Absorption in saturator liquor	Pyridine bases	154
Polish coals	(1) Caustic hydrolysis at 250° and 75 atm for 26 hr (2) Liquid product treated with mixed acids (3) Extraction with C_6H_6 and Et_2O	Methylpyridines	133

In view of the known differences in basicity among the various pyridine derivatives from coal and the fact that separations of bases can be effected as a result of these differences, several investigators have made studies to determine the base strengths of the various alkylpyridines. Ikekawa and co-workers determined base strengths of pyridines at 15° by titration with standard acid of 0.005 *M* solutions of the individual bases in 10% ethanol (130) (Table II-6). Brown and Mihm used ultraviolet absorption spectroscopy to determine the dissociation constants of alkyl-substituted pyridine bases in water at 25°. Their study showed that the introduction of an alkyl group into the 2-, 3-, or 4-position results in an increase in the pK_a value of 0.5 to 0.8 units. The increase in base strength on going from pyridine to 2-picoline (0.80 pK_a unit) is approximately the same as the increase from 2-picoline to 2,6-lutidine (0.78

unit). This points to the essential absence of any important steric effects in the addition of the proton to the latter base or in the solvation of the ion (129).

TABLE II-6. Physical Properties of Pyridine Derivatives

A. Data of Ikekawa *et al.* (130)

Base	$pK_a(15°)$	K_b	d^{15}	n_D^{15}
Pyridine	5.30	2.00×10^{-9}	0.9840	1.5100
3-Picoline	5.85	7.08×10^{-9}	0.9578	1.5052
2-Picoline	5.95	8.91×10^{-9}	0.9488	1.5035
4-Picoline	6.10	1.26×10^{-8}	0.9577	1.5043
3,5-Lutidine	6.34	2.19×10^{-8}	0.9432	1.5055
2,5-Lutidine	6.55	3.55×10^{-8}	0.9346	1.4966
2,3-Lutidine	6.56	3.63×10^{-8}	0.9569	1.5062
3,4-Lutidine	6.61	4.07×10^{-8}	0.9623	1.5127
2,6-Lutidine	6.72	5.25×10^{-8}	0.9278	1.5001
2,4-Lutidine	6.80	6.31×10^{-8}	0.9332	1.5022
3,4,5-Collidine	7.10	1.26×10^{-7}	0.9616	1.5103
2,3,5-Collidine	7.15	1.41×10^{-7}	0.9613	1.5110
2,4,5-Collidine	7.28	1.91×10^{-7}	0.9384	1.5095
2,3,4-Collidine	7.38	2.40×10^{-7}	0.9605	1.5185
2,3,6-Collidine	7.40	2.51×10^{-7}	0.9294	1.5058
2,4,6-Collidine	7.63	4.27×10^{-7}	0.9204	1.5012
2,3,4,5-Tetramethylpyridine	7.78	6.03×10^{-7}	–	–
2,3,5,6-Tetramethylpyridine	7.91			
2,3,4,6-Tetramethylpyridine	8.10	1.26×10^{-6}	0.9388	1.5132
2,3,4,5,6-Pentamethylpyridine	8.75	5.62×10^{-6}	–	–

B. Data of Brown and Mihm (129)

Base	$pK_a(25°)$	B.p., °C/mm Hg	n_D^{20}
Pyridine	5.17	114/743	1.5092
2-Picoline	5.97	127/740	1.5010
2-Ethylpyridine	5.97		
2-Propylpyridine	5.97		
2-Isopropylpyridine	5.83		
2-*t*-Butylpyridine	5.76		
3-Picoline	5.68	142/747	1.5058
3-Ethylpyridine	5.70		
3-Isopropylpyridine	5.72		
3-*t*-Butylpyridine	5.82		
4-Picoline	6.02	143/740	1.5051
4-Ethylpyridine	6.02		
4-Isopropylpyridine	6.02		
4-*t*-Butylpyridine	5.99		
2,6-Lutidine	6.75		

A number of reliable techniques have been developed over the years for the analysis of pyridine bases. When present in aqueous plant liquors, the mixed bases are most readily determined by absorption in strong acid and back-titration with standard aqueous ammonia (196). The effects of complicating impurities such as ammonia, acid gases, or complex ions can generally be counteracted by the use of special indicator systems (190), removal of impurities by distillation (200) or oxidation (173), and by the use of solution buffers (191). Dry bases or mixtures of bases are determined readily by titration with standard $HClO_4$ in glacial acetic acid or other suitable nonaqueous systems (164, 180, 183, 193, 201, 207, 210). Individual bases have been determined and base mixtures have been resolved through both infrared and ultraviolet spectrophotometry, where characteristic absorption patterns have been developed for the various bases (156, 156a, 171, 182). In addition, dyes and colored complexes can be formed between the bases and other organic compounds, and these can be determined photocolorimetrically to obtain quantitative estimates of the concentrations of bases in aqueous streams, in organic mixtures, and in the atmosphere (160, 163, 170, 172, 188, 199, 204). More recently, advances in pyridine base analysis have been made through use of gas-liquid chromatography. A considerable effort has been devoted to the area of separation and identification of individual bases by means of various combinations of solid supports and stationary phases (136, 158, 159, 161, 166, 175, 185, 209). Both liquid and paper chromatography, and more recently, thin-layer chromatography have proved to be viable supplemental techniques for resolution of pyridine bases (Table II-7). Another technique that offers promise for identification of bases is nuclear magnetic resonance, but owing to its relative infancy with regard to applications development, it has not yet attained a significant status in the coal tar analysis area (184, 205, 206). Cryoscopic analysis has been employed effectively for the determination of water in pyridine bases as well as for the determination of minor base impurities in 2- and 3-component systems (192, 197, 198). A related technique for assay of pyridine bases involves measurement of solution temperatures of salts of base fractions, notably the hydrochlorides (186, 187).

The largest part of the pyridine bases formed during the coking of coal remains in the coke-oven gas, and these are generally removed from the gas by scrubbing with acid. That portion that is condensed with the tar, or is retained by the crude benzene absorbed from the coke-oven gas, can also be extracted with acid, organic solvents (216, 238, 261), or ion-exchange resins (244, 259) (Table II-8).

Scrubbing of the gas is generally accomplished in saturators with aqueous H_2SO_4 (223, 229, 231, 233–235, 240, 247, 253, 254, 258). Industrial units are usually of a two-scrubber design in the first of which the major portion of NH_3 is absorbed to form $(NH_4)_2SO_4$. The NH_3-lean gas is then passed through the second scrubber where the pyridine bases are absorbed to form a pyridine-H_2SO_4 salt. At a given stage of saturation, the pyridine sulfate liquor is treated with NH_3 to spring the free bases which are steam-distilled to yield an aqueous pyridine base-rich overhead stream. The overhead fraction is then dehydrated and distilled fractionally to recover the product pyridine bases (218, 250, 256). Phosphoric acid or ammonium phosphate liquors have also been used to advantage for recovery of pyridine bases (215, 228, 243, 255), and in some cases, pickle liquors ($FeSO_4$ + H_2SO_4) from steel-processing operations have been utilized (213, 249).

TABLE II-7. Analytical Methods for Determination of Pyridine Bases

Method	Applicability	Ref.
Infrared or ultraviolet analysis	Wavelengths of characteristic peaks determined for 14 Me, 2 Et, 1 Pr, and 6 Et-Me derivatives, for fractions separated by distillation from pyridine bases boiling <185°	156
	2,3- and 2,5-Substituted pyridines differentiated by i.r. analysis on basis that former correspond to 1,2,3-trisubstituted benzenes and exhibit band at 1578–1588 cm^{-1}; latter correspond to 1,2,4-trisubstituted benzenes and exhibit band at 1599–1605 cm^{-1}	156a
	Principal i.r. absorption bands determined in the 1700–400 cm^{-1} region for the 3 picolines, *s*-collidine, and 3,5-lutidine	182
	Determination of individual bases in the picoline, lutidine, and collidine fractions of coal-tar bases	208
	Determination of pyridine bases in isooctane extract of high-temperature coal-carbonization liquor was made at 260 and 316.5 mμ	167
	Determination of 3- and 4-picoline and 2,6-lutidine in the crude β-picoline fraction of tar bases, with an accuracy of ±4.5%, was made from observation of the characteristic maxima at 240–300 mμ	171
	Determination of pyridine, 2-picoline and 2,3-lutidine in admixture, with an accuracy of ±2%, was made with Beckmann DK-2 spectrophotometer on isooctane extract of MeOH-H_2O solution of bases	181
	Crude bases from acid extraction process were extracted with isooctane and absorptivities measured at 260 and 316.5 mμ; total bases in solution given by: $p = \{[A_{260} - (A_{316.5} \times 23.24)]/21.25\} \times$ dilution factor	194
Photometric	Determination of pyridine bases in waste water (between 0.05 and 150 mg/1) made by adding liquor to HCl-KCN-chloramine mixture, then adding 1% barbituric acid solution; absorbance measured with Pulfrich photometer using Hg lamp and S57 filter	160

TABLE II-7. Analytical Methods for Determination of Pyridine Bases (*Continued*)

Method	Applicability	Ref.
	Pyridine and the three picolines resolved by photometric measurement of the polymethine dyes formed from the chloride-cyanide-base-barbituric acid reaction	204
	Small amounts of pyridine in aromatic hydrocarbons were detected by measuring absorbance in aq. ethanol of color complex from reaction of base with BrCN and benzidine	170
	Colorimetric determination of pyridine bases in the air of industrial establishments, with claimed sensitivity of 0.005 mg/3ml, is based on formation of colored complex by reaction with BrCN and benzidine	172
	Pyridine bases in the air were determined with accuracy of ± 5%, by photocolorimetric method based on color complex from reaction of bases with violuric acid; alcohols, ketones, hydrocarbons, and ethers do not interfere	163
	Non-α-substituted alkylpyridines determined in presence of α-substituted alkylpyridines since former displace *p*-$ONC_6H_4NMe_2$ from strongly colored ion $[Fe(CN)_5 \cdot ONC_6H_4NMe_2]^{3-}$, and green color disappears; color measured with Spekker absorptiometer and Ilford 608 filters	188
	Pyridine determined colorimetrically by measurement of colored products of reaction between base and epichlorohydrin or epoxyethylbenzene; substituents NH_2, CN, COOH, COOR, Br, and Cl retard color reaction	199
Gas chromatography	Retention volumes for 2-alkylpyridines (Me, Et, isoPr, *t*-Bu) on glycerol determined; elution from column occurs in order of decreasing steric hindrance	157
	Separation of three picolines and 2,6-lutidine accomplished most effectively on stationary phase of 15% 1,2,3-tris-(2-cyanoethoxy)propane on Celite 545 pretreated with 0.05% $CuCl_2$, and contained in 125 × 0.4 cm glass column	158
	Pyridine, picolines, lutidines, and collidines determined with stainless-steel column packed with solid support of Celite (60–72 mesh, acid-washed) or kieselguhr (60–72 mesh, acid- and base-washed) coated with 9% polyethylene glycol 1000 and 1% $(HOC_2H_4)_3N$ or 20% dioctyl sebacate and 1% $(HOC_2H_4)_3N$; detection by flame ionization	159
	Pyridine (b.p. 115.3°), 2-picoline (b.p. 129.4°), 2,6-lutidine (b.p. 144.0°), 3-picoline (b.p. 144.1°) and 4-picoline (b.p. 145.4°) eluted in order of b.p. from silica gel-glycerol columns; Celite, Gas Chrom Q and kieselguhr all caused reversal for first three members of series	161

TABLE II-7. Analytical Methods for Determination of Pyridine Bases (*Continued*)

Method	Applicability	Ref.
	26 Pyridine homologs separated essentially without tailings by Perkin-Elmer Fraktometer F 6/4 employing capillary column (50 m long, 0.25 mm diameter) packed with 5% *N*,*N*-bis(hydroxyethyl)trimethylenediamine in MeOH under pressure with N_2 gas	165
	Components of β-picoline fraction (pyridine, 3 picolines, and 2,6-lutidine) best resolved by use of 5% polyethylene glycol 400 or Chromosorb W (60–80 mesh)	166
	Aqueous mixture of pyridine, three picolines, and 2,6-lutidine almost completely separated by use of stationary liquid phase of 10% ethanolamine and 5% *o*-phenylphenol on carrier of C-22 refractory brick	169
	Specific retention volumes determined at 200° for six pyridine homologs on stationary phases of Reoplex 400, Apiezon L and silicone elastomer E301	174
	Elution sequence and relative retention volumes determined for the three picolines and 2,6-lutidine, in admixture with nicotinonitrile and isonicotinonitrile, on support of porous plate and stationary phases of 7% diglycerol or 10% poly(ethylene adipate)	175
	3- and 4-Picolines almost completely separated with triethanolamine used as stationary phase and oxygen-free nitrogen as carrier gas; complete resolution achieved with glycerol as stationary phase	185
	Pyridine homologs separated on carrier of Chromosorb P containing 6% NaOH and stationary phase of 20% poly(phenyl ether); at 110°, retention times (min.) were pyridine, 14.5; 2-picoline, 17.3; 2,6-lutidine, 20.5; 3-picoline, 30.5; 4-picoline, 33.7; 2,5-lutidine, 35.0; 2,4-lutidine, 40.5	209
	High-boiling collidine fractions (178–200°) separated according to b.p.'s with silicone DC-550 as stationary phase; each fraction further subdivided using polyethylene glycol or diglycerol	136
	5-Ethyl-2-methylpyridine determined (as low as 0.01% concn.) in commercial 2-methyl-5-vinylpyridine with alkaline Carbowax 400 column	168
Liquid chromatography	Collidine base fraction (b.p. 159–191.3°) resolved by distillation and percolation of fractions through chromatographic column made from $CuCl_2$-H_2O-glycerol-Al_2O_3; colored $CuCl_2$ complexes eluted and identified as 2,3- and 3,5-lutidines, 4-ethylpyridine, and 2,3,5-, 2,3,6-, and 2,4,6-collidines	162

TABLE II-7. Analytical Methods for Determination of Pyridine Bases (*Continued*)

Method	Applicability	Ref.
	Chromatographic color complexes determined for 2- and 3-picolines and 2,4- and 2,5-lutidines with $CuSO_4$, $CuSO_4$-HCl, $CuSO_4$-picric acid, Roth oxidation mixture, *p*-$Me_2NC_6H_4CHO$, *p*-$Me_2NC_6H_4CHO$·HCl and $K_4Fe(CN)_6$-HCl	179
	Chromatographic color complexes determined for pyridine, the three picolines, 2,3-, 2,4-, 2,5-, and 2,6-lutidines, 2,4,6-collidine, and 4-ethylpyridine; bases separated and analyzed to within 1–2%	189
	Impurities in technical β-picoline determined on Al_2O_3-$CuCl_2$ column; error was <3%	212
Paper chromatography	Nicotine, nornicotine, anabasine, anatabine, metanicotine, dihydrometanicotine, pseudooxynicotine, myosmine, and pyridine-1-oxide were resolved	178
	Method adapted to determination of color complexes resulting from opening of pyridine ring with ClCN to form glutaconic dialdehyde and subsequent reaction with barbituric acid; works on pyridine and 22 derivatives, but not with 2-picoline	202
Thin-layer chromatography	Pyridine, the three picolines, and 2,4- and 2,6-lutidines separated on silica gel G and alumina G by using 17 eluents; R_f values tabulated	211
Nuclear magnetic resonance	α-Picoline and water in reagent-grade pyridine determined using OH (δ = 4.7 ppm), and CH_3 peak intensities (δ = 2.45 ppm) of α-picoline; empirical relationship given for α-picoline as function of H_2O content, previously determined, and the peak intensities	184
	Study of Me-group signals for methylated pyridines and proton signals for deuterated pyridines indicated that *meta* and *para* protons more nearly magnetically equivalent than *ortho* and *para* protons	205
	Chemical shifts relative to water and spin-spin coupling constants listed for pyridine, the three picolines, 2,6-, 2,4-, and 2,3-lutidines and *s*-collidine	206
Titrimetric	Potentiometric titration method worked out, employing 0.1 *N* $HClO_4$ in Ac_2O, to determine pyridine bases and aromatic amines separately in one sample, with accuracy of 2–3% relative; method employs glass electrode and standard calomel electrode filled with satd. soln. of KCl in MeOH	210
	Pyridine bases in tar middle oil determined in presence of NH_3 and acidic constituents by dissolving sample in mixt. of glacial AcOH and HCHO and titrating potentiometrically	207

TABLE II-7. Analytical Methods for Determination of Pyridine Bases (*Continued*)

Method	Applicability	Ref.
	with 0.1 *N* $HClO_4$ in AcOH; Ag-AgCl electrode dipped in satd. soln. of KCl in AcOH and glass electrode were used	
	Pyridine bases determined by high-frequency titration with 0.5 *N* $HClO_4$ in AcOH; analysis time is 15 to 20 min and rel. error is $\leqslant 6\%$	193
	Results given for high-frequency titration of pyridine and 2-picoline in AcOH; mixtures of bases also titrated in mixed solvents of AcOH and 1,4-dioxane	183
	Accurate determination of tar bases possible by titration with $HClO_4$ in glacial AcOH	180
	Pyridine, the three picolines, and 2,6-lutidine quantitatively determined by potentiometric titration with $HClO_4$ in AcOH, with 1 to 2% error; pyridinecarboxylic acids determined by titration with Et_4NOH in mixed MeCOEt-MeOH	164
	Potentiometric titrations of pyridine and α-picoline carried out in C_6H_6-MeOH mixtures in ratios of 1:1 to 7:1 with 0.1 *N* $HClO_4$ in dioxane; C_6H_6-MeOH has advantages of high soly. for bases, lack of strong acid and basic properties, and dielectric constant that allows for constant potentials	201
	Pyridine determined in milligram quantities, usually within 2% error, by coulometric analysis in acetonitrile which is 0.05 *N* in $LiClO_4 \cdot 3H_2O$ (H_2O content 0.3%); salt serves as source of H_2O which is oxidized to H^+ anodically	203
	Pyridine bases in plant liquors determined by acidifying stream and boiling to remove H_2S, CO_2, and naphthalene, neutralizing cooled liquor to bromophenol blue with carbonate-free alkali, adding Na monophosphate buffer and distilling, and titrating using bromophenol blue as indicator	200
	Pyridine determined in aq. streams by acidifying with HCl or H_2SO_4 and titrating with standard NH_3 soln.; pyridine content obtained from inflection on titration curve without being influenced by NH_4^+ or CNS^-	196
	Use of buffer method (NaH_2PO_4, pH = 4.4) for determination of tar-base fractions (b.p. 114–122° and 122–138° in solns. produced good results (101.0 and 99.6% recoveries) and is superior to hypobromite method; results less satisfactory with higher-boiling fractions due to incomplete separation of pyridine homologs from buffer solution	191
	Pyridine bases–NH_3 mixtures determined by single titration if NH_3 : pyridine base molar ratio <1; α-naphtholphthalein	190

TABLE II-7. Analytical Methods for Determination of Pyridine Bases (*Continued*)

Method	Applicability	Ref.
	used as indicator for NH_3 and bromophenol blue for pyridine bases	
	Modifications given for analysis of coke-oven pyridine bases by hypobromite method whereby contained NH_3 is oxidized to N_2 with alk. hypobromite soln., and bases are distilled into 0.02 N H_2SO_4; increased gas flow rate shortens analysis time, and boiling of acidified samples reduces error to $\leqslant$7.7%	173
	Pyridine, α-picoline, and nicotine or mixtures thereof titrated thermometrically with stand. HCl or $NaNO_2$ and with no interference from alcs. or phenol; sharp break in titration curve occurs with HCl at addition of 1 meq. HCl/mmol; determinations have error of 0.62%	177
Miscellaneous	Up to 2 wt.% α-picoline in pyridine determined with uncertainty of about 0.1% by measuring lower soln. temp. of mixt. contg. the bases, H_2O, and KCl, in specified proportions	195
	Pyridine, the 3 picolines, 2,3-, 2,4-, 2,5-, and 2,6-lutidines, and 2,4,6-collidine determined in coal tar and crude benzene by measuring temps. of disappearance of crystals of hydrochlorides of fractions obtained by distn. of sample mixtures	187
	Pyridine bases determined during distn. by forming hydrochlorides, removing H_2O and excess HCl and measuring three temperatures corresponding to different crystalline states of the fractions	186
	Crystallization temperatures of 2,6-lutidine in presence of varying percentages of 3- and 4- picoline and mixtures of 3- and 4-picoline are tabulated and plotted	
	Cryoscopic method described for determination of pyridine, contains correction factor for water; f.p. depressions due to presence of $\leqslant$6 mole % of 2-, 3- or 4-picoline, 2,6-lutidine, or benzene are linear with concentration	197
	Cryoscopic curves developed for pure 3- and 4-picoline, 2,4-, 2,5-, and 2,6-lutidines; methods developed for determining 2,6-lutidine in mixture of isomeric picolines and mixture of 2,4- and 2,5-lutidines, with error of 0.6 to 0.8%; method also suitable for 3-component mixtures of pyridine bases	198
	Pyridine bases determined with average error of 1.2 to 2.1% by analysis with dioxane-SO_3; method also applicable to alkaloids, anabasine, and lupinine and to α, β′-bipiperidine and α, β′-bipyridine.	176

TABLE II-8. Methods of Recovery of Pyridine Bases

Technique	Details	Ref.
Extraction with H_2SO_4	Mechanism of trapping of pyridine bases from coke-oven gas with H_2SO_4 reviewed with graphs, tables, and equations	217
	Coal-tar distillates extracted with used 8% NaOH to remove phenols, then with 25% H_2SO_4, and finally with fresh 8% NaOH; oil then contains only traces of phenols and bases	221
	Concentration of $(NH_4)_2SO_4$ in saturator and acidity and temperature of mother liquor shown to have little effect on decomposition of pyridine sulfate; increased yield of pyridine bases achieved by reducing acidity of saturator to 4.5 to 5.0%	223
	Review monograph on production of pyridine bases at Russian coal-tar chemical plants	224
	Ammoniacal liquor treated in liming column to remove acid gas and phenols, and ammonia and pyridine vapors pass to acid saturator; technique allows for improved recovery of both phenol and pyridine	226
	Remodeled Krivoĭ Rog (U.S.S.R.) pyridine plant operates with following pyridine concns.: 0.280 and 0.016 g/m^3 in primary and exhaust gases, 0.18 and 0.043 g/l in NH_3 liquor and waste water, 10–11 g/l in mother liquor; overall pyridine recovery is 84%	227
	Recovery of pyridine doubled per ton of $(NH_4)_2SO_4$ produced, by maintaining saturator temperature at 55 to 60°, acidity of mother liquor at 6 to 7%, and concn. of bases in saturator at 8 to 11 g/l	229
	Pyridine recovered with NH_3 in system of two scrubbers equipped with spray nozzles; in first scrubber, acidulated $(NH_4)_2SO_4$ contg. no more than 6% H_2SO_4 is used to remove most of the NH_3 in the coke-oven gas while in second scrubber, liquor contg. no less than 8% H_2SO_4 is used to recover most of the pyridine bases	231
	Improved pyridine base recovery from raw benzene and naphthalene oils achieved by first separating phenols, then washing with spent H_2SO_4 from benzene refining, after dilution of acid to 25 to 30% concentration	232
	Pyridine bases and tar bases recovered from hot coke-oven gas by isolating pyridine bases in stream rich in NH_3 and free from acid gases	233
	Pyridine sulfate in saturator liquor hydrolyzed by neutralization of liquor with NH_3; vapors of pyridine, NH_3, and steam pass to dephlegmator, condenser, and separator, while	234

TABLE II-8. Methods of Recovery of Pyridine Bases (*Continued*)

Technique	Details	Ref.
	bottoms liquor contg. simple and complex cyanides is routed to separate plant	
	Pyridine bases (b.p. <160°) absorbed from coke-oven gas in acid liquor; liquor neutralized with NH_3 and pyridine bases distilled, p*H* being kept <9.0; separation of pyridine layer achieved by addition of NH_3 into steam distillate	235
	Review article on separation of pyridine and alkyl derivatives from coal tar; approximate percentages, b.p.'s, and separation methods discussed	236
	Amount of pyridine bases absorbed in saturator liquor increases with acidity of liquor, ranging from 0.300 to 0.425 g/m^3 at saturator acidities of 3.0 and 5.0%, respectively	240
	Method for extn. of pyridine bases from crude benzene is ineffective since 77% of bases are absorbed from coke-oven gas in H_2O and then lost in tar and in cooling tower; absorption of bases after separation of NH_3, but before separation of C_6H_6, is recommended	245
	Off gases from saturator passed through column contg. their aq. condensate, then through condenser where part of the H_2O is condensed; two phases condensed from gas in second condenser–aq. phase and phase contg. phenol, pyridine bases, and neutral oil	246
	Acidity of initial liquor increases degree of absorption of pyridine bases; presence of $(NH_4)_2SO_4$ does not affect absorption of pyridine; rise in temperature from 20 to 55° prevents absorption of pyridine due to variation in pressure of pyridine vapors over soln. contg. pyridine-H_2SO_4	247
	Pyridine bases recovered from coal-tar naphtha (b.p. 170–200°) by first washing with tar acid- and base-free naphtha, then with 2*N* H_2SO_4; recovery of bases was 92.9%	251
	Vapor pressure of pyridine over solns. of pyridine · H_2SO_4 contg. free H_2SO_4 and $(NH_4)_2SO_4$ determined from composition of vapor phase in equilibrium with solns. at 30 to 55°	252
	Pyridine recovered continuously by absorption in saturator liquor contg. $(NH_4)_2SO_4$ and 6% excess H_2SO_4; part of the liquor is neutralized countercurrently at 80° with excess NH_3, and pyridine enters gas stream from which it can be absorbed by C_6H_6	253
	Complete recovery of pyridine bases claimed in two-stage saturator whereby NH_3 is removed in the first; reduction	254

TABLE II-8. Methods of Recovery of Pyridine Bases (*Continued*)

Technique	Details	Ref.
	of NH_3 in the second-stage feed and the practical absence of $(NH_4)_2SO_4$ promoted absorption of pyridine bases	
	Dephenolized tar oils washed with saturator liquor from gas processing to remove tar bases; acid liquor then sprung with NH_3 to free bases which can then be separated	257
	Recovery of pyridine bases of better than 97% claimed in Russian process whereby bases are accumulated in saturator and are sprayed with acid solution of pyridine sulfate in separate scrubber	258
	Attempts to remove pyridine bases from phenolic oils unsuccessful due to formation of strong complexes of phenol-pyridine salts; necessity of phenol removal prior to acid extraction is indicated	260
	Plan for efficient recovery of pyridine bases requires two-stage extraction of absorber oil and one-stage extraction of naphthalene oil	239
	Nearly complete separation of pyridine bases from absorber oil and naphthalene oil achieved by maintaining coefficient of excess acid at 1.1 for former and 1.7 to 1.8 for latter; by so doing, free acid content was reduced from 6.0 to 2.8%	241
	Pyridine bases recovered from process liquors by steaming and enriching steam distillate by fractional distn. to 20% bases; two phases result with upper oily layer contg. ~ 65% bases that can be recovered with dil. H_2SO_4	248
Extraction with pickle liquor	Spent pickle liquor from steel finishing used to scrub coke-oven gas to recover pyridine bases; complex ferro-cyanides formed can be decomposed to release pyridine	213
	Pyridine and other bases removed from coke-oven gas by scrubbing with pickle liquor at 45 to 55° and p*H* 6.2 to 7.0, then neutralizing with NH_3 and steam-distilling to recover bases	249
Extraction with phosphoric acid	Pyridine bases recovered from NH_3-free coke-oven gas by scrubbing with soln. of phosphates of pyridine bases at up to 55° and neutralizing with NH_3; coefficient of extraction was 70 to 72% with 75 to 78% pyridine bases in crude separated product	215
	Empirical relationship developed between pyridine vapor pressure and salt concentration, solution composition, and absorption temperature	219
	Pyridine bases removed from coke-oven gas by first removing NH_3 in scrubber with aq. H_3PO_4 at p*H* 5 to 8, then contacting gas in second scrubber at p*H* <4	228

TABLE II-8. Methods of Recovery of Pyridine Bases (*Continued*)

Technique	Details	Ref.
	Optimum conditions for removal of pyridine bases from coke-oven gas with H_3PO_4 shown to be 50 to 60°, 4 to 5 moles/1 concentration of NH_4 phosphates and NH_3/H_3PO_4 ratio of 1.35 to 1.5:1; 80 to 85% of total bases were transferred to pyridine layer contg. 65 to 75% bases	243
	Pyridine bases recovered from crude benzene by extraction of 60 parts oil with 1 part by vol. of 15% aq. H_3PO_4, and neutralization of acid extract with NH_3	255
Extraction with mixed acids	Pyridine bases recovered from coke-oven gas by scrubbing with H_2SO_4–H_3PO_4; 95 to 97% yield and 200 g/l concentration of bases in absorbing solution were achieved	220
Other extraction techniques	Tar bases removed from coal-tar fractions by extraction with 50% solution of AcOH in water	216
	Pyridine bases and phenols removed from oils by counter-current extraction with 38 to 40% Na phenolate; aq. extract steam-distilled to yield pyridine–contg. vapors	222
	Pyridine bases removed from crude carbolic oil (b.p. 155–194°) by extraction at 15° with kogazine from Fischer-Tropsch synthesis	238
	Pyridine bases recovered from tar water of coking by multi-stage vacuum evaporation, extraction with organic solvents, and rectification	261
	Pyridine bases in crude benzene recovered by adsorption on ion-exchange resins; maximum recovery obtained on bed swollen to EtOH content of 14.5% by wt.	244
	Pyridine bases recovered from crude or distilled benzene by adsorption on sulfonated polystyrene-type cation-exchange resin preswollen with alcohol	259
Miscellaneous	Phenols must be removed from light-pyridine fraction since they form high-boiling azeotropes with the pyridines, which results in reduced yield of β-picoline and other bases	214
	Crude pyridine oil worked up by dephenolization with 15% NaOH, azeotropic distillation, dehydration with concentrated alkali solutions, and final rectification; recovery of pyridine bases represents 75% of content of starting oil	218
	Improved plan for recovery of pyridine bases from crude concentrates involves use of primary and secondary rectification systems	225
	Tar-base sulfate liquor sprung with NH_3, and mixture stirred at 80° and separated into two layers; upper one contains free pyridine bases	230

TABLE II-8. Methods of Recovery of Pyridine Bases (*Continued*)

Technique	Details	Ref.
	Recycling of NH_3 + CO_2 to distillation of neutralized liquor promotes salting out of crude pyridine bases, increases their yield, and reduces their water content	237
	Ammoniacal liquor from coking of coal (50 mg/1 of pyridine) passed at 3 atm through water jet pump together with benzene to give liquor contg. only traces of pyridine	242
	Fractional distillation of vapors from neutralized saturator liquor gives overhead of pyridine–H_2O azeotrope and pyridine-free H_2O as bottoms	250
	Technique for dehydration and fractionation of crude pyridine bases	256

The technical literature continues to list a large variety of techniques for the separation of individual pyridine-base isomers from crude base fractions (295, 323). These are based mainly on the ability of pyridine bases to form complexes with different reagents, particularly metal salts (Table II-9). The chlorides of copper, cobalt, nickel, zinc, manganese, and iron are especially useful in this respect (274–276, 287, 289, 298, 300, 301, 306, 309, 324). Adduct formation with organic molecules or organometallics also forms a basis for separation of pyridine base products (271–273, 280–282, 284, 291, 292, 298, 302, 309, 313, 315, 320). Preferential hydroxymethylation of certain substituted pyridines and conversion of these to higher-boiling products permits a separation from codistilling isomers (262, 264, 284, 288, 307, 322, 324). To some extent, isomers can be separated by steam-distillation of the crude base mixtures (266, 288, 293, 310). Continuous extraction with solvent pairs affords enrichment of certain of the pyridine-base isomers (290, 297, 321). Treatment with mineral acids produces pyridine salts which, because of differences in salt solubilities under different conditions, leads to separation of isomers (263, 267, 270, 273, 278, 280, 282, 294, 296, 298, 299, 309, 311, 312, 316). A novel isomer separation technique involves conversion of the bases to their *N*-oxides, separating these by distillation and, finally, reduction of the separated *N*-oxides with agents such as iron or phosphorous trichloride (285, 317).

The pyridine bases recovered commercially from saturator liquors are generally of sufficient purity that additional purification is not required. However, depending on the care exercised during extraction of the bases from the gas stream and subsequent processing by springing and distillation, the crude bases could be contaminated with neutral oils such as toluene, xylenes, and nitriles (329). These contaminants can be removed by adding water to the crude bases and careful distillation through a high-efficiency tower, whereby relatively low-boiling water azeotropes are removed initially, followed by the pyridine base-water azeotrope, and finally the anhydrous bases (Table II-10). Chemical treatments are also employed to remove impurities, and include reaction with an acetic anhydride-phthalic anhydride mixture (327) and oxidation with alkaline potassium permanganate (328). Hydrorefining with metal oxide or sulfide catalysts at elevated temperatures and pressures was found to be effective in reducing significantly the sulfur content of pyridine bases (325). A United States Bureau of Mines study has concluded that, for purification of pyridine bases, protection from water and oxygen is necessary (326).

B. *Petroleum*

A detailed study of Romashkino petroleum from Eastern Europe was made by Landa and co-workers. Identification of the various pyridine bases was effected on fractions that had been subjected to cracking, distillation over CaO and KOH, oxidation with dilute HNO_3, and treatment with various catalysts, such as Al_2O_3, MgO, Fe_2O_3, and CaO (332). The bases were isolated with the aid of H_2SO_4, $CuCl_2$, and Al_2O_3 impregnated with $CuCl_2$. Detection was carried out by Baudet's method and by paper chromatography (333).

As part of a study relating jet fuel composition to stability, determination was made of the nitrogen bases in a catalytically cracked California gas oil boiling above 350°F. Total nitrogen compounds in the oil amounted to 1 wt. %; of these, 40% were basic nitrogen compounds. Nitrogen bases in the 340 to 430°F fraction of the oil consisted of 35% pyridines, 25% quinolines, and 40% anilines (330).

Crystallizing mother liquor from petroleum naphthalene fractions was studied in detail with respect to both acidic and basic constituents. The bases were removed by washing with dilute H_2SO_4 and were identified (331).

TABLE II-9. Methods of Separation of Pyridine Bases

Bases	Method	Results	Ref.
3- and 4-Picolines and 2,6-lutidine	(1) Heat 3 hr with 38% HCHO at 150–160° (2) Steam distil	Separated 3-picoline of 95% purity	262
β-Picoline fraction	(1) Heat with HCHO at 110–250° (2) Distil	Separated 3-picoline	264
3- and 4-Picolines	(1) Boiled 15 hr with aq. HCHO (2) Steam distil	Separated 3-picoline	284
3- and 4-Picolines and 2,6-lutidine	(1) Heat with HCHO at 130–140° (2) Distil	Separated 3-picoline and 2,6-lutidine in purity of >97%	288
β-Picoline fraction	(1) Heat with 40% HCHO at 100–250° (2) Distil	Separated 3-picoline	307
Isomeric picolines	Treatment with aq. HCHO	Separation	322
3- and 4-Picolines	(1) Heat with catalyst and excess HCHO (2) Steam distil (3) Benzene extraction of steam distillate	Separated 3-picoline	324
Tar bases, b.p. 138–140°	Distil with H_2O until distillate cloud pt. drops <60°	Separated pyridine and 2-picoline	266
	Distil with H_2O until distillate cloud pt. drops <42° and starts to rise again	Separated 2,6-lutidine and 2-ethylpyridine	
β-Picoline and 2,6-lutidine	Distil with H_2O	Separation of each product in purity of >97%	288
5-Ethyl-2-methylpyridine and 2-methyl-5-vinylpyridine	Distil with H_2O in stripping column	Separation	293

Tar bases, b.p. 141–145°	(1) Distil with H_2O (2) Salt distillate with NaCl	Separated 3- and 4-picolines and 2,6-lutidine	310
β-Picoline fraction	Continuous extraction with pet. ether/water	Separated 2,6-lutidine product with purity of 70%	297
β-Picoline fraction	Continuous countercurrent extn. with naphtha (b.p. 78–98°)/water	Separated 2,6-lutidine of 69–77% purity	321
	Continuous countercurrent extn. with C_6H_6/aq. NaH_2PO_4 (pH 3.6)	Separated β-picoline	
5-Ethyl-2-methylpyridine and 2-methyl-5-vinylpyridine	Continuous countercurrent extn. with NH_3 (contg. 35 wt.% H_2O)	Separation of products with purities of 94–98 wt. %	290
3- and 4-Picolines and 2,6-lutidine	Distil (three parts bases: one part aq. $(NH_4)_2SO_4$ or NH_4OH)	Separated 2,6-lutidine	265
Tar bases, b.p. 155–160°	(1) Treat with HNO_3 (*d.* 1.52) at 60° (2) React with ice water (3) Fractional crystallization	Separated 2,4,6-collidine and 2,4-lutidine	263
3- and 4-Picolines and 2,6-lutidine	(1) Treat with conc. H_2SO_4 (2) Mix with BuOH (3) Decomp. BuOH-insolubles with strong NaOH	Separated 95% pure 2,6-lutidine in 85% yield	294
Tar bases, b.p. 170–180°	(1) Treat with conc. H_2SO_4 (2) Add higher alcohol	Separated 2,4,6-collidine and 3,5-lutidine	316

TABLE II-9. Methods of Separation of Pyridine Bases (*Continued*)

Bases	Method	Results	Ref.
Tar bases, b.p. 155–160°	(1) Treat with conc. H_3PO_4 (2) Decomp. ppt. with aq. alkali (3) Steam distil (4) Salt condensate with solid alkali	Separated 2,5-lutidine	267
Crude lutidines	(1) Dilute with EtOH (2) Add seed crystals of pure 2,3-lutidinium phosphate (3) Add H_3PO_4 at 70° (4) Cool and filter	Separated pure 2,3-lutidine in 45–50% yield (after 1 recrystallization from EtOH)	273
Tar bases, b.p. 165–175°	(1) Treat with 83% H_3PO_4 in MeOH (2) Filter ppt. and wash with MeOH	Separated 29% pure 2,4,6-collidine (after recrystallization from MeOH)	299
Tar bases, b.p. 150–170°	(1) Treat with anhyd. HCl (2) Filter (3) Add 85% H_3PO_4 in MeOH (4) Filter	Recovered 25% pure 2,3-lutidine from phosphate	311
Tar bases, b.p. 155–160°	(1) Treat with crystalline or 85% H_3PO_4 in MeOH (2) Dilute with MeOH and filter (3) Recrystallize ppt. from MeOH	Separated 98% pure 2,5-lutidine in 90% yield	312
Tar bases, b.p. 157–159°	Fractional crystallization of hydrochlorides	Separated pure 2,4-lutidine and eutectic of 2,4- and 2,5-lutidine	270
Crude lutidines	(1) Add 35% HCl and cool (2) Dehydrate by azeotroping with C_6H_6 (3) Filter insol. hydrochloride	Separated pure 2,4-lutidine in 50–60% yield (after recrystallization from EtOH)	273
Crude lutidines	(1) Treat with HCl (2) Dehydrate by azeotroping with C_6H_6 at	Separated 2,4-lutidine	278

	140–160 torr and up to 75° (3) Filter insol. hydrochloride (4) Decomp. hydrochloride with NaOH		
Tar bases, b.p. 143.5–145.5°	(1) Add anhyd. HCl (2) Cool to 15–20° and filter (3) Recrystallize salt from BuOH	Separated pure 2,6-lutidine	280
Tar bases, b.p. 155–160°	(1) Add anhyd. HCl at 120–130° (2) Filter	Separated 2,4-lutidine in 28% yield	282
Tar bases, b.p. 170–180°	(1) Add Ac_2O to remove aniline (2) Add dry HCl (3) Filter	Separated 2,4,6-collidine in 15–20% yield	282
Crude lutidines (b.p. 155-160°)	(1) Add 36% HCl (2) Add BuOH (3) Distil off H_2O (4) Cool and filter	Separated 2,4-lutidine	296
Crude collidines	Crystallization of hydrochlorides	Separated 2,4,6-collidine	298
3- and 4-Picolines and 2,6-lutidine	(1) Countercurrent extn. with C_6H_6 and 2 *N* HCl (2) Concn. of C_6H_6 layers	Enrichment of 3-picoline (83 wt. % max.) and 2,6-lutidine (73 wt. % max.)	303
Crude β-picoline fraction	Treatment with HCl	Separated 2,6-lutidine	309
Tar bases, b.p. 140–145°	(1) Dilute with <15% H_2O (2) Treat with Amberlite IR-120	Separated 3-picoline	304
....	....	....	....
3- and 4-Picolines and 2,6-lutidine	(1) Add anhyd. $(COOH)_2$ (2) Digest at 105–110° (3) Cool to 25–30° and filter	Separated 4-picoline in nearly theoretical yield	272
Tar bases, b.p. 142–145°	(1) Add anhyd. $(COOH)_2$	Separated β-picoline and 2,6-lutidine	284

TABLE II-9. Methods of Separation of Pyridine Bases (*Continued*)

Bases	Method	Results	Ref.
	(2) Heat to 80° and add absol. alc. (3) Cool and filter		
3- and 4-Picolines and 2,6-lutidine	(1) Add anhyd. $(COOH)_2$ (2) Distil off H_2O (3) Cool and filter	Separated 4-picoline	292
3- and 4-Picolines and 2,6-lutidine	Addn. of $(COOH)_2$	Separated 4-picoline	309
Tar bases, b.p. 143–146°	(1) Add cryst. $(COOH)_2$ (2) Heat to 80–90° (3) Cool and filter	Separated 4-picoline in 90% purity (m.p. −2°)	313
Tar bases, b.p. 142–144°	(1) Add anhyd. $(COOH)_2$ (2) Add absol. alc. (3) Cool and filter	Separated 3- and 4-picolines	320
....			
2-, 3-, and 4-Phenylpyridines	(1) Add metal salt in inert hydrocarbon diluent (2) Filter	Separated 2-phenylpyridine	268
2-, 3-, and 4-Phenylpyridines	Pass through Al_2O_3 chromatographic column impregnated with metal salt	Separated 2-phenylpyridine	269
Crude lutidines	(1) Add aq. $CuCl_2$ (2) Filter	Separated 2,3-lutidine and 2,3,6-collidine	274
Crude lutidines	Treat with Co or Cu chloride in dry MeOH	Separation	275
2,3,6- and 2,4,6-Collidines	Treat with $CuCl_2$ in dry MeOH	Separated 2,3,6-collidine	276
Crude collidines	Treat with $CuCl_2$, $NiCl_2$, or $CoCl_2$ in dry MeOH	Separated 3,5-lutidine	277

3- and 4-Picolines and 2,6-lutidine	(1) Add aq. Ni $(CNS)_2$ (2) Digest 0.5 hr at 70° (3) Filter	Separated 85% pure 4-picoline	279
Crude collidines	Treat with $ZnCl_2$	Separated 3,5-lutidine	281
	Treat with urea	Separated 2,3,6-collidine	
Middle-boiling bases	Treat with urea	Separated 2,3,4,6-tetramethylpyridine, 2,3-cyclopentenopyridine and 2,3-cyclopenteno-6-methylpyridine	281
	Pass through $CuCl_2$-Al_2O_3 chromatographic column treated with aq. glycerol	Separated 3,4,5-collidine and 2,3-cyclopenteno-6-methylpyridine	
Tar bases, b.p. 183–190°	Pass through $CuCl_2$-Al_2O_3 chromatographic column treated with aq. glycerol	Separated 2,3,5-collidine	281
Tar bases, b.p. 138–146°	(1) Treat with aq. urea (2) Dehydrate filtrate (3) Heat with aq. $CuCl_2$ (4) Filter	Separated 98% pure 3-picoline in 25% yield	282
Crude picolines	(1) Treat with $FeCl_2$, $CoCl_2$, $NiCl_2$, $MnCl_2$, or $CuCl_2$ in water or aliph. alc. (2) Filter	Separated 97–98% pure 4-picoline	287
Crude 3-picoline fraction	Treat with $FeCl_3$ in HCl	Separated 4-picoline	289
Crude collidines	Treat with $CuCl_2$	Separated 3,4- and 3,5-lutidines	298
Pyridine solvent	(1) Add HCl to p*H* 4–5 (2) Add satd. soln. of $ZnCl_2$ (3) Cool and filter	Separated pyridine	300
Crude 3-picoline fraction	(1) Add conc. HCl at 20–30° (2) Add finely ground CuCl and SO_2 (3) Filter	Separated 3-picoline	301

TABLE II-9. Methods of Separation of Pyridine Bases (*Continued*)

Bases	Method	Results	Ref.
Pyridine and 2-picoline	(1) Add Mn thiocyanate (2) Filter	Separated 2-picoline	305
Crude 4-picoline fraction	(1) Add HCl soln. of $CoCl_3$ (2) Filter	Separated 3-picoline	306
Crude 3- and 4-picolines	(1) Add aq. $CuSO_4$ (2) Add NaCl (3) Filter	Separated 94% 4-picoline	308
3- and 4-Picolines and 2,6-lutidine	Treat with CuCl or $CuCl_2$	Form addition compounds with all three bases	309
Tar bases, b.p. 170–180°	Treat with dichromate in dil. acid soln.	Separated 2,4,6-collidine in 15% yield	316
Crude base mixture	(1) Countercurrent extn. with PhMe/aq. $NiCl_2$ (2) Extract PhMe phase with H_2SO_4 (3) Spring bases with NaOH	Separated 89% pure 2-picoline	318
3- and 4-Picolines	(1) Add Li halide (2) Filter (3) Hydrolyze complex	Separated 85% pure 3-picoline	319
Crude picolines, b.p. 143–145°	Add $CuSO_4 \cdot 5H_2O$ or Cu_2Cl_2	Separation	322
Tar bases, b.p. 142–145°	(1) Treat with HCHO and steam distil (2) Add Cu_2Cl_2 to distillate (3) Filter complex (4) Decomp. complex with NaOH	Separated 3-picoline in nearly quantitative yield	324
Crude collidines	Treat with urea	Separated 2,3,6-collidine	281

Tar bases, b.p. 138–146°	(1) Treat with urea and H_2O at 100° (2) Filter (3) Steam-distil ppt.	Separated 2,6-lutidine of 94% purity (20 wt. % of steam distillate)	282
Tar bases, b.p. 169–173°	Treat with urea	Separated 2,4,6- and 2,3,6-collidines	315
Crude collidines	(1) Treat with $HCONH_2$ at −6 to −10° (2) Filter (3) Treat ppt. with conc. aq. NaCl (4) Distil oil phase	Separated pure 2,4,6-collidine	291
Crude 3-picoline fraction	(1) Treat with benzenesulfonic acid imide (2) Filter (3) Treat ppt. with alkali	Separated 2,6-lutidine	302
High-boiling bases	Treat with Na phenolate	Separated all three picolines and 2,4- and 2,5-lutidines	271
Crude lutidines	(1) Remove 2,4-lutidine with HCl (2) Add phenol and pet. ether (3) Cool to −20° with stirring (4) Filter adduct (5) Add NaOH and steam distil	Separated pure 2,5-lutidine in 50–55% yield	273
Crude collidines	Treat with *o*-cresol	Separated 2,3,5-collidine	298
Tar bases, b.p. 143.5–145.5°	(1) Remove 2,6-lutidine with HCl (2) Add salicylic acid and heat (3) Cool and filter	Separated 4-picoline	280
Tar bases, b.p. 148–158°	(1) Remove 2,4-lutidine with HCl (2) Remove 2,5-lutidine with phenol	Separated 2,3-lutidine (as *N*-oxide)	285

TABLE II-9. Methods of Separation of Pyridine Bases (*Continued*)

Bases	Method	Results	Ref.
	(3) Treat with AcOH and H_2O_2 (4) Cool to −12° and filter		
Crude collidines	(1) Treat with AcOH and H_2O_2 (2) Cool to −12° and filter	Separated 2,3,6-collidine (as *N*-oxide)	285
Tar bases, b.p. 143°	(1) Treat with AcOH and H_2O_2 (2) Fractionally distil *N*-oxides (3) Reduce distillates with PCl_3	Separated 3- and 4-picolines and 2,6-lutidine	317
Mixed 3- and 4-ethylpyridines	(1) Treat with AcOH and H_2O_2 (2) Fractionally distil *N*-oxide (3) Reduce distillates with PCl_3	Separation	317
Crude collidines	Treatment with picric acid	Separated 4-ethyl-2-methylpyridine	298
Tar bases, b.p. 169–173°	(1) Remove 2,4,6- and 2,3,6-collidines with urea (2) Distil raffinate (3) Treat with picric acid	Separated 4-ethylpyridine	315
4-Picoline and 2,6-lutidine	(1) Dissolve in pentane (2) Add tris(*o*-cresyl) borate (3) Filter borate complex (4) Distil borate complex	Separated 4-picoline of 96.5% purity	314

TABLE II-10. Methods for the Purification of Pyridine Bases

Bases	Treatment	Results	Ref.
Semi-refined pyridine fraction (98.3% pyridine)	(1) Add 1 part H_2O: 5 parts bases (2) Distil through tower with $\sim$ 50 theoretical plates	Removed neutral oil contaminants mainly as azeotropes with H_2O that are lower boiling than pyridine	329
Pyridine and 2-picoline	Protect from water and oxygen	Improved storage stability of products	326
Crude pyridine-base fractions	Hydrotreat at 250–320° and 5–50 atm with oxides of Mo, Ni, and W or sulfides of Ni and W	Product reduced in sulfur content from 0.38 to 0.07%	325
Crude pyridine fraction	(1) Add NaOH and $KMnO_4$ with stirring (2) Add C_6H_6 (3) Distil C_6H_6/H_2O azeotrope (4) Distil to recover bases	Produced base fractions that are clear and color-stable	328
Crude β-picoline fraction	Treatment with Ac_2O-phthalic anhydride mixture	Removed impurities	327

C. *Shale*

Nitrogen bases were recovered from the light tar fraction of Baltic oil shale and were identified. Separation of the bases was accomplished by treatment of the oil with a 100% excess of 40% H_2SO_4, after initial removal of phenols with 20% aqueous alkali. The recovered bases were distilled and their picrates formed and identified. This technique demonstrated that the light fraction contained 2,3- and 2,4-lutidine, 3- and 4-ethylpyridine, 2,3,6-collidine, 2-phenylpyridine, and quinoline; pyridine and picolines were absent (336).

Pyridine bases were identified in the chamber shale tar obtained by high-temperature gasification of Baltic bituminous shale. The chemical content of the bituminous shale tar was found to approach that of coal tar. The most valuable bases, such as pyridine, picolines, lutidines, and collidines, comprise 71.8 to 76.5% of the total bases. 3- and 4-Picoline amount to 35 to 40% of the bases in the 136 to 152° fraction (339).

From the <175° fraction of light oils from Fushun (China) shale oil, 12 basic nitrogen compounds were isolated. The <190° fraction of light oils from Maoming shale oil yielded 19 compounds, three of unknown structure. Analysis by ultraviolet spectroscopy in isooctane solution indicated the presence of pyridine, 2-ethylpyridine, 2,5-, 2,4-, 2,3-, and 3,5-lutidines, 6-ethyl-2-methylpyridine, 4-ethyl-2-methylpyridine, 5-ethyl-2-methylpyridine, 2-ethyl-4-methylpyridine, and 2,3,5-, 2,3,6-, and 2,4,5-collidines (337).

The crude heavy oil from NTU shale oil was extracted with dilute H_2SO_4 and the acid layer neutralized to yield a basic fraction rich in 2,4,6-trialkylpyridines (338).

A systematic study was made of the nitrogen compounds in shale-oil gas oil. A concentrate of the nitrogen compounds was made by Florisil adsorption, and this was further resolved by molecular distillation and thermal diffusion. Analysis by combinations of various techniques indicated that more than 50% of the nitrogen compounds in the gas oil were pyridines, dihydropyridines, indoles, and quinolines, with about twice as many pyridines as the others. Most of the remainder were compounds with one or more saturated rings condensed with the above compound types (334).

Tar water from the distillation of shale tar was found to contain 0.85% pyridine bases. These were separated and analyzed by gas chromatography employing a two-column technique. The first column comprised 20 wt. % of polyethylene glycol 4000 on diatomite; the second contained 15 wt. % of Apiezon L on Chromosorb W. By this means, the 80 to 165° fraction of recovered bases was shown to contain pyridine and its alkyl derivatives (335).

Examination of samples from the sandstone and shale beds of McMurray, Alberta was made by extraction with CH_2Cl_2. The extract was separated by elution chromatography on silica gel and alumina columns. Analysis of the eluate fractions revealed the presence of pyridines and pyrroles (340).

D. *Other Fuel Deposits*

Pyridine bases were recovered from coal tar derived from thermal decomposition of Dnieper Basin lignite by washing with 20% H_2SO_4, then with 13% NaOH (to destroy pyridine-phenol complexes) and finally with 20% H_2SO_4. The acid extract was washed with petroleum ether and neutralized with NH_3. From a 70 to 290° tar fraction were recovered crude bases containing 3.7% pyridine, 11.5% picolines, 24.7% lutidines, and 60.1% collidines and quinolines (342).

Tar distillates obtained from low-temperature carbonization of Texas lignite were found by Barrett analysis to contain 3 to 5% tar bases; this concentration applies for distillates of any boiling range and holds reasonably well for distillation residues. A study of these bases showed them to be heterocyclic, of pyridine and quinoline structures, with no evidence of aniline types (345). In a related study, a primary tar from the low-temperature carbonization of Texas lignite was found to contain 4% heterocyclic bases (pyridines, quinolines) (349).

Tar from lignite contains 200 versus 120 g of bases per ton in coal tar. A light oil fraction was extracted at 30° with 30% H_2SO_4, and the acid extract was purified by steam distillation and was neutralized with aqueous $(NH_4)_2CO_3$. Analysis by gas chromatography indicated that lignite-derived bases contain more pyridine, 3- and 4-picoline, and 2-ethylpyridine than those derived from coal tar (348).

A 230 to 270° fraction of lignite tar, containing approximately 1% organic bases, was extracted with 10 to 15% H_2SO_4 and water, the combined aqueous phases extracted with ketone to remove tar acids, and the acid sulfate liquor neutralized with aqueous NH_3. The bases recovered after dehydration represent a 90% yield and contain about 33% quinolines and 58% pyridines (346).

West, Abramovitch, and Pepper studied the composition of the basic components of tar obtained by a low-temperature Parry carbonization at 499° of Estevan (Saskatchewan) lignite. Gas chromatography was employed to separate and identify pyridines, quinolines, and other bases (346a).

The light tar from Lurgi gasification of Australian brown coal was examined by distillation to effect a primary separation, followed by extraction with 20% aqueous NaOH and 20% H_2SO_4 to isolate tar acids and tar bases. The portion of tar boiling $<380°$ was found to contain 1.5% nitrogen bases (341).

Tar bases present in the 114.6 to 167.0° fraction of light oil from brown coal tar were isolated by precipitation of their $HgCl_2$ double salts and identification of derivatives. Pyridine, the three picolines, all six lutidines and 2-, 3-, and 4-ethylpyridine were identified (343).

Tars produced by the pressure steam-oxygen gasification of Irsha-Borodinskoe (U.S.S.R.) brown coal contain 3.5% pyridine bases (347).

A light tar-base fraction from Indian gas tar, having a boiling range of 90 to 170°, was repeatedly fractionated to yield 36% of pyridine, 12% of 2-picoline, 16% of 3-picoline, 8% of 2,4-lutidine, and 16% of residue boiling >160°; total intermediate fractions amounted to 12% (344).

From a 200 to 250° fraction of sapropel tar, the picrates of a number of nitrogen bases were formed and identified. These included 3-isopropylpyridine, the three picolines, 2-ethyl-4-methylpyridine, 3-ethyl-4-methylpyridine, and 2-aminopyridine. From the distillation of sapropel tar below 250° a total of 30 pyridine bases have been isolated (350). In a related study, tar from the dry distillation of sapropels was found to contain 5% of basic compounds, of which only 50% of the bases was extractable with 10% H_2SO_4. A pyridine-picoline fraction isolated by acid extraction contained pyridine, the three picolines, and 2,6-lutidine (352). Extension of this study to higher-boiling fractions resulted in the identification of a number of pyridine bases (354).

The primary tar obtained by distillation of Latvian (Dobele region) sapropel was repeatedly distilled and the fractions extracted with 20% H_2SO_4 to remove the bases. The acid extract was freed of neutral oils, then neutralized with alkali. The crude bases were carefully distilled into small fractions and the picrates formed and fractionally precipitated. By this means pyridine, the three picolines, the three ethylpyridines, all six lutidines, 2,3,4-, 2,3,5-, 2,3,6-, and 2,4,6-collidines, 4-ethyl-2-methyl-, 5-ethyl-2-methyl-, 6-ethyl-2-methyl-, 2-ethyl-4-methyl-, and 3-ethyl-4-methylpyridines, 3-isopropylpyridine, 2,3,5,6-tetramethylpyridine, and 2-aminopyridine were identified (344).

Twenty bases were isolated from saprolic tar and identified: pyridine, the three picolines, the six lutidines, 2,3,5-, 2,3,6-, and 2,4,6-collidines, 4-ethyl-2-methyl-, 5-ethyl-2-methyl- and 6-ethyl-2-methylpyridines, and 2,3,5,6-tetramethylpyridine (351).

A study was made of the low-temperature carbonization of Mnevsk (U.S.S.R.) peat at 285 to 500°. Pyridine bases did not become prominent in the tar until the carbonization temperature was raised to 360°. A further increase in carbonization temperature to 420° resulted in a decrease in the concentration of bases (355).

Fifteen pyridine bases were identified in peat tar by passing benzene solutions of distillate fractions through a chromatographic column filled with Al_2O_3 and $CuCl_2$, isolating the colored zones, eluting each zone with ether, and forming the picrates of the recovered bases (356).

Tar bases were found in the pyrolysate from Gilsonite, a solid hydrocarbonaceous material found in the western United States. The bases were concentrated by extraction from the pyrolysis oil with 40% aqueous H_3PO_4 and subsequent neutralization with solid NaOH. A fraction boiling 160 to 220° was resolved by column chromatography, and the following were identified: 4-picoline, 2,5-, 2,6-, 3,4-, and 3,5-lutidines, and 2,3,5-collidine (357).

An investigation was made of the extractability of various materials from fractions of crude oil extracted from kerogen. Anthracene oil at 300° dissolved 2.04 to 3.05% of pyridine bases (as a percentage of the total organic matter). Cylinder oil extracted 1.56% bases; mazut removed 0.76% bases (358).

E. *Degradation of Alkaloids*

Complex alkaloids are very often degraded to simpler pyridine products by chemical, physical, and other means. This is generally a basis for the determination of the structure of the original alkaloids.

Zinc dust distillation of aspidospermine gave 3,5-diethylpyridine, together with 3-ethylindole and/or skatole (359). The same alkylpyridine was obtained by zinc dust distillation of quebrachamine (372).

The washed spores of *B. megaterium* were studied in detail with respect to production of dipicolinic acid. Mechanical breakage, boiling, and death of the spores caused the release of dipicolinic acid, as did also ultraviolet irradiation. Organic solvents (80% MeOH, 80% EtOH, and 80% acetone) extracted considerable amounts of the acid (361). In a separate study, the spores were disrupted by vibration in a Mickle tissue disintegrator and were extracted with hot water to recover dipicolinate (368). Addition of metal ions to bacterial spores, followed by treatment with surface-active agents, enabled the extraction of dipicolinic acid, probably as a chelate with the added metal (369, 370).

Oxidation and dehydrogenation techniques have been employed successfully to degrade alkaloids to simpler pyridine structures. Chromic acid oxidation of adenocarpine gave optically active $RCO(CH_2)_3NHCOCOOH{\cdot}H_2O$ where R = 2-piperidyl (363). Vigorous oxidation of gentianine in 2 *N* NaOH at 100° with aqueous $KMnO_4$ and purification by passage through Zeo-Karb 315 gave pyridine-3,4,5-tricarboxylic acid (364). Dehydrogenation of rhynchophylline gave 3,4-diethylpyridine (367).

Potassium fusion of the alkaloid voacorine produced 5-ethyl-3-methylpyridine (365). Alkali fusion of betaxanthins, or hot acid treatment, yielded 4-methylpyridine-2,6-dicarboxylic acid (371).

Pyrolysis of nicotine produced ten nitrogen bases that were subsequently identified (362). Microbial degradation of nicotine with species of *Pseudomonas* produced 3-succinoyl-6-hydroxypyridine and γ-oxo-γ-(3-pyridyl)butyric acid (366).

F. *Miscellaneous Sources*

Pyrolysis of ammoniated invert sugar at 600°, until the evolution of volatile material ceased, gave a steam-volatile fraction containing primarily 2-picoline (374).

Partial carbonization of reed grasses from Southeast Romania yields a tar which is separated into three fractions by distillation. The heavy oil fraction (b.p. 150–250°) contains about 1% pyridine bases, the recovery of which appears economical (375).

The dry distillation of fat-free soybeans yielded fractions that were analyzed by infrared spectroscopy and gas chromatography. The following nitrogen bases were found: pyridine; the three picolines, 2,6-, 2,4-, and 3,5-lutidines; 2- and 3-ethylpyridine, 2,4,6-collidine, and 2-ethyl-4-methylpyridine (376).

II. Pyridines by Synthetic Methods

1. From Other Ring Systems

A. *Carbocyclic Compounds*

Cyclohexanone and its derivatives reacted in the vapor phase with ammonia to form pyridines (Table II-11). It is believed that cyclohexenones are intermediates on the reaction (377).

TABLE II-11. Reaction of Cyclohexanones with Ammonia in the Vapor Phase (377)

Cyclohexanone	Catalyst	Product
Cyclohexanone	Cadmium phosphate acid clay	2-Picoline
2-Methylcyclohexanone	Cadmium phosphate acid clay	2,3-Lutidine
3-Methylcyclohexanone	Cadmium phosphate acid clay	2,6-Lutidine
4-Methylcyclohexanone	Cadmium phosphate acid clay	No pyridine compound
2-Cyclohexenone	Cadmium phosphate acid clay	2-Picoline
3,5-Dimethyl-2-cyclohexenone	Cadmium phosphate acid clay	2,4,6-Collidine

Passage of active nitrogen from a glow discharge tube into benzene for 25 to 50 hr yielded pyridine together with hydrogen cyanide, benzonitrile, terephthalonitrile, and phenylisocyanide (378). Free radical intermediates are postulated.

Cyclopentadiene, when ozonized in anhydrous ammonia at −50°, gave pyridine in 18% yield (379). Polymerization of cyclopentadiene under the reaction conditions lowered the pyridine yield.

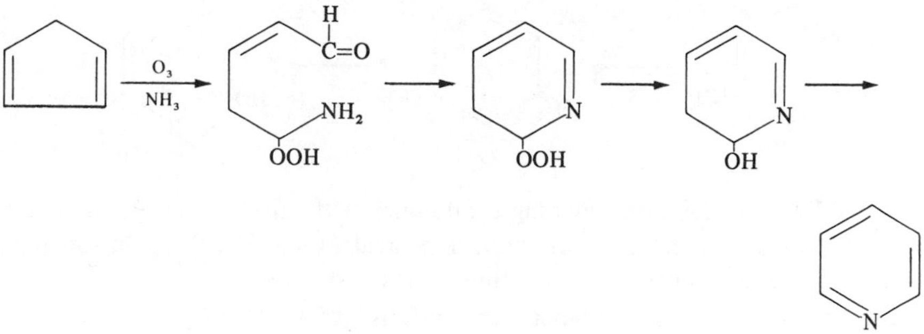

O$_3$ NH$_3$ → OOH, NH$_2$, C=O, H → N, OOH → N, OH → N

When heated with benzonitrile in a sealed tube at 270 to 300° anisylcyclone **(II-1)** gave a Diels-Alder adduct which eliminated carbon monoxide to form a highly substituted pyridine in 50 to 60% yield (380):

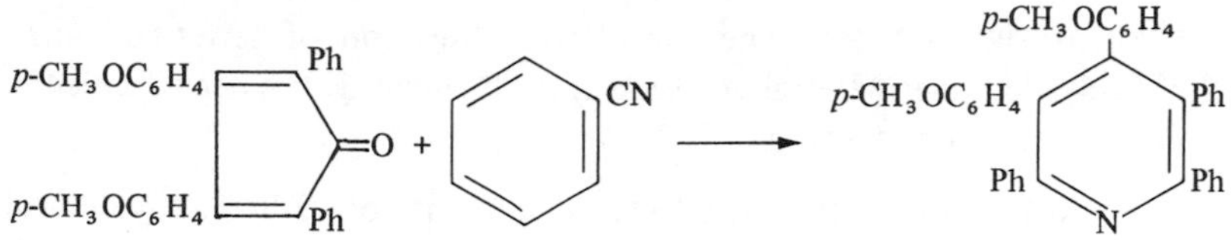

p-$CH_3OC_6H_4$, p-$CH_3OC_6H_4$, Ph, Ph, =O + CN → p-$CH_3OC_6H_4$, p-$CH_3OC_6H_4$, Ph, Ph, Ph, N

II-1

Similarly, tetracyclone and phenylcyanoformate to give phenyl 3,4,5,6-tetraphenylpicolinate (380):

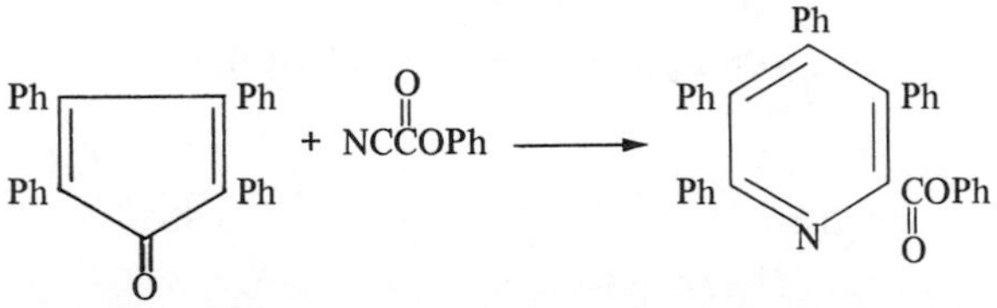

Ph, Ph, Ph, Ph, O + NCCOPh (O) → Ph, Ph, Ph, Ph, N, COPh (O)

When either the dimethyl **(II-2)** or diethylketal of 1,2,3,4-tetrachlorocyclopentadienone reacts with benzoylcyanide, a chlorine is eliminated to form the corresponding ester of 3,4,5-trichloro-6-benzoylpicolinic acid **(II-3)** (382):

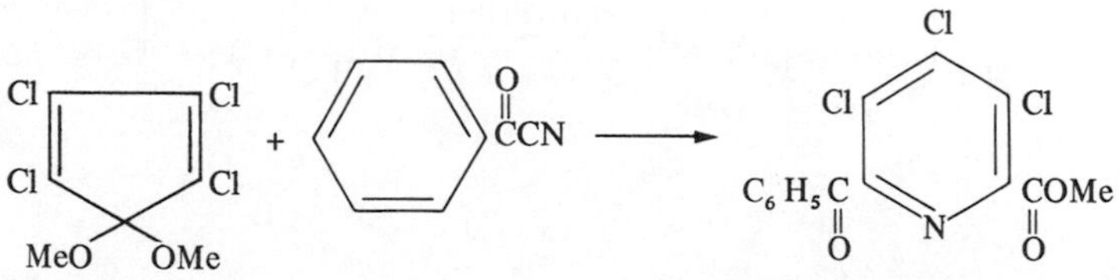

Cl, Cl, Cl, Cl, MeO, OMe + CCN (O) → Cl, Cl, Cl, C_6H_5C (O), N, COMe (O)

II-2 **II-3**

A new method for the conversion of 3-aminocatechols and related compounds to pyridine derivatives has been reported (383). The *o*-dihydroxy compounds are first oxidized to the corresponding *o*-quinones with silver oxide. Pyridines are formed by treating the quinones with a peracid (Table II-12). A ring-opening reaction to form a muconic acid intermediate is postulated.

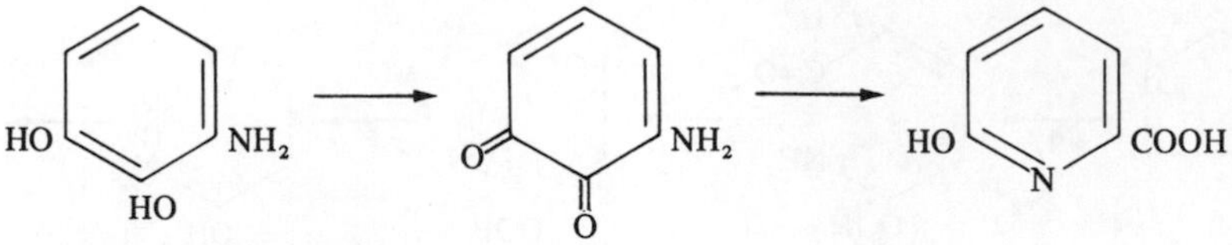

Flash vacuum pyrolysis of β-phenethylsulfonyl azide at 650° gives good yields of dihydro-[2,3]pyridine (383a).

B. *Furans*

a. TETRAHYDROFURANS. The vapor phase reaction of tetrahydrofurfuryl alcohol (2-hydroxymethyltetrahydrofuran) with ammonia has been studied with a number of catalyst systems (Table II-13).

b. DIHYDROFURANS AND FURFURAL. Treatment of 2-heptanoyl-2,5-dihydro-2,5-dimethoxy-furan (**II-4**) with hydroxylamine hydrochloride in methanol resulted in the formation of 1,5-dihydroxy-6-*n*-hexyl-2-pyridone (**II-5**):

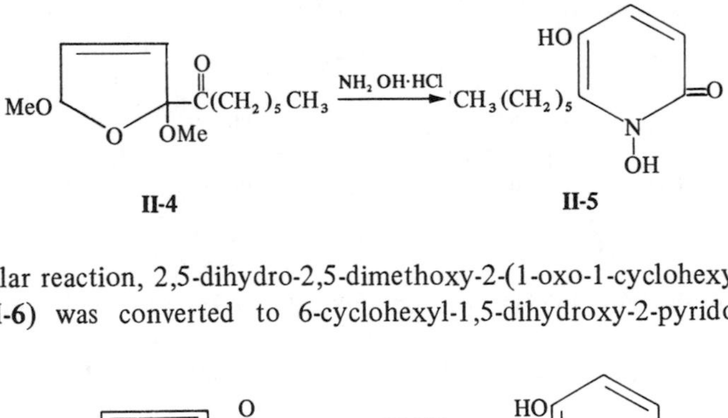

$NH_2OH{\cdot}HCl$

II-4 **II-5**

In a similar reaction, 2,5-dihydro-2,5-dimethoxy-2-(1-oxo-1-cyclohexylmethyl)-furan (**II-6**) was converted to 6-cyclohexyl-1,5-dihydroxy-2-pyridone (**II-7**)

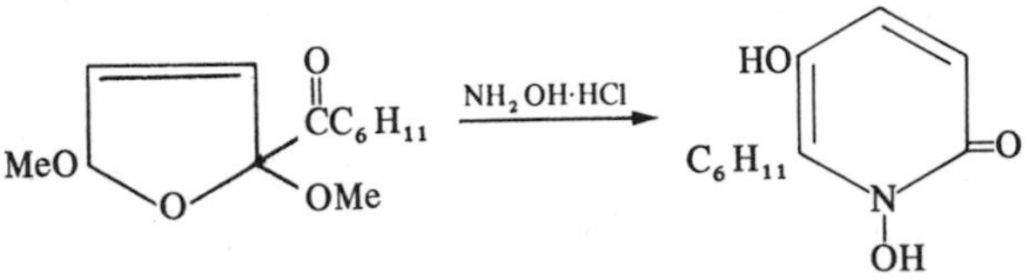

$NH_2OH{\cdot}HCl$

II-6 **II-7**

(394). When 2,5-dimethoxy-2-(α, α-dimethoxyethyl)-2,5-dihydrofuran (**II-8**) is treated with hydroxylamine hydrochloride, 1,5-dihydroxy-6-methyl-2-pyridone is formed (**II-9**) (395):

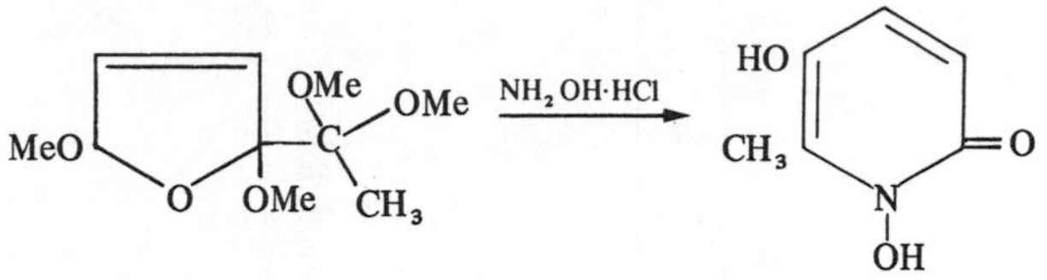

MeO, O, OMe, OMe, C, OMe, CH_3 —$NH_2OH \cdot HCl$→ HO, CH_3, N, OH, O

II-8 **II-9**

3-Hydroxypicolinic acid is formed from furfural by conversion to the cyanohydrin with sodium cyanide, followed by a treatment with urea to form the α-ureido-2-furanoacetonitrile in 32% yield. Chlorination of the ureido compound gave 3-hydroxypicolinic acid amide in 47% yield and 3-hydroxypicolinonitrile in 7% yield (396):

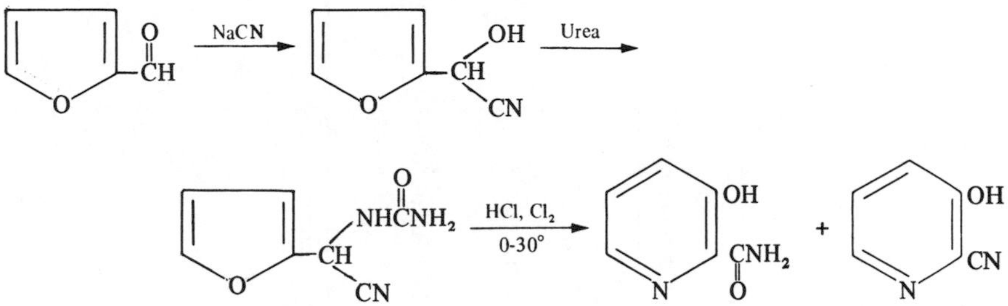

O, CH, O —NaCN→ O, OH, CH, CN —Urea→

O, NHCNH₂ (O), CH, CN —HCl, Cl_2, 0-30°→ N, OH, CNH_2 (O) + N, OH, CN

Furfural is converted to 2,3-dihydroxypyridine by the action of sodium hypochlorite at 0° followed by treatment with ammonium hydroxide at 95° (397).

c. FURFURYLAMINES AND ACYLFURANS. Furfurylamines can be rearranged to 3-pyridinols by treatment with an oxidizing agent such as hydrogen peroxide, a halogen, or sodium hypochlorite in the presence of acid (398) (Table II-14).

Acylfurans are converted to 3-pyridinols by heating in the presence of ammonia at elevated temperatures (Table II-15).

In the carbohydrate series, it was found that 5,6-diacetamido-5,6-dideoxy-L-idofuranose forms 2-aminomethyl-5-hydroxypyridine by mild hydrolysis, which leads to deacetylation followed by dehydration (407).

TABLE II-12. Conversion of *o*-Dihydroxyaromatics and *o*-Quinones to Pyridines (383)

Reactant	Peracid	Product	Yield (%)
3,4-Dihydroxyanthranilic acid	Peracetic	6-Hydroxyquinolinic acid	41.2
3,4-Dihydroxyanthranilic acid	Trifluoroperacetic	6-Hydroxyquinolinic acid	16.1
3-Amino-4-carboxy-1,2-benzoquinone	Paraperiodic	6-Hydroxyquinolinic acid	11.1
3-Aminocatechol	Perbenzoic	6-Hydroxypicolinic acid	12.3
3-Amino-1,2-benzoquinone	Peracetic, trifluoroacetic, or periodic	Pyridine 2-Pyridone	0–14 9.5–14.1
3,4-Diamino-5-methyl-1,2-benzoquinone	Peracetic	3-Amino-6-hydroxy-4-methylpicolinic acid	11.3

TABLE II-13. Formation of Pyridines from Tetrahydrofurfuryl Alcohol

Conditions	Product (yield %)	Ref.
NH_3, vapor phase, $(NH_4)_2MoO_4$ on alumina, 400°	Pyridine (69)	384
NH_3, vapor phase, H_2-reduced zinc molybdate on alumina, 520°	Pyridine (56) and dihydropyran	385
NH_3, vapor phase, 0.25% Pt on silica-alumina, 500°	Pyridine (48)	386
NH_3, vapor phase, cobalt molybdate on alumina, 525°	Pyridine (39)	387
NH_3, vapor phase, K-modified molybdate on alumina, 525°	Pyridine (45–57)	388
NH_3, vapor phase, vanadia-molybdena on alumina, 525°	Pyridine (44)	389
NH_3, vapor phase, molybdena-alumina, 500°	Pyridine (20), 3-picoline (5.6), 3-ethylpyridine (3.5)	390
NH_3, ethylamine, vapor phase, molybdena-alumina, 500°	Pyridine (8.8), 3-picoline (4.2), 3-ethylpyridine (24.1)	390
NH_3, ethanol, vapor phase, molybdena-alumina, 500°	Pyridine (7.5), 3-picoline (3.8), 3-ethylpyridine (19.8)	390
NH_3, methanol, vapor phase, molybdena-alumina, 500°	Pyridine (13.0), 3-picoline (17.7), 3-ethylpyridine (1.8)	390
NH_3, H_2, vapor phase, 0.3% Pd on alumina, 260–320°	Pyridine and piperidine	391
NH_3, vapor phase, Cr_2O_3, MoO_2, or reduced Cu on alumina, 400–550°	Pyridine	392
NH_3, vapor phase, catalyst	Pyridine, 3-picoline, piperidine	393

TABLE II-14. Formation of 3-Pyridinols from Furfurylamines

Starting material	Conditions	Product (yield %)	Ref.
Furfurylamine	HCl, 30% H_2O_2, reflux	3-Pyridinol (73)	398
2-(α-Aminoethyl)furan	HCl, 30% H_2O_2, reflux	2-Methyl-3-pyridinol (77)	398
2-(α-Aminoethyl)-3,4-bis(hydroxymethyl)furan	HCl, 30% H_2O_2, reflux	Pyridoxine (67)	398
2-Hydroxymethyl-5-aminomethylfuran	Electrolytic methoxylation, alkaline hydrolysis	6-Hydroxymethyl-3-pyridinol (74)	399
2-Hydroxymethyl-5-aminomethylfuran	Reflux 15 hr, *N* HCl	6-Methyl-3-pyridinol (88)	399

TABLE II-15. Formation of 3-Pyridinols from 2-Acylfurans

Starting material	Conditions	Product (yield %)	Ref.
2-Furyl propyl ketone	NH_3, methanol, NH_4Cl, 28% NH_4OH, autoclave at 180°	3-Hydroxy-2-propylpyridine	400–402
2-Acetylfuran	Alc. NH_3, NH_4Cl, sealed tube 204°, 2 hr	3-Hydroxy-2-methylpyridine	403
2-Acetylfuran	Liquid ammonia, NH_4Cl	3-Hydroxy-2-methylpyridine	404
2-Propionoylfuran	Liquid ammonia, NH_4Cl	2-Ethyl-3-hydroxypyridine	404
2-Benzoylfuran	NH_3, methanol, NH_4Cl, 28% NH_4OH, autoclave at 180°	3-Hydroxy-2-phenylpyridine (15–25)	405
2-(*p*-Toluoyl)furan	NH_3, methanol, NH_4Cl, 28% NH_4OH, autoclave at 180°	3-Hydroxy-2-(*p*-tolyl)pyridine (20–25)	405
2-(4-Methoxy-3-methylbenzoyl)furan	NH_3, methanol, NH_4Cl, 28% NH_4OH, autoclave at 180°	2-(4-Methoxy-3-methylphenyl)-3-hydroxypyridine (20–40)	405
3-(2-Furoyl)propionic acid	Aqueous conc. NH_3 sealed tube, 160–170°, 15–20 hr	3-(3-Hydroxy-2-pyridyl)propionic acid	406
4-(5-Methyl-2-furoyl)butyric acid	Aqueous conc. NH_3 sealed tube, 160–170°, 15–20 hr	4-(3-hydroxy-6-methyl-2-pyridyl)-propionic acid	406
8-(2-Furoyl)caprylic acid	Aqueous conc. NH_3 sealed tube, 160–170°, 15–20 hr	8-(3-Hydroxy-2-pyridyl)caprylic acid	406
8-(5-Methyl-2-furoyl)caprylic acid	Aqueous conc. NH_3 sealed tube, 160–170°, 15–20 hr	8-(3-Hydroxy-6-methyl-2-pyridyl) caprylic acid	406
3-(5-Acetyl-2-furyl)propionic acid	Aqueous conc. NH_3 sealed tube, 160–170°, 15–20 hr	3-(3-Hydroxy-6-methyl-2-pyridyl)propionic acid	406
8-(5-Acetyl-2-furyl)caprylic acid	Aqueous conc. NH_3 sealed tube, 160–170°, 15–20 hr	8-(5-Hydroxy-6-methyl-2-pyridyl) caprylic acid	406

TABLE II-16. Reactions of Pyran Derivatives with Ammonia

Pyran	Conditions	Products (yield %)	Ref.
2,6-Dialkoxy-3-(alkoxyalkyl)tetrahydropyran	Alumina or alumina with heavy metal salts at 300–350°	3-Alkylpyridine	408
2,6-Dialkoxy-3-(alkoxyalkyl)tetrahydropyran	Pt or Pd on alumina	3-Alkylpyridine	409
2,6-Diethoxy-3-(1-ethoxyethyl)tetrahydropyran	Alumina, 325–350°	3-Ethylpyridine, (50.5)	410
2,6-Dialkoxy-3-(1-alkoxyalkyl)tetrahydropyran	Autoclave at 220°	Alkylpyridines	411
2-Ethoxy-3,4-dihydro-2*H*-pyran	0.3% Pd on silica-alumina, steam, 300°	Pyridine (66)	412
2-Methoxy-3,4-dihydro-2*H*-pyran	HOAc, $Cu(OAc)_2$, NH_4OAc, air, 150 psi at 135°	Pyridine (93)	413
2,3-Dialkoxy-3,4-dihydro-2*H*-pyrans	Pt or Pd on alumina, steam, 200–250°	3-Alkoxypyridines	414
2-Ethoxy-3,4-dihydro-2*H*-pyran	Boric and phosphoric acids on silica-alumina, steam, 300°	Pyridine	415
2-Methoxy-5,6-dihydro-2*H*-pyran	Pt on silica-alumina, steam, 300°	Pyridine (36)	416

C. *Pyrans*

Pyridines can be formed from pyran compounds by reaction either with ammonia (Table II-16) or with hydroxylamine (Table 11-17).

D. *Pyrones*

a. 2-PYRONES. When treated with ammonium hydroxide and sodium hydroxide followed by acidification with concentrated hydrochloric acid, methyl coumalate (**II-10**) gave a 90% yield of methyl 6-hydroxynicotinate (**II-11**) (420, 421):

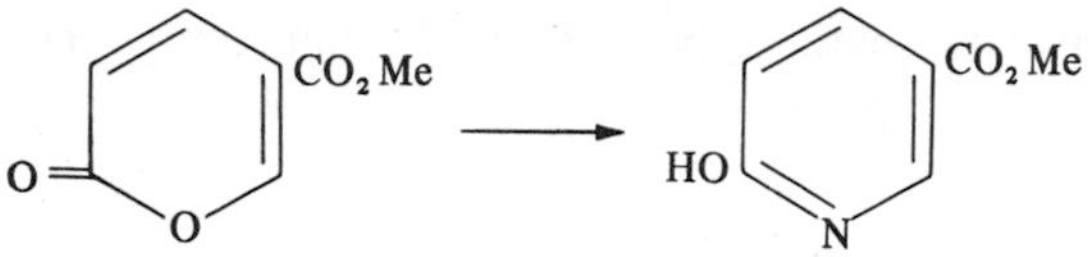

II-10 → **II-11**

b. 4-PYRONES. By heating 3-acetyl-2,6-dimethyl-4-pyrone with ammonium hydroxide at 75° for 1 hr, 3-acetyl-2,6-dimethyl-4-pyridone was formed (422). Treatment of 2-(*p*-chlorophenoxymethyl)-5-methoxy-4-pyrone with concentrated ammonium hydroxide gave 2-(*p*-chlorophenoxymethyl)-5-methoxy-4-pyridone. Similarly, 2-(6-chloro-2-methylphenoxymethyl)-5-methoxy-4-pyrone was converted to the corresponding 4-pyridone (423).

It was determined that when 3-[1-(carboxymethylamino)ethyl]-4-hydroxy-6-methyl-2-pyrone (**II-12**) was allowed to stand at room temperature for 80 days, it rearranged to 2,6-dimethyl-1,4-dihydro-4-oxo-1-pyridineacetic acid (**II-13**):

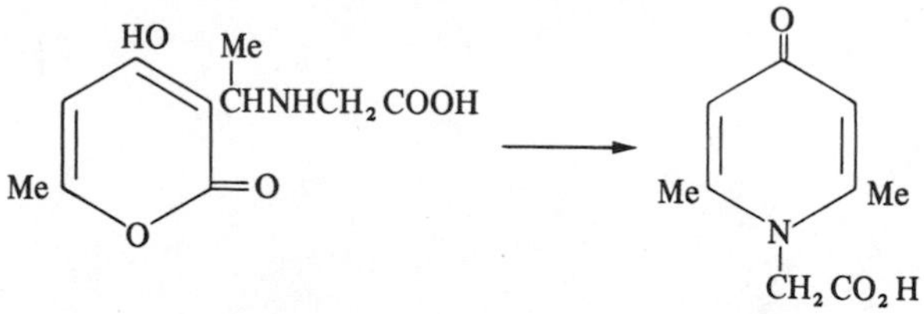

II-12 → **II-13**

The histidine derivative (**II-14**) also rearranged in a similar manner to a pyridone (**II-15**) (424):

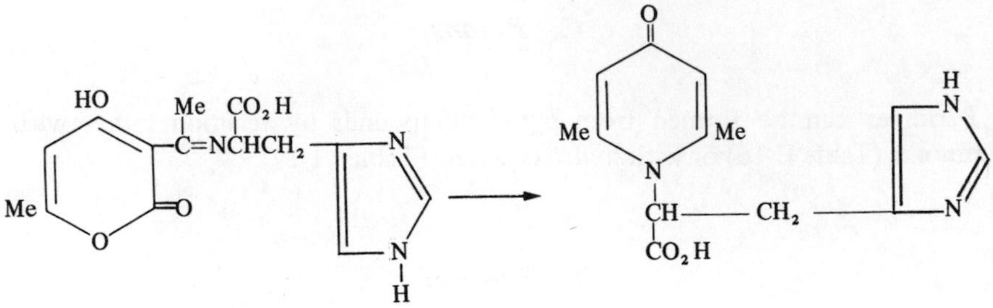

II-14 → **II-15**

E. *Pyrylium Salts*

The conversion of a number of pyrylium salts to pyridines is reported in Table II-18. Review articles on the synthesis of pyrylium compounds have appeared recently (437).

TABLE II-17. Reactions of Pyran Derivatives and Hydroxylamine

Pyran	Conditions	Product (yield %)	Ref.
2-Ethoxy-3,4-dihydro-2*H*-pyran	$NH_2OH \cdot HCl$, 130°, pressure	Pyridine (70–80)	417
2-Ethoxy-3,4-dihydro-2*H*-pyran	$NH_2OH \cdot HCl$, 100°	Pyridine (50–60)	417
2-Ethoxy-4-methyl-3,4-dihydro-2*H*-pyran	$NH_2OH \cdot H_2SO_4$, 105–112°	4-Picoline (88)	417
2-Ethoxy-4-ethyl-5-methyl-3,4-dihydro-2*H*-pyran	$NH_2OH \cdot HCl$, 100°	4-Ethyl-3-methylpyridine (61)	417
2-Butoxy-5-ethyl-3,4-dihydro-2*H*-pyran	$NH_2OH \cdot HCl$, AcOH, reflux	3-Ethylpyridine	418
2-Ethoxy-3-ethyl-3,4-dihydro-2*H*-pyran	$NH_2OH \cdot HCl$, AcOH, reflux	3-Ethylpyridine (60)	418
3-Butyl-2-ethoxy-6-methyl-3,4-dihydro-2*H*-pyran	NH_2OH, AcOH, reflux	5-Butyl-2-methylpyridine (37.5)	419
3-Butyl-2-ethoxy-4-methyl-3,4-dihydro-2*H*-pyran	NH_2OH, AcOH, reflux	3-Butyl-4-methylpyridine (80.5)	419
2-Ethoxy-3-hexyl-6-methyl-3,4-dihydro-2*H*-pyran	NH_2OH, AcOH, reflux	5-Hexyl-2-methylpyridine (42.3)	419

TABLE II-18. Reactions of Pyrylium Salts with Ammonia

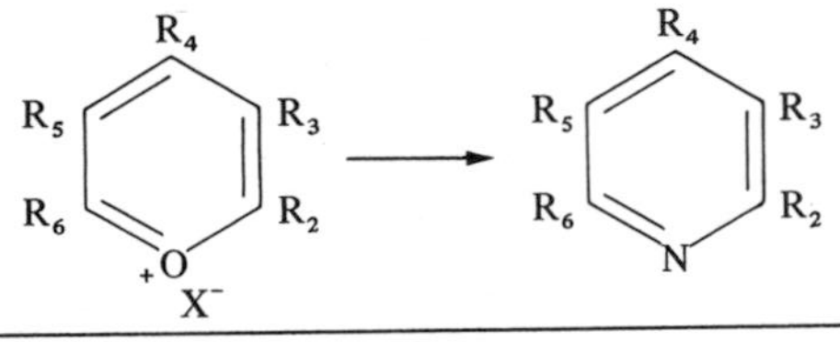

R_2	R_3	R_4	R_5	R_6	X^-	Conditions	Yield (%)	Ref.
Me	H	Me	H	Me	ClO_4^-	NH_3, NH_4OAc in HOAc, reflux		425, 434
Et	H	H	H	Me	ClO_4^-	NH_3, NH_4OAc in HOAc, reflux		425
Et	H	Me	H	Me	ClO_4^-	NH_3, NH_4OAc in HOAc, reflux		425
Me	Me	Me	H	Me	ClO_4^-	NH_3, NH_4OAc in HOAc, reflux		425, 434
Me	Me	Me	H	Me	$SnCl_5^-$	NH_3, NH_4OAc in HOAc, reflux		435
Me	H	Et	H	Me	ClO_4^-	NH_3, NH_4OAc in HOAc, reflux		425, 434, 435
Et	Me	Me	H	Et	ClO_4^-	NH_3, NH_4OAc in HOAc, reflux		425
Et	H	Et	H	Et	ClO_4^-	NH_3, NH_4OAc in HOAc, reflux		425
Et	H	Me	H	Et	ClO_4^-	NH_3, NH_4OAc in HOAc, reflux		425
Me	H	Me	Me	Me	ClO_4^-	NH_3, NH_4OAc in HOAc, reflux		425
Me	H	Ph	H	Me	ClO_4^-	NH_3, NH_4OAc in HOAc, reflux		425, 434
Me	Ph	Me	H	Me	ClO_4^-	NH_3		426
Ph	H	Ph	H	Ph	ClO_4^-	NH_4OH, cold		427
Ph	H	4-Acetoxy-2-buten-1-onyl	H	Ph	ClO_4^-	NH_4OH, cold	27	427
Ph	H	5-Iodofuryl	H	Ph	ClO_4^-	NH_4OH, cold	41	427
Ph	H	H	H	Ph	ClO_4^-	NH_4OAc, HOAc, reflux	96.4	428, 429
Ph	H	COOH	H	Ph	ClO_4^-	NH_4OAc, HOAc, reflux	94.1	428
Ph	H	Ph	H	1-$C_{10}H_7$	ClO_4^-	NH_4OAc, HOAc, reflux		429
1-$C_{10}H_7$	H	Ph	H	1-$C_{10}H_7$	ClO_4^-	NH_4OAc, HOAc, reflux		429

Ph	H	Ph	H	2-$C_{10}H_7$	ClO_4^-	NH_4OAc, HOAc, reflux		429
2-$C_{10}H_7$	H	Ph	H	2-$C_{10}H_7$	ClO_4^-	NH_4OAc, HOAc, reflux		429
Ph	H	Ph	H	*p*-PhC_6H_4	ClO_4^-	NH_4OAc, HOAc, reflux		429
p-PhC_6H_4	H	Ph	H	*p*-PhC_6H_4	ClO_4^-	NH_4OAc, HOAc, reflux		429
2-$C_{10}H_7$	H	Ph	H	*p*-PhC_6H_4	ClO_4^-	NH_4OAc, HOAc, reflux		429
Ph	H	Ph	H	*p*-BrC_6H_4	ClO_4^-	NH_4OAc, HOAc, reflux		429
Ph	H	Ph	H	*p*-PhCH=CH	ClO_4^-	NH_4OAc, HOAc, reflux		429
Ph	H	*p*-MeC_6H_4	H	Ph	ClO_4^-	NH_4OAc, HOAc, reflux		429
Ph	H	*p*-PhCH=CH	H	Ph	ClO_4^-	NH_4OAc, HOAc, reflux		429
Ph	H	*p*-FC_6H_4	H	Ph	ClO_4^-	NH_4OAc, HOAc, heat		430
Ph	H	*p*-ClC_6H_4	H	Ph	ClO_4^-	NH_4OAc, HOAc, heat		430
Ph	H	*p*-BrC_6H_4	H	Ph	ClO_4^-	NH_4OAc in HOAc, heat		430
Ph	H	*p*-IC_6H_4	H	Ph	ClO_4^-	NH_4OAc in HOAc, heat		430
p-$NO_2C_6H_4$	H	*p*-$NO_2C_6H_4$	H	*p*-$NO_2C_6H_4$	ClO_4^-	NH_4OAc in HOAc, heat		430
Ph	H	Ph	H	Me	BF_4^-	NH_3 in MeOH, reflux		431
Ph	H	Ph	H	Ph	BF_4^-	NH_3 in MeOH, reflux		431
Me	*n*-Pr	H	H	Me	ClO_4^-	NH_4OH		432
Me	*n*-Bu	H	H	Me	ClO_4^-	NH_4OH	30	432
Me	*n*-Hexyl	H	H	Me	ClO_4^-	NH_4OH	42	432
Me	*n*-Heptyl	H	H	Me	ClO_4^-	NH_4OH	23	433
t-Bu	H	*t*-Bu	H	*t*-Bu	BF_4^-	NH_3		433
Me	Et	H	H	Me	ClO_4^-	NH_3		434
Me	Me	H	Me	Me	ClO_4^-	NH_3		434
Me	isoPr	H	H	Me	ClO_4^-	NH_3		435
Me	Me	Et	Me	Me	ClO_4^-	NH_3	75	435
Me	H	Me	H	Et	ClO_4^-	NH_3		435
Me	Ph	H	H	Me	ClO_4^-	NH_3		436
Me	*p*-AcC_6H_4	H	H	Me	ClO_4^-	NH_3		436

F. *Pyrroles*

When *N*-methylpyrrole was pyrolyzed at 650°, it was converted primarily into 2-methylpyrrole (85%). Small amounts (2 to 5%) of pyrrole and pyridine were formed. Treatment of 2-methylpyrrole under similar conditions gave 85% of unreacted 2-methylpyrrole and 2 to 5% of pyrrole and pyridine. At 745°, both *N*-methylpyrrole and 2-methylpyrrole yielded significantly more pyridine and pyrrole, but the reaction was accompanied by as much as 40% of carbonization (438).

Pyrolysis of 1-(*n*-butyl)pyrrole at 500 to 570° gave mixture of isomers containing all three possible *n*-butyl compounds and from 1 to 9% of pyridine (439).

Generation of chlorocarbene by the treatment of methylene chloride with methyllithium, followed by reaction of the carbene with pyrrole, gave a 32% yield of pyridine (441):

$$CH_3Li + CH_2Cl_2 \longrightarrow :CHCl + CH_4 + LiCl$$

N-Li pyrrole + :CHCl ⟶ [bicyclic N-Li adduct with H, Cl] ⟶ pyridine + LiCl

Dichlorocarbene, which is formed by treatment of chloroform with potassium hydroxide, adds to 2,5-dimethylpyrrole to form a mixture of 3-chloro-2,6-dimethylpyridine, 2-dichloromethyl-2,5-dimethylpyrrolenine, and 2,5-dimethyl-3-pyrrolaldehyde. Similarly 2,3,4,5-tetramethylpyrrole gave a mixture of 3-chloro-2,4,5,6-tetramethylpyridine and 2-dichloromethyl-2,3,4,5-tetramethylpyrrolenine (441).

3,4-Dimethyl-2,5-diphenyl-1*H*-pyrrole reacts with dichlorocarbene, which is generated in neutral solution by the thermolysis of sodium trichloroacetate, to give 3-chloro-4,5-dimethyl-2,6-diphenylpyridine in 90% yield. By the same technique, 2,5-dimethyl-1*H*-pyrrole gave 70% of 3-chloro-2,6-dimethylpyridine, and 2,3,4,5-tetramethyl-1*H*-pyrrole yielded 55% of 3-chloro-2,4,5,6-tetramethylpyridine. Generation of dichlorocarbene from chloroform under basic conditions resulted in significantly lower yields because of competing reactions (441a).

The reaction of 2,5-dimethyl-1*H*-pyrrole with ethanolic sodium hydroxide and chloroform at 55° produces a mixture of 3-chloro-2,6-dimethylpyridine and 2-dichloromethyl-2,5-dimethyl-2*H*-pyrrole. It was determined that the dichloromethylated pyrrole compound is not an intermediate in the formation of the 3-chloropyridine derivative. When pure 2-dichloromethyl-2,5-dimethyl-2*H*-pyrrole (**II-15a**) is treated with ethanolic sodium ethoxide at 105°, an 82% yield of 2-ethoxymethyl-5-methylpyridine (**II-15b**) is obtained with no 3-chloropyridine derivative present. On the basis of this result, a mechanism has been proposed for this ring expansion reaction which involves attack by two equivalents of sodium ethoxide (441b)

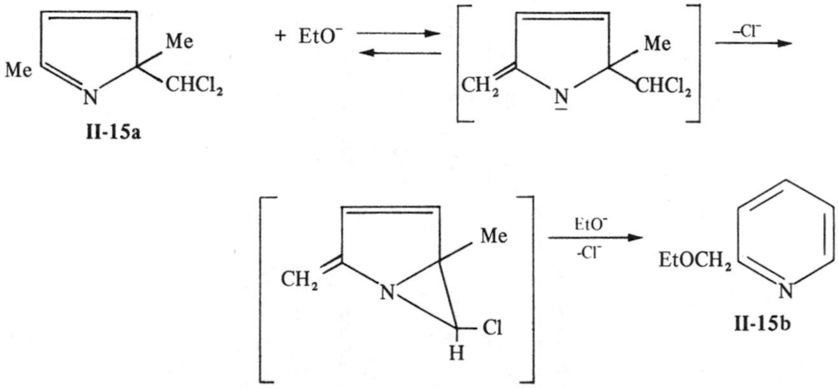

Me, Me, $CHCl_2$, N

II-15a

$+ EtO^-$ ⇌ [CH_2, N, Me, $CHCl_2$] $\xrightarrow{-Cl^-}$

[CH_2, N, Me, Cl, H] $\xrightarrow[-Cl^-]{EtO^-}$ $EtOCH_2$, N, Me

II-15b

TABLE II-19. Aromatization of Dihydropyridines

Starting material	Conditions	Product (yield %)	Ref.
1-Benzyl-2,3,4,5-tetracarbomethoxy-6-phenyl-1,6-dihydropyridine	Bromine, acetic acid, reflux	2-Phenyl-3,4,5,6-tetracarbomethoxypyridine (73)	442
3,5-Diacetyl-1,4-dihydropyridine	HNO_2, ethanol	3,5-Diacetylpyridine	443
3,5-Diacetyl-4-methyl-1,4-dihydropyridine	HNO_2, ethanol	3,5-Diacetyl-4-methylpyridine	443
3,5-Diacetyl-4-phenyl-1,4-dihydropyridine	HNO_2, ethanol	3,5-Diacetyl-4-phenylpyridine	443
3,5-Diacetyl-1,4-dihydropyridine	HNO_3, 20% aqueous, reflux	2,5-Pyridinedicarboxylic acid	444

TABLE II-20. Oxidative Dealkylation of Certain 1,4-Dihydropyridines (445)

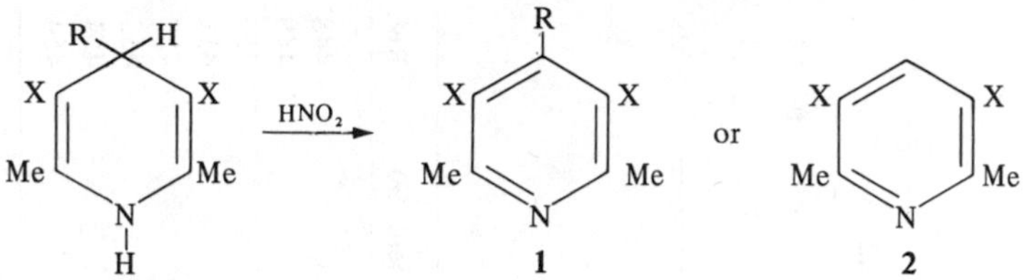

X = COOEt (R)	X = CN (R)	Product[a] (yield %)
Me		**1** (87)
	Me	**1** (46)
	Et	**1**
	isoPr	**1** (90)
	t-Bu	**2** (48)
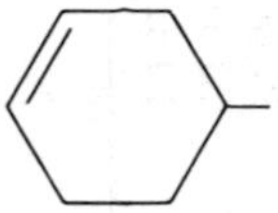		**2** (70)
C_6H_5		**1** (53)
$C_6H_5CH_2$		**2** (90)
	$C_6H_5CH_2$	**2** (48)
$C_6H_5CH{=}CH$		**1** (60)

[a] Mixtures of **1** and **2** were not observed.

TABLE II-21. Dehydrogenation of Tetrahydropyridines to Pyridines (447)

Starting material	Conditions	Pyridine (yield %)
1,2,5-Trimethyl-4-phenyltetrahydropyridine	K12 catalyst, 400–420°	92
1-Benzyl-2,5-dimethyl-4-phenyltetrahydropyridine	K12 catalyst, 400–420°	70
4-(α-Naphthyl)-1,2,5-trimethyltetrahydropyridine	K12 catalyst, 400–420°	62

TABLE II-22. Dehydrogenation of Piperidines to Pyridines

Piperidine	Product	Conditions	Yield (%)	Ref.
Piperidine	Pyridine	Pd on silica, H_2, 260°	53	448
	Pyridine	Ni-Al catalyst, H_2, 250–300°		449
	Pyridine	Cadmium phosphate, 450°	32	450
	Pyridine	Pt or Pd, H_2, 200–500°		451
N-Methylpiperidine	Pyridine	Pd on silica gel, H_2, 375°		452
3-(4-Acetylaminobutyl)piperidine	3-(4-Acetylaminobutyl)pyridine	Pd on C, 200°		453
2-Methyl-3-phenylpiperidine	2-Methyl-3-phenylpyridine	Pd on C, 310–315°	62	454
2-Methyl-3-phenylpiperidine	2-Methyl-3-phenylpyridine	Pd on C, nitrobenzene-toluene		454

G. *Hydropyridines*

a. DIHYDROPYRIDINES. The aromatization of various dihydropyridines with chemical oxidizing agents is summarized in Table II-19.

It has been reported (445) that when the dihydropyridine ring contains electron withdrawing substituents such as cyano or carboxylic ester, loss of a 4-substituent may occur in some cases during the oxidative aromatization reaction (see Table II-20). A carbonium ion mechanism has been postulated for the dealkylation, and a correlation has been noted between the ease of dealkylation and the ability of the leaving group to form a carbonium ion.

b. TETRAHYDROPYRIDINES. A palladium-on-alumina catalyst in the presence of nitrobenzene as a hydrogen acceptor has been used to dehydrogenate 4-aryltetrahydropyridines to the corresponding pyridines. The liquid phase system operated normally over a temperature range of 125° to 220°. Dehydrogenation of 4-phenyl-1,2,3,6-tetrahydropyridine at 135° gave a 79% yield of 4-phenylpyridine. Similarly, 4-(*p*-tolyl)pyridine and 4-(*p*-cumyl)pyridine were prepared from the corresponding tetrahydro-derivatives. Substitutents on nitrogen, if present, were eliminated during the reaction (446). In vapor phase dehydrogenations, substituents on nitrogen were also eliminated (see Table II-21).

c. PIPERIDINES. The results obtained on dehydrogenation of piperidines to pyridines in the presence of various catalysts are summarized in Table II-22.

H. *Quinolines and Condensed Pyridines*

Oxidation of quinoline by the classical method in which potassium permanganate is the oxidizing agent gave a 50% yield of quinolinic acid. This compound readily decarboxylated at 155 to 160° to form nicotinic acid (455). Treatment of 8-hydroxy-3-methylquinaldine **(II-16)** with fuming nitric acid yielded 5-methylquinolinic acid **(II-17)** (456):

II-16 **II-17**

The catalytic vapor phase oxidation of isoquinoline at 400° in the presence of a catalyst composed of vanadium pentoxide and stannic oxide on pumice stone gave a 47% yield of cinchomeronic acid. Other products produced were isonicotinic acid, phthalic acid, and pyridine. The best yield of cinchomeronic acid in a noncatalytic oxidation was 34%. When molybdenum trioxide was used to replace the stannic oxide, an increase was noted in the amount of pyridine produced (457). 4-Bromo-5-methylmerimine (**II-18**) gave a mixture of 5-bromo-6-methyl-3-cinchomeronamic acid (**II-19**) and 5-bromo-6-methyl-4-cinchomeronamic acid (**II-20**) when oxidized with potassium permanganate:

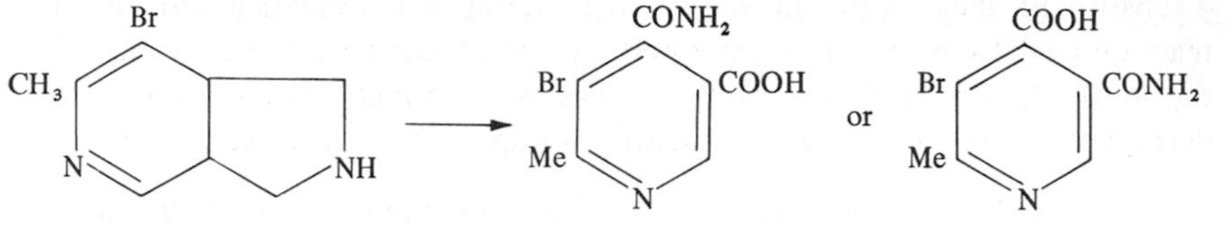

Br CH$_3$ N NH → Br Me N CONH$_2$ COOH or Br Me N COOH CONH$_2$

II-18 **II-19** **II-20**

The 5-chloro- and 5-iododerivatives were prepared similarly (458).

The reduction of quinolizinium bromide (**II-21**) with lithium aluminum hydride gave a butadiene derivative of pyridine (**II-22**) (459):

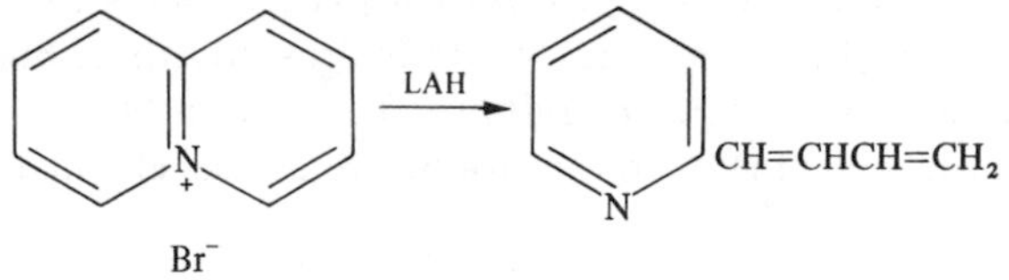

N$^+$ Br$^-$ —LAH→ N CH=CHCH=CH$_2$

II-21 **II-22**

Hydrogenation of 7-azaindole at 270° under pressure with a Raney nickel catalyst and decalin as the solvent gave 2-amino-3-ethylpyridine as the product (460).

I. *Oxazoles and Thiazoles*

Alkyl-substituted oxazoles will react with dienes in a Diels-Alder reaction to yield substituted pyridine derivatives after cleavage of the oxygen bridge of the adduct. For example, 2-methyl-5-ethoxyoxazole (**II-23**) reacts with maleic anhydride in boiling benzene to form 5-hydroxy-2-methylpyridine-3,4-dicarboxylic acid (**II-24**):

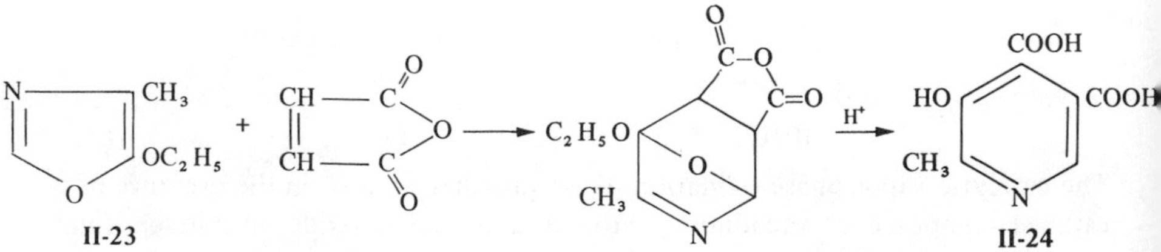

N CH$_3$ OC$_2$H$_5$ O + CH—C(=O) CH—C(=O) O → C$_2$H$_5$O CH$_3$ N O C–O C=O O —H$^+$→ HO CH$_3$ N COOH COOH

II-23 **II-24**

Additional examples of this versatile reaction are given in Table II-23. Thiazoles have been found to react in an analogous manner (Table II-24).

2. From Acyclic Compounds

A. *Cyclization of a Five-Carbon Chain*

a. 1,5-DIOXO COMPOUNDS AND DERIVATIVES. The conversion of 1,5-dialdehydes and 1,5-diketones to pyridines by reaction with a nitrogen source is a convenient synthetic method. The use of glutaraldehyde and its derivatives as starting materials has received increasing attention recently. The results obtained with a number of 1,5-dioxo compounds are given in Table II-25.

The lithium aluminum hydride reduction of 2-pyrone (**II-25**) gives a hemiacetal (**II-26**) that can tautomerize to an enol-aldehyde (**II-27**). Condensation of **II-27** with ammonia forms pyridine:

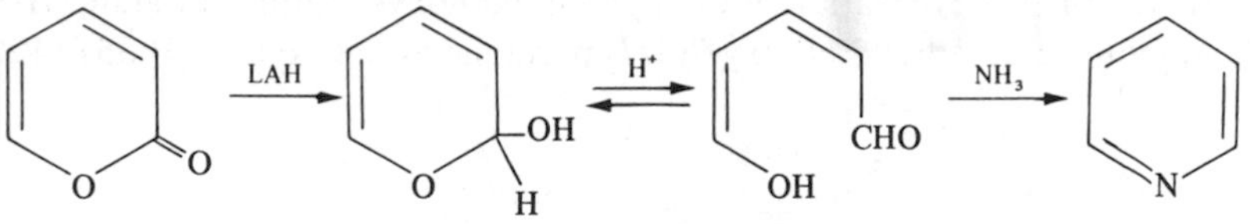

LAH; H^+; NH_3; OH; H; CHO; OH; O; N

II-25 **II-26** **II-27**

By means of this reaction, coumalic acid has been converted to nicotinic acid, and 2-pyrone-5,6-dicarboxylic acid converted to quinolinic acid in yields of up to 35% (494).

b. OXOCARBOXYLIC ACIDS AND DERIVATIVES. Condensation of acrylonitrile and γ-acetylpropyl acetate in the presence of alcoholic potassium hydroxide gave the acetate of 3-acetyl-3-(2-cyanoethyl)propanol (**II-28**). Cyclization of **II-28** was accomplished by boiling over alumina to give a 40% yield of 5-acetoxyethyl-6-methyl-3,4-dihydropyrid-2-one (**II-29**) (495):

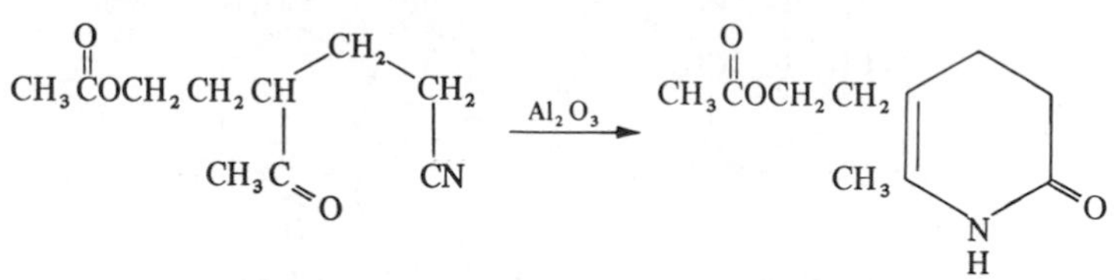

$CH_3\overset{O}{\overset{\|}{C}}OCH_2CH_2CH(COCH_3)CH_2CH_2CN \xrightarrow{Al_2O_3}$ $CH_3\overset{O}{\overset{\|}{C}}OCH_2CH_2$, CH_3, N, H, O

II-28 **II-29**

Pyridines have been prepared by catalytic dehydrogenation of an oxo nitrile with a metal or metal oxide (496). Heating 4-methyl-5-oxoheptanonitrile (**II-30**) at 200° in the presence of a palladium-on-alumina catalyst gave a 10% yield of 2-ethyl-3-methylpyridine (**II-31**):

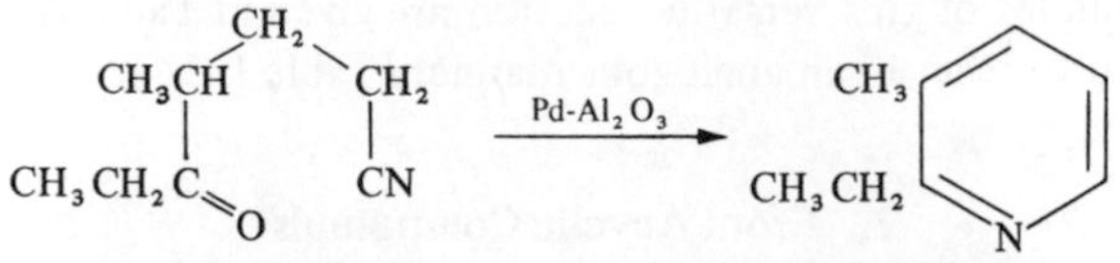

II-30 II-31

Treatment of 2-(2-cyanoalkyl)cyclohexanones (**II-32**) with 96% sulfuric acid at room temperature for several hours gave the corresponding cycloalkano [*e*]-2-pyridones (**II-33**) by means of a consecutive cyclization-aromatization reaction (497):

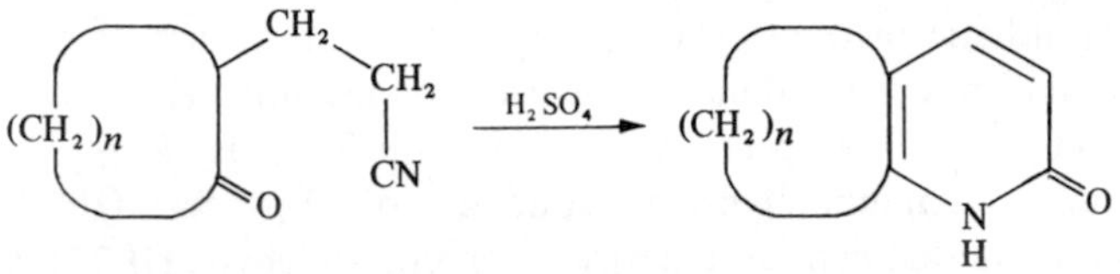

II-32 II-33

It was found that cyclization of diethyl 4,4-diethoxy-2-formylglutarate (**II-34**) with ammonia gave ethyl 3-ethoxy-2(1*H*)-pyridone-5-carboxylate (**II-35**) (498):

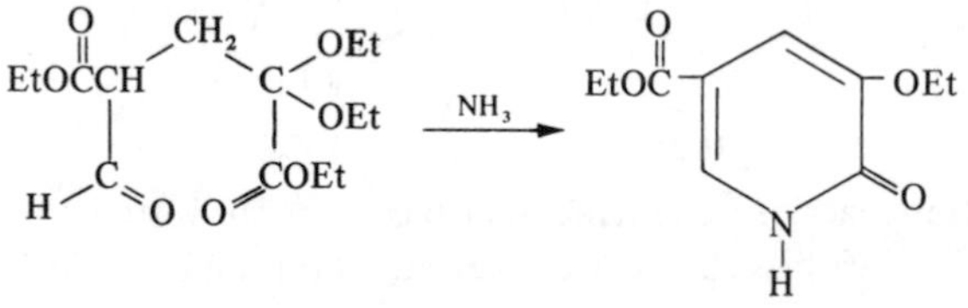

II-34 II-35

c. 1,5-DICARBOXYLIC ACIDS AND DERIVATIVES. Citrazinic acid (**II-37**) was prepared in 43.7% yield by boiling citric acid (**II-36**) in methanol in the presence of *p*-toluenesulfonic acid, followed by a treatment with ammonia under pressure (499):

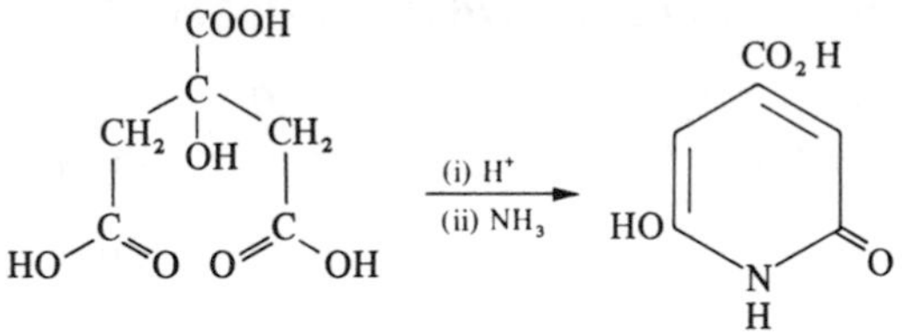

II-36 II-37

In a similar reaction, it was found that if citric acid is treated under pressure in the presence of either excess ammonia or urea, the citrazinic acid amide is formed (500). When trimethyl citrate was heated under pressure in the presence of ammonium hydroxide, a 62% yield of citrazinic acid was obtained (501). The cyclization of citric acid triamide with sodium carbonate in ethylene glycol gave citrazinic acid amide, which was then hydrolyzed to citrazinic acid in an 80.5% overall yield (502).

A nitrile-ester cyclization of ethyl γ-cyanobutyrate (**II-38**) in the presence of ethyl α-bromoisobutyrate (**II-39**), zinc, and a trace of mercuric chloride yielded 2,6-bis(1-carbethoxy-1-methylethyl)pyridine (**II-40**) (503):

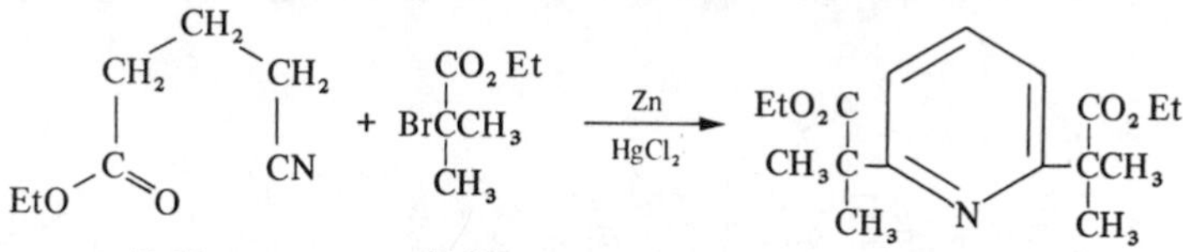

$$\text{EtO}_2\text{C(CH}_2)_3\text{CN} + \text{BrC(CH}_3)_2\text{CO}_2\text{Et} \xrightarrow[\text{HgCl}_2]{\text{Zn}} \text{II-40}$$

II-38 **II-39** **II-40**

Evidence obtained in the reaction suggests that a dihydropyridine intermediate is formed, which disproportionates into the pyridine compound and the corresponding piperidine derivative. Ring closure of diethyl methylenebiscyanoacetate (**II-41**) in the presence of sodium ethoxide gave a 38% yield of ethyl 5-cyano-2-ethoxy-6-hydroxynicotinate (**II-42**) (504):

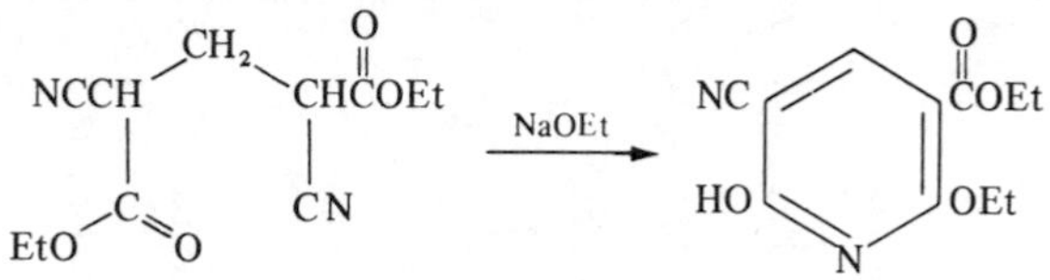

$$\text{II-41} \xrightarrow{\text{NaOEt}} \text{II-42}$$

II-41 **II-42**

Diethyl acetonedicarboxylate and ethyl cyanoacetate when condensed in the presence of diethylamine form a nitrile-ester (**II-43**). This compound in the presence of sulfuric acid undergoes cyclization to yield ethyl 2,6-dihydroxy-3-ethoxycarbonyl-4-pyridylacetate (**II-44**) (505):

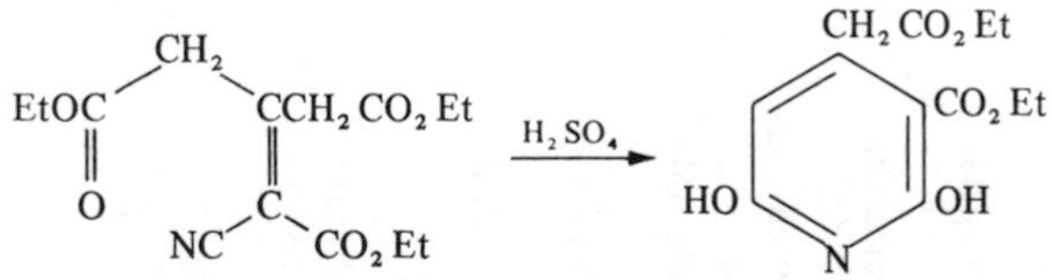

$$\text{II-43} \xrightarrow{\text{H}_2\text{SO}_4} \text{II-44}$$

II-43 **II-44**

A series of trisubstituted cyanoglutaconic esters and the analogous cyanocrotonic esters have been subjected to acid and alkaline cyclization procedures, and the results are tabulated in Table II-26 (506).

The cyclization of 1,5-dinitriles with halogen acids to form pyridines appears to be a method of general utility. For example, the salts of substituted tetracyanopropenes (**II-47**) react in the following manner:

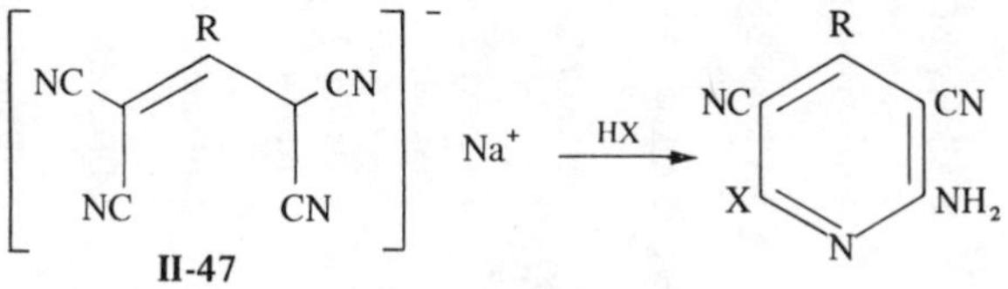

$$\text{II-47 } (\text{Na}^+) \xrightarrow{\text{HX}} \text{2-amino-6-X-4-R-3,5-dicyanopyridine}$$

II-47

TABLE II-23. Reaction of Oxazoles to Form Pyridines

Oxazole	Reactant	Conditions	Product (yield %)	Ref.
5-Ethoxy-2-methyloxazole	Maleic anhydride	Benzene, reflux	5-Hydroxy-2-methylpyridine-3,4-dicarboxylic acid (54)	461
5-Alkoxy-2-ethyloxazole	Maleic anhydride	Benzene, reflux	2-Ethyl-5-hydroxypyridine-3,4-dicarboxylic acid (47)	461
5-Alkoxy-2-propyloxazole	Maleic anhydride	Benzene, reflux	5-Hydroxy-2-propylpyridine-3,4-dicarboxylic acid (31.5)	461
5-Alkoxy-2-butyloxazole	Maleic anhydride	Benzene, reflux	2-Butyl-5-hydroxypyridine-3,4-dicarboxylic acid (42.3)	461
5-Alkoxy-2-amyloxazole	Maleic anhydride	Benzene, reflux	2-Amyl-5-hydroxypyridine-3,4-dicarboxylic acid (40)	461
5-Alkoxy-2,4-dimethyloxazole	Maleic anhydride	Benzene, reflux	2,6-Dimethyl-5-hydroxypyridine-3,4-dicarboxylic acid (36)	461
5-Alkoxy-2-ethyl-4-methyl-oxazole	Maleic anhydride	Benzene, reflux	2-Ethyl-5-hydroxy-6-methyl-pyridine-3,4-dicarboxylic acid (25)	461
5-Alkoxy-2-amyl-4-methyl-oxazole	Maleic anhydride	Benzene, reflux	2-Amyl-5-hydroxy-6-methyl-pyridine-3,4-dicarboxylic acid (15.8)	461
4,5-Dimethyloxazole	Maleimide	Benzene, reflux	5,6-Dimethylpyridine-3,4-dicarboximide (85)	462
2,4-Dimethyloxazole	Maleimide	Benzene, reflux	2,6-Dimethylpyridine-3,4-dicarboximide	462
2,5-Dimethyloxazole	Maleimide	Benzene, reflux	2,5-Dimethylpyridine-3,4-dicarboximide	462
2,4,5-Trimethyloxazole	Maleimide	Benzene, reflux	2,5,6-Trimethylpyridine-3,4-dicarboximide	462
2-Methyl-4,5-tetramethylene-oxazole	Maleimide	Benzene, reflux	2-Methyl-5,6-tetramethylenepyridine-3,4-dicarboximide	462

5-Ethoxy-2-methyloxazole	Maleimide	Benzene, reflux	5-Hydroxy-2-methylpyridine-3,4-dicarboximide (54)	462
5-Ethoxy-2-ethyloxazole	Maleimide	Benzene, reflux	2-Ethyl-5-hydroxypyridine-3,4-dicarboximide	462
5-Ethoxy-2-propyloxazole	Maleimide	Benzene, reflux	5-Hydroxy-2-propylpyridine-3,4-dicarboximide	462
2-Butyl-5-ethoxyoxazole	Maleimide	Benzene, reflux	2-Butyl-5-hydroxypyridine-3,4-dicarboximide	462
2-Amyl-5-ethoxyoxazole	Maleimide	Benzene, reflux	2-Amyl-5-hydroxypyridine-3,4-dicarboximide	462
5-Ethoxy-2-phenyloxazole	Maleimide	Benzene, reflux	5-Hydroxy-2-phenylpyridine-3,4-dicarboximide	462
2,4-Dimethyl-5-ethoxy-oxazole	Maleimide	Benzene, reflux	2,6-Dimethyl-5-hydroxypyridine-3,4-dicarboximide	462
5-Ethoxy-4-ethyl-2-methyl-oxazole	Maleimide	Benzene, reflux	6-Ethyl-5-hydroxy-2-methyl-pyridine-3,4-dicarboximide	462
5-Ethoxy-4-ethyl-4-methyl-oxazole	Diethyl maleate	110°, 2 hr	Diethyl 3-hydroxy-2-methyl-pyridine-3,4-dicarboxylate (85)	463, 473
5-Ethoxy-4-methyloxazole	Fumaronitrile	Methanol, reflux, 5 hr	4,5-Dicyano-3-hydroxy-2-methyl-pyridine (75)	463
5-Ethoxy-4-methyloxazole	2,5-Dihydrofuran	Pressure vessel, 175°, 3 hr, 1% trichloroacetic acid	7-Hydroxy-6-methyl-1,3-dihydrofuro-[3,4-*b*] pyridine (R.I. #1309)	463
5-Acetyl-4-methyloxazole	Maleimide		5-Acetyl-6-methylpyridine-3,4-dicarboximide (45)	464
5-Ethoxy-4-(ethoxycar-bonylmethyl)oxazole	Diethyl maleate	130°, 3 hr	Diethyl 2-(ethoxycarbonylmethyl)-3-hydroxypyridine-4,5-dicarboxylate	465
5-Ethoxyoxazole	Dimethyl maleate	110°, 2 hr	Dimethyl 3-hydroxycinchomeronate (43.5)	466

TABLE II-23. Reaction of Oxazoles to Form Pyridines (*Continued*)

Oxazole	Reactant	Conditions	Product (yield %)	Ref.
5-Cyano-4-methyloxazole	4,7-Dihydro-1,3-dioxepin	Sealed tube, 150°, 20 hr; acid hydrolysis	5-Hydroxy-3,4-bis(hydroxymethyl)-6-methylpyridine	467
2,5-Dimethyloxazole	Maleic anhydride		2,5-Dimethylcinchomeronic acid (78)	468
4,5-Dimethyloxazole	Maleic anhydride		5,6-Dimethylcinchomeronic acid (85)	468
2,4,5-Trimethyloxazole	Maleic anhydride		2,5,6-Trimethylcinchomeronic acid (52)	468
2-Methyl-4,5-tetramethylene-oxazole	Maleic anhydride		2-Methyl-5,6-tetramethylene-cinchomeronic acid (35)	468
4-Methyloxazole	Dimethyl fumarate	110–120°, $PhNO_2$	4,5-Dicarboxymethyl-3-hydroxy-2-methylpyridine	469
4-Methyloxazole	Diethyl fumarate	110–120°, $PhNO_2$	4,5-Dicarboxyethyl-3-hydroxy-2-methylpyridine	469
4-Methyloxazole	Acrylonitrile	110–120°, $PhNO_2$	5-Cyano-3-hydroxy-2-methylpyridine	469
4-Methyloxazole	Fumaronitrile	110–120°, $PhNO_2$ (or MeOH, reflux)	4,5-Dicyano-3-hydroxy-2-methyl-pyridine	469, 473, 475
4-Methyloxazole	Acrylonitrile	150°, PhMe, sealed tube, 4 hr	2-Methyl-3-pyridinol; 2-methylisonicotinonitrile	470
4-Phenyloxazole	Acrylonitrile	150°, AcOH, sealed tube, 30 hr	2-Phenyl-3-pyridinol	470

2,4-Dimethyloxazole	Acrylonitrile	110°, AcOH, sealed tube, 20 hr	2,6-Dimethyl-3-pyridinol	470
4-Methyloxazole	Maleic anhydride	90°, AcOH, sealed tube, 20 hr; hydrolysis	6-Methylpyridine-3,4-dicarboxylic acid	471, 473
4-Methyloxazole	Maleimide	AcOH, 5 hr	6-Methylpyridine-3,4-dicarboximide	471
4-Methyloxazole	Fumaronitrile	90°, AcOH, sealed tube, 20 hr; hydrolysis	3-Hydroxy-2-methyl-nicotinitrile	471
4-Methylozaxole	$MeOCH_2CH=C(CN)CH_2OMe$	95°, AcOH, 40 hr	4,5-Bis(methoxymethyl)-2-methyl-3-pyridinol	472
4-Methylozaxole	Ethyl crotonate	90°, AcOH, 70 hr	Ethyl 5-hydroxy-4,6-dimethylnicotinate	472
4-Methyloxazole	2,5-Dihydrofuran	Heat, pressure	7-Hydroxy-6-methyl-1,3-dihydrofuro[3,4-*b*]pyridine	476
5-Ethoxy-4-methyloxazole	2-Butene-1,4-diol	100°, pressure	Pyridoxine	473, 474
5-Ethoxy-4-methyloxazole	1,4-Dimethoxy-2-butene		4,5-Dimethoxymethyl-3-hydroxy-2-methylpyridine	473, 474

TABLE II-24. Reaction of Thiazoles to Form Pyridines (477)

Thiazole	Reactant	Conditions	Product (Yield %)
5-Ethoxy-4-methylthiazole	Diethyl maleate	200° sealed tube 8 hr; hydrolysis	5-Ethoxy-6-methyl-3,4-dicarboxylic acid (63.5)
5-Ethoxy-4-methylthiazole	2-Butene-1,4-diol	100°, $PhNO_2$, $AlCl_3$	Pyridoxine

TABLE II-25. Conversion of Dioxo Compounds and Derivatives to Pyridines

Starting compound	Conditions	Product (yield %)	Ref.
Glutaraldehyde	Malachite green, $(NH_4)_2SO_4$ in aq. H_2SO_4	Pyridine (53)	478
	Rosaniline, $(NH_4)_2SO_4$ in aq. H_2SO_4	Pyridine (33)	478
	Crystal violet, $(NH_4)_2SO_4$ in aq. H_2SO_4	Pyridine (60)	478
	Indophenol blue, $(NH_4)_2SO_4$ in aq. H_2SO_4	Pyridine (50)	478
	Methylene blue, $(NH_4)_2SO_4$ in aq. H_2SO_4	Pyridine (80)	478
	Indigo carmine, $(NH_4)_2SO_4$ in aq. H_2SO_4	Pyridine (60)	478
	Aurin, $(NH_4)_2SO_4$ in aq. H_2SO_4	Pyridine (30)	478
	$Cu(OAc)_2$, NH_4OAc, AcOH, O_2, 80°, 3 hr	Pyridine (68)	479
	Pd, Fe^{+3}, $(NH_4)_2SO_4$, O_2, 100°, pressure	Pyridine (20)	479
	Sulphur, $(NH_4)_2SO_4$, $HOCH_2CH_2OH$, aq. H_2SO_4	Pyridine (10)	480
	$CuCl_2$, NH_4Cl, aq. HOAc	Pyridine (44) Chloropyridine (15)	481
	Pt on silica-alumina, NH_3, H_2O, 300°	Pyridine (28)	482
	Tetrachlorobenzoquinone, $(NH_4)_2SO_4$, aq. H_2SO_4, dioxane, reflux 2 hr	Pyridine (92)	483

$Me_2NCH{=}C(CHO)CH{=}CHO$	NH_4Cl, H_2O, heat	Nicotinaldehyde	484, 485
Glutaconaldehyde enol	$(NH_4)_2CO_3$	Pyridine	486
α-Methylglutaconaldehyde enol	$(NH_4)_2CO_3$	2-Picoline	486
α, α′-Dimethylglutaconaldehyde enol	$(NH_4)_2CO_3$	3,5-Dimethylpyridine	486
α, α′-Dimethylglutaraldehyde	$NH_2OH \cdot HCl$, AcOH, reflux	3,5-Dimethylpyridine	487
α,β-Dimethylglutaraldehyde	$NH_2OH \cdot HCl$, AcOH, reflux	3,4-Dimethylpyridine	487
α-Ethylglutaraldehyde	$NH_2OH \cdot HCl$, AcOH, reflux	3-Ethylpyridine	487
α, α′-Diethylglutaraldehyde	$NH_2OH \cdot HCl$, AcOH, reflux	3,5-Diethylpyridine	487
1,1,3,5,5,-Pentaethoxypentane	NH_4Cl, aq. H_2SO_4, NH_3	Pyridine (54)	488, 489
Glutaraldehyde	2,4-Dinitrophenylhydrazine HCl	Pyridine	490
α-Methylglutaraldehyde	2,4-Dinitrophenylhydrazine HCl	3-Methylpyridine	490
α, α′-Dimethylglutaraldehyde	2,4-Dinitrophenylhydrazine HCl	3,5-Dimethylpyridine	490
5-Butyl-2-oxohexanol	NH_2OH, AcOH, reflux	5-Butyl-2-methylpyridine (37.5)	491
5-Hexyl-2-oxohexanol	NH_2OH, AcOH, reflux	5-Hexyl-2-methylpyridine (42.3)	491
2-Butyl-3-methylglutaraldehyde	NH_2OH, AcOH, reflux	3-Butyl-4-methylpyridine (80.5)	491
β-Phenylglutaraldehyde	NH_2OH, AcOH, reflux	4-Phenylpyridine (75.6)	492
α-Methyl-β-phenylglutaraldehyde	NH_2OH, AcOH, reflux	3-Methyl-4-phenylpyridine (40)	492
3-Carbethoxy-2,6-heptanedione	NH_3, EtOH, air oxidation	Ethyl 2,6-dimethylnicotinate	493
2,6-Heptanedione	NH_3, EtOH, air oxidation	2,6-Lutidine	493

TABLE II-26. Cyclization of Trisubstituted Cyanoglutaconic Esters (**II-45**) or Trisubstituted Cyanocrotonic Esters (**II-46**) (506)

$$EtO_2C{-}C(R){=}C(Me){-}C(R')(CO_2Et)(CN) \quad \textbf{II-45}$$

$$EtO_2C{-}C(R){=}C(Me){-}C(R')(H)(CN) \quad \textbf{II-46}$$

Ester	R	R′	Cyclization method[a]	Product (yield %)
II-45	Me	Me	*A*	2,6-Dihydroxy-3,4,5-trimethylpyridine (53)
II-45	Et	Me	*B*	3-Ethyl-2,6-dihydroxy-4,5-dimethylpyridine (9)
II-45	Et	Me	*A*	3-Ethyl-2,6-dihydroxy-4,5-dimethylpyridine (16)
II-46	Me	Et	*A*	3-Ethyl-2,6-dihydroxy-4,5-dimethylpyridine (37)
II-46	Et	Me	*B*	3-Ethyl-2,6-dihydroxy-4,5-dimethylpyridine (13)
II-46	Et	Me	*A*	3-Ethyl-2,6-dihydroxy-4,5-dimethylpyridine (6)
II-45	Et	Et	*B*	3,5-Diethyl-2,6-dihydroxy-4-methylpyridine (30)
II-45	Et	Et	*A*	3,5-Diethyl-2,6-dihydroxy-4-methylpyridine (48)
II-46	Et	Et	*A*	3,5-Diethyl-2,6-dihydroxy-4-methylpyridine (13)

[a] Method *A*: boil with concentrated HCl; method *B*: boil with MeOH-KOH.

Examples of this reaction are given in Table II-27. The same general reaction has also been found to take place in an alkaline medium (508).

d. COMPOUNDS HAVING TERMINAL UNSATURATION. Perchloropentadienonitrile has been cyclized to pyridine compounds by two different types of reactions. In the first technique, sodium metal and an alcohol are used to affect the reaction. The second technique utilizes a Grignard reagent to bring about the cyclizations. Results obtained with both reagents are given in Table II-28.

It was found that 1-cyano-1,3-butadiene could be converted to pyridine in 35% yield by passing it over a silica-alumina catalyst at 510°(516).

Passage of ammonia and dimethyl vinylethynylcarbinol (**II-47a**) over a cadmium oxide-on-alumina catalyst at 360° gave a mixture of tar bases containing mainly 2,4-lutidine, 2,6-lutidine, and 2,4,6-trimethylpyridine. Similar

II-47a $\xrightarrow{NH_3}$ 2,4-lutidine + 2,6-lutidine + 2,4,6-trimethylpyridine

results were obtained with a chromia-magnesium oxide-on-alumina catalyst at 420° (517). When methyl vinylethynylcarbinol was used as the starting material, the tar base product was a mixture of 2-picoline, 4-picoline, and 2,4-lutidine (518, 519). Methyl phenyl vinylethynylcarbinol and ammonia when passed over a cadmium phosphate-on-alumina catalyst at 400° gave a 14.4% yield of 2-methyl-6-phenylpyridine (520). In these reactions it is postulated that the alcohol is converted to an amine as the first step in the reaction.

e. MISCELLANEOUS 1,5-BIFUNCTIONAL COMPOUNDS. It was found that 1,1,3,5-tetramethoxyhexane could be passed over a dehydrating catalyst such as chromium oxide-on-alumina the presence of ammonia at 325° to give a mixture of pyridines which consisted mainly of 2-picoline with small amounts of lutidines and collidines (521).

Pyridine derivatives were obtained in high yield by cyclizing unsaturated 1,5-amino-aldehydes and 1,5-amino-ketones (Table II-29).

2. 4-1 Condensations

Pyridines can be formed in a 4-1 condensation by the reaction of dienes with nitriles:

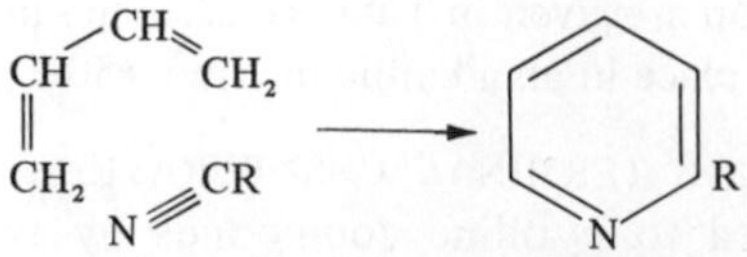

CH=CH₂ / CH=CH₂ + N≡CR → 2-R-pyridine

a. DIENES WITH NITRILES. Janz and Monahan continued their study of the condensation of "cyanogen-like" compound with dienes. The reaction of CF_3CF_2CN, $CF_3CF_2CF_2CN$, CF_2ClCN, and $CFCl_2CN$ with butadiene in the gas phase at 350 to 450° C gave the corresponding pyridines. When fluorine was the only halogen present, near quantitative yields based upon the nitrile were obtained (523). Isoprene and pentadiene were also studied as the diene (524). When competitive rate measurements for the reaction of butadiene with CF_3CN and CF_3CF_2CN, CF_3CN and $CF_3CF_2CF_2CN$, and CF_3CF_2CN and $CF_3CF_2CF_2CN$ were made, there were marked differences in the reaction rates (525). A summary of this work has been published (526).

Iminochlorides add in a Diels-Alder fashion to dienes to give 1,2-dihydropyridines. The iminochlorides were prepared *in situ* from the reaction of formamide or acetamide with phosphorus oxychloride. The condensation was carried out in nitrobenzene at 170 to 180°. The compounds prepared are given in Table II-30. The dihydropyridines were dehydrogenated by heating with sulfur (527).

Alkylpyridines are the reported products obtained when 2-aza-1,3-butadiene or 3-alkenyl-1,3,5-triazines were heated with acrylonitriles at 300 to 400° over an alumina-potassium catalyst (528).

b. MISCELLANEOUS CONDENSATIONS. A Diels-Alder reaction takes place between 2,3-dimethyl-1,3-butadiene or isoprene and *N*-carbobutyloxymethylene-*p*-toluenesulfonamide in boiling benzene (10 hr) (II-48). Alkaline hydrolysis of the tetrahydropyridine derivative produced the corresponding methyl substituted pyridinecarboxylic acids (529):

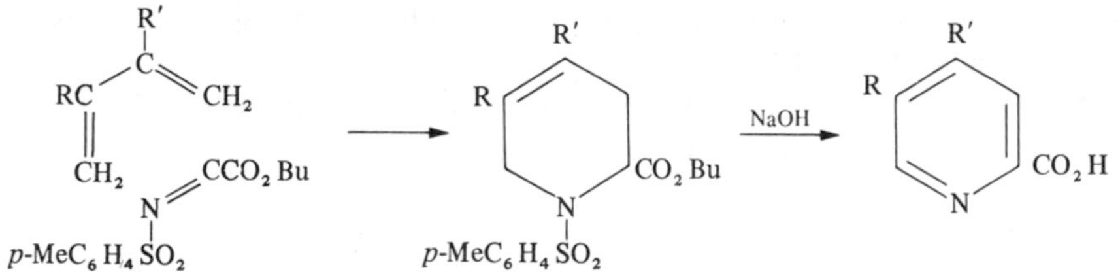

RC(=CH₂)–C(R′)=CH₂ + p-MeC₆H₄SO₂–N=CCO₂Bu → tetrahydropyridine (N-SO₂C₆H₄Me-p, R, R′, CO₂Bu) —NaOH→ pyridine (R, R′, CO₂H)

R = Me R′ = Me

R = H R′ = Me

II-48

2-Methyl-1-buten-3-yne was heated with acetone (1:3 mole ratio) and ammonia at 400° over a cadmium phosphate-alumina catalyst to give unreported yields of tri- and tetramethylpyridines (530). In a similar manner, 2-methyl-4-phenyl-1-buten-3-yne gave 37% of 2,4,6-trimethylpyridine and 26% of 2,6-dimethyl-3-phenylpyridine (531).

C. *3-2 Condensations*

Reactions involving a three-carbon moiety with a two-carbon fragment to form the pyridine ring are most common. There are many variations of such condensations and three general schemes are shown in **II-49**.

CO, –CH, –CO, CH_2–, CO–, N; C–, –C, –CO, CH_2–, CO–, N; C–, –C, –CO, C=C=, CO–, N

II-49

a. 1,3-DICARBONYL COMPOUNDS AND DERIVATIVES WITH METHYLENIC COMPOUNDS. This method is illustrated in **II-49**. The simplest example of a 3-2 synthesis is the reaction of a 1,3-diketone or β-ketoaldehyde with ammonia where the dioxo compound also serves as the methylenic reactant (**II-50**).

CO, CH_2, –CO, CH_2 CO–, CO–, NH_3 → –CO– (pyridine, N)

II-50

However, depending on the methylenic compound, 1,2-dihydropyridines often result. The synthesis of 2-pyridones from 1,3-diketones and cyanoacetamides is listed in Table II-31 (see also Ch. XII).

A series of 4,6-disubstituted 1,2-dihydro-3-cyano-2-dicyanomethylenepyridines were prepared (532) as depicted in **II-51**. The reaction medium was either

R′, CO, CH_2, RCO, CH_2 CN, C=C(CN)$_2$, H_2N → R′, CN, R, =C(CN)$_2$, N, H

II-51

TABLE II-27. Cyclization of Nitriles to Pyridines

Nitrile	Conditions	Product (yield %)	Ref.
1,1,3,3-Tetracyanopropene anion, sodium salt	Anhyd. HCl, acetone	2-Amino-6-chloro-3,5-dicyanopyridine (90)	507
1,1,3,3-Tetracyanopropene anion, sodium salt	Anhyd. HBr, acetone	2-Amino-6-bromo-3,5-dicyanopyridine (93.5)	507
2-Ethoxy-1,1,3,3-tetracyanopropene anion, sodium salt	Anhyd. HCl, acetone	2-Amino-6-chloro-3,5-dicyanopyridine (77.7)	507
1,1,2,3,3-Pentacyanopropene anion, sodium salt	Aq. HCl, acetone, reflux	2-Amino-6-chloro-3,4,5-tricyanopyridine (69.5)	507
1,1,2,3,3-Pentacyanopropene anion, sodium salt	Aq. HBr, acetone, reflux	2-Amino-6-bromo-3,4,5-tricyanopyridine (90)	507
2-Phenyl-1,1,3,3-tetracyanopropene anion, sodium salt	Anhyd. HCl, dioxane	2-Amino-6-chloro-3,5-dicyano-4-phenylpyridine (87.6)	507
2-(*p*-Dimethylaminophenyl)-1,1,3,3-tetracyanopropene anion, sodium salt	Anhyd. HCl, dioxane	2-Amino-6-chloro-3,5-dicyano-4-(*p*-dimethylaminophenyl) pyridine (86)	507
2-Dimethylamino-1,1,3,3-tetracyanopropene anion, sodium salt	Anhyd. HCl, acetone	2-Amino-6-chloro-3,5-dicyano-4-dimethylaminopyridine (10)	507
2-Amino-1,1,3,3-tetracyanopropene anion, sodium salt	Aq. HBr, reflux	2,4-Diamino-6-bromo-3,5-dicyanopyridine (82)	507
2-Bromo-1,1,3,3-tetracyanopropene anion, sodium salt	Anhyd. HBr, acetone	2-Amino-4,6-dibromo-3,5-dicyanopyridine (93)	507
2-Amino-1,1,3,3-tetracyanopropene anion, sodium salt	Aq. HI, reflux	2,4-Diamino-3,5-dicyano-6-iodopyridine (72)	507
2-Amino-1,1,3,3-tetracyanopropene anion, sodium salt	Anhyd. HCl, acetonitrile	2,4-Diamino-6-chloro-3,5-dicyanopyridine (52)	507
1,1,2,3,3-Pentacyanopropene anion, sodium salt	Conc. H_2SO_4, EtOH, reflux	2-Amino-3,4,5-tricyano-6-ethoxypyridine (75)	507

1,1,2,3,3-Pentacyanopropene anion, sodium salt	Conc. H_2SO_4, isoPrOH, reflux	2-Amino-6-isopropoxy-3,4,5-tricyano-pyridine (31)	507
1,1,3,3-Tetracyanopropene anion, sodium salt	Aq. KOH, 2-methyl-2-thiopseudourea sulfate	2-Amino-3,5-dicyano-6-methylthio-pyridine (90)	508
1,1,3,3-Tetracyanopropene anion, sodium salt	Aq. KOH, ethyl mercaptan	2-Amino-3,5-dicyano-6-ethylthio-pyridine (80)	508
1,1,3,3-Tetracyanopropene anion, sodium salt	Aq. KOH, 2-benzyl-2-thiopseudourea hydrochloride	2-Amino-6-benzylthio-3,5-dicyano-pyridine (45)	508
1,1,3,3-Tetracyanopropene anion, sodium salt	Aq. KOH, MeOH, reflux	2-Amino-3,5-dicyano-6-methoxy-pyridine (85)	508
1,1,3,3-Tetracyanopropene anion, sodium salt	Aq. KOH, EtOH, reflux	2-Amino-5-cyano-6-ethoxy-3-pyridinecarboxamide (69)	508
2-Cyanomethyl-1,1,3,3-tetracyano-propene anion, sodium salt	Aq. HCl, acetone, heat	2-Amino-6-chloro-4-cyanomethyl-3,5-dicyanopyridine	509
2-Cyanomethyl-1,1,3,3-tetracyano-propene anion, sodium salt	Aq. HBr, acetone	2-Amino-6-bromo-4-cyanomethyl-3,5-dicyanopyridine	509
2-Cyanomethyl-1,1,3,3-tetracyano-propene anion, sodium salt	HI, acetone	2-Amino-4-(1-cyano-4-hydroxy-2,4-dimethylpent-1-en-1-yl)-6-iodo-3,5-dicyanopyridine (24)	509
1,1,3,3-Tetracyanopropene anion, sodium salt	$SOCl_2$, THF	2-Amino-6-chloro-3,5-dicyanopyridine	510
Glutacononitrile	HBr, acetic acid	2-Amino-6-bromopyridine	511
Glutacononitrile	HI, acetic acid	2-Amino-6-iodopyridine	511
3-Hydroxyglutaronitrile	HBr, ethyl ether	2-Amino-6-bromopyridine (55)	511
3-Hydroxyglutaronitrile	HBr, acetic acid	2-Amino-6-bromopyridine (70)	511
3-Hydroxyglutaronitrile	HI, ethyl ether	2-Amino-6-iodopyridine (60)	511
3-Hydroxyglutaronitrile	HI, acetic acid	2-Amino-6-iodopyridine (90)	511

TABLE II-27. Cyclization of Nitriles to Pyridines (*Continued*)

Nitrile	Conditions	Product (yield %)	Ref.
3-Hydroxy-3-methylglutaronitrile	HBr, ethyl ether	2-Amino-6-bromo-4-methylpyridine (87.5)	511
3-Hydroxy-3-methylglutaronitrile	HI, acetic acid	2-Amino-6-iodo-4-methylpyridine (88)	511
3-Ethyl-3-hydroxyglutaronitrile	HBr, ethyl ether	2-Amino-6-bromo-4-ethylpyridine (65)	511
3-Ethyl-3-hydroxyglutaronitrile	HI, acetic acid	2-Amino-4-ethyl-6-iodopyridine (85)	511
3-Hydroxy-3-phenylglutaronitrile	HBr, ethyl ether	2-Amino-6-bromo-4-phenylpyridine (75)	511
3-Hydroxy-3-phenylglutaronitrile	HI, acetic acid	2-Amino-6-iodo-4-phenylpyridine (54)	511
Tris (cyanomethyl) carbinol	HBr, acetic acid, MeOH	Methyl 2-amino-6-bromo-4-pyridylacetate	511
3-Hydroxyglutaronitrile	HBr, nitromethane	2-Amino-5,6-dibromopyridine	511
3-Hydroxyglutaronitrile	HBr, aniline, 180°	2,6-Dianilinopyridine	512
3-Hydroxy-2-methlyglutaronitrile	HBr, methylene chloride	2-Amino-6-bromo-3-methylpyridine 2-Amino-6-bromo-5-methylpyridine	511
4-Amino-1,3-dicyano-2-methylpenta-1,3-diene	Heated to 170°	2-Amino-5-cyano-4,6-dimethylpyridine	513

TABLE II-28. Conversion of Perchloropentadienonitrile to Pyridine

Reactant	Conditions	Product (yield %)	Ref.
Methanol	Na, R.T. 2 hr, boil 1 hr	2,6-Dimethoxy-3,4,5-trichloropyridine	514
Ethanol	Na, R.T. 2 hr, boil 2 hr	2,6-Diethoxy-3,4,5-trichloropyridine	514
Benzyl alcohol	Na in C_6H_6, R.T. 0.5 hr, boil 1 hr	2,6-Dibenzyloxy-3,4,5-trichloropyridine	514
Cyclohexanol	Na in C_6H_6	2,6-Dicyclohexyloxy-3,4,5-trichloropyridine	514
Phenylmagnesium bromide	Et_2O, –60°, 1 hr	2-Phenyl-3,4,5,6-tetrachloropyridine	515
p-Tolylmagnesium bromide	Et_2O, –60°, 1 hr	2-(*p*-Tolyl)-3,4,5,6-tetrachloropyridine	515
p-Fluorophenylmagnesium bromide	Et_2O, –60°, 1 hr	2-(*p*-Fluorophenyl)-3,4,5,6-tetrachloropyridine	515
p-Anisylmagnesium bromide	Et_2O, –60°, 1 hr	2-(*p*-Anisyl)-3,4,5,6-tetrachloropyridine (29.9)	515
α-Naphthylmagnesium bromide	Et_2O, –60°, 1 hr	2-(α-Naphthyl)-3,4,5,6-tetrachloropyridine (41.5)	515
2-Thienylmagnesium bromide	Et_2O, –60°, 1 hr	2-(2-Thienyl)-3,4,5,6-tetrachloropyridine (41.1)	515
Methylmagnesium bromide	Et_2O, –60°, 1 hr	2-Methyl-3,4,5,6-tetrachloropyridine (40.2)	515

TABLE II-29. Conversion of Unsaturated 1,5-Amino-aldehydes and 1,5-Amino-ketones to Pyridines (522)

Starting material	Conditions	Product (yield %)
$H_2NCMe{=}C(CO_2Et)CH{=}CHCHO$	Heat, vacuum, 130°	Ethyl 2-methylnicotinate (85)
$H_2NCMe{=}CAcCH{=}CHCHO$	Heat, vacuum, 120°	3-Acetyl-2-methylpyridine (90)
$H_2NCMe{=}C(CN)CH{=}CHAc$	AcOH, reflux, sublimed	2,6-Dimethylnicotinonitrile (90)
$H_2NCPh{=}C(CN)CH{=}CHAc$	Heat, vacuum, 140–150°	6-Methyl-2-phenylnicotinonitrile (90)
$H_2NCMe{=}C(CO_2Et)CH{=}CHBz$	Heat, vacuum, 150–170°	Ethyl 2-methyl-6-phenylnicotinate (90)
p-$MeC_6H_4CNH_2{=}C(CN)CH{=}CHAc$	Heat, vacuum, 170°	6-Methyl-2-(*p*-tolyl)nicotinonitrile
$H_2NCMe{=}C(CO_2Et)CH{=}CHCOC_7H_{15}$	Heat, vacuum, 150°	Ethyl 6-heptyl-2-methylnicotinate (92)

TABLE II-30. Condensation of Iminochlorides with Dienes (527)

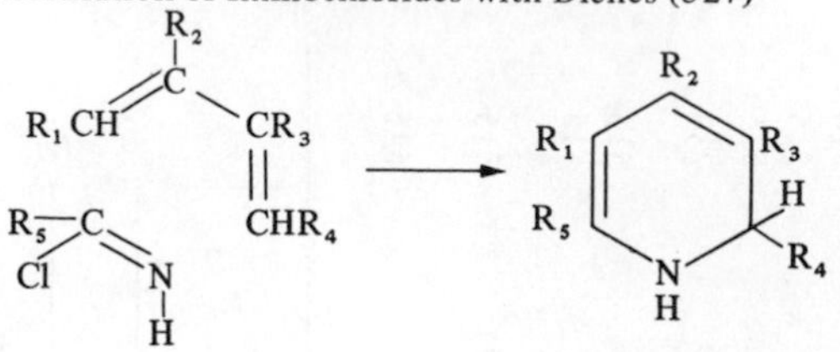

R_1	R_2	R_3	R_4	R_5	Products
H	Me	Me	H	H	3,4-Dimethyl-1,2-dihydropyridine
Me	Me	H	Me	H	2,4,5-Trimethyl-1,2-dihydropyridine
Me	Me	H	Ph	H	4,5-Dimethyl-2-phenyl-1,2-dihydropyridine
Me	H	Me	H	H	Mixture: 3,5-dimethyl- and 2,4-dimethyl-1,2-dihydropyridine
Me	H	Me	H	Me	Mixture: 3,5,6-trimethyl- and 2,4,6-trimethyl-1,2-dihydropyridine

10% aqueous sodium hydroxide or sodium-ethanol. The 4,6-dimethyl derivative was prepared in near quantitative yield. When the diketone was 2-acetylcyclohexanone, 3-cyano-2-dicyanomethylene-4-methyl-1,2,5,6,7,8-hexahydroquinoline (**II-52**) was prepared in 48% yield (532).

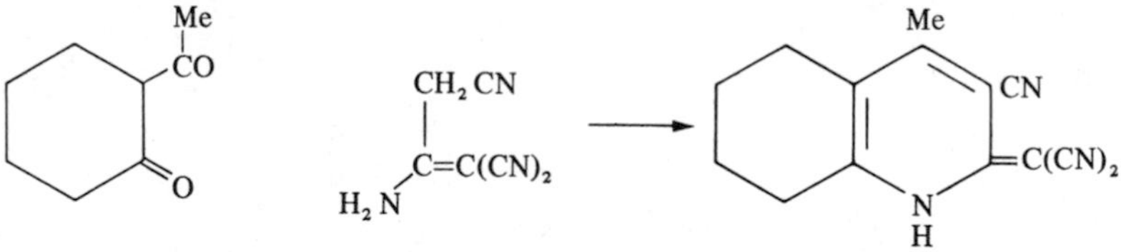

II-52

When malonamide was heated with excess ethyl 2,4-dioxopentanecarboxylate at 130 to 140°, the imide of 2-hydroxy-6-methylpyridine-3,4-dicarboxylic acid (**II-53**) was formed in unreported yield (535).

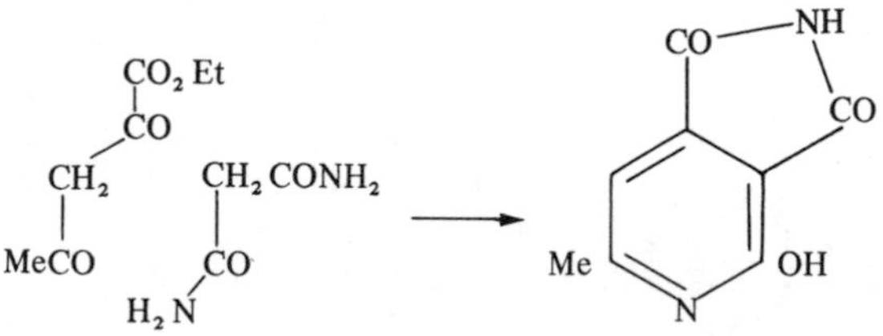

II-53

Two moles of acetylacetone condense with *N,N'*-bis(cyanoacetyl)hydrazine to give an *N,N'*-bipyridone **II-54** (539).

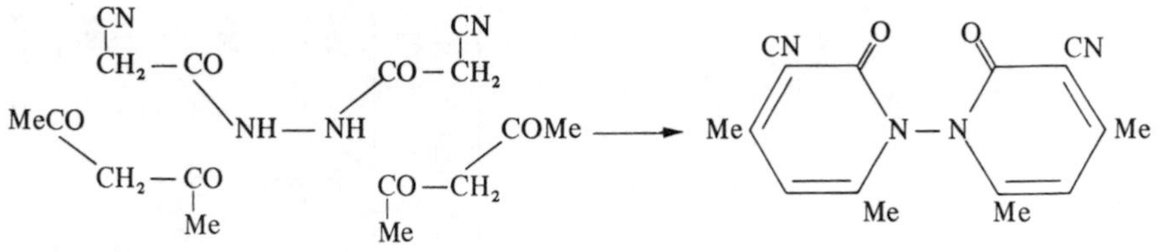

II-54

A similar reaction can occur with *N,N'*-bis(cyanoacetyl)diamides (**II-55**) (539).

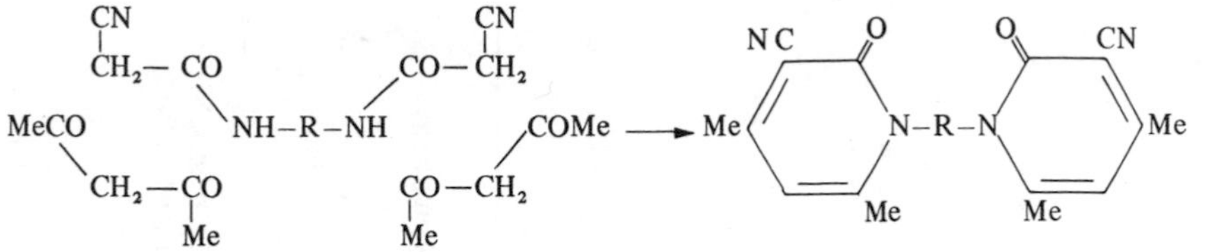

II-55

where R = CO or malonyl but not oxalyl.

In a reaction similar to that of a diketone, the sodium salt of 2-hydroxymethylcyclopentanone reacts with malonamide amidine hydrochloride to give

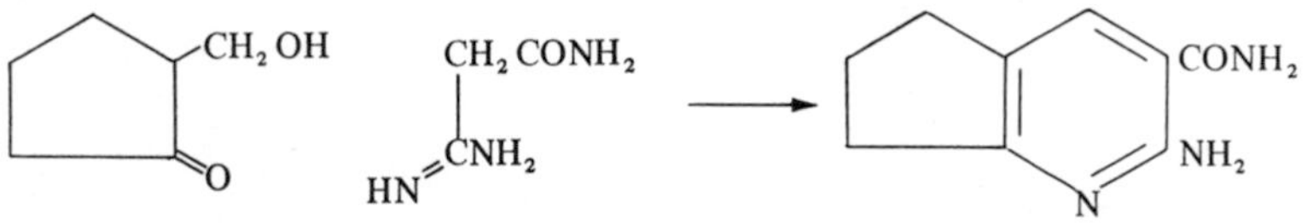

II-56

2-amino-3-carboxamido-5,6-cyclopentenopyridine (**II-56**) in a 28% yield. A similar reaction was also reported for 2-hydroxymethylcyclohexanone (545).

The reaction of 1-ethoxy-2,4-dioxopentane with ethyl β-amino-β-ethoxyacrylate and ammonia is reported to give ethyl 2-amino-4-ethoxymethyl-6-methylni-

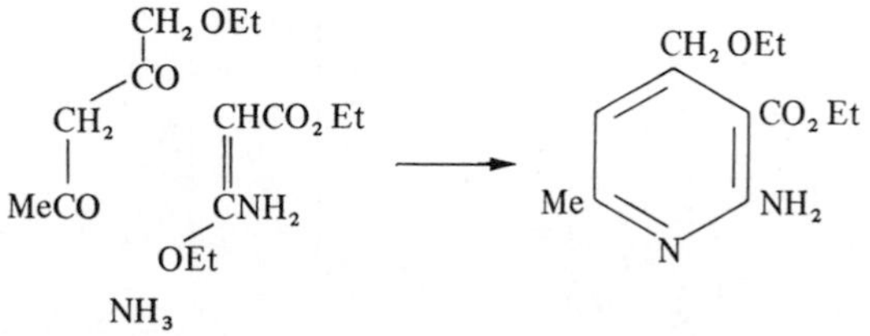

II-57

TABLE II-31. Condensation of 1,3-Diketones with Cyanoacetamide and *N*-Substituted Cyanoacetamides

R_1	R_2	R_3	R_4	Conditions	Yield (%)	M.p. (°C)	Ref.
Me	H	Me	H	Basic ion exchange resin	66	298	533
Me	H	Me	H	Piperidine	97	294	538
Me	H	Me	Phenyl			261	539
Me	H	Me	4-Antipyryl			245	539
Me	H	Me	Amino	EtOH; Et_2NH	81	174	540
Me	H	Me	Amino	EtOH; $NaNH_2$	43		540
Me	H	Me	Anilino			214	539
Me	H	Me	Tosylamino			255	539
Me	H	Me	Phenacetylamino			224	239
Me	H	Me	*p*-Nitrobenzylidenamino			258	540

Me	H	Me	Acetylamino	EtOH; Et_2NH	85	208	540
Me	H	Me	Chloroacetylamino	EtOH; Et_2NH	64	209	540
Me	H	$CH(OEt)_2$	H	Piperidine; 24 hours	80	147	537
Me	H	CO_2Et	H	$(COOEt)_2$; NaOMe	35	215	544
Me	H	CH_2CO_2H	H	NaOMe		228	533
Me	H	$CH_2CH_2CO_2H$	H	NaOMe		217	533
Me	H	Carbamyl	H	NaOMe		300	533
Me	H	Phenyl	Amino		8	206	540
CF_3	H	Me	H	NaOEt	17	232	536
Et	H	CO_2Me	H		80	240	542
Et	H	CO_2Et	H	NaOEt; 5 hr at 50°	78	216	542
Et	H	CH_2CO_2H	H	NaOMe	58	205	533
Et	H	$CH_2CH_2CO_2H$	H	NaOMe		211	533
n-Propyl	H	CO_2Et	H		80		542
$PhCH_2$	H	CO_2Et	H		35	166	542
Phenyl	H	Me	Amino	EtOH; $NaNH_2$; 3 days	44	238	540
Phenyl	H	Phenyl	Amino		43	177	540
2,3,4-$(MeO)_3C_6H_2$	H	CO_2Et	H	Piperidine, 10 hr; boil in EtOH			543
HO	H	Me	Amino		86	196	540
HO	Me	Me	H	MeOH; Piperidine; 29 hr reflux	54	294	541

cotinate in 35% yield (**II-57**) (546). The 4-carbethoxy and 4-carbamoyl derivatives were prepared in a similar manner.

Triethyl 3-carboxamido-2-methyl-4,5,6-pyridinetricarboxylate (**II-58**) has been prepared in 75% yield by heating ethyl α-oxo-β-ethoxyaconitate with β-methyl-β-aminoacrylonitrile (547).

CO_2Et, COEt, EtO_2C-C, CHCN, EtO_2C-CO, CMe, H_2N → CO_2Et, EtO_2C, $CONH_2$, EtO_2C, Me, N

II-58

Ethyl *n*-butylmalonate was condensed in the presence of potassium butoxide with ethyl 2-aminoacrylate to give 2-methyl-5-*n*-butylpyridine (548).

Acetoacetic ester has served as both of the components in a 3-2 condensation with either loss of the carbethoxyl group or of its conversion to an amide function (**II-59**) (549).

Me, CO, CH_2, $CHCO_2Et$, CO_2Et, COMe, NH_3 → Me, $CONH_2$, HO, Me, N

II-59

The β-chlorovinyl ketones, RCOCH=CHCl, react with methylenic compounds and excess ammonia or enamine compounds with elimination of chloride ion. Thus benzylvinyl chloride and ethyl β-aminocrotonate, when boiled in benzene for 8 hr gave a 51% yield of ethyl 2-methyl-6-phenylnicotinate (**II-60a**) (550).

CHCl, CH, $CHCO_2Et$, RCO, CMe, H_2N → CO_2Et, R, Me, N

II-60a R = Ph(51%)
II-60b R = *p*-ClC_6H_4(63%)
II-60c R = *o*-ClC_6H_4 (55%)
II-60d R = *p*-$O_2NC_6H_4$ (66%)

Diethyl 3-acetyl-2-methyl-4,5-pyridinedicarboxylate has been prepared by the condensation of diethyl 3-formyl-2-oxosuccinate with the imine of acetylacetone (**II-61**) (551).

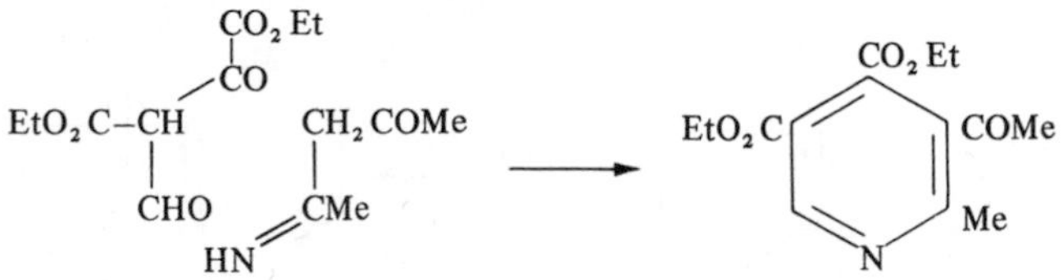

II-61

b. α,β-UNSATURATED CARBONYL COMPOUNDS WITH METHYLENIC COMPOUNDS. Another type of 3-2 condensation is the combination of an α,β-unsaturated carbonyl compound with an active methylenic compound. The nitrogen may be attached to either component but is often supplied separately as ammonia or an amine. A general example is **II-62**. The products are typically

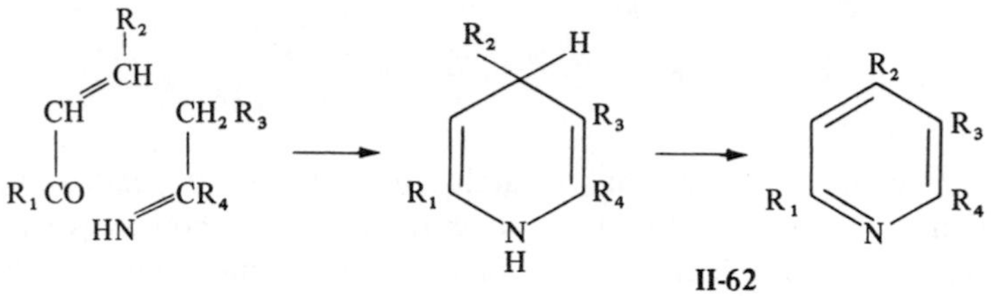

II-62

dihydropyridines, but under many reaction conditions (high temperatures over a contact catalyst) pyridines are the isolated products. The yields of pyridines are generally low, and those reported in the patent literature are subject to interpretation.

Sometimes the synthesis is conducted in two steps. The carbonyl compound and the methylenic compound are condensed, and the reaction product cyclized with a nitrogen compound. For instance, a Russian patent (552) claims the reaction of acrolein, ethylacrolein, methyl vinyl ketone, methacrolein, or crotonaldehyde with butyl vinyl, butenyl ethyl, or propenyl ethyl ethers. This is followed by cyclization by the action of hydroxylamine, although no details are given in *Chemical Abstracts.* Cinnamaldehyde is also reported to condense with an alkenyl alkyl ether to give a product that, after saponification, was cyclized to a 4-phenylpyridine (553).

The synthesis of 2-pyridones by the condensation of an α,β-unsaturated carbonyl compound, an aldol, or the Mannich base of an aldehyde or ketone with an *N*-(carbamoylmethyl)cycloammonium compound is summarized in Table II-32.

4-Methylthionaphtheno[2,3-*b*]pyridine (R.I. #8470) and 4-methylthionaphtheno[3,2-*b*]pyridine were reported (557) to be the products of heating the tin double salts of 2- or 3-aminobenzo[*b*]thiophene hydrochloride (R.I. #1353) with methyl vinyl ketone in the presence of iron and zinc chlorides (II-63).

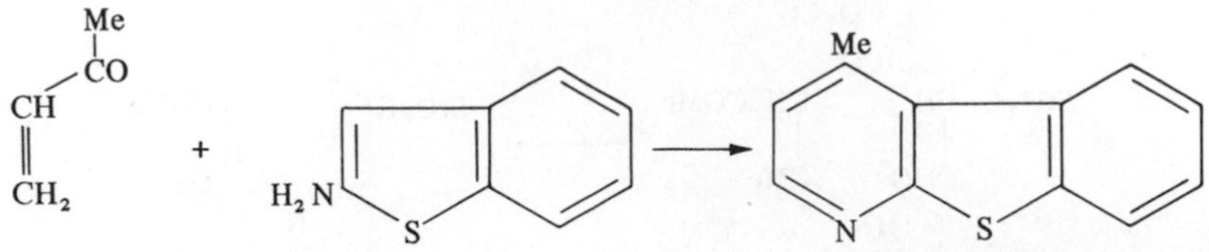

II-63

The reaction of α-carbonylacetylenes with enamines followed by ring-closure in the presence of a dehydration catalyst produces substituted pyridines **(II-64)** (556, 557). Other such addition products are given in Table II-33.

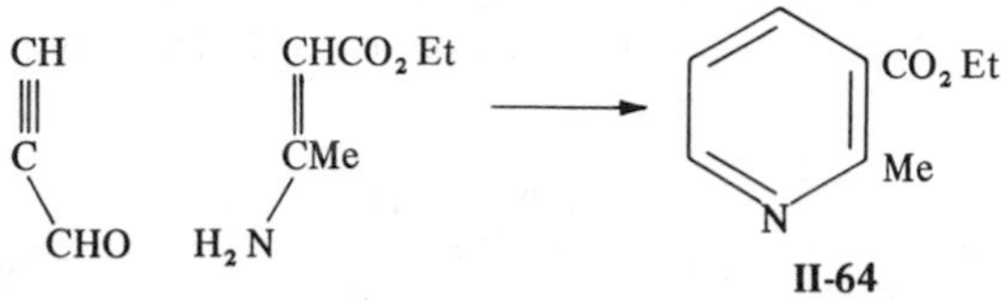

II-64

As described later, considerable research effort has been expended to develop an inexpensive commercial synthesis of pyridine and its homologs from small molecules. The 3-2 condensation with ammonia as the nitrogen source has been studied and is the subject of many patents.

When an equimolar mixture of acrolein and acetaldehyde is passed over a silica-alumina-lead fluoride catalyst at 400°, the respective yields of pyridine, 2-, 3-, and 4-picolines were 28, 8.5, 26.5, and 8.5% (558). The addition of oxygen to the reaction is purported to improve the yield of pyridine (559). By changing reaction conditions and catalysts, 3-picoline can be recovered as the major pyridine product (560). Silica-alumina catalysts modified with boron and phosphorus result in the predominant formation of picolines (561).

Crotonaldehyde and a ketone, such as acetone, gives alkyl substituted pyridines, such as collidines, methylethylpyridines, and lutidines (562).

The product from the condensation of crotonaldehyde and acetaldehyde with ammonia over a silica-alumina catalyst at 400° is reported to be mainly 3,4-lutidine (563). With formaldehyde instead of acetaldehyde, a 55 to 67% yield of pyridine was reported (564).

Two unsaturated ketones with additional functional groups have been condensed as in **II-65** (565).

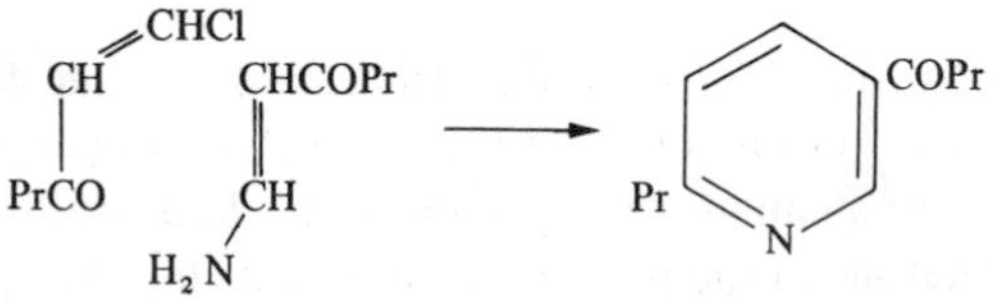

II-65

TABLE II-32. Condensation of α,β-Unsaturated Carbonyl Compounds and Related Compounds with *N*-(Carbamoylmethyl)pyridinium Salts (554)[a]

Carbonyl compound	Pyridinium compound	Product	M.p. (°C)	Yield (%)
BzCH=CHPh	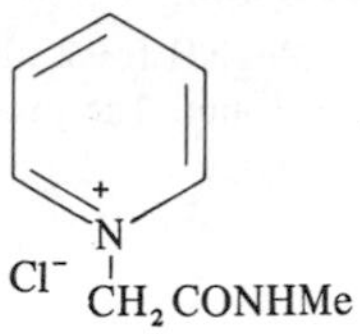 N^+, Cl^- $CH_2CONHMe$	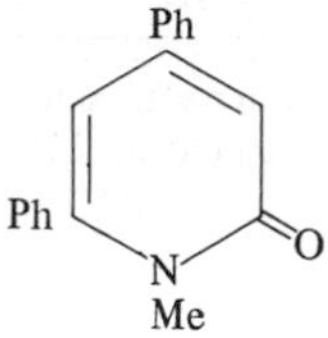 Ph, Ph, N, Me, O	91	45
PhCH=CHCHO	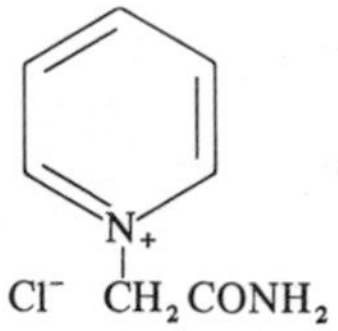 N^+, Cl^- CH_2CONH_2	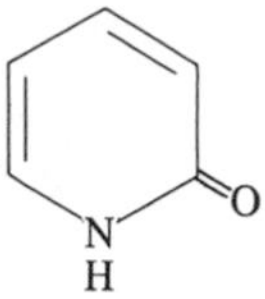 N, H, O	227–228	47
BzCH=CHPh	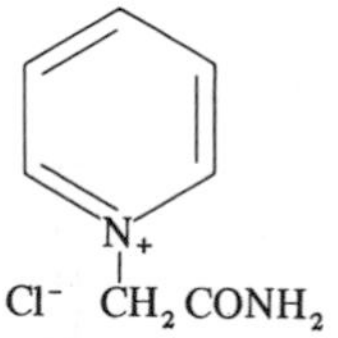 N^+, Cl^- CH_2CONH_2	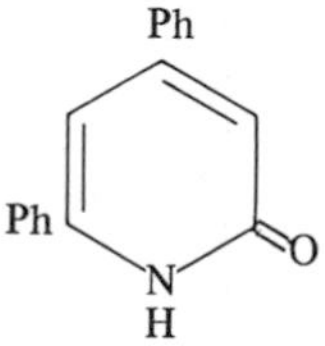 Ph, Ph, N, H, O	266–268	22
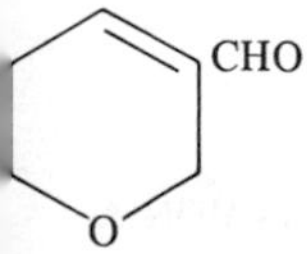 CHO, O	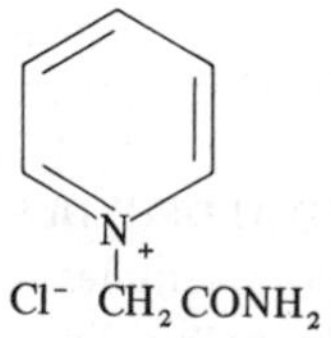 N^+, Cl^- CH_2CONH_2	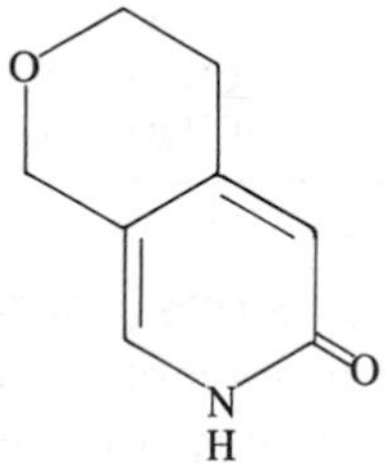 O, N, H, O	237	45
3-PyCH=CHCO-3-Py	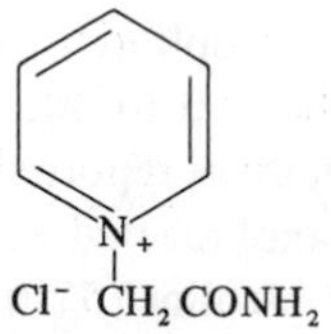 N^+, Cl^- CH_2CONH_2	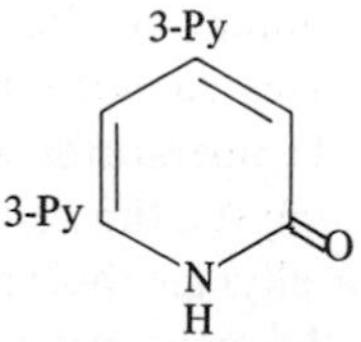 3-Py, 3-Py, N, H, O	282	80

$BzCH_2CH_2NMe_2$	(pyridinium, N^+–CH_2CONH_2 Cl^-)	Bz (2-pyridone, N–H)	197	50

[a]The reactants were boiled under reflux in MeOH with Me_2NH for 2 hr and evaporated to dryness, and then the residue was heated at 210 to 220° for 5 min. The product was extracted into benzene.

TABLE II-33. Pyridines by Reaction of α-Carbonylacetylenes and Enamines (556)

(Pyridine: R_1 at C-2, R_2 at C-3, R_3 at C-6)

R_1	R_2	R_3	Yield (%)	B.p./mm or m.p.
Me	CO_2Et	Me	90	60°/0.1
Me	CO_2Et	OH	–	160–180°/0.02
Me	COMe	H	80	–
Me	COMe	Me	90	–
Me	CN	Me	90	M.p. 82°
Me	CN	H	80	M.p. 38°
Me	CO_2Et	Ph	80	M.p. 45°
Ph	CN	Me	90	M.p. 70°
p-Tolyl	CN	Me	90	M.p. 102°
Me	CO_2Et	Heptyl	92	125°/0.1
Me	CO_2Et	*p*-$CH_3OC_6H_5$	80	M.p. 70°
Pr	CO_2Et	*p*-$CH_3OC_6H_5$	80	M.p. 84°

c. SELF–CONDENSATION OF α,β-UNSATURATED ALDEHYDES WITH AMMONIA. Although the self-condensation of α,β-unsaturated aldehydes with ammonia in a 3-2 synthesis is historically significant in the work of Chichibabin, little work has been done in recent times in this area. Thus acrolein and ammonia were passed in the vapor phase over silica-alumina or fluoride-containing alumina catalyst (566) or supported chromium oxide, molybdenum oxide or tungsten oxide (667), or silica-alumina treated with boric or phosphoric acid (568) to give pyridine and 3-picoline. Temperatures were in the 350 to 400° range. At 200°, over alumina-group VI catalysts, the major product was reported to be 2-picoline (569). When acrolein (or allyl alcohol) and propanol reacted with ammonia over a silica-alumina catalyst, the major product was 3-picoline (570).

The condensation of crotonaldehyde and ammonia over basic catalysts was studied (571). Alumina impregnated with calcium oxide gave better yields of pyridine bases and less tar formation than when potassium hydroxide was the base. Thermal cracking of crotonaldehyde limits the usefulness of the synthesis. In the presence of water vapor, the main pyridine product was 5-ethyl-2-picoline.

d. MISCELLANEOUS 3-2 CONDENSATIONS. Some 3-2 condensations are difficult to classify further because of the small number of examples. For instance, 2-carbethoxycycloheptanone was condensed with cyanoacetamide to give 3-cyano-6,7,8,9-tetrahydro-5*H*-cyclohepta[*e*]-2-pyridone (**II-66**) (R.I. #1782) in 50% yield (572).

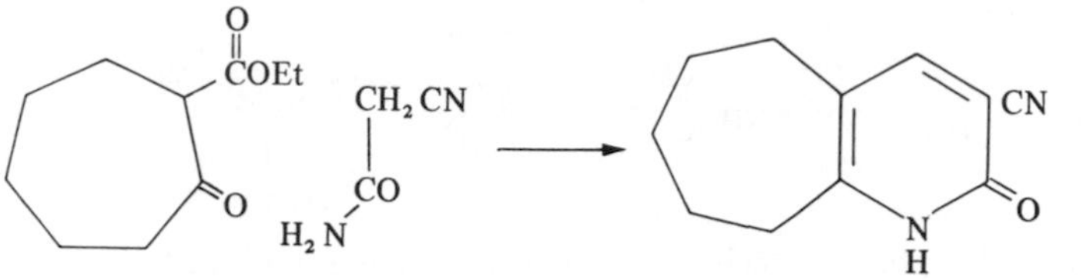

COEt CH$_2$CN CO H$_2$N O CN O N H

II-66

In a photosynthesis, malonyl chloride condenses with acetonitrile at room temperature to give 6-chloro-4-hydroxy-2-pyridone (**II-67**) (573).

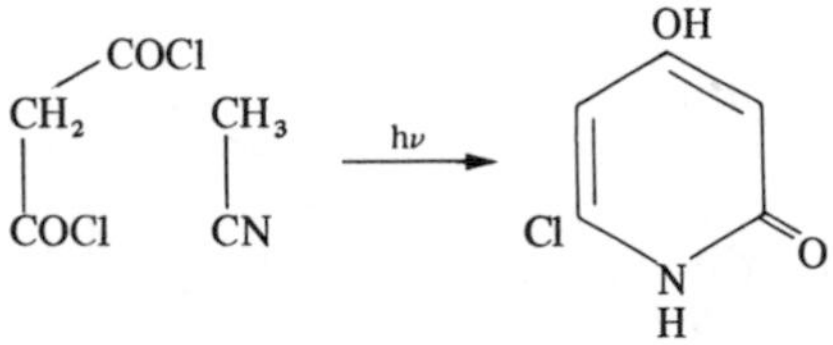

COCl CH$_2$ COCl CH$_3$ CN $h\nu$ OH Cl O N H

II-67

Ethyl β-aminocrotonate and propargyl aldehyde in benzene gave ethyl 5-aminohexa-2,4-dien-1-al-4-carboxylate which, when heated at 130°, gave ethyl 2-methylpyridine-3-carboxylate (**II-68**) (574).

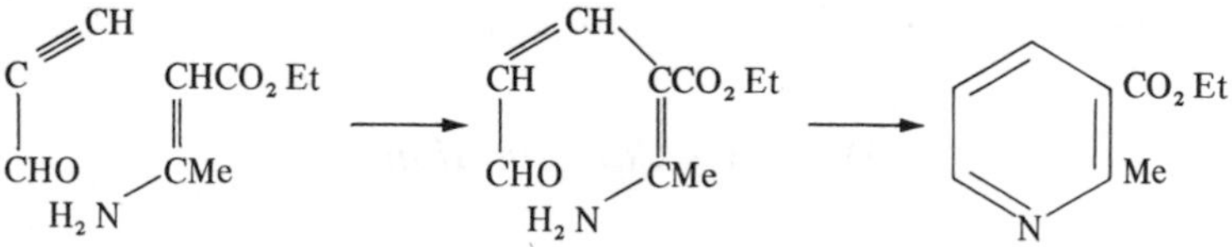

CH C CHO CHCO$_2$Et CMe H$_2$N → CH CH CHO CCO$_2$Et CMe H$_2$N → CO$_2$Et Me N

II-68

Ethyl nitromalonate was heated at 100° with ethyl β-aminomethacrylate and sodium to give ethyl 2-methyl-4,6-dihydroxy-5-nitropyridine-3-carboxylate (**II-69**) (575).

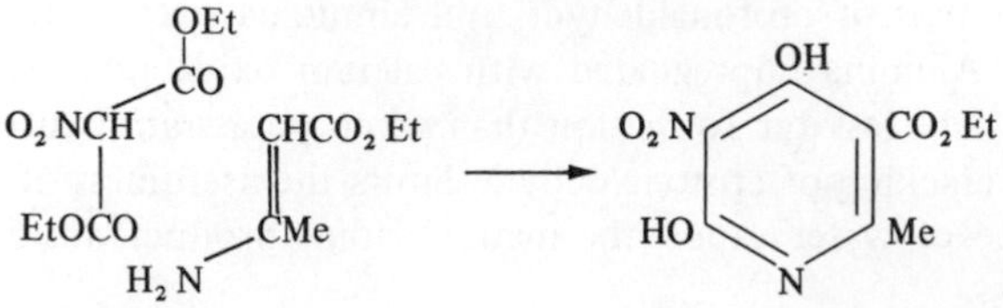

II-69

Condensation of α-phenyl-α-acetylacetamide and benzyl methyl ketone gave 4,6-dimethyl-3,5-diphenyl-2-pyridone (**II-70**) (578).

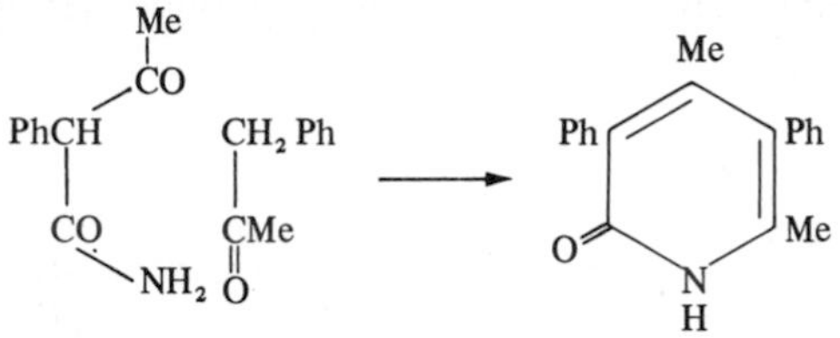

II-70

Cadmium phosphate or calcium phosphate were found to be catalysts for the synthesis of pyridine bases in the vapor phase from ammonia and various compounds in 3-2 condensations. Allyl alcohol and benzyl methyl ketone reacted with ammonia at 375° to give a 20% yield of 2-methyl-3-phenylpyridine (**II-71**) (577).

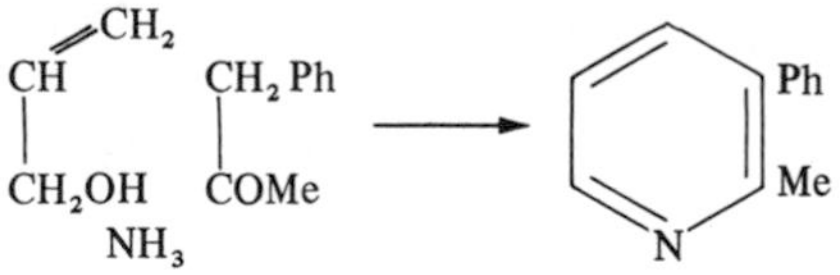

II-71

A mixture of ammonia, 1,1,3-triethoxybutane and various ketones reacted at 350 to 400° to give 2,3-disubstituted-4-picolines (578). These reactions are summarized in Table II-34. Similarly allyl alcohol and ketones gave 2-methyl-3-alkylpyridines (579) (Table II-35).

D. *1-3-1 Condensations*

The 1-3-1 condensation is not well known and is limited to a few examples. α-Methylstyrene can be condensed with formaldehyde and ammonium chloride or amine hydrochlorides to give 1,2,3,6-tetrahydropyridines that can be dehydrogenated to pyridines. These products are summarized in Table II-36.

TABLE II-34. Condensation of 1,1,3-Triethoxybutane, Ammonia and Ketones (578)

$$\mathrm{MeCH(OEt)CH_2CH(OEt)_2} + \mathrm{CH_2R_2COR_1} + \mathrm{NH_3} \longrightarrow \text{4-Me-3-}R_2\text{-2-}R_1\text{-pyridine}$$

R_1	R_2	Yield (%)	Conditions
Me	Et	35	1 : 1 : 4–6 mole ratio of triethoxybutane, ketone, and ammonia at 350–400° over $Cd_3(PO_4)_2$ catalyst.
Me	H	31	
Ph	Me	26	
Me	Ph	33	
Me	$PhCH_2$	7	

TABLE II-35. Condensation of Allyl Alcohol, Ammonia, and Ketones (579)

$$\mathrm{CH_2{=}CHCH_2OH} + \mathrm{CH_2R_2COR_1} + \mathrm{NH_3} \longrightarrow \text{3-}R_2\text{-2-}R_1\text{-pyridine}$$

R_1	R_2	Yield (%)	Conditions
Me	*n*-Pr	15	Mole ratio, ketone alcohol ammonia of 1:1.25–1.5: 4–5 at 375–400°, over $Cd_3(PO_4)_2$ catalyst.
Me	*n*-Bu	18	
Me	isoPr	5	
Me	*sec*-Bu	5	
Me	isoBu	16	
Me	isoAm	16	

E. *2-2-1 Condensations*

a. ALDEHYDES WITH AMMONIA. The condensation of aldehydes and ammonia can proceed either by a 2-2-1 or a 2-1-2 route, depending on whether the aldehyde contributing only one carbon atom to the ring forms the 2- or 4-carbon of the pyridine ring. The reaction of a single aldehyde favors a 2-2-1 condensation **(II-72)** and is discussed in this section. Reactions of mixtures of

TABLE II-36. Reactions of α-Methylstyrenes with Formaldehyde and Ammonia or Amines

R_1	R_2	Conditions	B.p. (°C)	Ref.
H	H	20–40°; 20 hr	M.p. 97–112°	580
m-Br	H	20–40°; 20 hr	–	580
p-F	H	20–40°; 20 hr	139–141°/4 mm	580
p-Cl	H	20–40°; 20 hr	157–160°/8 mm	580
p-CH_3	H	20–40°; 20 hr	162–170°/10 mm	580, 582
p-isoPr	Me	60° for 2 hr; mixed with MeOH	125–135°/0.8 mm	581
p-isoPr	H	60° for 2 hr; mixed with MeOH	110–125°/0.3 mm	582
p-CH_3	Me		110–115°/0.9 mm	582

II-72

aldehydes or aldehydes and ketones tend to give 2-1-2 condensation and are dealt with in the following section. Although the mechanism for forming pyridines from aldehydes and ammonia has been of concern since Chichibabin's time, recent work has done little to elucidate the mechanism (583, 584).

(1) *Acetaldehyde and Ammonia*

(a) *Liquid Phase.* The reaction of acetaldehyde and ammonia in water at 150 to 250° under pressure gives predominately 5-ethyl-2-methylpyridine and is a commercial process. Although 5-ethyl-2-methylpyridine is the major product, some thirteen other pyridine derivatives have been identified as by-products (585, 586). Paraldehyde has been reported to give better yields of 5-ethyl-2-methylpyridine than acetaldehyde (587, 588). In the system paraldehyde, ammonia, and aqueous ammonium acetate, the temperature should be higher than 220°. The concentration of ammonia did not affect the yield but at least 6 moles of ammonia per mole of paraldehyde were needed (590). Pressures used were in the 20 to 100 atm range.

A summary of conditions and yields is given in Table II-37.

(b) *Vapor Phase.* The vapor phase condensation of acetaldehyde and ammonia, usually over a modified alumina or silica-alumina catalyst, can result in picolines as the major products rather than 5-ethyl-2-methylpyridine. The loss of activity with use of the catalyst has been studied (603). A summary of reaction conditions is given in Table II-38.

(2) ***Other Aldehydes and Ammonia.*** In the search for methods of manufacture of pyridine and pyridine derivatives aldehydes other than acetaldehyde have been condensed with ammonia. The results of these reactions are summarized in Table II-39.

b. KETONES WITH AMMONIA. Although ketones do not generally give good yields of pyridines, acetone and ammonia have been vaporized over a silica-alumina catalyst between 300 and 600° to give a 21 to 31% yield of 2,4,6-trimethylpyridine (614, 615).

c. ALCOHOLS WITH AMMONIA. The reaction of ammonia with ethanol, allyl alcohol, and derivatives of vinyl alcohol produced alkylpyridines (Table II-40). This method of synthesis generally gives low yields of pyridine derivatives.

d. ACETYLENES WITH AMMONIA. A commercial method for the manufacture of pyridine from acetylene and ammonia continued to be sought. However, most work has led to product mixtures and to low yields of any single product (Table II-41).

e. MISCELLANEOUS. Isobutylene, formaldehyde, and ammonium chloride were condensed at 107° and 232 psig to give a red oil containing various tetrahydro-4-picolines. Dehydrogenation of the oil over a chromia-alumina catalyst at 475° gave 4-picoline. Substituting $MeNH_2 \cdot HCl$ for ammonium chloride gave 4-picoline and 2,4-lutidine (634).

When acetophenone and benzylidenediurea were heated at 137° for 7 hr, the major product was 4,6-diphenyl-2-oxodihydropyridine, but about a 10% yield of 2,4,6-triphenylpyridine was also obtained (**II-73**) (635).

Ph
CO
CH_3 CH_3
H_2NCONH
Ph–CH COPh
$NHCONH_2$
→
Ph
Ph Ph
N

II-73

TABLE II-37. Liquid Phase Condensation of Acetaldehyde (Paraldehyde) and Ammonia

Catalyst	Temperature (°C)	Pressure (atm.)	Yield of 5-ethyl-2-methylpyridine (%)	Ref.
NH_4HF_2-Na hexametaphosphate	260	–	81	581, 588
Guanidine-HF	260	–	74	591
Various metal fluorides	260	–	56–77	592
NH_4HF_2-undisclosed co-catalyst	230–260	<80	67	593
MoO_3	190–250	45	70	594
NH_4OAc	220	50	–	595
NH_4OAc	225	–	70	596
NH_4 salts	200	–	65	597
NH_4OAc	255–265	110–120	70	598
NH_4 cation exchange resin	230	–	75	599
$NH_4OAc + AcNH_2$	230	–	79	600
NH_4 molybdate	195–230	–	65	601
KHF_2	230–250	100	70	602

TABLE II-38. Vapor Phase Condensation of Acetaldehyde and Ammonia

Catalyst	Temperature (°C)	Product (yield %)	Ref.
Al_2O_3-SiO_2-Cd (or ZnO)	–	4-Picoline	604
Al_2O_3-CuO (or PbO)	250–600	2- and 4-Picoline 5-Ethyl-2-methylpyridine	605
Al_2O_3-SiO_2-Metal Oxide	400	2-Picoline (21) 4-Picoline (21)	606
Al_2O_3-SiO_2-Metal Oxide	400	2-Picoline (26–29) 4-Picoline (24–27)	607
Al_2O_3-SiO_2-ThO_2	420–430	2- and 4-Picoline (39–43)	608
Al_2O_3-SiO_2	<200	5-Ethyl-2-methylpyridine (major product)	609
Al_2O_3-SiO_2	<200	2- and 4-Picolines (major)	609

When treated with carbon monoxide (150 atm) at 280° in benzene and in the presence of metal carbonyls, allylamine has given pyridines. When the metal carbonyl was cobalt octacarbonyl, 3,5-dimethyl-2-ethylpyridine and 3,5-dimethyl-4-ethylpyridine were obtained in yields as high as 19 and 9%, respectively, depending on the weight ratio of allylamine to benzene. The major product was pyrrolidone (636). With nickel carbonyl at 230 to 280° and 300 atm, only traces of pyrrolidone and of the pyridines were obtained, but ruthenium carbonyl yielded 27 and 14% of the respective pyridines (637).

F. *2-1-2 Condensations*

The reaction of an aldehyde with ammonia and a second aldehyde or ketone to produce pyridine and alkylpyridines has been the subject of numerous patents. The product is often a complex mixture of compounds with the actual yield obscured by a nondisclosed definition of the term. The condensation of ammonia and an activated ketone such as ethyl acetoacetate takes place at modest temperatures to give pyridinediols or dihydropyridines.

a. MIXTURES OF ALDEHYDES AND KETONES WITH AMMONIA. As discussed in the previous section, the condensation of an aldehyde and ammonia with a second aldehyde (or a ketone) may occur by a 2-2-1 or a 2-1-2 condensation. A 2-1-2 path is generally favored in the vapor phase and may be dependent on the nature of the R groups in **II-74**.

R–CHO, $R'CH_2$–COR″, CH_2R'–COR″, NH_3 ⟶ pyridine (R at 4; R′ at 3,5; R″ at 2,6) **II-74**

TABLE II-39. Condensation of Various Aldehydes with Ammonia

Aldehyde	Catalyst	Conditions	Product (Yield)	Ref.
EtCHO	$Co_3Al_2(PO_4)_3$	350°, vapor phase, ~ 2 sec contact time	2-Et-3,5-$Me_2C_5H_2N$(65%)	611
n-PrCHO	$Co_3Al_2(PO_4)_3$	350°, vapor phase, ~ 2 sec contact time	3,5-Et_2-2-Pr-C_5H_2N(74%)	611
n-PrCHO	$Co_3Al_2(PO_4)_4$	350°, vapor phase; space velocity, 742 hr^{-1}	2-Et-3,5-$Me_2C_5H_2N$(65%)	612
n-$PrOCH_2CH_2CHO$	SiO_2-Al_2O_3-ZnO	400°, vapor phase, presence of PrOH	3-Picoline	610
n-$PrOCH_2CH_2CH_2CHO$	SiO_2-Al_2O_3-ZnO	400°, vapor phase, presence of PrOH	3,4-Lutidine	610
R-CHO	A sulfonic acid	150–315°, liquid phase	Alkylpyridines (improved)	612

TABLE II-40. Condensation of Alcohols and Ammonia

Alcohol	Conditions	Product (yield %)	Ref.
Ethanol	Vapor phase, 300–400°, various metal oxide catalysts	2- and 4-Picoline, 5-Et-2-MePyridine and β-collidine	616
Allyl alcohol	Vapor phase, 280–350° Pt, Pd, or Os catalyst	3-Picoline (18–21), 3,5-Me_2Pyridine (7) and 3,5-Me_2 4-EtPyridine (2.6)	617
Allyl alcohol	Vapor phase, 310°, Pd-Al_2O_3 catalyst	36% of the recovered product were basic compounds	618
Ethyl vinyl ether	Liquid phase, NH_4OH, $AcONH_4$, 2 hr at 155°	5-Et-2-MePyridine	619
Vinyl acetate	Liquid phase, (MeOH); 3 hr at 130°	5-Et-2-MePyridine (9)	620
Vinyl acetate	Vapor phase, 300–450°, Al_2O_3 catalyst	2- and 4-Picoline	620
Vinyl acetate	Liquid phase, NH_4OH, 2 hr at 230°	5-Et-2-MePyridine (68)	621
Methyl vinyl ether	Liquid phase, (MeOH); 1 hr at 230°, NH_4Cl catalyst	Picoline and 5-Et-2-MePyridine	622

TABLE II-41. Condensation of Acetylenes and Ammonia

Acetylene	Conditions	Products (yield %)	Ref.
Acetylene	Vapor phase NH_3 and HCHO, 400–420°	Pyridine, 2-, 3-, and 4-picoline	623
Acetylene	Vapor phase, NH_3, clay-ZnO catalyst	MeCN (75), pyridine bases	624
Acetylene–C_3H_8 mixture	Liquid phase, NH_4OH, 8 hr 185°, 200 psig NH_4HF_2 catalyst	5-Et-2-MePyridine (88)	625
Acetylene	Vapor phase, NH_3, 100°, Ag-zeolite catalyst	2-Picoline	626
Acetylene	Vapor phase, NH_3, 420°, CdF_2 floridin catalyst	Pyridine (28), condensate containing 50% pyridine homologs	627
Acetylene	Vapor phase, NH_3, 300°, SiO_2 $CdHPO_4$ catalyst	Fraction (51) that could be oxidized to pyridine and isonicotinic acid	628
Acetylene	Liquid phase, H_2O, NH_4OAc, 200°, $Cd(OAc)_2$ catalyst	5-Et-2-MePyridine	629
Vinylacetylene	Vapor phase, NH_3, MeCOEt, 400°, $Cd_3(PO_4)_2$-Al_2O_3 catalyst	Crude product (97) containing 6-Et-2-Me-Pyridine, 2,3,6-$(Me)_3$Pyridine and 2,3,5,6-$(Me)_4$Pyridine	630
Vinylacetylene	Vapor phase, NH_3, acetophenone, 350–400° $Cd_3(PO_4)_2$-Al_2O_3	6-Et-2-MePyridine	632
Vinylacetylene	Vapor phase, NH_3, Me_2CO, 340–350°, Al-Cd catalysts	Lutidines	632
Dimethylvinylethynyl carbinol	Vapor phase, NH_3, 370°, CdO-Al_2O_3 catalyst	2,4-Me_2Pyridine, 2,6-Me_2Pyridine, and 2,4,6-Me_3Pyridine	633

The condensations (Tables II-42 and II-43) are generally carried out in the vapor phase at about 400° over a silica-alumina catalyst or over silica-alumina modified by a metal fluoride. With formaldehyde and acetaldehyde, the products contain predominantly pyridine (20 to 30% yield) and picolines in lesser amounts. Replacement of acetaldehyde by crotonaldehyde gives higher yields of pyridine (638). Methylal has been substituted for formaldehyde and this results in an increase in the yield of 3-picoline. Higher homologs of the two aldehydes give mostly lutidines, collidines, or other alkylpyridines, as do aliphatic ketones.

b. ALDEHYDES OR KETONES WITH ACTIVE METHYLENIC COMPOUNDS

(1) *With Methylenic Ketones.* The usual methylenic ketone that is condensed with ammonia and an aldehyde in a 2-1-2 condensation is acetoacetic ester. The reaction takes place under mild conditions, such as in boiling ethanol, and the product is most often a dihydropyridine **II-75**.

R–CHO + EtO_2CCH_2COMe + H_2CCO_2Et(OCMe) + NH_3 ⟶ (4-R,4-H-1,4-dihydro-2,6-dimethyl-3,5-bis(CO_2Et)pyridine)

II-75

Haley and Maitland demonstrated that the yield was optimum at a weakly basic p*H* in the reaction of acetaldehyde, ethyl acetoacetate, and ammonia (**II-75**, R = Me) (663). The yield at a p*H* of 6 was only 3%. However, Ihsan reported yields of 53% or better at p*H* 3.25 to 5.0 (664). The reaction was studied with a number of aldehydes in an aqueous medium and the optimum p*H* was 9.2 when R = Me; Pr; Ph; 2-ClC_6H_4; 3-HOC_6H_4; 3,4-$(MeO)_2C_6H_3$; 2-, 3-, and 4-$NO_2C_6H_4$; styryl; 2-naphthyl; vinyl; and propenyl. However when R = 2-furyl, the optimum p*H* was 7.1 (665).

Hexamethylenetetramine and ethyl acetoacetate, when boiled in ethanol for 20 to 30 min in the presence of ammonium phosphate gave diethyl 2,6-dimethyl-1,4-dihydropyridine-3,5-dicarboxylate. In the absence of the phosphate, 4 to 5 hr of boiling under reflux were required. However, when equal parts of ethyl acetoacetate, hexamethylenetetramine, and acetic acid were heated at 100° for 30 min diethyl 2,6-dimethylpyridine-3,5-dicarboxylate was the product (666).

In other examples, piperonal was condensed with ammonia and ethyl acetoacetate to give a 77% yield of the dihydropyridine **II-76** (667), which was oxidized to the pyridine in 80% yield with nitric acid.

TABLE II-42. Condensation of Formaldehyde, Acetaldehyde, and Ammonia in the Vapor Phase

Catalyst	Temperature (°C)	Products (yield %)	Comments	Ref.
Al_3O_3 or SiO_2	500	Pyridine and 3-picoline	HCHO in form of CH_2(OH)OMe	639
SiO_2-Al_2O_3-PbF_2	400	Pyridine (0.28 mole/mole AcH); 2-picoline (0.01 mole/mole AcH); 3- and 4-picoline (0.06 mole/mole AcH)	4.9 sec contact time; steam; air	640
SiO_2-Al_2O_3-ZnO	440	Pyridine (14.7); 2-picoline (18.9), 4-picoline (17.5)	Trioxane or acetaldehyde can replace HCHO	641
SiO_2-Al_2O_3-ZnO	435	Pyridine (38); 2-picoline (6), 3- and 4-picoline (12)	1:2–5 mole ratio of HCHO and AcH	642
SiO_2-Al_2O_3	300–500	Pyridine (28); 3-picoline (16.7)	1:2 mole ratio of NH_3 to aldehydes	643
SiO_2-Al_2O_3	200–400	Pyridine (21.6)	200-psig pressure	644
SiO_2-Al_2O_3	425	Pyridine and 3-picoline (60)	Paraldehyde and trioxane as feed	645
SiO_2-Al_2O_3-PbF_2	400	Pyridine (0.1 mole/mole AcH); 3-picoline (0.22 mole/mole AcH)	1.0:1.09:0.73 AcH–CH_2O–PrOH and 1.11:1 NH_3–AcH ratios	646

TABLE II-43. Condensation of an Aldehyde and Ammonia with Another Aldehyde or Ketone

Aldehydes	Conditions	Products (yield %)	Ref.
HCHO; Crotonaldehyde	Vapor phase; MeOH; SiO_2-CdF_2; 500°	Pyridine (28); 3-picoline (4)	647
	Vapor phase; SiO_2-Al_2O_3-PbF_2; 400°	Pyridine (56)	648
HCHO; ethylmethylacrolein	Vapor phase; MeOH; SiO_2-Al_2O_3-PbF_2; 400°	3,5-Lutidine, 3-picoline	649
HCHO; ethylpropylacrolein	Vapor phase; MeOH; SiO_2-Al_2O_3-PbF_2; 400°	3,5-Diethylpyridine (46)	649
HCHO; butyraldehyde	Vapor phase; SiO_2-Al_2O_3-PbF_2; 400°	Pyridine; 3,5-diethylpyridine, 3-picoline, 3-ethylpyridine	650
Allyl alcohol, 4-methylcyclohexanone	Vapor phase; $Cd_3(PO_4)_2$-Clay; 350°	6-Methyl-1,2,3,5-tetrahydroquinoline (38)	651
Allyl alcohol, cyclopentanone	Vapor phase	Cyclopenteno-2,3-pyridine	652
Allyl alcohol, Et_2CO	Vapor phase; $Cd_3(PO_4)_2$; 400°	2-Ethyl-3-methylpyridine (25)	653
Allyl alcohol, Pr_2CO	Vapor phase; $Cd_3(PO_4)_2$; 400°	3-Methyl-2-propylpyridine (30)	653
Crotonaldehyde, butyraldehyde	Vapor phase; SiO_2-Al_2O_3-PbF_2, 450°	Pyridine, 2-picoline, 4-picoline, 2-propylpyridine, 3-ethyl-4-methylpyridine	654
3-Methoxybutyraldehyde, crotonaldehyde, propionaldehyde	Vapor phase; SiO_2-Al_2O_3, 400°	3,4-Lutidine (14); 3,5-lutidine (6); 2-ethylpyridine (3) trace of picolines	655
HCHO, butyraldehyde, Me_2CO	Vapor phase, SiO_2-Al_2O_3-PbF_2, 400°	5-Ethyl-2-methylpyridine; 2,6-lutidine; 3,5-diethylpyridine	656
HCHO, EtMeCO, EtCHO	Vapor phase; SiO_2-Al_2O_3-PbF_2, 400°	2-Ethyl-5-methylpyridine; 3,5-lutidine; 2,3,5-trimethylpyridine	656
HCHO, PrCHO, Me_2CO	Vapor phase; SiO_2-Al_2O_3-PbF_2, 400°	5-Ethyl-2-methylpyridine; 2,6-lutidine; 3,5-diethylpyridine	657
HCHO, EtCHO, MeCOEt	Vapor phase; SiO_2-Al_2O_3-PbF_2, 400°	2-Ethyl-5-methylpyridine; 3,5-lutidine	657
HCHO, EtCHO, Me_2CO	Vapor phase; SiO_2-Al_2O_3-PbF_2, 400°	2,5-Lutidine; 3,5-lutidine, 2,6-lutidine	657

HCHO, MeCHO, Me_2CO	Vapor phase; SiO_2-Al_2O_3-PbF_2, 400°	2-Picoline; 2,6-lutidine; pyridine; 4-picoline; 2,4-lutidine; 2,4,6-collidine	657
HCHO, MeCHO, EtCHO	Vapor phase, SiO_2-Al_2O_3-PbF_2, 400°	Pyridine (0.14 mole/mole MeCHO); 3-picoline (0.15 mole/mole MeCHO)	657
Ethylideneacetone, mesityl oxide, HCHO, MeOH	Vapor phase, SiO_2-Al_2O_3-PbF_2, 400°	2-Picoline and 2,4-lutidine (major); 2,6-lutidine and 2,4,6-collidine (trace)	651
Ethylideneacetone, mesityl oxide, MeCHO, MeOH	Vapor phase, SiO_2-Al_2O_3-PbF_2, 400°	2,4-Lutidine (20); 2,6-lutidine (3.7)	662
Ethylideneacetone, mesityl oxide, *n*-PrCHO, MeOH	Vapor phase, SiO_2-Al_2O_3-PbF_2	6-Methyl-2-propylpyridine; 5-ethyl-2,4-dimethylpyridine	662
Ethylideneacetone, mesityl oxide, *n*-PrCHO, MeOH	Vapor phase, SiO_2-Al_2O_3-PbF_2	2-Isopropyl-6-methylpyridine	662
Mesityl oxide, MeOH, HCHO	Vapor phase, SiO_2-Al_2O_3-PbF_2	2,4-Lutidine (17); 2,6-lutidine; 2- and 4-picoline	663
Mesityl oxide, MeOH, MeCHO	Vapor phase, SiO_2-Al_2O_3-PbF_2	2,4,6-Collidine (major); 2,4-lutidine; 2,6-lutidine; 2- and 4-picoline	663
$PhCH_2CHO$, MeCHO	Liquid phase, EtOH 225°, 6 hr	3,5-Diphenylpyridine; 2-methyl-5-phenylpyridine; 2-methyl-3,5-diphenylpyridine	659
$PhCH_2CHO$, EtCHO	Liquid phase, EtOH 225°, 6 hr	3,5-Diphenylpyridine; ethyldimethylpyridines; 2-ethyl-3-methyl-5-phenylpyridine; 3-methyl-5-phenylpyridine; 2-ethyl-3,5-diphenylpyridine; 2-benzyl-3,5-diphenylpyridine	659
Cinnamaldehyde; butyraldehyde	Liquid phase, $(NH_4)_2CO_3$, 61°, 25 min	3-Benzylidene-5-ethyl-2-propyl-2,3-dihydropyridine (major)	660
Cinnamaldehyde; CH_3CHO	Liquid phase, $(NH_4)_2CO_3$, 61°, 25 min	3-Benzylidene-2-methyl-2,3-dihydropyridine 3-Benzylidene-2-(1-propenyl)-2,3-dihydropyridine	660

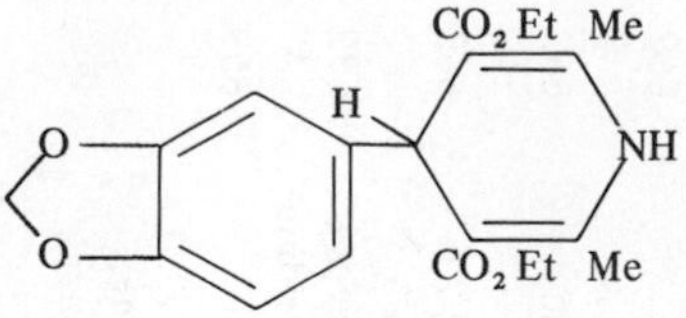

II-76

Penta-*O*-acetylaldehydo-D-glucose was used as the aldehyde in the condensation with ethyl acetoacetate to give diethyl 1,4-dihydro-2,6-dimethyl-4-D-arabino-1,2,3,4,5-pentaacetoxypentylpyridine-3,5-dicarboxylate (668).

Ethyl acetoacetate, ammonia, and substituted benzaldehydes can condense to give 4-pyridones in a 1-3-1 condensation as shown in **II-77** (669) (Section II **2D**, p. 272).

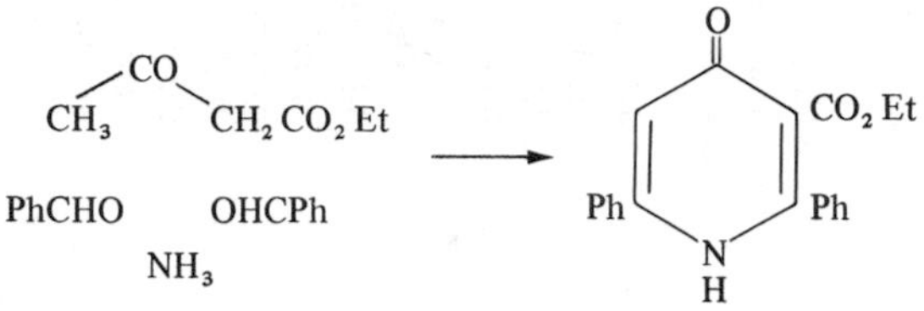

II-77

II-77

(2) *With Methylenic Esters and Amides.* When cyanoacetamide, ammonium hydroxide, and acetaldehyde were condensed at room temperature, the product was 3,5-dicyano-2,6-dihydroxy-4-methylpyridine (**II-78**). The ethyl derivative was the product from propionaldehyde (670).

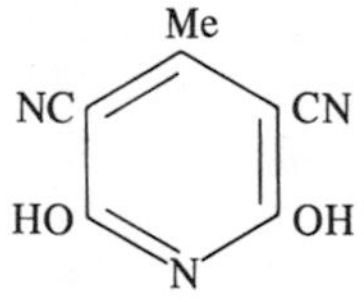

II-78

The products obtained from the condensation of aldehydes or ketones with 2-phenyl-2-iminopropionitriles are listed in Table II-44 (671).

G. *Cyclization Not Involving the Ring Nitrogen*

Although the majority of pyridine syntheses from acyclic compounds require the formation of at least one carbon-nitrogen bond, a few preparations start with acyclic compounds that are capable of being cyclized through the formation of carbon-to-carbon bonds. Most of these compounds are secondary or tertiary amines or substituted amides.

TABLE II-44. Condensation of Iminonitriles with Aldehydes and Ketones (671)

$$R_2R_1CO + 2\ R_3C_6H_4-C(NH_2)=CH-CN \longrightarrow \text{4-}R_1\text{-4-}R_2\text{-3,5-dicyano-2,6-bis}(R_3C_6H_4)\text{-1,4-dihydropyridine}$$

R_3	R_1	R_2	Yield (%)
H	H	H	23
p-Me	H	H	13
p-MeO	H	H	61
p-Cl	H	H	20
m-Cl	H	H	8
H	Me	H	49
p-Me	Me	H	44
p-MeO	Me	H	59
p-Cl	Me	H	34
m-Cl	Me	H	31
H	Me	Me	43
p-Me	Me	Me	56
p-MeO	Me	Me	42
p-Cl	Me	Me	28
m-Cl	Me	Me	23

a. CYCLIZATION OF SECONDARY AND TERTIARY AMINES. Pyridine was obtained in 20% yield by passing diethanolamine and methanol over a molybdenum-vanadium-alumina catalyst at 500°. A 7% yield of 2- and 3-picoline also resulted. Substituting paraformaldehyde for methanol gave a 13.5% yield of pyridine (672).

Aliphatic imino compounds were converted to pyridine bases in the presence of iodine at 400°. Compounds tested were propyl ethylideneamine, ethyl *n*-propylideneamine, *n*-propyl *n*-propylideneamine, and *n*-propyl isopropylideneamine; yields of pyridine bases ranged from about 1% to over 24% (673).

N,N-Bis(2-chloro-2-butenyl)benzylamine, when treated for 6 hr at 80° with sulfuric acid, yielded 3-acetyl-*N*-benzyl-4-methyl-1,2,5,6-tetrahydropyridine. Thermal decomposition of this base gave 2,3,4-trimethylpyridine in 36% yield (674).

9-Methylacridine has been prepared by passing diphenylamine and acetic acid over a silica-alumina catalyst at 388° (675).

N,N-Bis(2-cyanoethyl)alkylamines and *N*-methylaniline react with sodamide to give 1-alkyl-4-amino-3-cyano-1,2,5,6-tetrahydropyridines. The respective yields with various alkyl groups were: Me, 87%; Et, 85%; and Ph, 72% (676).

b. CYCLIZATION OF *N*-SUBSTITUTED AMIDES. The cyclization of amides to pyridine generally requires the participation of the acyl group. Thus the amides **II-79** were cyclized to pyridines by the action of lithium aluminum hydride (677).

R = Ph, Me

II-79

Cyclization of *N*-(3-benzyloxy-4-methoxybenzoyl)-2-(2-furyl)ethylamine by the action of phosphorus oxychloride gave 4-(3-benzyloxy-4-methoxyphenyl)-6,7-dihydrofuro[3,2-*c*]pyridine (R.I. #1306) (678).

N-Benzoyl-2-(2-pyrrolo)ethylamine was cyclized to 4-phenyl-6,7-dihydro-1*H*-pyrrolo[3,2-*c*]pyridine (R.I. # 1269) by the action of phosphorus oxychloride. The product was readily dehydrogenated to 1-phenylpyrrolo[3,2-*c*]pyridine (679). A similar reaction was reported for *N*-acyl-2-(2-furyl)-ethylamines (680).

III. References

1. Waller and Henderson, *Biochem. Biophys. Research Commun.*, **5**, 5 (1961).
2. Schenk and Shütte, *Naturwissenschaften,* **48**, 223 (1961).
3. McKennis, Bowman, and Turnbull, *J. Amer. Chem. Soc.*, **82**, 3974 (1960).
4. McKennis, Bowman and Turnbull, *Proc. Soc. Expt. Biol. Med.*, **107**, 145 (1961); *Chem. Abstr.*, **55**, 20197g (1961).
5. Coper, Hadass and Salazar, *Z. Krebsforsch.*, **70**, 138 (1967); *Chem. Abstr.*, **68**, 11062b (1968).
6. Senoh and Sakan, *Biol. Chem. Aspects Oxygenases, Proc. Kyoto,* 93 (1966).
7. Fukuda and Gilvarg, *J. Biol. Chem.*, **243**, 3871 (1968).
8. Shadani, Hathaway, and Kelly, *J. Dairy Sci.*, **45**, 833 (1962); *Chem. Abstr.*, **59**, 4477b (1963).
9. Auricchio and Quagliariello, *Boll. Soc. Ital. Biol. Sper.*, **33**, 159 (1957); *Chem. Abstr.*, **51**, 16628e (1957).
10. Oyama, Nakamura, and Tanabe, *Bull. Agr. Chem. Soc. Jap.*, **24**, 743 (1960); *Chem. Abstr.*, **55**, 5648h (1961).
11. Portellada, *Anais Microbiol. Univ. Brasil,* **7**, 77 (1959); *Chem. Abstr.*, **55**, 9551g (1961).

12. Hankes and Henderson, *J. Biol. Chem.*, **225**, 349 (1957); *Chem. Abstr.*, **51**, 9855c (1957).
13. Hankes and Henderson, *J. Biol. Chem.*, **222**, 1069 (1956); *Chem. Abstr.*, **51**, 2153h (1957).
14. Oyama, Nakamura, and Tanabe, *Kogyo Gijutsum Hakko Kenkyusko Kenkyu Hokoku*, **19**, 75 (1961); *Chem. Abstr.*, **55**, 27535g (1961).
15. Youatt, *Aust. J. Chem.*, **14**, 308 (1961); *Chem. Abstr.*, **55**, 23671f (1961).
16. Kitahara and Hayano, Japanese Patent 29,811 (1964); *Chem. Abstr.*, **63**, 5946a (1965).
17. Martin and Foster, *J. Bacteriol.*, **76**, 167 (1958); *Chem. Abstr.*, **52**, 17386e (1958).
18. Olson, *U. S. At. Energy Comm. BNL*, **512 (C-28)**, 316 (1958); *Chem. Abstr.*, **53**, 22201d (1959).
19. Franke, Frehse, and Heinen, *Getreide Mehl*, **8**, 81 (1958); *Chem. Abstr.*, **53**, 20596b (1959).
20. Hollander, Hollander, and Brown, *Endocrinology*, **64**, 621 (1959); *Chem. Abstr.*, **53**, 20390c (1959).
21. Oyama and Miyachi, *Biochem. Biophys. Acta*, **34**, 202 (1959); *Chem. Abstr.*, **53**, 8308d (1959); also *Chem. Abstr.*, **53**, 19037i (1959).
22. Bishop, Rankine, and Talbot, *J. Biol. Chem.*, **234**, 1233 (1959); *Chem. Abstr.*, **53**, 15282e (1959).
23. Ricci and Ivoldi, *Bull. Soc. Chim. Biol.*, **40**, 1163 (1958); *Chem. Abstr.*, **51**, 6738i (1957); also *Chem. Abstr.*, **53**, 538f (1959).
24. Harper, Dendt, and Elverjen, *Amer. J. Physiol.*, **105**, 175 (1958); *Chem. Abstr.*, **53**, 1484f (1959).
25. Preiss and Handler, *J. Biol. Chem.*, **233**, 488 (1958); *Chem. Abstr.*, **53**, 1439c (1959).
26. Caspe and Pitcoff, *Clin. Chem.*, **4**, 374 (1958); *Chem. Abstr.*, **53**, 1452d (1959).
27. Dietrich, Friedland, and Kaplan, *J. Biol. Chem.*, **233**, 964 (1958); *Chem. Abstr.*, **53**, 2417h (1959).
28. Shuster, Langan, and Kaplan, *Nature*, **182**, 512 (1958).
29. Schmidt, *Tabak-Forsch. Wiss. Beilage Deut. Tobakzig*, **25**, 85 (1958); *Chem. Abstr.*, **53**, 4660f (1959).
30. Schmeltz and Stedman, *Chem. Ind.* (London), 1244 (1962).
31. Quin and Pappas, *J. Agri. Food Chem.*, **10**, 79 (1962); *Chem. Abstr.*, **59**, 12917 (1963).
32. Avundzhyan, *Dokl. Akad. Nauk. Arm. SSR*, **34**, 211 (1962); *Chem. Abstr.*, **58**, 759b (1963).
33. Morselli, Ong, Bowan, and McKennis, *J. Med. Chem.*, **10**, 1033 (1967); *Chem. Abstr.*, **67**, 115024m (1967).
34. Neurath, Duenger, Geive, Leuttich, and Wechern, *Beitr. Tabakforsch*, **3**, 563 (1966); *Chem. Abstr.*, **67**, 8792p (1967).
35. Il'in, *Izvest. Akad. Nauk. SSR, Ser. Biol.*, 206 (1959); *Chem. Abstr.*, **53**, 15233d (1959).
36. Il'in, *Tabak*, **19**, 11 (1958); *Chem. Abstr.*, **53**, 19050c (1959).
37. Il'in and Lovkova, *Biokhimiya*, **24**, 274 (1959); *Chem. Abstr.*, **53**, 20305a (1959).
38. Jeney and Nemeth, *Dohanykutato Intezet Kozlemenyei*, **1**, 9 (1958); *Chem. Abstr.*, **53**, 20705a (1959).
39. Ligete, *Dohanykutato Intezet Kozlemenyei*, **1**, 12 (1958); *Chem. Abstr.*, **53**, 20704d (1959).
40. Wada, Kisaki, and Ihida, *Arch. Biochem. Biophys.*, **80**, 258 (1959); *Chem. Abstr.*, **53**, 14130d (1959).
41. Quagliariello, Auricchio, Rinaldi, and Violante, *Clin. Chim. Acta.*, **3**, 441 (1958); *Chem. Abstr.*, **53**, 12434a (1959).

42. Ackerman and List, *Z. Physiol. Chem.*, **313**, 30 (1958); *Chem. Abstr.*, **53**, 14362g (1959).
43. McGuire and Tompkins, *J. Biol. Chem.*, **234**, 791 (1959); *Chem. Abstr.*, **53**, 14273i (1959).
44. Moline, Walker, and Schweigert, *J. Biol. Chem.*, **234**, 880 (1959); *Chem. Abstr.*, **53**, 14273i (1959).
45. Onishi and Yamasaki, *Bull. Agr. Chem. Soc. Jap.*, **21**, 177 (1957); *Chem. Abstr.*, **53**, 13513f (1959).
46. Perlusz and Nemeth, *Elelmegesi Ipar*, **11**, 220 (1957); *Chem. Abstr.*, **53**, 22755g (1959).
47. Glock and Wright, *Anal. Chem.*, **35**, 246 (1963); *Chem. Abstr.*, **58**, 11692c (1963).
48. Veselinov, *Nauch. Tr. Vizzh. Inst. Khranitana Vkusova Prgm. Plovdiv.*, **9**, 195 (1962); *Chem. Abstr.*, **53**, 1972h (1959).
49. Matsuyama and Ishitoya, *Nippon Negeihagoka Keishi*, **34**, 659 (1960); *Chem. Abstr.*, **53**, 1972e (1959).
50. Gwynn, *Tobacco Sci.*, **7**, 1 (1963); *Chem. Abstr.*, **58**, 10526b (1963).
51. Weybrew and Mann, *Tobacco Sci.*, 7, 28 (1963); *Chem. Abstr.*, **58**, 14452c (1963).
52. Chakraborty and Weybrew, *Tobacco Sci.*, **7**, 122 (1963); *Chem. Abstr.*, **59**, 7868b (1963).
53. McKennis, *Ann. N.Y. Aca. Sci.*, **90**, 36 (1960); *Chem. Abstr.*, **58**, 4433e (1963).
54. Il'in, *Faziol Rast.*, **10**, 79 (1963); *Chem. Abstr.*, **59**, 5511h (1963).
55. Il'in, *Tabak*, **23**, 48 (1962); *Chem. Abstr.*, **59**, 3093f (1963).
56. Griffith, Hellman, and Byerrum, *Biochemistry*, **1**, 336 (1962); *Chem. Abstr.*, **59**, 5993f (1963).
57. Pyriki and Müller, *Ber. Inst. Tabakforsch.* (Dresden), 5, 127 (1958); *Chem. Abstr*, **53**, 3609g (1959).
58. Phillips and Bacot, *U.S. Dept. Agr. Tech. Bull.*, 1186 (1958); *Chem. Abstr.*, **53**, 3610d (1959).
59. Solt and Dawson, *Plant Physiol.*, **33**, 375 (1958); *Chem. Abstr.*, **53**, 3400h (1959).
60. Koelle, *Zuechter*, **31**, 346 (1961); *Chem. Abstr.*, **59**, 2662c (1963).
61. Kobashi, Hoshaku, and Watanabe, *Nippon Nogei Kagaku Kaishi*, **37**, 766 (1963); *Chem. Abstr.*, **59**, 18190d (1963).
62. Jarboe and Rosene, *J. Chem. Soc.*, **2455** (1961).
63. Onishi, *Koryo*, **55**, 9 (1960); *Chem. Abstr.*, **55**, 12778b (1961).
64. Caba-Torres and Tina, *Kongr. Pharm. Wiss. Vertr. Originalmitt*, **23**, 245 (1963); *Chem. Abstr.*, **62**, 10838d (1965).
65. Truhaut and de Clercq, *Bull. Soc. Chim. Biol.*, **41**, 1693 (1959); *Chem. Abstr.*, **55**, 4783d (1961).
66. Waltz, Haeusermann, and Moser, *Beitr. Tabakforsch.*, **2**, 283 (1964); *Chem. Abstr.*, **62**, 8134b (1965).
67. Choman, Abdel-Ghaffar, Reid, and Cone, *Bacteriol. Proc.*, **54**, 21 (1954); *Chem. Abstr.*, **51**, 15071b (1957).
68. Bonnet and Neukomm, *Helv. Chim. Acta*, **40**, 113 (1957).
69. Schmeltz, Stedman, Chamberlain, and Burdick, *J. Sci. Food Agr.*, **15**, 774 (1964); *Chem. Abstr.*, **62**, 6815h (1965).
70. Aronoff, *Plant Physiol.*, **31**, 355 (1956); *Chem. Abstr.*, **51**, 2129h (1957).
71. Dewey, *J. Amer. Chem. Soc.*, **80**, 1634 (1958).
72. Hasse and Berg, *Naturwissenschaften*, **44**, 584 (1957); *Chem. Abstr.*, **52**, 11848g (1958).
73. Schöpf, Braun, and Komzak, *Chem. Ber.*, **89**, 1821 (1956).

74. Pyriki, *Ber. Inst. Tabakforsch.* (Wohlsdorf-Biendorf), **1**, 62 (1954); *Chem. Abstr.*, **52**, 15842i (1958).
75. Izwa and Kobashi, *Bull. Agr. Chem. Soc. Jap.*, **21**, 357 (1957); *Chem. Abstr.*, **52**, 9524a (1958).
76. Brouwer, *Chem. Weekblad*, **58**, 530 (1962); *Chem. Abstr.*, **58**, 9547 (1963).
77. Hegarty, Court, and Thorne, *Aust. J. Agr. Res.*, **15**, 168 (1964); *Chem. Abstr.*, **60**, 16190b (1964).
78. Kuffner and Faderl, *Montash. Chem.*, **87**, 71 (1965); *Chem. Abstr.*, **50**, 15528g (1956).
79. Bose, De, and Dalal, *J. Indian Chem. Soc.*, **33**, 131 (1956); *Chem. Abstr.*, **50**, 13377g (1956).
80. Jeffrey and Eoff, *Anal. Chem.*, **27**, 1903 (1955).
81. Kemula and Chodkowski, *Rocz. Chem.*, **29**, 839 (1955); *Chem. Abstr.*, **50**, 6256f (1956).
82. Cundiff and Markunas, *Anal. Chem.*, **27**, 1650 (1955).
83. Bose, De, and Dalal, *Current Sci.* (India), **24**, 196 (1955); *Chem. Abstr.*, **50**, 2928f (1956).
84. Sandri, *Mikrochim. Acta*, **2**, 221 (1959); *Chem. Abstr.*, **54**, 25591g (1960).
85. Torres and Nunez, *Ars. Pharm.*, **4**, 303 (1963); *Chem. Abstr.*, **61**, 4691b (1964).
86. Biglino, *Ann. Chem.*, **49**, 1294 (1959).
87. Gopalachari and Sriramamierty, *Current Sci.* (India), **33**, 113 (1964); *Chem. Abstr.*, **60**, 12355g (1964).
88. Stephens and Weybrew, *Tobacco Sci.*, **3**, 48 (1959); *Chem. Abstr.*, **53**, 14425f (1959).
89. Bose, Gupta, and Mohamed, *J. Indian Chem. Soc.*, **35**, 81 (1958); *Chem. Abstr.*, **53**, 8325f (1959).
90. Schroeter, *Abhandl. Deut. Akad. Wiss. Berlin Kl. Chem. Geol. Biol.*, **99** (1963); *Chem. Abstr.*, **61**, 2193f (1964).
91. Yamashita, Kobashi, and Watanabe, *Nippon Nogeikagaku Kaishi*, **37**, 291 (1963); *Chem. Abstr.*, **61**, 2198c (1964).
92. Kobashi, Watanabe, and Hoshaku, *Nippon Nogeikagaku Kaishi*, **36**, 589 (1962); *Chem. Abstr.*, **61**, 3428c (1964).
93. Osman and Barson, *Phytochemistry*, **3**, 587 (1964); *Chem. Abstr.*, **61**, 13643e (1964).
94. Moser, U.S. Patent 3,151, 118 (1964); *Chem. Abstr.*, **61**, 16459d (1964).
95. Mors, Gottlieb, and Djerassi, *J. Amer. Chem. Soc.*, **79**, 4507 (1957).
96. Kusumoto and Ohara, *Bull. Chem. Soc. Jap.*, **30**, 195 (1957); *Chem. Abstr.*, **52**, 1550i (1958).
97. Kotake, Kusumoto, and Ohara, *Ann. Chem.*, **606**, 148 (1957).
98. Hasse and Maisack, *Naturwissenschaften*, **42**, 627 (1955); *Chem. Abstr.*, **52**, 1549c (1958).
99. Fairbairn and Challen, *Biochem. J.*, **72**, 556 (1959); *Chem. Abstr.*, **53**, 22258h (1959).
100. Robertson and Marion, *Can. J. Chem.*, **37**, 1043 (1959).
101. Swanholm, St. John, and Scheuer, *Pacific Sci.*, **13**, 295 (1959); *Chem. Abstr.*, **53**, 20317e (1959).
102. Gonzalez and Rodríguez, *Anales Real. Soc. Espan. Fis. Quim., Ser. B.*, **58**, 431 (1962); Chem. Abstr., **58**, 368g (1963).
103. Kazimierz, *Rocz. Nauk Rolniczych, Ser. A*, **86**, 527 (1962); *Chem. Abstr.*, **58**, 4811b (1963).
104. Volgina, *Biol. Aktivn. Veshchestvam, Plodov i Yayod, Sverdlovsk*, 186 (1961); *Chem. Abstr.*, **58**, 12861a (1963).
105. Bouillene-Walrand and Bouillene, *Bull. Soc. Roy. Sci.* (Liege), **30**, 258 (1961); *Chem. Abstr.*, **58**, 8232f (1963).

106. Smith and Abashian, *Amer. J. Botany,* **50**, 435 (1963); *Chem. Abstr.,* **59**, 9094b (1963).
107. Harashima, Takaichi, and Nakamura, *Agr. Biol. Chem.,* **25**, 39 (1961); *Chem. Abstr.,* **55**, 10537c (1961).
108. Cromwell, *Biochem. J.,* **64**, 259 (1956); *Chem. Abstr.,* **51**, 1381a (1957).
109. Friedman and Leete, *J. Amer. Chem. Soc.,* **85**, 2141 (1963).
110. Frank, Baker, and Sobotka, *Nature,* **197**, 490 (1963).
111. Owen, Reddon, and Whiting, *Can. J. Animal Sci.,* **42**, 182 (1962); *Chem. Abstr.,* **58**, 10562g (1963).
112. Labik, *Voeding,* **24**, 138 (1963); *Chem. Abstr.,* **59**, 4323c (1963).
113. Summers and Fisher, *Z. Ernachrungswiss,* **3**, 40 (1962); *Chem. Abstr.,* **59**, 4323c (1963).
114. Leete, *J. Amer. Chem. Soc.,* **80**, 4393 (1958).
115. Bernauer, Berlage, Schmid, and Karrer, *Helv. Chim. Acta.,* **41**, 1202 (1958).
116. List, *Planta Med.,* **6**, 424 (1958); *Chem. Abstr.,* **53**, 10393i (1959).
117. Griffith and Byerrum, *Biochem. Biophys. Res. Commun.,* **10**, 293 (1963); *Chem. Abstr.,* **58**, 14358c (1963).
118. Takahaski, Suguki, and Tamura, *J. Amer. Chem. Soc.,* **87**, 2066 (1965).
119. Dreyer, *J. Org. Chem.,* **33**, 3658 (1968).
120. Hill, *J. Amer. Chem. Soc.,* **80**, 1609 (1958).
121. Appelgren, Hansson, and Schmiterlow, *Acta Physiol. Scand.,* **56**, 249 (1962); *Chem. Abstr.,* **58**, 14585g (1963).
122. Ogg, Bates, Cogbill, Harrow, and Peterson, *J. Assoc. Offic. Agr. Chem.,* **42**, 305 (1959); *Chem. Abstr.,* **53**, 15491a (1959).
123. Suzue and Ryokuero, *Bull. Inst. Chem. Res., Kyoto Univ.,* **42**, 288 (1964).
124. Sodykov and Ostroshchenko, "Pyridine Derivatives from Natural Products" *in* "Khim. Tekhnol i Primenenie Proizvodnylch Piridina i Khinolina," Riga, 1957, pp. 89–96; *Chem. Abstr.,* **55**, 16583f (1961).
125. Werle and Schievelbein, *Arzneim. Forsch.,* **12**, 202 (1962); *Chem. Abstr.,* **57**, 2606 (1963).
126. Dariev, *Khim. Tekhnol. Topliv Masel,* **4**, 24 (1959); *Chem. Abstr.,* **53**, 18446d (1959).
127. Kubicka, Kvapil, and Huml, Czech. Patent 122, 956 (1967); *Chem. Abstr.,* **68**, 61421p (1968).
128. Andre, Dath, Mahieu, and Grand'Ry, *Brennst. Chem.,* **48**, 369 (1967); *Chem. Abstr.,* **68**, 31869y (1968).
129. Brown and Mihm, *J. Amer. Chem. Soc.,* **77**, 1723 (1955); *Chem. Abstr.,* **50**, 2586f (1956).
130. Ikekawa, Sato, and Maeda, *Pharm. Bull.* (Japan), **2**, 205 (1954); *Chem. Abstr.,* **50**, 994e (1956).
131. Tsuda and Maruyama, *Pharm. Bull.* (Japan), **1**, 149 (1953); *Chem. Abstr.,* **50**, 13896a (1956).
132. Dariev and Mar'yasin, *Khim. Tekhnol. Masel,* **3**, 60 (1958); *Chem. Abstr.,* **53**, 19350i (1959).
133. Nadziakiewicz and Heilpern, *Koks, Smola, Gaz,* **4**, 211 (1959); *Chem. Abstr.,* **59**, 8493i (1963).
134. Kuznetsov, Fadeicheva, and Kigel, *Akad. Nauk Ukr. SSR* 102 (1962); *Chem. Abstr.,* **58**, 10010b (1963).
135. Tanaka, Arakawa, and Yoshimura, *Coal Tar* (Japan), **7**, 576 (1955); *Chem. Abstr.,* **50**, 17382d (1956).
136. Funasaka and Kojima, *Bunseki Kagaku,* **9**, 741 (1960); *Chem. Abstr.,* **56**, 9393g (1962).

137. Arnall, *J. Chem. Soc.*, 1702 (1958).
138. Gepshtein, *Vost. Nauchn.–Issled. Uglekhim. Inst., Sb. Statei,* **2**, 92–6, (1964); *Chem. Abstr.*, **62**, 7548c (1965).
139. Wojdylo, *Przemysl. Chem.*, **13**, 708 (1957); *Chem. Abstr.*, **52**, 8505c (1958).
140. Yeh and Kalechits, *Jan Liao Hsileh Pao,* **2**, 146 (1957); *Chem. Abstr.*, **52**, 20145b (1958).
141. Saha, *J. Indian Chem. Soc.*, **34**, 79 (1957); *Chem. Abstr.*, **51**, 10873c (1957).
142. Werle, *Przemysl. Chem.*, **45**, 324 (1966); *Chem. Abstr.*, **66**, 39691p (1967).
143. Gepshtein, *Nauchn. Tr. Vost. Nauchn.–Issled. Uglekhim. Inst.*, **16**, 30 (1963); *Chem. Abstr.*, **61**, 8091d (1964).
144. Tanaka and Arakawa, *Bunseki Kagaku,* **6**, 281 (1957); *Chem. Abstr.*, **52**, 11662g (1958).
145. Bertling and Peters, *Erdoel Kohle*, **18**, 286 (1965); *Chem. Abstr.*, **63**, 392f (1965).
146. Dariev and Babueva, *Izv. Sibirsk. Otd. Akad. Nauk SSSR, Ser. Khim. Nauk,* 131 (1963); *Chem. Abstr.*, **60**, 14295d (1964).
147. Werle, *Przemysl. Chem.*, **42**, 545 (1963); *Chem. Abstr.*, **60**, 1493f (1964).
148. Karr and Chang, *J. Inst. Fuel,* **31**, 522 (1958); *Chem. Abstr.*, **53**, 3663e (1959).
149. Potashnikov, *Zhur. Priklad. Khim.*, **33**, 1381 (1960); *Chem. Abstr.*, **54**, 20160b (1960).
150. Tóth-Sarudy, *Magyar Kém. Lapja,* **13**, 401 (1958); *Chem. Abstr.*, **53**, 12639d (1959).
151. Yamada, *Kôgyô Kagaku Zasshi,* **60**, 1310 (1957); *Chem. Abstr.*, **53**, 13552c (1959).
152. Thewalt, *Brennst.-Chem.*, **40**, 88 (1959); *Chem. Abstr.*, **53**, 10707h (1959).
153. Funasaka, Kojima, and Igaki, *Bunseki Kagaku,* **16**, 1026 (1967); *Chem. Abstr.*, **69**, 37755a (1968).
154. Panfilov, El'bert, and Revva, *Probl. Podzemn. Gazif. Kuzbasse,* No. 2, 91 (1967); *Chem. Abstr.*, **69**, 37757c (1968).
155. Suzumura, *Bull. Chem. Soc. Jap.*, **34**, 1097 (1962); *Chem. Abstr.*, **57**, 13718f (1962).
156. Cook and Church, *Anal. Chem.*, **28**, 993 (1956).
156a. Abramovitch, Giam, and Notation, *Can. J. Chem.*, **38**, 624 (1960).
157. Chumakov, Filippovich, and Degtyarev, *Reakt. Sposobnost Org. Seedin., Tartu. Gos. Univ.*, **4**, 302 (1967); *Chem. Abstr.*, **69**, 2387e (1968).
158. Turkova, Vitenberg, and Belen'kii, *Neftekhimiya,* **7**, 458 (1967); *Chem. Abstr.*, **68**, 65462a (1968).
159. Van der Meeren and Verhaar, *Anal. Chim. Acta*, **40**, 343 (1968); *Chem. Abstr.*, **68**, 26791n (1968).
160. Voigt and Stech, *Fortschr. Wasserchem. Ihrer Grenzgeb.*, **2**, 242 (1965); *Chem. Abstr.*, **68**, 15919j (1968).
161. Mitra, Ghosh, Saha, and Sinha, *Technology* (Sindri), **4**, 105 (1967); *Chem. Abstr.*, **69**, 92746y (1968).
162. Kawasaki and Funakubo, *Koru Taru* (Japan), **16**, 236 (1964); *Chem. Abstr.*, **61**, 11805a (1964).
163. Lugovkin and Zemlyanitskaya, *Nauchn. Raboty Inst. Okhrany Truda Vses. Tsentr. Sov. Prof. Soyuzov,* 121 (1965); *Chem. Abstr.*, **65**, 15972c (1966).
164. Kondratov, Rus'yanova, and Malysheva, *Zh. Analit. Khim.*, **21**, 996; *Chem. Abstr.*, **65**, 19312i (1966).
165. Heyns, Stute, and Winkler, *J. Chromatog.*, **21**, 302 (1966).
166. Ruzhentseva, Pervacheva, Toropov, and Shcherbina, *Med. Prom. SSR,* **19**, 34 (1965); *Chem. Abstr.*, **64**, 10405c (1966).
167. Sinha and Mitra, *Technology* (Sindri), **2**, 46 (1965); *Chem. Abstr.*, **64**, 485f (1966).
168. Takayama, *Kogyo Kagaku Zasshi,* **62**, 658 (1959); *Chem. Abstr.*, **57**, 7913c (1962).

169. Kametani and Kubota, *Yakugaku Zasshi,* **82**, 659 (1962); *Chem. Abstr.*, **57**, 6588h (1962).
170. Pribyl, *Collect. Czech. Chem. Commun.*, **27**, 1330 (1962).
171. Bragilovskaya, Kogan, and Neimark, *Koks i Khim.*, No. 4, 44 (1962); *Chem. Abstr.*, **57**, 4041h (1962).
172. Dolgopolova and Ogloblina, *Sb. Nauchn. Rabot Inst. Okhramy Truda Vses. Tsentr. Soveta Prof. Soyuzov,* No. 1, 103 (1961); *Chem. Abstr.*, **57**, 3744i (1962).
173. Gruzdeva and Khokhlova, *Vost. Nauchn-Issled. Uglekhim. Inst., Sb. Statei,* 81 (1960); *Chem. Abstr.*, **57**, 1205i (1962).
174. Janak and Hrivnac, *Collect. Czech. Chem. Commun.*, **25**, 1557 (1960).
175. Rezl and Stajgr, *Chem. Prumysl.*, **11**, 413 (1961); *Chem. Abstr.*, **56**, 9i (1962).
176. Ostroshchenko, Sadykov, and Salit, *Zhur. Priklad. Khim.*, **34**, 2768 (1961); *Chem. Abstr.*, **56**, 10915i (1962).
177. Daftary and Haldar, *Anal. Chim. Acta,* **25**, 538 (1961); *Chem. Abstr.*, **56**, 9423e (1962).
178. Kuffner and Kirchenmayer, *Monatsh. Chem.*, **92**, 701 (1961); *Chem. Abstr.*, **56**, 1424e (1962).
179. Albesmeyer and Krampitz, *Naturwissenschaften,* **44**, 466 (1957); *Chem. Abstr.*, **52**, 4646e (1958).
180. Westbunk, *Brennst.-Chem.*, **38**, 375 (1957); *Chem. Abstr.*, **52**, 4962b (1958).
181. Chang and Karr, *Anal. Chem.*, **29**, 1617 (1957).
182. Bohon, Isaac, Hoftiezer, and Zellner, *Anal. Chem.*, **30**, 245 (1958).
183. Riolo and Notarianni, *Ann. Chim.* (Rome), **49**, 1981 (1959); *Chem. Abstr.*, **55**, 3301i (1961).
184. Kameo, Hirashima, Manabe, and Hiyama, *Kagaku To Kogyo* (Osaka), **39**, 22 (1965); *Chem. Abstr.*, **63**, 6316c (1965).
185. Brooks and Collins, *Chem. Ind.* (London), 1021 (1956).
186. Swietoslawski, Rostafinska, and Janek, *Przemysl. Chem.*, **33**, 212 (1954); *Chem. Abstr.*, **51**, 2253i (1957).
187. Bylicki, Rostafinska, and Wnek, *Przemysl. Chem.*, **34**, 565 (1955); *Chem. Abstr.*, **51**, 5387e (1957).
188. Biddiscombe and Herington, *Analyst,* **81**, 711 (1956); *Chem. Abstr.*, **51**, 4880i (1957).
189. Lindner, *Arkiv. Kemi,* **10**, 483 (1957) *Chem. Abstr.*, **51**, 11177e (1957).
190. Neimark and Kogan, *Sb. Nauchn. Tr. Ukr. Nauchn.-Issled. Uglekhim. Inst.*, No. 13, 172 (1962); *Chem. Abstr.*, **61**, 1278d (1964).
191. Neimark and Bragilevskaya, *Koks i Khim.*, No. 12, 38 (1957); *Chem. Abstr.*, **52**, 16725h (1958).
192. Gepshtein, *Koks i Khim,* No. 3, 53 (1960); *Chem. Abstr.*, **54**, 11812h (1960).
193. Noskov, *Nauchn. Tr. Vost. Nauchn.-Issled. Uglekhim. Inst.*, **16**, 265 (1963); *Chem. Abstr.*, **61**, 10049c (1964).
194. Chang and Karr, *Anal. Chem.*, **30**, 971 (1958).
195. Adey and Cox, *Analyst,* **84**, 414 (1959); *Chem. Abstr.*, **54**, 1183g (1960).
196. Hori, Sono, and Hirabayashi, *Nagoya Kogyo Gijutsu Shikensho Hokoku,* **8**, 804 (1959); *Chem. Abstr.*, **54**, 2094f (1960).
197. Adey, *Analyst,* **88**, 359 (1963); *Chem. Abstr.*, **60**, 29f (1964).
198. Gepshtein, *Pererabotka Vydelenie i Analizy Koksokhim. Produktov Sb.*, 83 (1961); *Chem. Abstr.*, **60**, 7437e (1964).
199. Roth and Schrimpf, *Arch. Pharm.* (Weinheim) **293**, 22 (1960); *Chem. Abstr.*, **54**, 16286e (1960).

200. Neimark and Kogan, *Zavodskaya Lab.*, **26**, 430 (1960); *Chem. Abstr.*, **54**, 15088f (1960).
201. Riolo and Molteni, *Ann. Chim.* (Rome), **50**, 220 (1960); *Chem. Abstr.*, **54**, 18198g (1960).
202. Nielsch and Giefer, *Z. Anal. Chem.*, **171**, 401 (1960); *Chem. Abstr.*, **54**, 8428f (1960).
203. Streuli, *Anal. Chem.*, **28**, 130 (1956).
204. Asmus and Kurandt, *Z. Anal. Chem.*, **149**, 3 (1956); *Chem. Abstr.*, **50**, 5465i (1956).
205. Bernstein and Schneider, *J. Chem. Phys.*, **24**, 469 (1956).
206. Baker, *J. Chem. Phys.*, **23**, 1981 (1955).
207. Minczewski and Lada, *Rocz. Chem.*, **29**, 948 (1955).
208. Tsuda, Maruyama, and Ikekawa, *J. Pharm. Soc. Jap.*, **75**, 1309 (1955); *Chem. Abstr.*, **50**, 3957a (1956).
209. Anwar, Hanson, and Patel, *J. Chromatogr.*, **34**, 529 (1968); *Chem. Abstr.*, **69**, 8284r (1968).
210. Kondratov, Rus'yanova, Gepshtein, and Golenkova, *Koks i Khim.*, No. 4, 38 (1968); *Chem. Abstr.*, **69**, 20972w (1968).
211. Kanamuru, Suzuki, and Sengoku, *Showa Yakka Daigaku Kiyo*, No. 4, 1 (1966); *Chem. Abstr.*, **69**, 24344r (1968).
212. Gepshtein, *Vost. Nauchn.-Issled. Uglekhim. Inst. Sb. Statei,* No. 4, 102 (1967); *Chem. Abstr.*, **69**, 24382b (1968).
213. Muns and Berkebile, U. S. Patent 2,880,061 (1959); *Chem. Abstr.*, **53**, 15505e (1959).
214. Spitsyn, *Koks i Khim.*, No. 4, 43 (1958); *Chem. Abstr.*, **52**, 13231b (1958).
215. Bunakov, Kagasov, Khlebnikov, Timofeev, Voronin, Kutsoglidi, Kharlampovich, and Yushina, *Vost. Nauchn.-Issled. Uglekhim. Inst. Sb. Statei,* No. 4, 3 (1967); *Chem. Abstr.*, **68**, 88884d (1968).
216. Waddington, Brit. Patent 969,013 (1961); *Chem. Abstr.*, **61**, 13094e (1964).
217. Kagasov and Kharlampovich, *Koks i Khim.*, **5**, 37 (1966); *Chem. Abstr.*, **65**, 8613c (1966).
218. Wiszniowski, *Koks, Smola, Gaz*, **10**, 240 (1965); *Chem. Abstr.*, **65**, 2018h (1966).
219. Bunakov and Kharlampovich, *Koks i Khim.*, No. 2, 40 (1966); *Chem. Abstr.*, **65**, 1806d (1966).
220. Kagasov, Timofeev, Voronin, and Khlebnikov, *Koks i Khim.*, No. 11, 38 (1965); *Chem. Abstr.*, **64**, 4817a (1966).
221. British Patent 898, 519 (1962); *Chem. Abstr.*, **57**, 6216i (1962).
222. Eisenlohr and Grob, U. S. Patent 2,997,477 (1961); *Chem. Abstr.*, **56**, 15731g (1962).
223. Petrenko, Goritskaya, and Shapiro, *Koks i Khim.*, No. 2, 31 (1962); *Chem. Abstr.*, **56**, 13172e (1962).
224. Petrenko, "Production of Pyridine Bases at Coal-Tar Chemical Plants", 1961, 176 pp.; *Chem. Abstr.*, **56**, 7620d (1962).
225. Spitsyn and Smirnov, *Koks i Khim.*, No. 7, 41 (1958); *Chem. Abstr.*, **52**, 16725i (1958).
226. Udovichenko and Gomel'skii, *Koks i Khim.*, No. 9, 40 (1957); *Chem. Abstr.*, **52**. 1590i (1958).
227. Petrenko, *Koks i Khim.*, No. 6, 40 (1958); *Chem. Abstr.*, **52**, 15030b (1958).
228. Langguth, U. S. Patent 2,849, 288 (1958); *Chem. Abstr.*, **52**, 21006e (1958).
229. Popov and Karvatskaya, *Koks i Khim.*, No. 2, 41 (1958); *Chem. Abstr.*, **52**, 9566i (1958).
230. Gonzalez and Muricio, Spanish Patent 241,513 (1958); *Chem. Abstr.*, **55**, 952a (1961).
231. Otto, Brit. Patent 768,345 (1957); *Chem. Abstr.*, **51**, 11698g (1957).

232. Wiszniowski, *Koks, Smola, Gaz,* **1**, 68 (1956); *Chem. Abstr.*, **51**, 1587f (1957).
233. Sweeney, U. S. Patent 2,799,678 (1957); *Chem. Abstr.*, **51**, 15926g (1957).
234. Cherkasova, Yaroslavskaya, and Gogel, *Koks i Khim.*, No. 8, 40 (1956); *Chem. Abstr.*, **51**, 6122d (1957).
235. Masek and Vojtovic, *Chem. Prum.*, **17**, 535 (1967); *Chem. Abstr.*, **67**, 118982h (1967).
236. Moll, *Chem. Tech.* (Berlin), **19**, 528 (1967); *Chem. Abstr.*, **67**, 110349m (1967).
237. Slavicek, Czechoslovakian Patent 119,215 (1966); *Chem. Abstr.*, **66**, 106875s (1967).
238. Kowalski and Szczurek, *Koks, Smola, Gaz,* **3**, 11 (1958) *Chem. Abstr.*, **52**, 13230i (1958).
239. Karlinskii and Chaiskii, *Koks i Khim.*, No. 4, 38 (1963); *Chem. Abstr.*, **59**, 2548d (1963).
240. Oganesyan, Chakhoyan, Azatyan, Vardanyan, and Simkin, *Tr. Nauchn.-Issled. Inst. Avtomatiz. Proizy. Protsessov Khim. Prom. Tsvetn. Met.*, No. 10, 19 (1962); *Chem. Abstr.*, **59**, 1295g (1963).
241. Oshurkova, Kulakov, and Mykol'nikov, *Koks i Khim.*, No. 1, 42 (1960); *Chem. Abstr.*, **54**, 10286c (1960).
242. Wolfskehl, German Patent 1,020,978 (1957); *Chem. Abstr.*, **54**, 869e (1960).
243. Kharlampovich and Kudryashova, *Koks i Khim.*, No. 2, 31 (1964); *Chem. Abstr.*, **60**, 11804f (1964).
244. Roy, Banerjee, and Basu, *Indian J. Technol.*, **2**, 59 (1964); *Chem. Abstr.*, **60**, 11805c (1964).
245. Kagasov and Yurkina, *Koks i Khim.*, No. 8, 32 (1964); *Chem. Abstr.*, **61**, 14420c (1964).
246. Lohrmann and Schroeder, German Patent 942,629 (1956); *Chem. Abstr.*, **53**, 4705g (1959).
247. Kuz'minykh and Kuznetsova, *Nauch. Doklady Vysshei Shkoly, Khim. Khim. Tekhnol.*, 799 (1958); *Chem. Abstr.*, **53**, 6551h (1959).
248. Nádasy and Ráskay, *Nehézvegyipari Kutató Intézet*, **1**, 113 (1958); *Chem. Abstr.*, **53**, 2578i (1959).
249. Hoak, U. S. Patent 2,712,980 (1955); *Chem. Abstr.*, **50**, 558e (1956).
250. Trefny and Haarmann, Ger. Patent 1,030,832 (1958); *Chem. Abstr.*, **54**, 14274h (1960).
251. Holdsworth and Holmes, Brit. Patent 992,319 (1965); *Chem. Abstr.*, **63**, 4063c (1965).
252. Kuz'minykh and Kuznetsova, *Zhur. Priklad. Khim.*, **33**, 865 (1960); *Chem. Abstr.*, **54**, 14878d (1960).
253. Schmidt and Ostrowsky, Ger. Patent 1,028,579 (1958); *Chem. Abstr.*, **54**, 16469d (1960).
254. Kharlampovich and Kagasov, *Koks i Khim.*, No. 5, 30 (1959); *Chem. Abstr.*, **54**, 18937h (1960).
255. Brit. Patent 827,636 (1960); *Chem. Abstr.*, **54**, 18942d (1960).
256. Geller and Ratte, U. S. Patent 2,717,232 (1955); *Chem. Abstr.*, **50**, 4237g (1956).
257. Sweeney, U. S. Patent 2,720,526 (1955); *Chem. Abstr.*, **50**, 2150a (1956).
258. Kagasov, Kholoptsev, Nemirovskii, Loparev, and Kharlampovich, *Koks i Khim.*, No. 6, 32 (1960); *Chem. Abstr.*, **54**, 25704e (1960).
259. Banerjee, Roy, and Lahiri, *J. Sci. Ind. Research* (India), **19A**, 174 (1960); *Chem. Abstr.*, **54**, 25705c (1960).
260. Bellet, *Industrie Chim. Belge,* **20**, Spec. No. 13–19 (1955); *Chem. Abstr.*, **50**, 11645f (1956).

261. Ivanov, Kozak, and Sharonova, U.S.S.R. Patent 114,863 (1958); *Chem. Abstr.*, **53**, 10713c (1959).
262. Kubička, Czechoslovakian Patent 83,777 (1955); *Chem. Abstr.*, **50**, 7874d (1956).
263. Achremowicz and Skrowaczewska, Polish Patent 52,039 (1966); *Chem. Abstr.*, **68**, 95698z (1968).
264. Chumakov, Shapavalova, Filippovich, Chernyavskii, Purik, and Nizhnyakovskii, U.S.S.R. Patent 196,844 (1967); *Chem. Abstr.*, **68**, 105004m (1968).
265. French Patent 1,484,583 (1967); *Chem. Abstr.*, **68**, 87185q (1968).
266. Neuhaeuser, Wolf, and Brand, German (East) Patent 57,852 (1967); *Chem. Abstr.*, **69**, 59106s (1968).
267. Gepshtein, U.S.S.R. Patent 133,887 (1960); *Chem. Abstr.*, **55**, 15516a (1961).
268. Chumakov and Lugovskoi, U.S.S.R. Patent 180,600 (1966); *Chem. Abstr.*, **65**, 12176h (1966).
269. Chumakov and Lugovskoi, U.S.S.R. Patent 180,601 (1966); *Chem. Abstr.*, **65**, 12177a (1966).
270. Rostafińska, *Rocz. Chem.*, **29**, 803 (1955).
271. Oberkobusch, *Brennst.-Chem.*, **40**, 145 (1959); *Chem. Abstr.*, **53**, 14469i (1959).
272. Williams and Hensel, U.S. Patent 2,728,771 (1955); *Chem. Abstr.*, **50**, 10796i (1956).
273. Vymetal and Rubicek, *Chem. Prumysl.*, **14**, 426 (1964); *Chem. Abstr.*, **61**, 10506c (1964).
274. Gepshtein, *Nauchn. Tr. Vost. Nauchn.-Issled. Uglekhim. Inst.*, **16**, 26 (1963); *Chem. Abstr.*, **61**, 13275c (1964).
275. Chumakov and Lugovskaya, U.S.S.R. Patent 179,318 (1966); *Chem. Abstr.*, **65**, 2232f (1966).
276. Chumakov and Lugovskaya, U.S.S.R. Patent 177,890 (1966); *Chem. Abstr.*, **65**, 694h (1966).
277. Chumakov and Lugovskaya, U.S.S.R. Patent 178,377 (1966); *Chem. Abstr.*, **64**, 19570f (1966).
278. Smirnov and Markus, U.S.S.R. Patent 176,899 (1965); *Chem. Abstr.*, **64**, 12651d (1966).
279. Koennecke, Gawalek, Kemula, and Sybilska, East Ger. Patent 39,197 (1965); *Chem. Abstr.*, **64**, 2069a (1966).
280. Nadasy, *Acta Chim. Acad. Sci. Hung.*, **30**, 255 (1962); *Chem. Abstr.*, **57**, 4953e (1962).
281. Suzumura, *Bull. Chem. Soc. Jap.*, **34**, 1846 (1961); *Chem. Abstr.*, **57**, 2185h (1962).
282. Gepshtein, *Vost. Nauchn.-Issled. Uglekhim. Inst. Sb. Statei,* 22 (1960); *Chem. Abstr.*, **57**, 16546e (1962).
283. Ustavshchikov and Koshel, *Uch. Zap. Yaroslavsk, Tekhnol. Inst.*, **6**, 23 (1961); *Chem. Abstr.*, **57**, 620e (1962).
284. Podkletnov and Podkopaeva, *Soobshch. Sakhalinsk. Kompleksn. Nauchn.-Issled. Inst., Akad. Nauk SSSR,* No. 8, 121 (1959); *Chem. Abstr.*, **57**, 11152h (1962).
285. Evans and Kynaston, *J. Chem. Soc.*, 5556 (1961).
286. Vymetal, *Chem. Listy*, **55**, 1444 (1961); *Chem. Abstr.*, **56**, 5036h (1962).
287. Danish Patent 90,478 (1957); *Chem. Abstr.*, **56**, 3746i (1962).
288. Kubicka and Zeman, *Chem. Prumysl.*, **7**, 169 (1957); *Chem. Abstr.*, **52**, 2377d (1958).
289. Chumakov and Chumakova, U.S.S.R. Patent 106,597 (1957); *Chem. Abstr.*, **52**, 2088h (1958).
290. Findlay, U. S. Patent 2,799,677 (1957); *Chem. Abstr.*, **52**, 450g (1958).
291. Pfeiffer, Ger. Patent 931,473 (1955); *Chem. Abstr.*, **52**, 18473a (1958).
292. Dimond, U.S. Patent 2,818,411 (1957); *Chem. Abstr.*, **52**, 4964h (1958).

293. Dilbert, U.S. Patent 2,995,500 (1961); *Chem. Abstr.*, **56**, 2430f (1962).
294. Danish Patent 88,934 (1960); *Chem. Abstr.*, **55**, 18080h (1961).
295. Ivashchenko, *Inst. Khim., Akad. Nauk Latv. SSR, Riga,* 13 (1957) (publ. 1960); *Chem. Abstr.*, **55**, 26407d (1961).
296. Chumakov and Gangrskii, U.S.S.R. Patent 130,515 (1960); *Chem. Abstr.*, **55**, 4932i (1961).
297. Kostyuk, *Inst. Khim., Akad. Nauk Latv. SSR, Riga,* 69 (1957) (publ. 1960); *Chem. Abstr.*, **55**, 12822a (1961).
298. Vymetal and Rubicek, *Chem. Prumysl.*, **15**, 202 (1965); *Chem. Abstr.*, **63**, 2808c (1965).
299. Gepshtein, *Vost. Nauchn.-Issled. Uglekhim. Inst., Sb. Statei,* No. 2, 87 (1964); *Chem. Abstr.*, **62**, 7548b (1965).
300. Wang, *Hua Hsueh Tung Pao,* No. 10, 597, 633 (1964); *Chem. Abstr.*, **62**, 13119f (1965).
301. Chumakov and Gangrskii, U.S.S.R. Patent 128,021 (1960); *Chem. Abstr.*, **55**, 1661b (1961).
302. Dykhanov, U.S.S.R. Patent 104,875 (1957); *Chem. Abstr.*, **51**, 9713e (1957).
303. Yamamoto, Japanese Patent 1517 (1956); *Chem. Abstr.*, **51**, 8809d (1957).
304. Fukuoka, Nagamawari, and Matsumoto, Japanese Patent 7576 (1955); *Chem. Abstr.*, **51**, 18556c (1957).
305. Chumakov, U.S.S.R. Patent 101,691 (1955); *Chem. Abstr.*, **51**, 12986i (1957).
306. Chumakov, U.S.S.R. Patent 105,811 (1957); *Chem. Abstr.*, **51**, 14829e (1957).
307. Sugido and Oodo, Japanese Patent 2984 (1955); *Chem. Abstr.*, **51**, 13366a (1957).
308. Shrader and Dimond, U.S. Patent 2,767,187 (1956); *Chem. Abstr.*, **51**, 9710c (1957).
309. Stajgr and Vymetal, *Chem. Prumysl.*, **14**, 182 (1964); *Chem. Abstr.*, **61**, 1670g (1964).
310. Swiderski, Szuchnik, and Wasiak, *Rocz. Chem.*, **38**, 1145 (1964); *Chem. Abstr.*, **61**, 16046g (1964).
311. Gepshtein, *Nauchn. Tr. Vost. Nauchn.-Issled. Uglekhim. Inst.*, **16**, 36 (1963); *Chem. Abstr.*, **61**, 13275g (1964).
312. Gepshtein, *Zh. Prikl. Khim.*, **37**, 1162 (1964); *Chem. Abstr.*, **61**, 4114a (1964).
313. Oberkobusch, German Patent 1,012,603 (1957); *Chem. Abstr.*, **54**, 11057a (1960).
314. Von Dohlen and Tully, U.S. Patent 2,924,602 (1960); *Chem. Abstr.*, **54**, 13148d (1960).
315. Arnall, *Chem. Ind.* (London), 1145 (1959); *Chem. Abstr.*, **54**, 864c (1960).
316. Profft and Melichar, *J. Prakt. Chem.*, **2**, 87 (1955).
317. Jerchel and Hippchen, German Patent 1,019,653 (1957); *Chem. Abstr.*, **54**, 1557i (1960).
318. Waddington, British Patent 943,193 (1963); *Chem. Abstr.*, **60**, 4115c (1964).
319. Fotis and Fields, U.S. Patent 3,112,322 (1963); *Chem. Abstr.*, **60**, 10657b (1964).
320. Podkletnov, *Soobshcheniya Sakhalin, Kompleks. Nauch.-Issledovatel. Inst.*, **3**, 131 (1957); *Chem. Abstr.*, **54**, 21084h (1960).
321. Kostyuk and Grigor'ev, *Tr. Nauchn.-Tekhn. Soveshch.*, (Leningrad), 326 (1961) (publ. 1963); *Chem. Abstr.*, **60**, 346d (1964).
322. Fu and Chiang, *Hua Hsüeh Shih Chieh,* **13**, 458 (1958); *Chem. Abstr.*, **54**, 23121f (1960).
323. Gepshtein, *Koks i Khim.*, No. 3, 49 (1959); *Chem. Abstr.*, **53**, 13145b (1959).
324. Bartz, *Bull. Acad. Polon. Sci., Cl. III,* **2**, 395 (1954); *Chem. Abstr.*, **50**, 332e (1956).
325. Smeykal and Moll, British Patent 1,039,856 (1966); *Chem. Abstr.*, **65**, 15338b (1966).
326. Helm, Lanum, Cook, and Ball, *J. Phys. Chem.*, **62**, 858 (1958).

327. Kyte, Jeffery, and Vogel, *J. Chem. Soc.*, 4454 (1960).
328. Sisco and Wiederecht, U. S. Patent 2,708,653 (1955); *Chem. Abstr.*, **50**, 2958c (1956).
329. British Patent 1,108,686 (1968); *Chem. Abstr.*, **69**, 43808m (1968).
330. Nixon and Thorpe, *Amer. Chem. Soc., Div. Petrol. Chem., Preprints,* **1**, No. 3, Gen. Papers, 265 (1956); *Chem. Abstr.*, **53**, 15539a (1959).
331. Lisicki and Sosnkowska, *Przemysl. Chem.*, **38**, 24 (1959); *Chem. Abstr.*, **53**, 15540b (1959).
332. Landa and Galik, *Conf. Chem. Chem. Process. Petrol. Natur. Gas, Plenary Lect., Budapest* (publ. 1968), 880 (1965); *Chem. Abstr.*, **69**, 88498b (1968).
333. Galik and Landa, *Sb. Vys. Sk. Chem. Technol. Praze, Technol. Paliv,* **13**, 87 (1967); *Chem. Abstr.*, **68**, 88739k (1968).
334. Dinneen, Cook, and Jensen, *Anal. Chem.*, **30**, 2026 (1958).
335. Eizen, Khallik, and Klesment, *Eesti NSV Teaduste Akad. Toimetised, Fuusik.-Mat.-jaTehnikateaduste Seer.*, 230 (1966); *Chem. Abstr.*, **66**, 4593n (1967).
336. Raudsepp and Degtereva, *Goryuchie Slantsy. Khim. i Tekhnol. Sbornik,* 131 (1956); *Chem. Abstr.*, **53**, 13569c (1959).
337. Ku, T'sêng, and Kalechits, *K'o Hsüeh T'ung Pao*, No. 16, 507 (1957); *Chem. Abstr.*, **55**, 25220g (1961).
338. Iida, *Yakugaku Zasshi,* **82**, 144 (1962); *Chem. Abstr.*, **57**, 2489i (1962).
339. Kobyl'skaya, *Tr. Vses. Nauch.-Issled. Inst. Pererabotki Ispol'zovan Topliva,* 198 (1959); *Chem. Abstr.*, **57**, 4943h (1962).
340. Nagy and Gagnon, *Geochim. Cosmochim. Acta,* **23**, 155 (1961); *Chem. Abstr.*, **62**, 1464h (1965).
341. Brown, *Aust. J. Appl. Sci.*, **10**, 294 (1959); *Chem. Abstr.*, **53**, 22839e (1959).
342. Kuznetsov and Fadeicheva, *Inst. Khim. Akad. Nauk Latv. SSR, Riga,* 37 (1957) (publ. 1960); *Chem. Abstr.*, **55**, 14879c (1961).
343. Naumann and Leibnitz, *J. Prakt. Chem.*, [4] **4**, 43 (1956); *Chem. Abstr.*, **51**, 10873a (1957).
344. Saha, *J. Indian Chem. Soc.*, **34**, 79 (1957); *Chem. Abstr.*, **51**, 10873c (1957).
345. Kahler, Rowlands, and Ellis, *Am. Chem. Soc., Div. Gas Fuel Chem., Preprints,* **1**, 81 (1959); *Chem. Abstr.*, **57**, 6233a (1962).
346. Grossmann and Strobach, East German Patent 37,502 (1965); *Chem. Abstr.*, **63**, 11198g (1965).
346a. West, M. Sc. Thesis, University of Saskatchewan (1960).
347. Makovetskii and Kigel, *Gazovaya Prom.*, **4**, No. 4, 16 (1959); *Chem. Abstr.*, **53**, 12637h (1959).
348. Strobach and Langosch, *Freiberger Forschungsh.*, **340A**, 109 (1964); *Chem. Abstr.*, **63**, 6756f (1965).
349. Kahler, Rowlands, Brewer, Powell, and Ellis, *J. Chem. Eng. Data,* **5**, 94 (1960); *Chem. Abstr.*, **54**, 16794i (1960).
350. Fedotova and Vanags, *Latvijas PSR Zinātnu Akad. Vēstis,* No. 2, 75 (1959); *Chem. Abstr.*, **54**, 526f (1960).
351. Fedotova and Vanags, *Latvijas PSR Zinātnu Akad. Vēstis,* No. 7, 81 (1958); *Chem. Abstr.*, **53**, 11367d (1959).
352. Fedotova and Vanags, *Latvijas PSR Zinātnu Akad. Vēstis,* No. 5, 101 (1956); *Chem. Abstr.*, **50**, 17387i (1956).
353. Fedotova and Vanags, *Trudy Inst. Khim., Akad. Nauk Latv. SSR,* **2**, 53 (1958); *Chem. Abstr.*, **53**, 21941f (1959).
354. Fedotova and Vanags, *Latvijas PSR Zinātnu Akad. Vēstis,* No. 5, 93 (1958); *Chem. Abstr.*, **53**, 1348b (1959).

355. Kuznetsov, Fadeicheva, Govorova, Chernykh, and Kigel, *Akad. Nauk Ukr. SSR, Inst. Teploenerget., Zbirnik Prats,* No. 17, 46 (1959); *Chem. Abstr.*, **55**, 7806h (1961).
356. Evdokimova and Rakovskii, *Vestsi Akad. Navuk Belarusk. SSR, Ser. Khem. Navuk,* 124 (1965); *Chem. Abstr.*, **64**, 10975f (1966).
357. Sugihara and Sorensen, *J. Amer. Chem. Soc.*, **77**, 963 (1955).
358. Vimba, Brakšs, and Kalninš, *Latvijas PSR Zinātnu Akad. Vēstis,* No. 6, 69 (1957); *Chem. Abstr.*, **52**, 11397i (1958).
359. Conroy, Brook, Rout, and Silverman, *J. Amer. Chem. Soc.*, **79**, 1763 (1957).
360. Woese, *J. Bacteriol.*, **77**, 38 (1959); *Chem. Abstr.*, **53**, 7317i (1959).
361. Rode and Foster, *J. Bacteriol.*, **79**, 650 (1960); *Chem. Abstr.*, **55**, 3719g (1961).
362. Truhaut and de Clercq, *Bull. Assoc. Franc. Etude Cancer,* **44**, 426 (1957); *Chem. Abstr.*, **53**, 17317d (1959).
363. Vega, Dominguez, and Ribas, *Anal. Real. Soc. Españ. Fís. y Quím.*, **50B**, 895 (1954); *Chem. Abstr.*, **51**, 441e (1957).
364. Govindachari, Nagarajan, and Rajappa, *J. Chem. Soc.*, 551 (1957).
365. Janot, Percheron, Chaigneau, and Goutarel, *C. R. Acad. Sci. Paris, Ser. C,* **244**, 1955 (1957).
366. Wada, *Arch. Biochem. Biophys.*, **64**, 244 (1956); *Chem. Abstr.*, **51**, 553e (1957).
367. Nozoye, *Ann. Rept. ITSUU Lab.* (Toyko), No. 8, 10 (1957); *Chem. Abstr.*, **51**, 16504c (1957).
368. Young, *Can. J. Microbiol.*, **5**, 197 (1959); *Chem. Abstr.*, **53**, 15203i (1959).
369. Halvorson and Howitt, *Spores Symp., 2nd, Allerton Park, Illinois,* 149 (1960) (publ. 1961); *Chem. Abstr.*, **59**, 6657e (1963).
370. Cooney and Lundgren, *Can. J. Microbiol.*, **8**, 823 (1962); *Chem. Abstr.*, **58**, 9440f (1963).
371. Piattelli, Minale, and Trota, *Phytochemistry,* **4**, 121 (1965); *Chem. Abstr.*, **62**, 13120a (1965).
372. Witkop, *J. Amer. Chem. Soc.*, **79**, 3193 (1957).
373. Wilcox, Wyler, Mabry, and Dreiding, *Helv. Chim. Acta,* **48**, 252 (1965).
374. Davison and Wiggins, *Chem. Ind.* (London), 982 (1956); *Chem. Abstr.*, **51**, 4389f (1957).
375. Blum, Bolchi, Heller, and Stancu, *Acad. Rep. Populare Romîne, Inst. Energet. Studii Cercetări Energet.*, 8, 257 (1958); *Chem. Abstr.*, **53**, 11834i (1959).
376. Negishi and Nishi, *Yakugaku Zasshi,* **84**, 879 (1964); *Chem. Abstr.*, **62**, 15071g (1965).
377. Ishiguro, Morita, and Ikushima, *Yakugaku Zasshi,* **80**, 83 (1960); *Chem. Abstr.*, **54**, 12131i (1960).
378. Arnovich, Bel'skii and Mikhailov, *Izvest. Akad. Nauk S.S.S.R., Otdel. Khim. Nauk,* 696 (1956); *Chem. Abstr.*, **51**, 1893b (1957).
379. Fremery and Fields, U.S. Patent 3,166,556 (1965); *Chem. Abstr.*, **62**, 13130d (1965).
380. Peláez, *Coloq. problemas sintesis Diels-Alder, Soc. Españ. Fis. Quim.* (Madrid) 109 (1953); *Chem. Abstr.*, **51**, 8708e (1957).
381. Jaworski, and Korybut-Daszkiewicz, *Rocz. Chem.*, **41**, 1521 (1967); *Chem. Abstr.*, **69**, 2822t (1968).
382. Jaworski and Polaczkowa, *Rocz. Chem.*, **31**, 1337 (1957); *Chem. Abstr.*, **52**, 11037g (1958).
383. Boyer and Morgan, *J. Amer. Chem. Soc.*, **83**, 919 (1961).
383a. Abramovitch and Holcomb, unpublished results (1974).
384. The Netherlands Appl. Patent 6,409,131 (1965); *Chem. Abstr.*, **63**, 584a (1965).
385. Belgian Patent 628,646 (1963); *Chem. Abstr.*, **61**, 13288h (1964).
386. Bartholomew and Campbell, Brit. Patent 1,088,706 (1967); *Chem. Abstr.*, **68**, 95693u (1968).

387. Denton, U.S. Patent 2,963,484 (1960); *Chem. Abstr.*, **55**, 15516c (1961).
388. Lane, Brit. Patent 911,524 (1962); *Chem. Abstr.*, **58**, 11333g (1963).
389. Denton, U.S. Patent 2,979,510 (1961); *Chem. Abstr.*, **55**, 18778g (1961).
390. Butler, Dodsworth and Groom, *Chem. Commun.*, 54 (1966).
391. Manly, O'Halloran and Rice, U.S. Patent 3,238,214 (1966); *Chem. Abstr.*, **64**, 17551h (1966).
392. Ishigaki, Kashiro, Tsuda, Shono, and Hachihama, *Kogyo Kagaku Zasshi,* **67**, 399 (1964); *Chem. Abstr.*, **61**, 639d (1964).
393. Ishigaki, Kashiro, Tsuda, Shono, and Hachihama, *Kogyo Kagaku Zasshi,* **66**, 1886 (1963); *Chem. Abstr.*, **60**, 14465f (1964).
394. Clauson-Kaas, Elming, and Nielsen, U.S. Patent 2,748,142 (1956); *Chem. Abstr.*, **51**, 2881a (1957).
395. Clauson-Kaas, U.S. Patent 2,801,252 (1957); *Chem. Abstr.*, **52**, 1260c (1958).
396. French Patent 1,372,758 (1964); *Chem. Abstr.*, **62,** 532e (1965).
397. French Patent 1,477,996 (1967); *Chem. Abstr.*, **69,** 2761x (1968).
398. Elming, Carlsten, Lennart, and Ohlsson, Brit. Patent 862,581 (1961); *Chem. Abstr.*, **56**, 11574g (1962).
399. Elming and Clauson-Kaas, *Acta Chem. Scand.*, **10**, 1603 (1956); *Chem. Abstr.*, **52**, 13728f (1958).
400. Sugiyama, Sugiyama, and Aso, *Tôhoku J. Agr. Research,* **10**, 409 (1959); *Chem. Abstr.*, **54**, 21083i (1960).
401. Sugisawa, *Tôhoku J. Agr. Research,* **11**, 389 (1960); *Chem. Abstr.*, **55**, 19917i (1961).
402. Sugisawa and Aso, *Chem. Ind.* (London), 781 (1961); *Chem. Abstr.*, **55**, 27306d (1961).
403. Williams, Kaufmann, and Mosher, *J. Org. Chem.*, **20**, 1139 (1955); *Chem. Abstr.*, **50**, 5658f (1956).
404. Sugisawa and Aso, *Tôhoku J. Agr. Research,* **10**, 137 (1959); *Chem. Abstr.*, **54**, 11015b (1960).
405. Sugisawa, Sugiyama, and Aso, *Tôhoku J. Agr. Research,* **12**, 245 (1961); *Chem. Abstr.*, **57**, 16535b (1962).
406. Gruber, *Chem. Ber.*, **88**, 178 (1955).
407. Paulsen, *Angew. Chem.*, **74**, 585 (1962); *Chem. Abstr.*, **57**, 16724g (1962).
408. Chumakov and Martynova, U.S.S.R. Patent 172,805 (1965); *Chem. Abstr.*, **64**, 716b (1966).
409. Chumakov, Martynova, and Marchenko, U.S.S.R. Patent 196,845 (1967); *Chem. Abstr.*, **68**, 105005n (1968).
410. Chumakov, Martynova, Zinov'eva and Khimchenko, *Zh. Obshch. Khim.*, **34**, 3511 (1964); *Chem. Abstr.*, **62**, 4003c (1965).
411. Chumakov, Zinov'eva, and Khimchenko, U.S.S.R. Patent 164,283 (1964); *Chem. Abstr.*, **61**, 16050e (1964).
412. The Netherlands Patent 6,408,149 (1965); *Chem. Abstr.*, **63**, 2961b (1965).
413. Colchester, Brit. Patent 1,110,865 (1968); *Chem. Abstr.*, **69**, 59100k (1968).
414. Chumakov and Sherstyuk, U.S.S.R. Patent 201,408 (1967); *Chem. Abstr.*, **69**, 19027k (1968).
415. Campbell and Corran, Brit. Patent 1,008,455 (1965); *Chem. Abstr.*, **64**, 3496f (1966).
416. Bartholomew, Campbell, Oliver, and Dutton, Brit. Patent 1,087,279 (1967); *Chem. Abstr.*, **68**, 87183n (1968).
417. Pasedach and Seefelder, *Festschrift Carl Wurster zum 60 Geburtstag,* 113 (1960); *Chem. Abstr.*, **56**, 10090f (1962).
418. Chumakov and Sherstyuk, *Metody Polucheniya Khim. Reaktivov Preparatov,* No. 13, 134 (1965); *Chem. Abstr.*, **65**, 3829c (1966).

419. Chumakov and Sherstyuk, *Tetrahedron Lett.*, 129 (1965).
420. Gault, Gilbert, and Briaucort, *C. R. Acad. Sci., Paris, Ser. C*, **266**, 131 (1968); *Chem. Abstr.*, **69**, 10331x (1968).
421. Boyer and Schoen, *J. Amer. Chem. Soc.*, **78**, 423 (1956).
422. Woods, *J. Org. Chem.*, **22**, 341 (1957).
423. British Patent 781,413 (1957); *Chem. Abstr.*, **52**, 2929c (1958).
424. Inoue and Iguchi, *Chem. Pharm. Bull.* (Tokyo), **12**, 382 (1964); *Chem. Abstr.* **60**, 14466h (1964).
425. Praill and Whitear, *Proc. Chem. Soc.*, 312 (1959); *Chem. Abstr.*, **54**, 7703d (1960).
426. Balaban, Mateescu, and Nenitzescu, *Acad. Rep. Populare Romine, Studii Cercetari Chim.*, **9**, 211 (1961); *Chem. Abstr.*, **57**, 15065g (1962).
427. Dorofeenko, Nazarova, and Novikov, *Zh. Obshch. Khim.*, **34**, 3918 (1964); *Chem. Abstr.*, **62**, 9099c (1965).
428. Siemiatycki and Fugnitto, *Bull. Soc. Chim. Fr.*, 538 (1961).
429. Simalty-Siemiatycki, *Bull. Soc. Chim. Fr.*, 1944 (1965).
430. Dorofeenko, Krivun, and Mezheritskii, *Zh. Obshch. Khim.*, **35**, 632 (1965); *Chem. Abstr.*, **63**, 2947d (1965).
431. Durden and Crosby, *J. Org. Chem.*, **30**, 1684 (1965).
432. Dulenko and Dorofeenko, *Dopovidi Akad. Nauk Ukr. RSR*, No. 1, 78 (1963); *Chem. Abstr.*, **59**, 6359d (1963).
433. Dimroth and Mach., *Angew. Chem. Int. Ed. Engl.*, 7, 460 (1968).
434. Praill and Whitear, *J. Chem. Soc.*, 3573 (1961).
435. Balaban and Nenitzescu, *J. Chem. Soc.*, 3553 (1961).
436. Balaban, Nenitzescu, Gavat, and Mateescu, *J. Chem. Soc.*, 3564 (1961).
437. Balaban, Schroth, and Fischer, *Advan. Heterocycl. Chem.*, **10**, 241 (1969); Dorofeenko, Sadekova, and Kuzneksov, "Preparative Chemistry of Pyrilyum Salts," in "Preparative Chemistry of Heterocyclic Compounds, Vol. 1", Izd. Rostov. Univ., Rostov-on-Don, U.S.S.R., 1972; *Chem. Abstr.*, **78**, 43238 m (1973).
438. Patterson and Drenchko, *J. Org. Chem.*, **27**, 1650 (1962).
439. Jacobson and Jensen, *J. Phys. Chem.*, **66**, 1245 (1962).
440. Closs and Schwartz, *J. Org. Chem.*, **26**, 2609 (1961).
441. Nicoletti and Forcellese, *Gazz. Chim. Ital.*, **95**, 83 (1965).
441a. Jones and Rees, *J. Chem. Soc. C*, 2249 (1969).
441b. Jones and Rees, *J. Chem. Soc. C*, 2255 (1969).
442. Huisgen and Herbig, *Ann. Chem.*, **688**, 98 (1965).
443. Inoue, Sugiyama, and Ozawa, *Nippon Kagaku Zasshi*, **82**, 1272 (1961); *Chem. Abstr.*, **57**, 15067a (1962).
444. Kuthan and Palecek, *Collect. Czech. Chem. Commun.*, **31**, 2618 (1966); *Chem. Abstr.*, **65**, 3828a (1966).
445. Loev and Snader, *J. Org. Chem.*, **30**, 1914 (1965).
446. Schmidle and Mansfield. U.S. Patent 2,847,414 (1958); *Chem. Abstr.*, **53**, 15098h (1959).
447. Prostakov and Gaivoronskaya, *Zh. Obshch. Khim.*, **32**, 76 (1962); *Chem. Abstr.*, **57**, 12426f (1962).
448. Oberrauch, German Patent 1,192,648, (1965); *Chem. Abstr.*, **63**, 8323g (1965).
449. Shuikin and Brusnikina, *Zh. Obshchei Khim.*, **29**, 438 (1959); *Chem. Abstr.*, **53**, 21931d (1959).
450. Ishiguro and Otsuka, *Yakugaku Zasshi*, 78, 1383 (1958); *Chem. Abstr.*, **53**, 5544g (1959).
451. Horrobin, U.S. Patent 2,765,310 (1956); *Chem. Abstr.*, **51**, 2877g (1957).
452. Silverstone, Brit. Patent 1,011, 322 (1965); *Chem. Abstr.*, **64**, 5052b (1966).

453. Baizer, U.S. Patent 3,246,000 (1966); *Chem. Abstr.*, **64**, 19568h (1966).
454. Julia, Siffert, and Bagot, *Bull. Soc. Chim. Fr.*, 1007 (1968).
455. Wang and Sung, *Nan Ching Yo Hsüeh Yüan Hsüeh Pao,* No. 3, 32 (1958); *Chem. Abstr.*, **55**, 530e (1961).
456. Oakes and Rydon, *J. Chem. Soc.*, 4433 (1956).
457. Komatsu, *Muroran Kogyô Hôkoku,* **3**, 61 (1958); *Chem. Abstr.*, **53**, 13149d (1959).
458. Gadekar and Frederick, *J. Heterocycl. Chem.*, **5**, 125 (1968).
459. Miyadera and Kishida, *Tetrahedron Lett.*, 905 (1965).
460. Robison, Butler, and Robison, *J. Amer. Chem. Soc.*, **79**, 2573 (1957).
461. Kondrat'eva and Huang, *Dokl. Akad. Nauk SSSR*, **141**, 861 (1961); *Chem. Abstr.*, **56**, 14229e (1962).
462. Kondrat'eva and Huang, *Dokl. Akad. Nauk SSSR,* **141**, 628 (1961); *Chem. Abstr.*, **56**, 14229h (1962).
463. Harris, Firestone, Pfister, Boettcher, Cross, Curie, Monaco, Peterson, and Reuter, *J. Org. Chem.*, **27**, 2705 (1962).
464. Huang and Kondrat'eva, *Izv. Akad. Nauk SSSR, Otd. Khim. Nauk,* 525 (1962); *Chem. Abstr.*, **57**, 15064i (1962).
465. Kawazu, Japanese Patent 18627 (1967) *Chem. Abstr.*, **69**, 10366n (1968).
466. Florent'ev, Drobinskaya, Ionova, Karpeiskii, and Turchin, *Dokl. Akad. Nauk SSSR* **177**, 617 (1967); *Chem. Abstr.*, **69**, 27198g (1968).
467. Kimel and Leimgruber, French Patent 1,384,099 (1965); *Chem. Abstr.*, **63**, 4263b (1965).
468. Kondrat'eva, *Izv. Akad. Nauk SSSR*, *Otdel. Khim. Nauk* 484 (1959); *Chem. Abstr.*, 21940d (1959).
469. French Patent 1,343,270 (1963); *Chem. Abstr.*, **60**, 11991g (1964).
470. Naito, Yoshikawa, Ishikawa, Isoda, Omura, and Takamura, *Chem. Pharm. Bull.* (Tokyo), **13**, 869 (1965); *Chem. Abstr.*, **63**, 11488h (1965).
471. Yoshikawa, Ishikawa, Omura, and Naito, *Chem. Pharm. Bull* (Tokyo), **13**, 873 (1965); *Chem. Abstr.*, **63**, 11489c (1965).
472. Yoshikawa, Ishikawa and Naito, *Chem. Pharm. Bull* (Tokyo), **13**, 878 (1965); *Chem. Abstr.*, **63**, 11489g (1965).
473. Pfister, Harris, and Firestone, U. S. Patent 3,227,722 (1966); *Chem. Abstr.*, **65**, 16949g (1966).
474. Pfister, Harris, and Firestone, U. S. Patent 3,227,724 (1966); *Chem. Abstr.*, **64**, 8149h (1966).
475. Pfister, Harris, and Firestone, U. S. Patent 3,227,721 (1966); *Chem. Abstr.*, **64**, 9689g (1966).
476. The Netherlands Patent 6,403,004 (1964); *Chem. Abstr.*, **62**, 7733h (1965).
477. French Patent 1,400,843 (1965); *Chem. Abstr.*, **63**, 9922d (1965).
478. Colchester and Corran, British Patent 1,077,573 (1967); *Chem. Abstr.*, **68**, 87171g (1968).
479. Belgian Patent 669,512 (1966); *Chem. Abstr.*, **65**, 7153f (1966).
480. Colchester, British Patent 1,074,944 (1967); *Chem. Abstr.*, **68**, 87189u (1968).
481. Colchester, British Patent 1,102,261 (1968); *Chem. Abstr.*, **69**, 27266c (1968).
482. Bartholomew and Campbell, British Patent 1,088,707 (1967); *Chem. Abstr.*, **68**, 95694v (1968).
483. The Netherlands Patent 6,511,861 (1966); *Chem. Abstr.*, **65**, 5446e (1966).
484. Arnold, *Collect. Czech. Chem. Commun.*, **25**, 1308 (1960); *Chem. Abstr.*, **54**, 17235a (1960).
485. Arnold, *Experientia,* **15**, 415 (1959); *Chem. Abstr.*, **54**, 14247b (1960).

486. Yanovskaya and Kucherov, *Izv. Akad. Nauk S.S.S.R., Otdel. Khim. Nauk,* 2184 (1960); *Chem. Abstr.,* **55**, 14452e (1961).
487. Chumakov, Sherstyuk, and Dzygun, *Ukr. Khim. Zh.,* **31**, 597 (1965); *Chem. Abstr.,* **63**, 11492b (1965).
488. MacLean and Parker, U.S. Patent 2,740,789 (1956); *Chem. Abstr.,* **51**, 496c (1957).
489. Kobayashi and Sakata, *Yakugaku Zasshi,* **82**, 447 (1962); *Chem. Abstr.,* **58**, 4552e (1963).
490. Cavill, Ford, and Solomon, *Aust. J. Chem.,* **13**, 468 (1960).
491. Chumakov and Sherstyuk, *Tetrahedron Lett.,* 129 (1965).
492. Chumakov and Lugovskaya, *Zh. Obshch. Khim.,* **34**, 3515 (1964); *Chem. Abstr.,* **62**, 2760c (1965).
493. Tanaka and Murata, *Kôgyô Kagaku Zasshi,* **59**, 1181 (1956); *Chem. Abstr.,* **52**, 14607c (1958).
494. Morgan, *J. Org. Chem.,* **27**, 343 (1962).
495. Kadyrov and Aliev, *Uzbeksk Khim. Zh.,* **8**, 30 (1964); *Chem. Abstr.,* **61**, 10588e (1964).
496. Simpson, Schnitzer, and Cobb, U. S. Patent 3,007,931 (1959); *Chem. Abstr.,* **56**, 4738h (1962).
497. Meyers and Garcia-Muñoz, *J. Org. Chem.,* **29**, 1435 (1964).
498. Mangoni, *Gazz. Chim. Ital.,* **90**, 941 (1960); *Chem. Abstr.,* **55**, 2115g (1961).
499. Bazier, Dub, Gister, and Steinberg, *J. Amer. Pharm. Assoc.,* **45**, 478 (1956); *Chem. Abstr.,* **51**, 408a (1957).
500. De Maldé and Alneri, *Chemica Industria,* **38**, 473 (1956); *Chem. Abstr.,* **51**, 408d (1957).
501. Rubstov, Mikhlina, and Furshtatova, *Zhur. Priklad. Khim.,* **29**, 946 (1956); *Chem. Abstr.,* **50**, 14752d (1956).
502. British Patent 741,712 (1955); *Chem. Abstr.,* **50**, 16875e (1956).
503. Lapin, Arsenijevic, and Horeau, *Bull. Soc. Chim. Fr.,* 1700 (1960).
504. Hellmann and Seegmuller, *Chem. Ber.,* **91**, 2420 (1958).
505. Ferrier and Campbell, *J. Chem. Soc.,* 3513 (1960).
506. Bailey and Brunskill, *J. Chem. Soc.,* 2554 (1959).
507. Little, Middletown, Coffman, Engelhart, and Sausen, *J. Amer. Chem. Soc.,* **80**, 2832 (1958).
508. Cottis and Tieckelmann, *J. Org. Chem.,* **26**, 79 (1961).
509. Atkinson and Johnson, *J. Chem. Soc.,* 1252 (1968).
510. Little and Middletown, U. S. Patent 2,790,805 (1957); *Chem. Abstr.,* **51**, 15598g (1957).
511. Johnson, Panella, Carlson, and Hunneman, *J. Org. Chem.,* **27**, 2473 (1962).
512. Johnson, U.S. Patent 3,225,041 (1965); *Chem. Abstr.,* **64**, 6623f (1966).
513. Bullock and Gregory, *Can. J. Chem.,* **43**, 332 (1965).
514. Roedig, Grohe, and Klatt, *Chem. Ber.,* **99**, 2818 (1966).
515. Roedig, Grohe, Klatt, and Kleppe, *Chem. Ber.,* **99**, 2813 (1966).
516. Tamano and Kawai, Japanese Patent 23,175 (1965); *Chem. Abstr.,* **64**, 3502f (1966).
517. Kotlyarevskii and Vereshchagin, *Izv. Akad. Nauk S.S.S.R., Otdel. Khim. Nauk,* 715 (1959); *Chem. Abstr.,* **53**, 219371 (1959).
518. Kotlyarevskii and Vereshchagin, *Izv. Akad. Nauk. S.S.S.R., Otdel. Khim. Nauk,* 1629 (1960); *Chem. Abstr.,* **55**, 8404g (1961).
519. Vereshchagin, *Materialy k Konf. Molodykh Nauchn. Sotrudn. Vost.-Sibirsk. Filiala Sibirsk. Otd. Akad. Nauk. S.S.S.R., Blagoveshchensk, Sbornik,* No. 3, 23 (1960); *Chem. Abstr.,* **57**, 11154a (1962).

520. Vershchagin and Kotlyarevskii, *Izv. Akad. Nauk S.S.S.R., Otdel. Khim. Nauk,* 1440 (1960); *Chem. Abstr.*, **55**, 533c (1961).
521. MacLean and Hobbs, U. S. Patent 2,780,627 (1957); *Chem. Abstr.*, **51**, 15597c (1957).
522. Bohlmann and Rahtz, *Chem. Ber.*, **90**, 2265 (1957).
523. Janz and Monahan, *J. Org. Chem.*, **29**, 569 (1964).
524. Janz and Monahan, *J. Org. Chem.*, **30**, 1249 (1965).
525. Janz and Monahan, *J. Phys. Chem.*, **69**, 1070 (1965).
526. G. J. Janz, "1,4-Cycloaddition Reactions," Hamer, (Ed.)., Chapter 4, Academic, New York, 1967.
527. Lora-Tamayo, Garcia-Muñoz, and Madronero, *Bull. Soc. Chim. Fr.*, 1331 (1958).
528. Kazanskii, Kondrateva, and Dolskaya, U.S.S.R. Patent 199,890 (1967); *Chem. Abstr.*, **68**, 95710 (1968).
529. Albrecht and Kresze, *Chem. Ber.*, **98**, 1431 (1965).
530. Kotlyarevskii and Vasil'eva, *Izv. Akad. Nauk S.S.S.R., Otdel. Khim. Nauk,* 1834 (1961); **56**, 11565 (1962).
531. Vasil'eva, Kotlyarevskii and Faiershtein, *Izv. Akad. Nauk S.S.S.R., Ser. Khim.*, 332 (1965); *Chem. Abstr.*, **62**, 14619 (1965).
532. Junek, *Monatsh. Chem.*, **95**, 1201 (1964); *Chem. Abstr.*, **62**, 1629 (1965).
533. Cuiban, Cilianu-Bibian, Popescu, and Rogozea, French Patent 1,366,064 (1964); *Chem. Abstr.*, **61**, 14643 (1964).
534. Trivedi, *Current Sci.*, **28**, 322 (1959); *Chem. Abstr.*, **54**, 24724 (1960).
535. Papini, Ridi, and Checchi, *Gazz. Chim. Ital.*, **90**, 1399 (1960).
536. Portnoy, *J. Org. Chem.*, **30**, 3377 (1965).
537. British Patent 852,398 (1960); *Chem. Abstr.*, **55**, 10477 (1961).
538. Wibaut, Uhlenbroek, Kooijman, and Kettrenes, *Rec. Trav. Chim. Pays-Bas,* **79**, 481 (1960).
539. Reid and Schleimer, *Ann. Chem.*, **626**, 106 (1959).
540. Reid and Meyer, German Patent 1,189,994 (1965); *Chem. Abstr.*, **63**, 588 (1965).
541. Kutney and Selby, *J. Org. Chem.*, **26**, 2733 (1961).
542. Fuchs, Senkariuk, Nemes, Lazar, and Somogyi, Hungarian Patent 150,475 (1961); *Chem. Abstr.*, **60**, 2911 (1964).
543. Kametani, Ogasawara, Kozuka, and Nyu, *Yakugaku Zasshi,* **87**, 1189 (1967); *Chem. Abstr.*, 68, 9222 (1968).
544. Yokoyama, Toyoshima, Hamano, and Kanai, Japanese Patent 1665 (1964); *Chem. Abstr.*, **60**, 15843 (1964).
545. Dornow and Neuse, *Arch. Pharm.* (Weinheim), **287**, 361 (1954); *Chem. Abstr.*, **51**, 16488 (1957).
546. Naito and Ueno, Japanese Patent 26,386 (1963); *Chem. Abstr.*, **60**, 11991 (1964).
547. Jones, U. S. Patent 2,748,135 (1956); *Chem. Abstr.*, **51**, 2877 (1957).
548. Schreiber and Adam, *Chem. Ber.*, **93**, 1948 (1960).
549. Buckles, Langsjoen, Mueller, and Svendsen, *Proc. Minn. Acad. Sci.* **25–26**, 257 (1957–58); *Chem. Abstr.*, **54**, 24724 (1960).
550. Kochetkov, Khomutova, and Likhosherstov, *Zh. Obshch. Khim.*, **29**, 1657 (1959); *Chem. Abstr.*, **54**, 8815 (1960).
551. Jones, U. S. Patent 2,724,714 (1955); *Chem. Abstr.*, **50**, 10797 (1956).
552. Chumakov, Shertstyuk, and Dzygun, U.S.S.R. Patent 172,329 (1965); *Chem. Abstr.*, **63**, 16315, (1965).
553. Chumakov and Lugovaskaya, U.S.S.R. Patent 163,284 (1964); *Chem. Abstr.*, **62**, 534, (1965).

554. Thesing, German Patent 1,092,016 (1960); *Chem. Abstr.*, **55**, 19957 (1961).
555. Zhiryakov, Abramenko, and Kurepina, U.S.S.R. Patent 159,908 (1964); *Chem. Abstr.*, **60**, 11992 (1964).
556. Bohlman, German Patent 1,190,464 (1965); *Chem. Abstr.*, **63**, 584 (1965).
557. Bohlman and Rahtz, *Chem. Ber.*, **90**, 2265 (1957).
558. Hargrave, British Patent 963,887 (1964); *Chem. Abstr.*, **61**, 9472 (1964).
559. Distillers Co. Ltd., The Netherlands Patent 6,512,937 (1966); *Chem. Abstr.*, **65**, 8884 (1966).
560. Hargrave, British Patent 887,688 (1962); *Chem. Abstr.*, **56**, 12859 (1962).
561. Campbell, Campbell, and Corran, British Patent 1,005,984 (1965); *Chem. Abstr.*, **64**, 2063 (1966).
562. Hargrave, British Patent 924,527 (1963); *Chem. Abstr.*, **64**, 708 (1966).
563. Hargrave, British Patent 898,869 (1962); *Chem. Abstr.*, **57**, 12443 (1963).
564. Hall, Millidge, and Young, British Patent 1,038,537 (1966); *Chem. Abstr.*, **65**, 16949 (1966).
565. Hardegger and Nikles, *Helv. Chim. Acta.*, **40**, 1016 (1957).
566. Moll and Uebel, East German Patent 58,960 (1967); *Chem. Abstr.*, **69**, 9983 (1969).
567. Toi, Hayashi, Amano, and Hachihama, *Kogyo Kagaku Zasshi,* **63**, 828 (1960); *Chem. Abstr.*, **56**, 8686 (1962).
568. Campbell and Corran, British Patent 1,020,857 (1966); *Chem. Abstr.*, **64**, 14171 (1966).
569. Yahama and Hayashi, Japanese Patent 179 (1961); *Chem. Abstr.*, **56**, 7285 (1962).
570. Hargrave, British Patent 896,049 (1962); *Chem. Abstr.*, **57**, 13736 (1963).
571. Butler, *J. Catalysis,* **6**, 26 (1966); *Chem. Abstr.*, **65**, 10558 (1966).
572. Godar and Mariella, *J. Amer. Chem. Soc.*, **79**, 1402 (1957).
573. Davis, Elvidge, and Foster, *J. Chem. Soc.*, 3638 (1962).
574. Pasedach and Seefelder, German Patent 1,207,930 (1965); *Chem. Abstr.*, **64**, 9694 (1966).
575. Dornow and Von Plessen, *Chem. Ber.*, **99**, 244 (1966).
576. Wajon and Arens, *Rec. Trav. Chim. Pays-Bas,* **76**, 65 (1957).
577. Ishiguro, Japanese Patent 6382 (1960); *Chem. Abstr.*, **55**, 6501 (1961).
578. Ishiguro, Morita, and Ikushima, *Yakugaku Zasshi*, **80**, 784 (1960); *Chem. Abstr.*, **54**, 24712 (1960).
579. Morita and Ikushima, *Yakugaku Zasshi,* **78**, 216 (1958); *Chem. Abstr.*, **52**, 11846 (1958).
580. Denton, U.S. Patent 2,972,615 (1961); *Chem. Abstr.*, **55**, 15514 (1961).
581. Schmidle, Locke, and Mansfield, *J. Org. Chem.*, **21**, 1195 (1956).
582. Schmidle, Locke, and Mansfield, *J. Org. Chem.*, **21**, 1194 (1956).
583. Gielas, *Bull. Soc. Chim. Fr.,* 3093 (1967).
584. Herzenberg and Boccato, *Chim. Ind.* (Paris), **80**, 248 (1958).
585. Mahan, Turk, Schnitzer, Williams, and Sammons, *Ind. Eng. Chem., Chem. Eng. Data Ser.,* **2**, 76 (1957); *Chem. Abstr.*, **52**, 15516 (1958).
586. Motoda, Omae, Yamamoto, and Yoshie, *Kogyo Kagaku Zasshi,* **65**, 354 (1962).
587. Mahan and Becker, U. S. Patent 2,766,248 (1956); *Chem. Abstr.*, **51**, 5843 (1957).
588. Stoops and Becker, U.S. Patent 2,745,833 (1956); *Chem. Abstr.*, **51**, 493 (1957).
589. Farberov, Ustavshchikov, Kut'in, Vernova, and Yarosh, *Izvest. Vysshykh Ucheb. Zavedenii, Khim. Khim. Tekhnol.*, No. 5, 92 (1958); *Chem. Abstr.*, **53**, 11364 (1959).
590. Kudo *Repts. Statist. Appl. Research, Union Japan. Scientists and Engrs.*, **6**, 13 (1959); *Chem. Abstr.*, **53**, 21934 (1959).
591. Mahan, U. S. Patent 2,877,228 (1959); *Chem. Abstr.*, **53**, 13182 (1959).

592. Mahan, U. S. Patent 2,775,596 (1956); *Chem. Abstr.*, **51**, 6704 (1957).
593. Pin-Ts'ai Li, K'ai-Kuo Wu and Yu-Tsan Hu, *Chung-Kuo K'o Hsueh Yuan Ying Yung Hua Hsueh Yen Chin So Chi K'an,* 1 (1963); *Chem. Abstr.*, **63**, 14804 (1965).
594. Takeba, Terada, and Sato, U. S. Patent 2,935,513 (1960); *Chem. Abstr.*, **55**, 573 (1961).
595. British Patent 749,718 (1956); *Chem. Abstr.*, **51**, 11396 (1957).
596. Narasaki and Suzuki, *Tokyo Kogyo Shikensho Hokoku,* **52**, (1957); *Chem. Abstr.*, **51**, 12089 (1957).
597. Bamford, U. S. Patent 2,769,007 (1956); *Chem. Abstr.*, **51**, 6704 (1957).
598. Dunn, British Patent 742,268 (1955); *Chem. Abstr.*, **50**, 16874 (1956).
599. Tsukamoto, Hirashiro and Takebe, Japanese Patent 5873 (1959); *Chem. Abstr.*, **54**, 14275 (1960).
600. Tsukamoto, Nonaka, and Suzuki, Japanese Patent 6119 (1959); *Chem. Abstr.*, **54**, 14275 (1960).
601. Takeba and Sato, Japanese Patent 6118 (1959); *Chem. Abstr.*, **54**, 14274 (1960).
602. Farberov, Kut'in, Ustavshchikov, and Shemyakina, *Zh. Prikl. Khim.*, **37**, 661 (1964); *Chem. Abstr.*, **60**, 14467 (1964).
603. Ferraiolo, Donetti, and Peloso, *Ann. Chim.* (Rome), **57**, 250 (1967); *Chem. Abstr.*, **67**, 94345 (1967).
604. Yameda, Tamano, Hashimoto, and Wada, Japanese Patent 21,536 (1961); *Chem. Abstr.*, **57**, 13735 (1962).
605. Schwarz and Bayer, German Patent 1,105,871 (1961); *Chem. Abstr.*, **56**, 11576 (1962).
606. Cooper, British Patent 900,799 (1962); *Chem. Abstr.*, **58**, 513 (1963).
607. Kurabayashi, Yanagiya, and Kamakura, *Tokyo Kogyo Shikensho Hokoku,* **58**, 448 (1963); *Chem. Abstr.*, **61**, 9460 (1964).
608. Donetti and Merli, French Patent 1,396,796 (1965); *Chem. Abstr.*, **63**, 8326 (1965).
609. Fabbri and Campazzi, *Chim. Ind.* (Milan), **49**, 458 (1967); *Chem. Abstr.*, **67**, 90627 (1967).
610. Hargrave and Cooper, British Patent 923,348 (1963); *Chem. Abstr.*, **59**, 9996 (1963).
611. Adams and Falbe, *Brennstoff Chem.*, **47**, 184 (1966); *Chem. Abstr.*, **65**, 10557 (1966).
612. The Netherlands Patent 6,512,874 (1966); *Chem. Abstr.*, **65**, 3846 (1966).
613. Mahan and Stoops, U. S. Patent 2,749,348 (1956); *Chem. Abstr.*, **51**, 1296 (1957).
614. Aries, British Patent 817,038 (1959); *Chem. Abstr.*, **54**, 2367 (1960).
615. Zellner, U. S. Patent 2,796,421 (1957); *Chem. Abstr.*, **52**, 447 (1958).
616. Dorafeev, *Trudy Khar'kov. Politekh. Inst.*, **8**, 89 (1956); *Referat. Zhur., Khim.*, Abstr. No. 14823 (1957); *Chem. Abstr.*, **53**, 15071 (1959).
617. Jezo and Tihlarik, Czechoslovakian Patent 90,632 (1959); *Chem. Abstr.*, **54**, 17425 (1960).
618. Jezo and Tihlarik, *Chem. Zvesti,* **12**, 558 (1958); *Chem. Abstr.*, **53**, 2229 (1959).
619. Kobayashi, Japanese Patent 1134 (1956); *Chem. Abstr.*, **51**, 5845 (1957).
620. Tsutsumi and Hasegawa, *Technol. Rept. Osaka Univ.*, **7**, 185 (1957); *Chem. Abstr.*, **52**, 10081 (1958).
621. Tsukamoto and Suzuki, Japanese Patent 3321 (1959); *Chem. Abstr.*, **54**, 13148 (1960).
622. Kobayashi, Sakata, Nishino, and Nakajima, Japanese Patent 3322 (1959); *Chem. Abstr.*, **54**, 13148 (1960).
623. Cislak and Wheeler, U. S. Patent 2,934,537 (1960); *Chem. Abstr.*, **54**, 17425 (1960).
624. Yamamoto, *Yuki Gosei Kagaku Kyokai Shi,* **21**, 196 (1963); *Chem. Abstr.*, **59**, 5019 (1963).

625. Mahan and Turk, U. S. Patent 2,995,558 (1957); *Chem. Abstr.*, **56**, 4737 (1962).
626. Jones and Landis, U. S. Patent 3,264,307 (1966); *Chem. Abstr.*, **65**, 15347 (1966).
627. Scheuber, *Promotionsarb.* (Zurich), 3422 (1964); *Chem. Abstr.*, **63**, 5596 (1965).
628. Utsumi, Japanese Patent 8285 (1954); *Chem. Abstr.*, **50**, 14001 (1956).
629. Ban and Yamamoto, Japanese Patent 3790 (1960); *Chem. Abstr.*, **55**, 1661 (1961).
630. Kotlyarevskii and Vereshchagin, *Izv. Akad. Nauk S.S.S.R., Otdel. Khim. Nauk*, 1272 (1960); *Chem. Abstr.*, **54**, 24714 (1960).
631. Kotlyarevskii, Vasil'ev and Vereshchagin, *Izv. Sibir. Otdel., Akad. Nauk S.S.S.R.*, No. 9, 52 (1959); *Chem. Abstr.*, **54**, 9915 (1960).
632. Kotlyarevskii and Vereshchagin, U.S.S.R. Patent 121,794 (1959); *Chem. Abstr.*, **54**, 5701 (1960).
633. Vereshchagin, Vasil'ev Nakhmanovich, and Kotlyarevskii, *Izv. Sibir. Otdel., Akad. Nauk S.S.S.R.*, No. 6, 89 (1959); *Chem. Abstr.*, **54**, 9924 (1960).
634. Boettner, U.S. Patent 2,744,903 (1956); *Chem. Abstr.*, **51**, 495 (1957).
635. Mamaev, *Biol. Aktivn. Soedin., Akad. Nauk S.S.S.R.*, 38 (1965); *Chem. Abstr.*, **63**, 18081 (1965).
636. Falbe, Korte, Bechstedt, Foerster, Hack, Hoerhold, and Mandelartz, *Chem. Ber.*, **98**, 1928 (1965).
637. Falbe, Weitkamp, and Korte, *Tetrahedron Lett.*, 2677 (1965).
638. Mahan, U.S. Patent 2,742,474 (1956); *Chem. Abstr.*, **50**, 16876 (1956).
639. Cislak and Wheeler, U.S. Patent 2,807,618 (1957); *Chem. Abstr.*, **52**, 2932 (1958).
640. The Netherlands Patent 6,600,923 (1966); *Chem. Abstr.*, **66**, 18678 (1967).
641. Shimizu, Igarashi, and Hashimoto, U.S. Patent 3,284,456 (1966); *Chem. Abstr.*, **66**, 37773 (1967).
642. Shimizu, Igarashi, and Hasimoto, U. S. Patent 3,272,825 (1966); *Chem. Abstr.*, **65**, 18563 (1966).
643. Aries, British Patent 790,994 (1958); *Chem. Abstr.*, **52**, 16375 (1958).
644. Bradshaw, Parkes, and Ford, British Patent 742,643 (1955); *Chem. Abstr.*, **50**, 16875 (1956).
645. Aries, U. S. Patent 2,700,042 (1955); *Chem. Abstr.*, **50**, 1924 (1956).
646. Cooper and Hargrave, French Patent 1,319,193 (1963); *Chem. Abstr.*, **60**, 509 (1964).
647. French Patent 1,398,750 (1965); *Chem. Abstr.*, **63**, 13224 (1965).
648. Cooper and Hargrave, French Patent 1,332,907 (1963); *Chem. Abstr.*, **60**, 508 (1964).
649. Cooper and Hargrave, British Patent 975,995 (1964); *Chem. Abstr.*, **62**, 5258 (1965).
650. Cooper and Hargrave, French Patent 1,332,908 (1963); *Chem. Abstr.*, **60**, 509 (1964).
651. Ishiguro, Morita, and Ikushima, *Yakugaku Zasshi*, **79**, 1073 (1959); *Chem. Abstr.*, **54**, 4587 (1960).
652. Ishiguro, Japanese Patent 8319 (1959); *Chem. Abstr.*, **54**, 17425 (1960).
653. Ishiguro, Morita, and Ikushima, *Yakugaku Zasshi*, **77**, 660 (1957); *Chem. Abstr.*, **51**, 16463 (1957).
654. Cooper and Hargrave, French Patent 1,332,909 (1963); *Chem. Abstr.*, **60**, 509 (1964).
655. Hargrave, British Patent 907,059 (1962); *Chem. Abstr.*, **58**, 2438 (1963).
656. Cooper and Hargrave, French Patent 1,332,906 (1963); *Chem. Abstr.*, **60**, 508 (1964).
657. Hargrave and Cooper, British Patent 966,264 (1964); *Chem. Abstr.*, **61**, 10660 (1964).
658. Hargrave, French Patent 1,321,085 (1963); *Chem. Abstr.*, **61**, 6995 (1964).
659. Farley and Eliel, *J. Amer. Chem. Soc.*, **78**, 3477 (1956).
660. Levin, U.S. Patent 2,922,785 (1960); *Chem. Abstr.*, **54**, 9962 (1960).
661. Cooper and Hargrave, British Patent 968,945 (1964); *Chem. Abstr.*, **61**, 13286 (1964).
662. Cooper and Hargrave, British Patent 968,946 (1964); *Chem. Abstr.*, **61**, 13289 (1964).
663. Haley and Maitland, *J. Chem. Soc.*, 3155 (1951).

664. Ehsan and Karimullah, *Pakistan J. Sci. Ind. Res.*, **11** (1968); *Chem. Abstr.*, **69**, 96403 (1968).
665. Kamal and Oureshi, *Pakistan J. Sci. Res.*, **15**, 35 (1963); *Chem. Abstr.*, **60**, 1689 (1964).
666. Checchi, *Gazz. Chim. Ital.*, **89**, 2151 (1959); *Chem. Abstr.*, **55**, 5499 (1961).
667. Furdik and Gvozdjakova, *Acta Fac. Rerum Nat. Univ. Comenianae, Chimia*, 8, 581 (1964); *Chem. Abstr.*, **61**, 13277 (1964).
668. Micheel and Moeller, *Ann. Chem.*, **670**, 63 (1963).
669. Baliah, Gopalakrishnan, and Govindarajan, *J. Indian Chem. Soc.*, **31**, 832 (1954); *Chem. Abstr.*, **50**, 998 (1956).
670. Lukes and Kuthan, *Collect. Czech. Chem. Commun.*, **25**, 2173 (1960); *Chem. Abstr.*, **55**, 1605 (1961).
671. Kuthan and Hakr, *Collect. Czech. Chem. Commun.*, **32**, 1438 (1967); *Chem. Abstr.*, **67**, 64207 (1967).
672. Belgian Patent 660,964 (1965); *Chem. Abstr.*, **64**, 2065 (1966).
673. The Netherlands Patent 6,516,712 (1966); *Chem. Abstr.*, **65**, 16946 (1966).
674. Lůkes, Hudlicky, and Zdenek, *Chem. Listy*, **50**, 258 (1956); *Chem. Abstr.*, **50**, 7796 (1956).
675. Erner, U.S. Patent 3,019,227 (1962); *Chem. Abstr.*, **56**, 12864 (1962).
676. Colonge, Descotes, and Frenay, *Bull. Soc. Chim. Fr.*, 2264 (1963).
677. Fujisawa and Sugasawa, *Tetrahedron*, **7**, 185 (1959).
678. Kametani, Nomura, and Morita, *J. Pharm. Soc. Jap.*, **76**, 652 (1956); *Chem. Abstr.*, **51**, 403 (1957).
679. Herz and Tocker, *J. Amer. Chem. Soc.*, **77**, 6353 (1955).
680. Herz and Tocker, *J. Amer. Chem. Soc.*, **77**, 3554 (1955).

CHAPTER III

Quaternary Pyridinium Compounds

OSCAR R. RODIG

Department of Chemistry, University of Virginia
Charlottesville, Virginia

Introduction

Since the appearance of the first chapter on quaternary pyridinium compounds, literally thousands of preparations, reactions, and uses of these substances have been reported. Since it is neither the purpose nor the desire of this chapter to present an all-inclusive treatise on this subject, it covers what are, in the opinion of the author, the major new developments in this area. The emphasis has been placed on a survey of new reactions and novel applications of previously known reaction types, along with an overview of recent studies on spectroscopic and physical properties of these compounds. In a number of cases, only accomplishments reported in the more recent literature have been stressed with the intent that the work cited acts as a lead reference from which previous endeavors in the area may be elicited. Not included are pyridine oxides and metal complexes,* because these substances are treated in separate chapters. Furthermore, since pyridinium compounds of biological interest are also reviewed in a separate chapter, only a cursory coverage of their uses in medicinal chemistry and in the study of enzymic reactions is presented herein.

*This chapter, unfortunately, was not received (Ed.).

I. Preparation

1. By Displacement of Halogen or Similar Leaving Group

A. *Alkyl Halides and Related Alkylating Agents*

Perhaps the most common method to prepare quaternary pyridinium compounds (**III-1**) continues to be the Menschutkin reaction, the reaction of a pyridine derivative (**III-2**) with an organic halide (**III-3**) (1). As expected, strong

R^1 ... N + R^2–X ⟶ R^1 ... N^+–R^2 X^-

III-2 **III-3** **III-1**

electron donor substituents on the pyridine ring increase the reaction rate, but the pyridine nitrogen atom does cause deviations of substituent effects over those observed in the benzene series. Thus from a study of the thermodynamic dissociation constants of the pyridinium ion in water, Fischer and co-workers concluded that while there is the expected resonance interaction of electron-donating substituents at C-4 with the pyridine ring nitrogen, electron attracting substituents appear to exert only an inductive effect (2). Furthermore, a substituent at C-3 may appear to exert a resonance effect (methoxyl) or it may not (amino). In a study of the *N*-methylation of *ortho*-substituted pyridines, Deady and Zoltewicz found that both steric and electronic effects come into play (2a).

It was reported earlier that 2,6-di-*t*-butylpyridine does not react with methyl iodide, presumably because the nitrogen atom is hindered sterically by the *t*-butyl groups (3). However, Okamoto and Shimagama (4) found that the pyridinium methyl iodide **III-4** can be prepared under high pressure (~5000 atm) and that it is surprisingly stable, subliming at 250° without decomposition. To account for this unexpected stability the authors suggest that the *t*-butyl groups may act as a kind of claw which holds the methyl group to the nitrogen atom. A more systematic study of this reaction under pressure has been reported by le Noble and Ogo (5) who investigated the rate of alkylations of various 2,6-dialkylpyridines with methyl, ethyl, and isopropyl iodide. It has also been possible to prepare the previously elusive tritylpyridinium chloride (**III-5**) by the reaction of trityl chloride with pyridine under high pressure (6). It is not highly

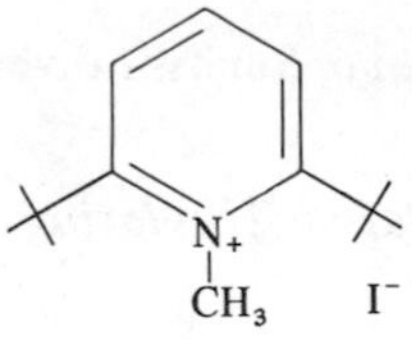

CH_3 I^- C_6H_5 C_6H_5 Cl^- C_6H_5

III-4 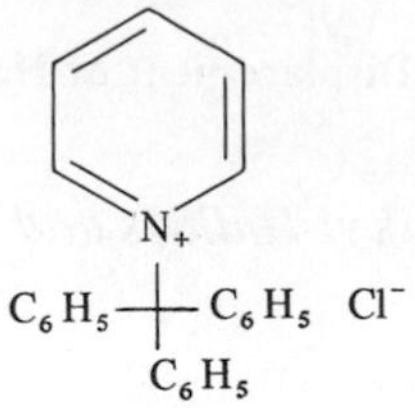 **III-5**

stable, decomposing to triphenylcarbinol and pyridinium chloride in the presence of moisture. Interestingly, tritylpyridinium bromide can be prepared at room temperature and atmospheric pressure (7, 8). An explanation for this difference based on the reactivities of the two halides in the solvents employed has been advanced (6).

Since neutral reactants form ionic products in the Menschutkin reaction, it provides a good example for studying solvent effects on reaction rate. Thus, in the formation of pyridinium compounds, correlations of the rates of reaction with solvent polarizability (9), Z values (10), volumes of activation (11), and protic-aprotic solvents (12) have been reported.

Mixed pyridinium salts of type **III-6** have been prepared because of their potential interest as antagonists for alkyl phosphate intoxication, as occurs with certain chemical warfare agents and insecticides (13).

Activated halides ordinarily react readily with pyridines to form the corresponding pyridinium compounds. The chloroacetyl derivatives of phenolic ethers (**III-7**) were converted to ketomethylpyridinium chloride on heating with

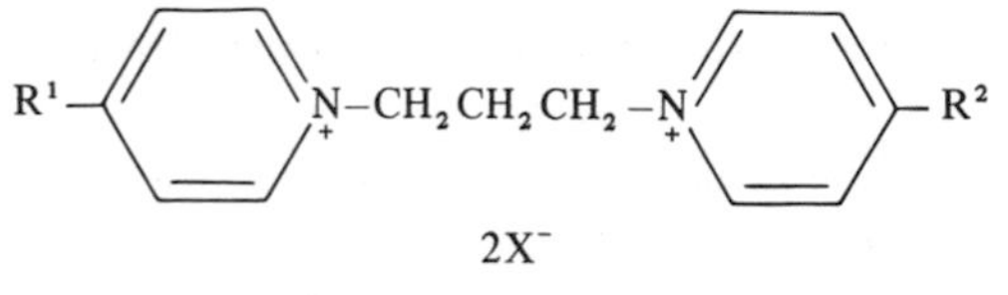

$COCH_2Cl$

R^1 $N^+-CH_2CH_2CH_2-N^+$ R^2 OR

$2X^-$

III-6 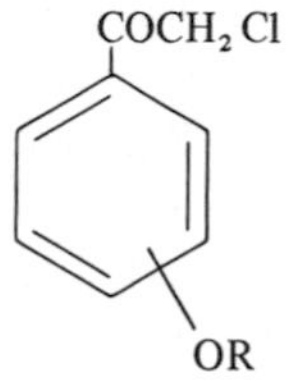 **III-7**

pyridine for several minutes. In the process, the aromatic ether group is dealkylated (14). The pyridinium products could be cleaved with alkali to give the corresponding hydroxybenzoic acids (15). A series of furanobromoketones have been condensed with 2-aminopyridine to give the pyridinium derivatives **III-8**, which were cyclized to the imidazopyridines **III-9**. Some of the latter compounds showed antimicrobial activity (16). RSO_2CH_2Cl does not react with pyridines to give the corresponding salts. The latter can be obtained from RSO_2CHN_2, or better $RSO_2CH_2OSO_2CF_3$, and the pyridine (16a).

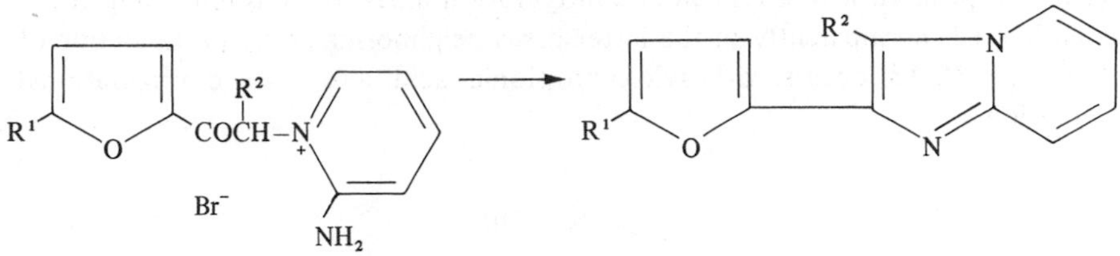

III-8 III-9

The reaction of 2,2′-bipyridyl or its methylated derivatives with aliphatic or aromatic α-haloketones gives high yields of the pyridinium salts **III-10**. When these, in turn, are treated with either bromine or *p*-toluenesulfonyl chloride in pyridine cyclization to the dipyridoimidazolium salts **III-11** occurs (17). It is postulated that the mechanism of cyclization probably proceeds by bromination at the methylene group since precedent for such a reaction exists (18, 19). In the case of the tosyl chloride-initiated cyclization, the mechanism has been tentatively postulated as proceeding *via* the intermediate **III-12**, formed from the 1-phenacyl-2-(2-pyridyl)pyridinium halide (**III-10**) as shown.

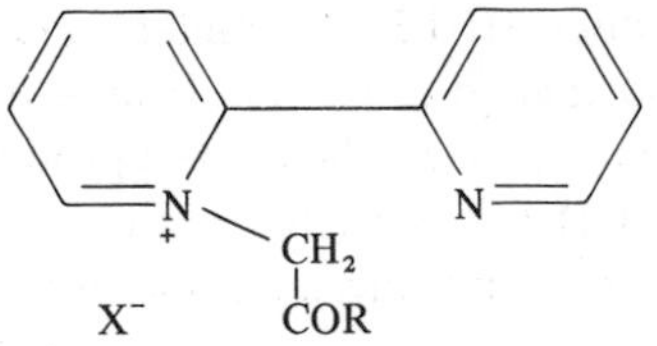

III-10 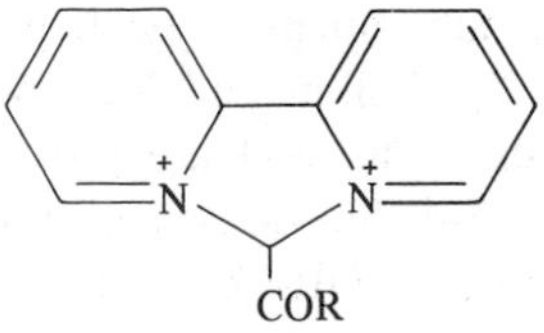 III-11

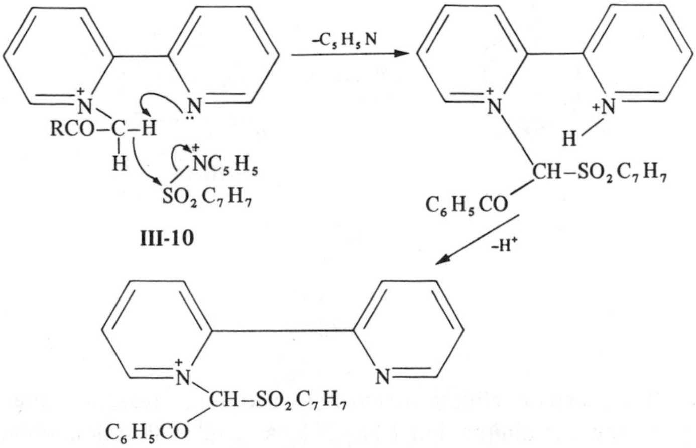

III-10

III-12

The kinetics of reactions of substituted pyridines with phenacyl bromide have been reported (20). The reaction of 3-hydroxypyridine with α-bromo-acids is found to proceed with inversion of configuration unless there is branching at the β carbon (21). Apparently in the latter cases neighboring group participation of the type **III-13** occurs. α-Tosyloxypropionic acid also gives configurational retention.

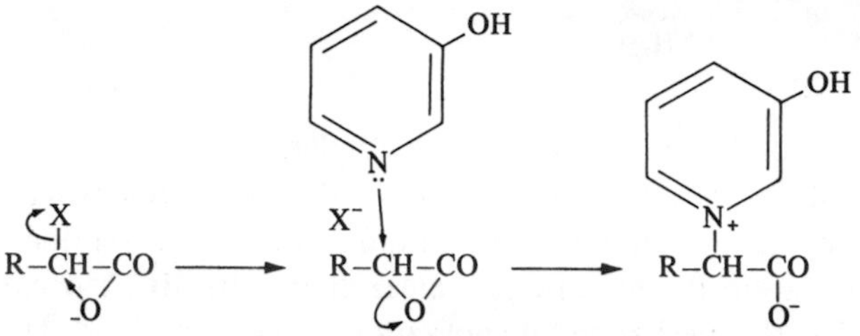

III-13

When α-bromo acids are heated in the presence of pyridine the formation of the pyridinium bromide may be accompanied by spontaneous decarboxylation (22). It has been suggested that the ylid **III-14** is a key intermediate in this reaction, subsequently yielding the product **III-15**. A similar reaction was reported by Phillips and Ratts who found that ylids formed from α-halo acids and pyridine can condense with aldehydes to give decarboxylated products (23).

When 2,3-dichloro-1,4-naphthoquinone (**III-16**) is heated with pyridine, the pyridinium salt (**III-17**) is formed. The remaining nuclear chlorine now shows

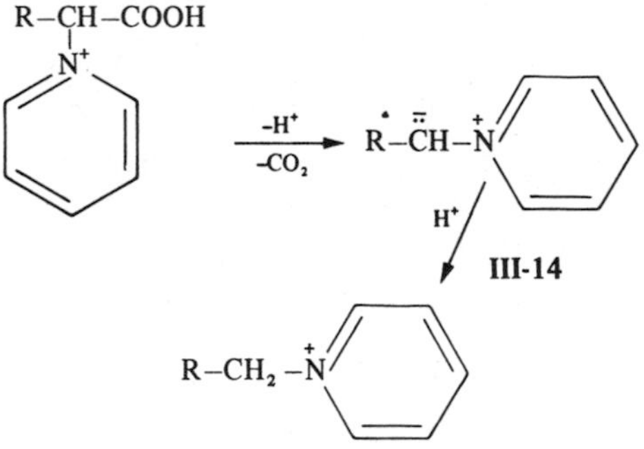

III-14

III-15

enhanced displacement ability toward nucleophilic reagents due to the proximity of the pyridinium ion (24, 25). A similar situation obtains with *N*-substituted 2,3-dichloromaleimides (26).

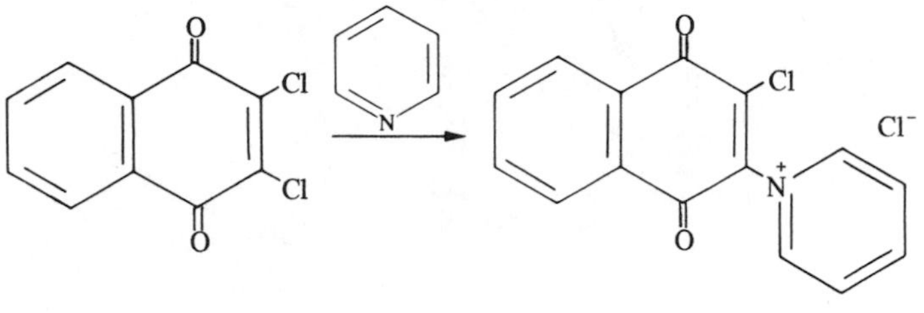

III-16 III-17

Neilands and co-workers have reported the preparation of pyridine betaines of type **III-21** by treating β-dicarbonyl compounds (**III-18**) with phenyliodosoacetate. The intermediate adducts (**III-19**) which form lead to **III-21** when heated in the presence of pyridine, presumably (but not necessarily) by way of the carbene **III-20** (27, 28) (see also ref. 28a).

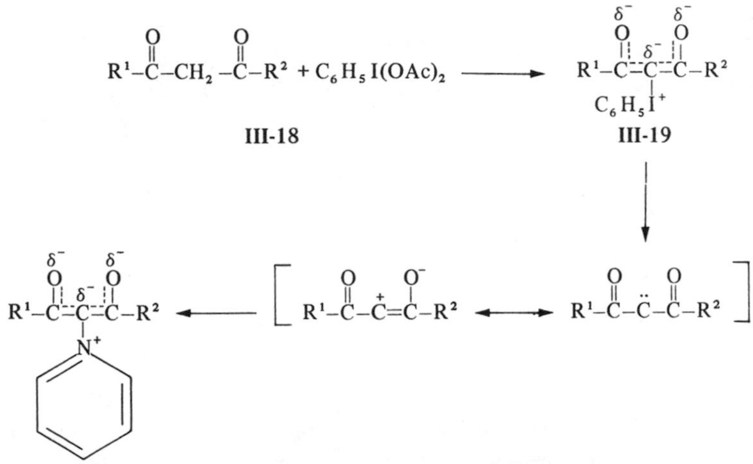

$$R^1\text{-}CO\text{-}CH_2\text{-}CO\text{-}R^2 + C_6H_5I(OAc)_2 \longrightarrow$$

III-18 III-19

III-21 III-20

Kinetic studies on the formation of pyridinium salts from α-chloroethers show the reaction to be first order, which is in accord with the dissociation of the α-chloro ether to a carboxonium ion as the rate determining step (29). The action of pyridine and picolines on 2,3,4,6-tetraacetyl-α-D-glucopyranosyl bromide (**III-22**) yielded the salts **III-23** with a free hydroxyl group at C-2 (30). A mechanism to explain the product could be that shown in **III-24** which involves neighboring group participation of the C-2 acetyl group.

Pyridinium compounds of type **III-26** have resulted from the reaction of pyridine and picolines with *N*-chloromethyl-4-nitrophenoxyacetamide **III-25**

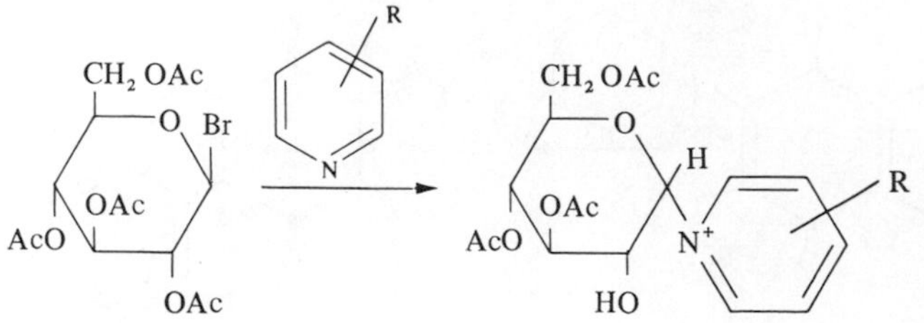

III-22 III-23

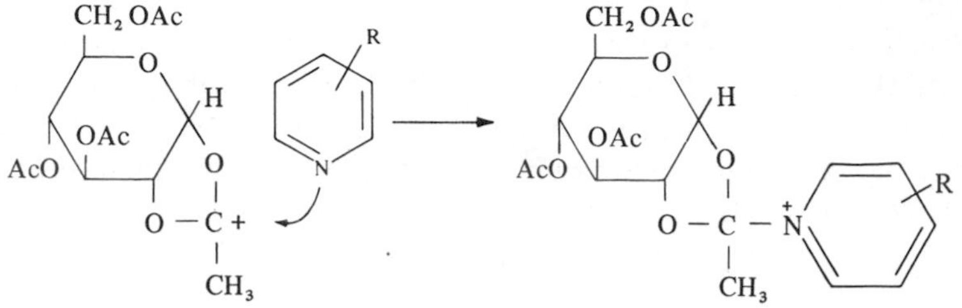

III-24

(31), while β-chloroethyl sulfonates (**III-27**; Z = RSO_3-) (32) and β-bromoethyl esters (**III-27**; Z = RCOO–) (33) have also been used to prepare pyridinium salts.

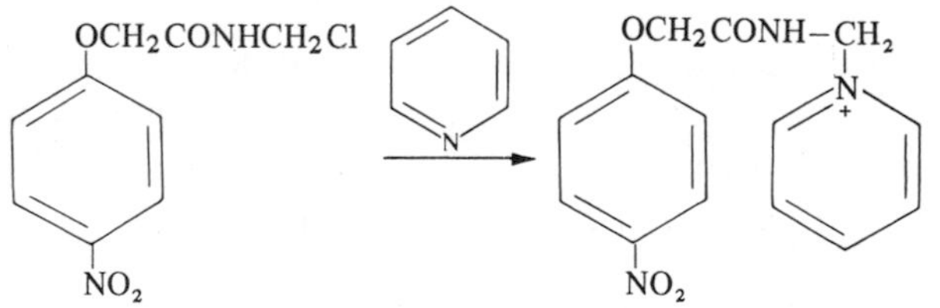

ZCH_2CH_2X

III-25 III-26 III-27

A general procedure for the synthesis of pyridinium compounds from alcohols, presumably *via* the corresponding alkyl chloride intermediates, has been reported (34). The method makes use of the dimethylformamide-thionyl chloride reagent, which, in turn, has been investigated actively to determine the reactive species (35). Pyridine has also been *N*-methylated using cyclic acyl phosphates (35a).

a. QUATERNIZATION BY HALOGENATION IN PYRIDINE. The reaction of olefins with halogens and pyridine derivatives to form α-haloalkylpyridinium halides reported previously for only bromine and iodine monochloride has been extended to the use of chlorine (36), and the chlorination of allyl chloride in a benzene-pyridine mixture gave about 20% of the pyridinium salt **III-28** as well as the trichloropropane **III-29** (37). The product **III-28** is in keeping with the formation of the most stable α-halo carbonium ion intermediate.

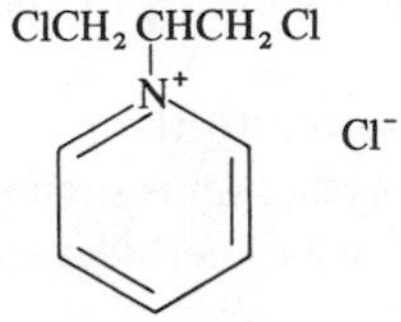

$ClCH_2CHCH_2Cl$ / N^+ (pyridinium) Cl^- $ClCH_2CHCH_2Cl$ / Cl

III-28 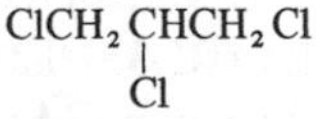 III-29

Diner and Lown obtained iodoalkane pyridinium nitrates (**III-30**) or alkene pyridinium iodides (**III-31**), depending on the substrate, when iodonium nitrate (INO_3, prepared by adding ICl to $AgNO_3$) in chloroform/pyridine was added to alkenes (38). The reaction with 2,3-dihydropyran yields the *trans*-diequatorial isomer **III-32**. The addition is subject to steric effects and occurs frequently in anti-Markovnikov 1,2-addition to give the corresponding iodoalkane pyridinium nitrates.

R–CH–CH$_2$I / N^+ NO_3^- R–C=CH$_2$ / N^+ I^- H O I H N NO_3^-

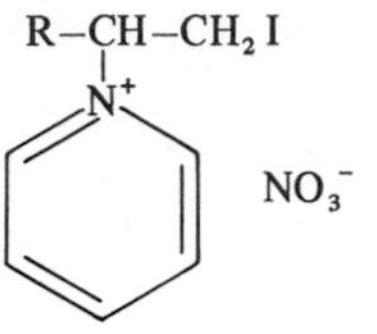

III-30 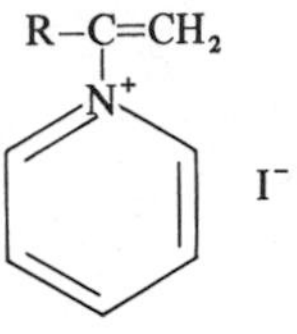 III-31 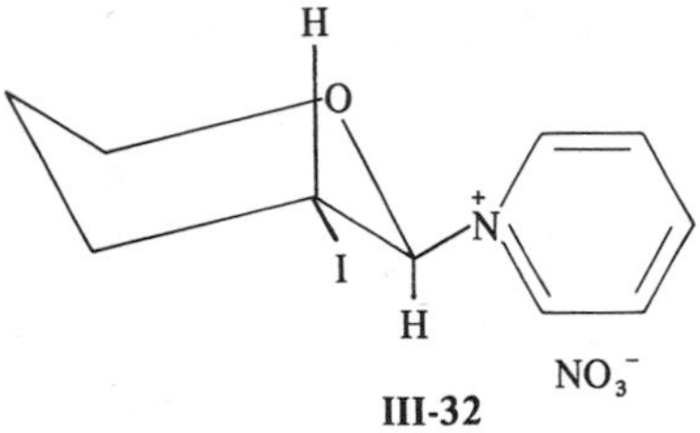 III-32

Ferrocene (**III-33**) reacts with chlorine, bromine, *N*-chloro- or *N*-bromosuccinimide in pyridine to yield the tetrahalocyclopentadienylides **III-34** (39). The

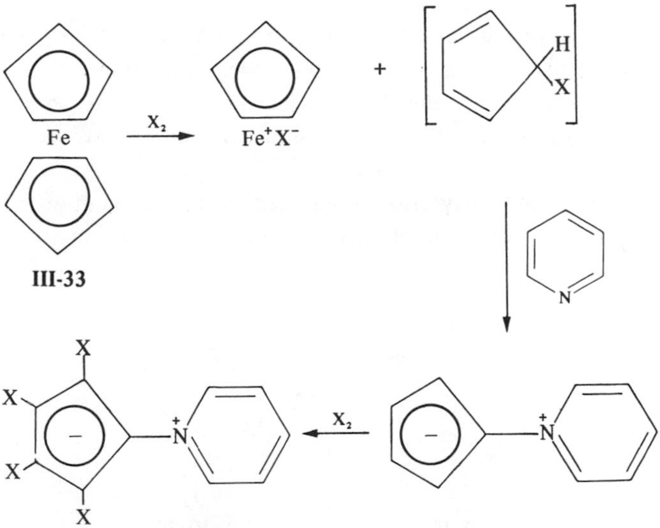

Fe $\xrightarrow{X_2}$ Fe^+X^- + [H X]

III-33

N

X X X X N $\xleftarrow{X_2}$ N

III-34

mode of formation of these products presumably occurs by the route shown; no ferricinium compounds were detected.

4-Methylpyrimidine (**III-35**) has been converted to the pyridinium salt **III-36** by the action of iodine in pyridine (40). This reaction is similar to that reported for the conversion of 6-methylpurine to purine-6-methylenepyridinium iodide (**III-37**) (41).

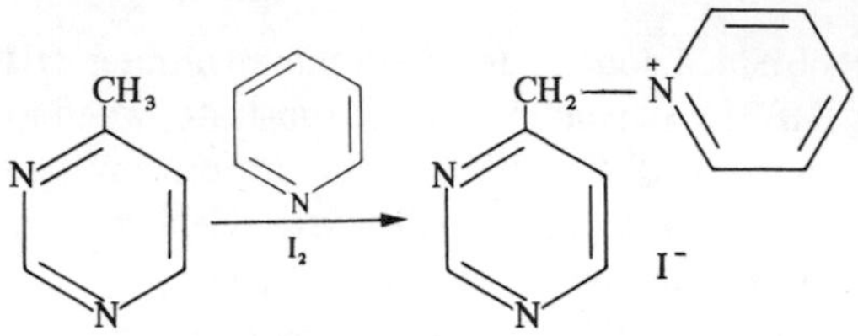

CH_3 N N N I_2 CH_2—$\overset{+}{N}$ N N I^-

III-35 **III-36**

CH_2 $\overset{+}{N}$ N N N N H I^-

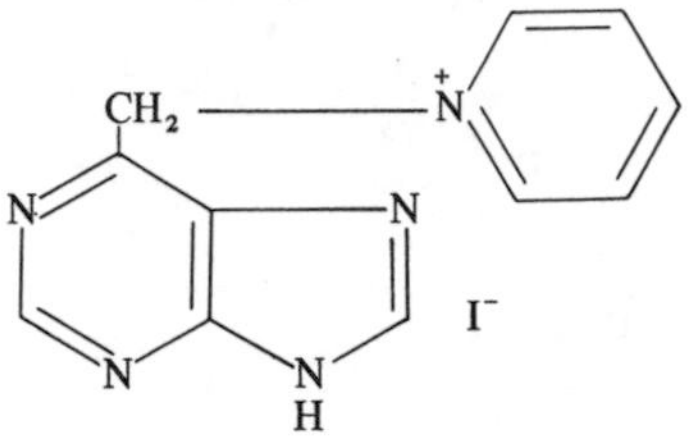

III-37

b. QUATERNIZATION WITH DIAZOMETHANE, TOSYLATES, SULTONES, AND OTHER ALKYLATING AGENTS. When 4-phenylacetylnicotinic acid (**III-38**) was treated with diazomethane, the *N*-methylbetaine **III-39** was obtained, while 2-aryl-5-azaindan-1,3-diones (**III-40**) and dimethyl sulfate gave betaines **III-41** (42).

Pyridinium compounds may also be prepared by displacement of the mesylate or tosylate group. For example, when the mesylation or tosylation of

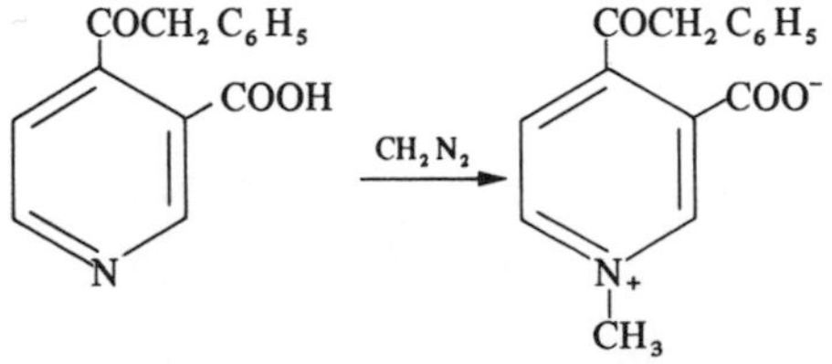

$COCH_2C_6H_5$ COOH N CH_2N_2 $COCH_2C_6H_5$ COO^- N_+ CH_3

III-38 **III-39**

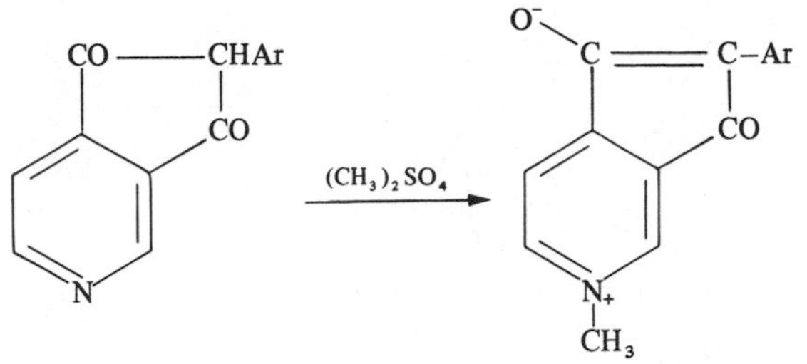

CO——CHAr | CO (pyridine fused) —$(CH_3)_2SO_4$→ O^-–C═C–Ar | CO (N^+–CH_3)

III-40 III-41

6-hydroxymethyluracil (**III-42**) was attempted, the pyridinium salt (**III-43**) was formed instead (43). It was rationalized that the initially formed sulfonyl esters undergo facile ionization, yielding a resonance stabilized benzylic type cation which can then undergo S_N1 attack by pyridine.

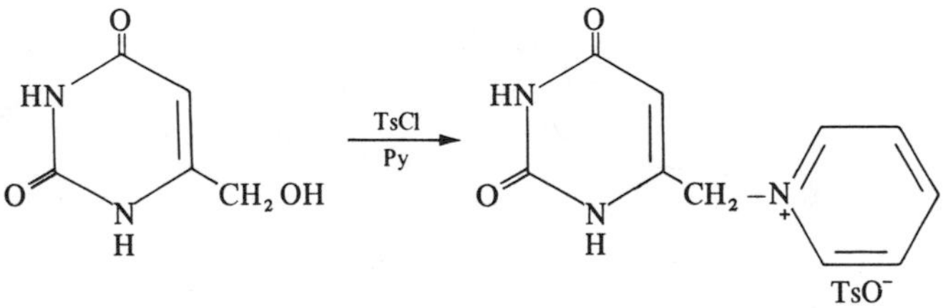

HN, O, N H, CH_2OH —TsCl, Py→ HN, O, N H, CH_2-N^+ (pyridine) TsO^-

III-42 III-43

An interesting observation concerning tosyl group displacement was reported by Jones and co-workers (44). When the pyranoside tosylate **III-44** was treated with hydroxylamine hydrochloride in pyridine, a 91% yield of **III-45** could be isolated, displacement of the tosylate group having occurred with retention of configuration. On the other hand, when **III-44** was treated with sodium benzoate in dimethylformamide a 90% yield of the oxime benzoate **III-46** was obtained. The retention of configuration observed in the formation of the pyridinium salt has been explained as follows. The oximino *p*-toluenesulfonate **III-47** is formed initially which eliminates *p*-toluenesulfonic acid in the pyridine medium, yielding an α-nitrosoolefin **III-48**. Attack of this substance by pyridine gives, after protonation, the most stable product **III-45**.

The use of the tolyl sultone **III-49** in forming quaternary pyridinium salts of type **III-50** was reported by Bradsher and co-workers (45). Subsequent

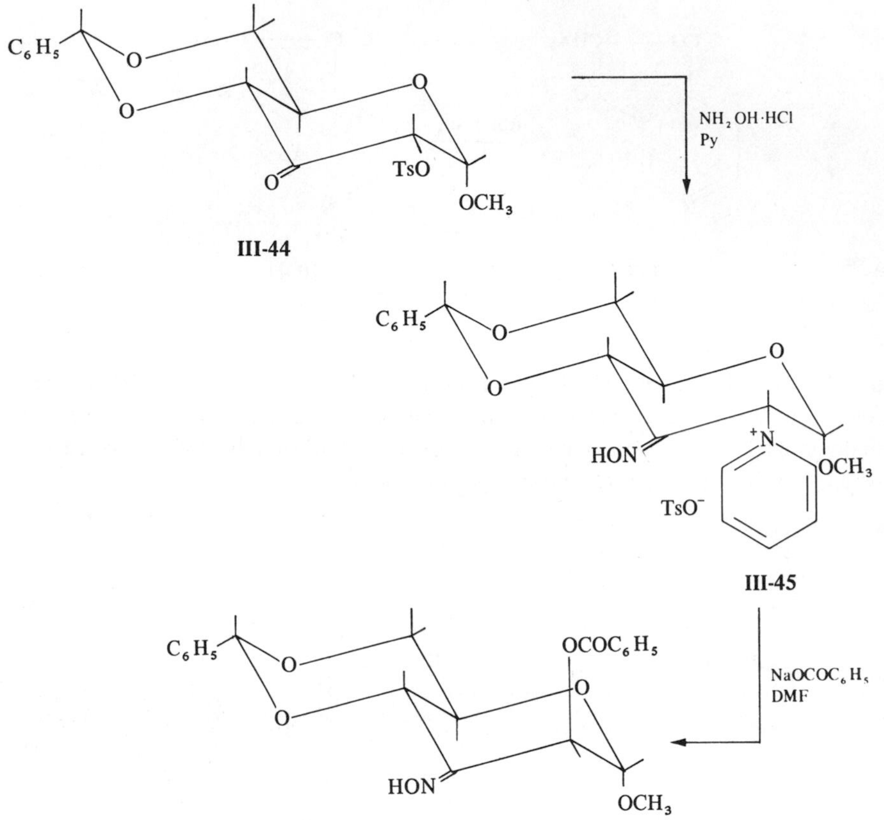

C_6H_5 O O O O TsO OCH_3

III-44

$NH_2OH \cdot HCl$
Py

C_6H_5 O O O HON ^+N OCH_3 TsO^-

III-45

$NaOCOC_6H_5$
DMF

C_6H_5 O O $OCOC_6H_5$ O HON OCH_3

III-46

O H H–O–N OTs —TsOH→ O O=N H

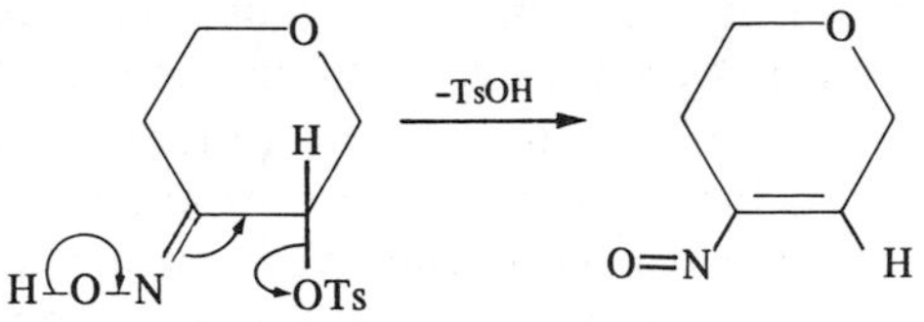

III-47 III-48

cyclization of compounds **III-50** to the corresponding acridizinium betaines was generally found difficult or not possible to accomplish under the acidic conditions studied. The opening of an epoxide ring to form the corresponding hydroxypyridinium derivative has been observed (45a).

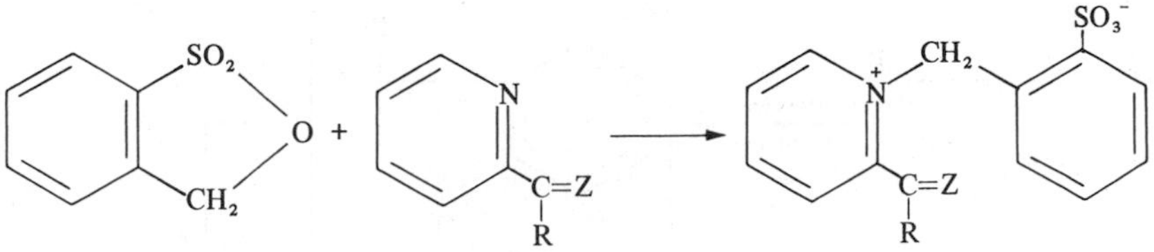

III-49 $Z = O, OCH_2CH_2O$ III-50
$R = H, CH_3, Ar$

The alkylation of the sodium salt of 2-pyridone (**III-51**) using triethyloxonium tetrafluoroborate has been studied (46) (see also Chap. XII). The reagent reacts with the pyridone within 5 min to yield three products: 2-ethoxypyridine (21%), 1-ethyl-2-pyridone (40%), and 1-ethyl-2-ethoxypyridinium tetrafluoroborate (**III-52**) (29%). The product percentages were found to vary somewhat with reaction time, and the dialkylated product (**III-52**) arises from further alkylation of either of the other two monoalkylation products. The dialkylated product can also act as an alkylating agent, one of much higher selectivity than triethyloxonium tetrafluoroborate. Substituted pyridine- and quinoline-1-oxides yield the corresponding 1-ethoxypyridinium and quinolinium tetrafluoroborates when treated with triethyloxonium tetrafluoroborate (47).

Finally, a methylpyridinium salt has been prepared using either the mono- or dimethoxynitropyrimidine **III-53**; a methoxy methyl group transfers readily to the pyridine ring nitrogen atom (48).

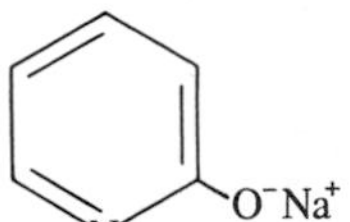

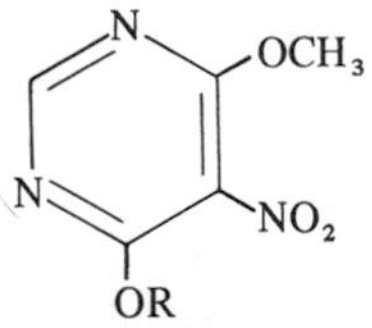

$R = H, CH_3$

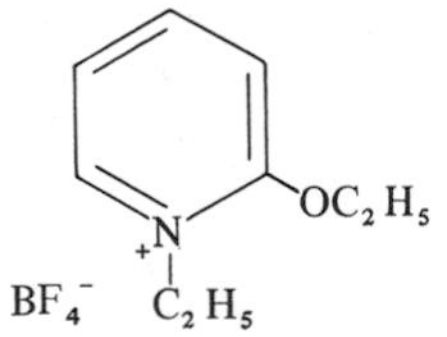

III-51 III-52 III-53

B. *Aryl Halides and Related Arylating Agents*

A method for converting indoles (**III-54**) to their 2-pyridinium derivatives (**III-57**) using *N*-bromosuccinimide or dioxane dibromide in pyridine has been reported (49, 50). The products serve as intermediates in the preparation of oxindoles, and the suggested mechanism of their formation proceeds *via* **III-55** and **III-56** (50).

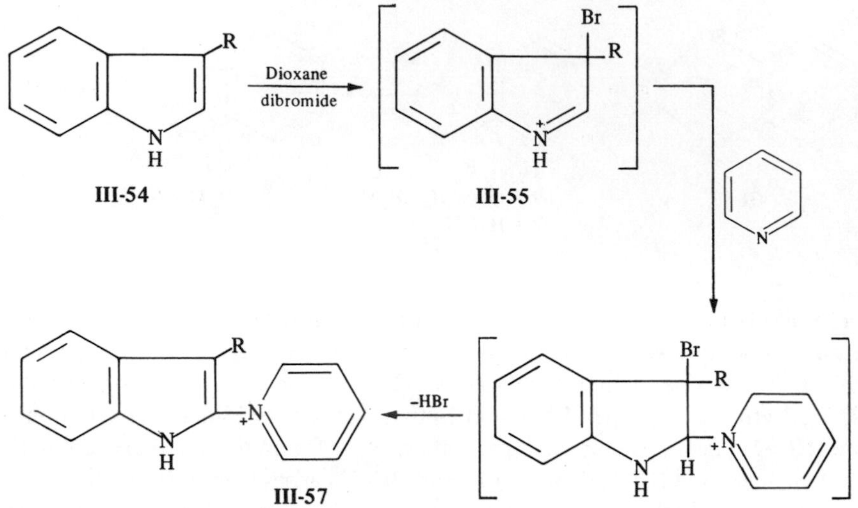

III-54 III-55

Dioxane dibromide

III-57 III-56

-HBr

When 2-chlorobenzimidazole is heated with pyridine, the pyridinium salt **III-58** is formed, which, on treatment with ammonia, is converted to the bright yellow crystalline *N*-pyridinium 2-benzimidazolide **(III-59)** (51). This compound was found to be more stable than the parent cyclopentadiene derivative **III-60** prepared earlier from dibromocyclopentadiene and pyridine (52).

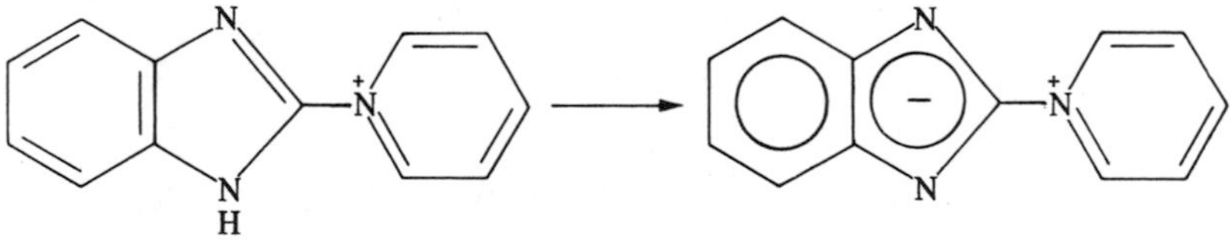

III-58 III-59

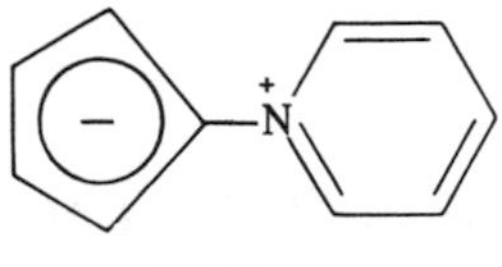

III-60

Nozoe and co-workers obtained the pyridinium salt **III-62** when they attempted to prepare the tosylate of the hydroxyazulene **III-61** (53). The same compound could be prepared from the corresponding chloroazulene.

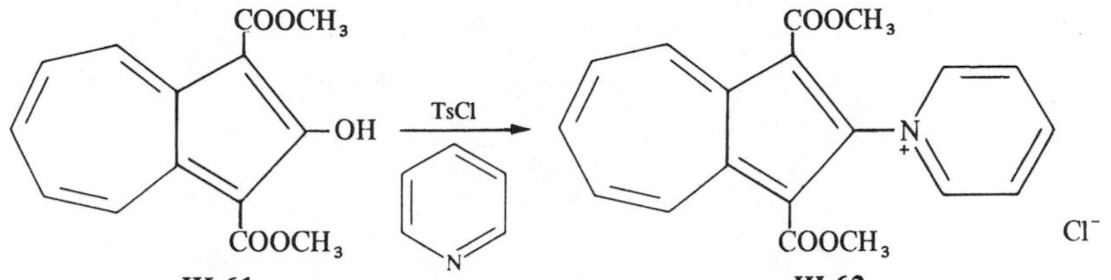

COOCH$_3$ ... OH —TsCl, pyridine→ ... COOCH$_3$ ... $\overset{+}{N}$ ... Cl$^-$

III-61 III-62

Picryl- and 2,4-dinitrobenzylpyridinium halides have been prepared at room temperature or below from the corresponding chloro compounds, demonstrating the ease of displacement of activated halogen in aromatic systems (54). Similar activation of the tosylate group also causes its displacement by pyridine or its derivatives. Steric factors were found to be important, and under the conditions studied, 2,6-lutidine yielded no product (55). The further displacement of the pyridinium or picolinium group in such substances by fluoride ion constitutes a method of synthesis for aromatic fluorine compounds (56).

Letzinger and co-workers (57, 58) found that pyridine could displace a nitro group at low temperature from *p*-nitroanisole or *p*-nitrophenyl phosphate **(III-63)** if the reaction is photoinduced, yielding 1-arylpyridinium salts **(III-64)**. Under similar conditions, *m*-nitroanisole, nitrobenzene, and *p*-dinitrobenzene fail

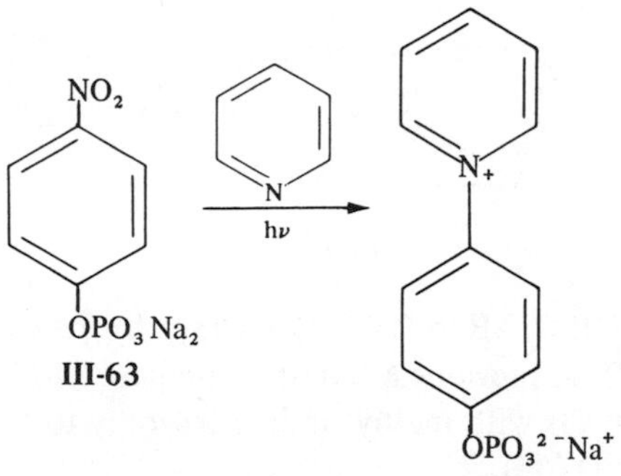

NO_2 ... OPO_3Na_2 —pyridine, hν→ N^+ ... $OPO_3^{2-}Na^+$

III-63 III-64

to react, which led the authors to infer that, in an excited state, *p*-phosphoryl and *p*-methoxyl may act as activating substituents in aromatic nitro group displacement reactions.

In later work, it was found that electron-donating *meta* and *para* substituents in compounds of type **III-65** markedly retard the photoinduced nucleophilic displacement of the nitro group by pyridine, as does also the methoxyl group in the series **III-66** (59). The authors suggest that such deactivation is due to interaction of the photoexcited nitroaromatic ring with the ring bearing the

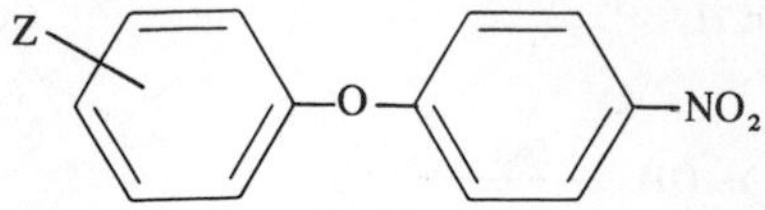

III-65

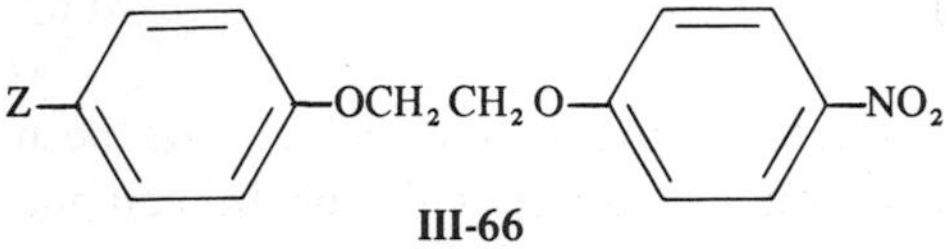

III-66

electron-donating group to yield a transient species of low reactivity toward nucleophilic substitution.

C. *Pyridylpyridinium Salts*

A detailed procedure for the preparation of *N*-(4-pyridyl)pyridinium chloride hydrochloride (**III-67**; R = H), originally reported by Koenigs and Greiner (60), is now available (61). This compound has recently been converted to the

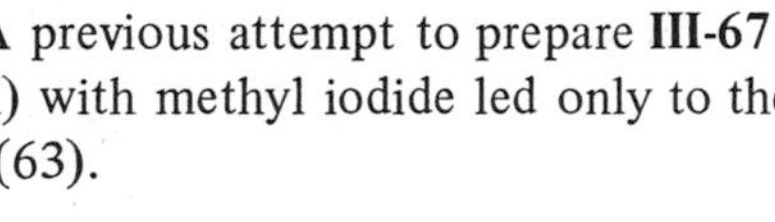

III-67

diquaternary salt (**III-67**; R = CH_3, CH_2CH_3) by reaction with the respective dialkyl sulfate (62). A previous attempt to prepare **III-67** (R = CH_3) by treating **III-67** (R = H, X = Cl) with methyl iodide led only to the corresponding iodide (**III-67**; R = H, X = I) (63).

D. *Selective Quaternization*

Garmaise and Paris (64) conducted an interesting study on the selective quaternization of systems having both a ring nitrogen and a dimethylamino group, such as **III-68** and **III-69**. They found that normally both monoquaternary products are formed which can exist in equilibrium. Thus the relative proportions depend on the reaction conditions, with the ring alkylated product being the most thermally stable.

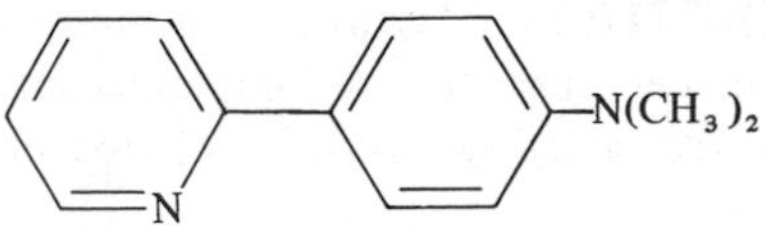

III-68

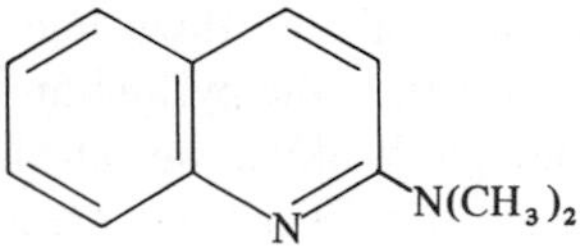

III-69

Methyl iodide reacts exclusively with the 3-amino group of 3-amino-2-methylaminopyridine, in contrast with 2-methylaminopyridine and 3-aminopyridine where only ring *N*-methylation occurs. Further, the ratio of ring to 3-amino methylation in 2,3-diaminopyridine is solvent dependent, the observed results being attributed to a combination of hydrogen bonding and steric effects (64a).

Methylation and *N*-oxide formation of the 2-, 3-, and 4-dimethylaminopyridines was studied by Frampton, Johnson, and Katritzky who found that each of the substances behaves differently toward the two reactions (65). The 3-isomer forms the methiodide **(III-70)** and the *N*-oxide **(III-71)**, while both reactions occur on the exocyclic nitrogen with the 2-isomer **(III-72)** and both on the ring nitrogen with the 4-isomer **(III-73)**. Selective quaternization of *S*-(-)-nicotine **(III-74)** was carried out by Jarboe and co-workers (66, 67). With alkyl halides they were able to quaternize selectively the pyridine nitrogen using acetic acid as solvent and subsequently quaternize the pyrrolidine nitrogen in methanol.

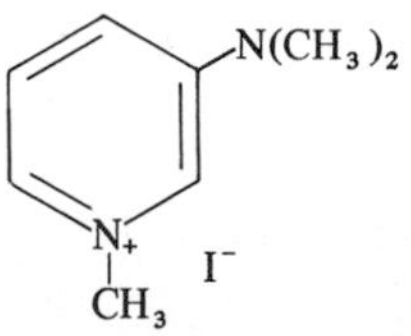

III-70

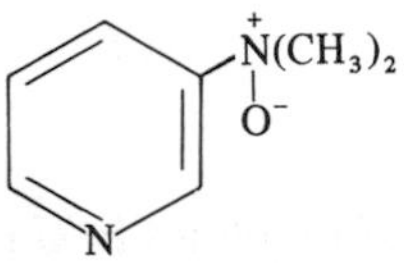

III-71

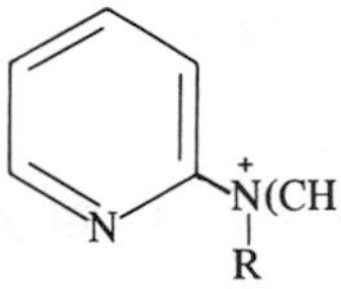

$R = O^-, CH_3$

III-72

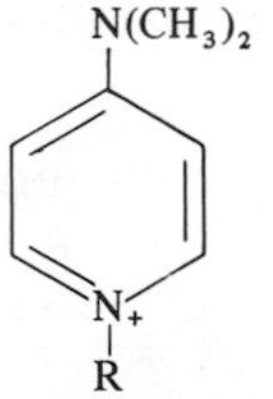

$R = O^-, CH_3$

III-73

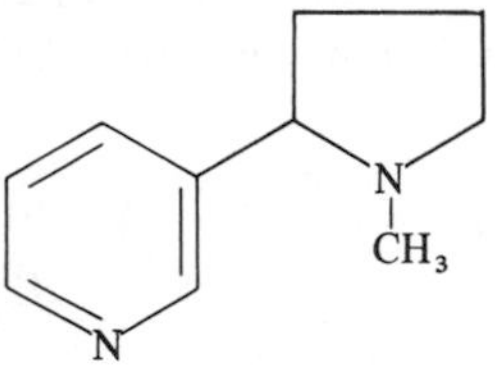

III-74

Alkylation of the thiolactam **III-75** first yields the thioalkylated product **(III-76)** and then the pyridinium salt **(III-77)** (68). 3-Hydroxy-2-bromopyridines also undergo *N*-alkylation. *O*-Alkylation becomes the predominant route (over pyridinium salt formation) when bulky alkylating agents such as iodoacetate are used.

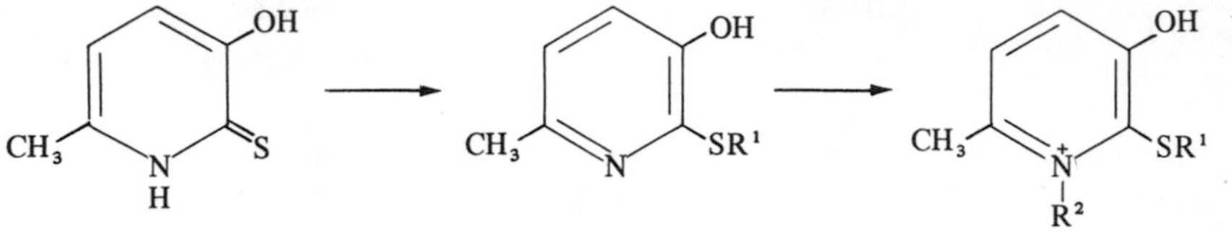

III-75 **III-76** **III-77**

The quaternization of quinazoline **(III-78)** with methyl iodide in the absence of solvent or in ethanol gives both the 3-methyl- **(III-79)** and the 1-methyl- **(III-80)** isomer in the ratio of 5 : 1, respectively (69). Previous work seemed to indicate that the 3-alkylquinazolinium salt was the only one formed. The selective quaternization of the 1-methyl-1*H*-imidazo[4,5-*b*]pyridine system also has been studied (69a).

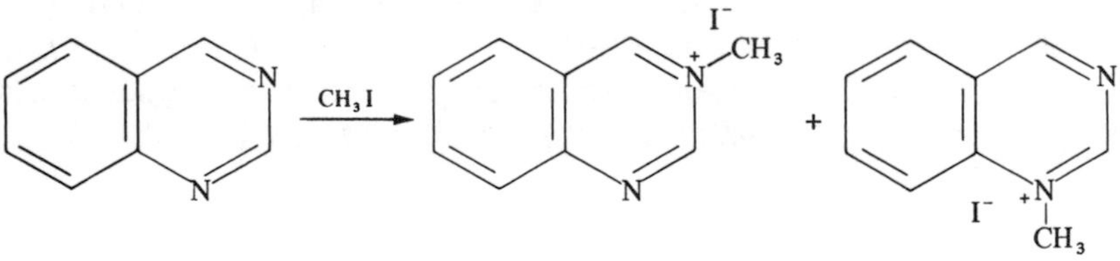

III-78 **III-79** **III-80**

2. By Addition of Pyridines to Unsaturated Systems

Some unusual preparations of pyridinium compounds by the addition of the pyridine system to alkenes have recently been reported. Dipyridine bromine(I) perchlorate **(III-81)** (70) was found to add to styrenes and to vinyl ethers to give the brominated pyridinium perchlorate salts **III-82** and **III-83** in good yield (71).

$[Br(C_5H_5N)_2]ClO_4$

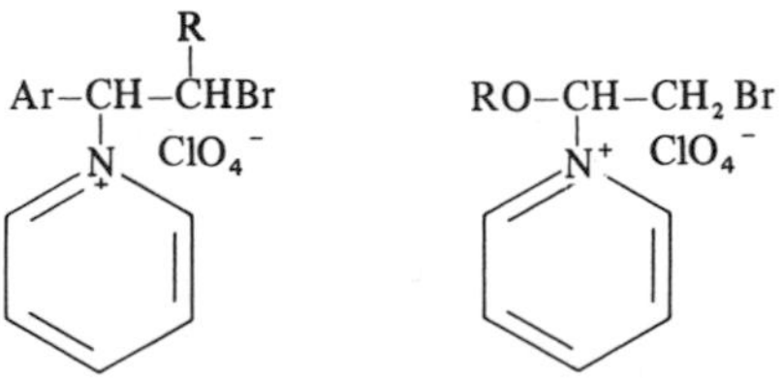

III-81 **III-82** **III-83**

The addition was found to be regiospecific and, where possible, stereospecific in all of the cases studied. Fonken and Mackellar reported the preparation of 1-dodecylpyridinium dodecyl sulfate (**III-84**) by treating 1-decanol with *N*-bromoacetamide and sulfur dioxide in pyridine (72). Since these are conditions which can cause the dehydration of alcohols (73), the reaction may proceed by way of 1-decene as the intermediate. The reported conversion of the chloroaldehyde **III-85** to the quaternary salt **III-86** probably involves the initial addition of pyridine to the double bond system, followed by the loss of chloride ion (74).

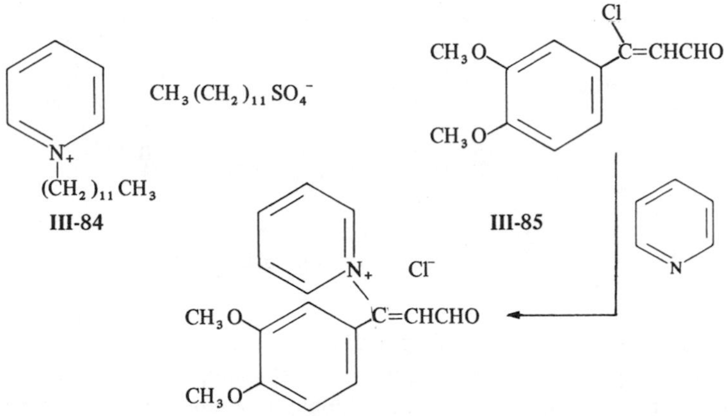

$CH_3(CH_2)_{11}SO_4^-$; $(CH_2)_{11}CH_3$

III-84

CH_3O, CH_3O, Cl, C=CHCHO

III-85

N

N_+ Cl^-, CH_3O, CH_3O, C=CHCHO

III-86

The addition of pyridine and α- and β-picoline to propargyl aldehyde or ethynyl ketones (**III-87**) to yield vinylpyridinium salts (**III-88**) has been reported to give the *trans* isomer (75). The salts are stable crystalline substances and may find potential use in replacing β-chlorovinyl ketones (76) in ring closure reactions.

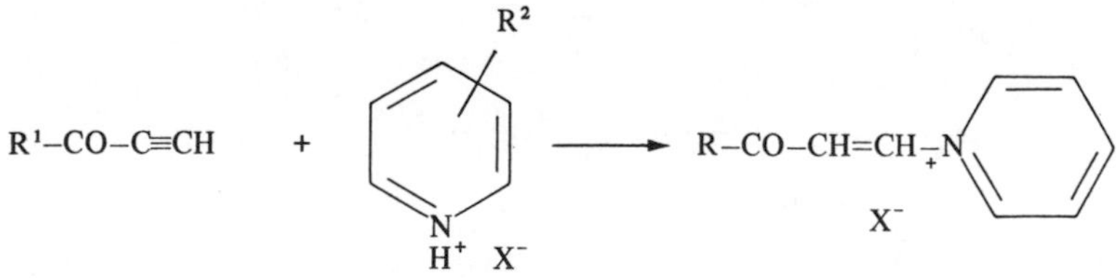

R^1–CO–C≡CH + (pyridine, R^2, N, H^+ X^-) ⟶ R–CO–CH=CH–$\overset{+}{N}$ X^-

R^1 = H, Alkyl, Aryl

III-87

R^2 = H, CH_3

III-88

Pyridine has been added to 3,4-dimethoxypropenylbenzene (**III-89**) electrochemically to give a nearly quantitative yield of the dipyridinium salt

III-90 on a preparative scale (77). The preparation of pyridinium salts *via* the addition of iodonium nitrate to alkenes has been mentioned previously (38) (see Section I.1.A.a).

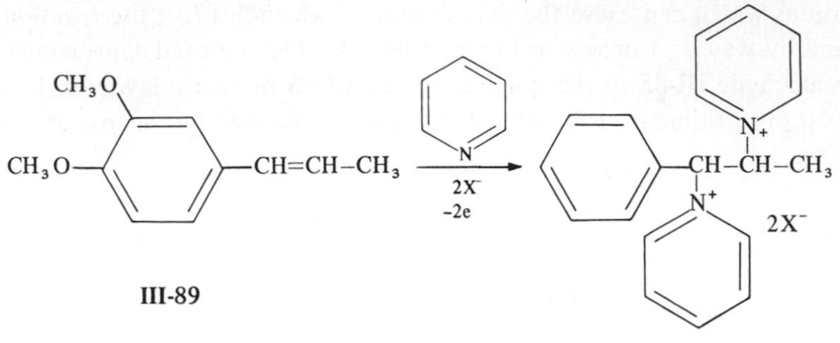

CH_3O CH_3O —CH=CH—CH_3 → (pyridine, $2X^-$, −2e) —CH—CH—CH_3 N^+ N^+ $2X^-$

III-89 **III-90**

3. From Pyrylium Salts

A series of preparations of pyridinium salts from the reaction of 2,4,6-trimethylpyrylium perchlorate with aliphatic and aromatic primary amines have been reported (78). In the case of aliphatic amines, *N*-alkyl-3,5-xylidines **(III-89a)** were formed as well as the expected pyridinium salts, and xylidine formation was favored by a bulkier R group and higher solvent polarity. A third product isolated in some cases was assigned the structure **III-90a**. Such compounds are probably intermediates in the conversion of pyrylium salts to pyridines (79).

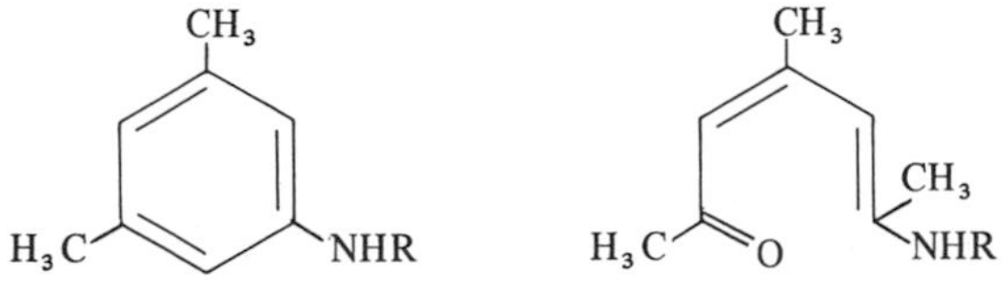

CH_3 H_3C NHR CH_3 H_3C O CH_3 NHR

III-89a **III-90a**

The reaction of 2,4,6-triphenylpyrylium perchlorate with methylamine yields two unexpected products in addition to 1-methyl-2,4,6-triphenylpyridinium perchlorate: these products are 1-methyl-2,4,6-triphenylpyridinium chloride (in carbon tetrachloride solvent) and 2,4,6-triphenylpyridine (80). The first of these substances is thought to arise by a displacement of perchlorate ion by chloride arising from the solvent, while the latter is postulated as being formed by an S_N2 attack of methylamine on 1-methyl-2,4,6-triphenylpyridinium perchlorate with consequent formation of dimethylamine.

Pyrylium salts have been condensed with amino acids, their esters, and peptides to yield quaternary pyridinium salts (81-84a). For example, 2,4,6-trimethylpyrylium perchlorate (**III-91**) reacts with amino acids to give the complex **III-92** which can be converted to the monomer **III-93** on further treatment with perchloric acid (82). It has been reported that the reaction of

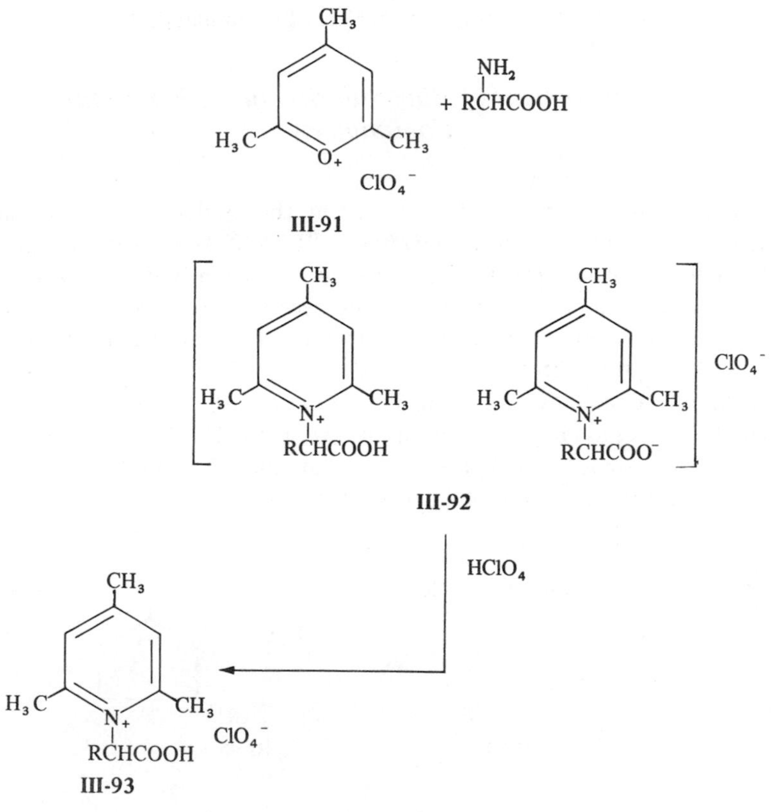

CH₃ / H₃C / O+ / CH₃ / ClO₄⁻ + NH₂ / RCHCOOH

III-91

[CH₃ / H₃C / N+ / CH₃ / RCHCOOH   CH₃ / H₃C / N+ / CH₃ / RCHCOO⁻] ClO₄⁻

III-92

$HClO_4$

CH₃ / H₃C / N+ / CH₃ / RCHCOOH   ClO₄⁻

III-93

pyrylium salts with di- and triaminopyrimidines occurs only at the most reactive amino group (position 5), with no reaction occurring with the relatively inactive amino groups at positions 2-, 4-, and 6- (84).

Trimethylpyrylium perchlorate (**III-91**) was used recently for the modification of amino groups of the enzymes chymotrypsin and acetoacetate carboxylase (85). The authors state that the method shows considerable promise as selective reagents for the chemical modification of protein amino groups.

The reactions of pyrylium salts with amines to form pyridinium compounds have been reviewed (86, 87), and their reactions with a series of substituted anilines to produce solvent sensitive dyes have been reported (88-91). The reactions of pyrylium salts with hydroxylamine and with substituted hydrazines have also been investigated (92).

4. By Transformation of Other Pyridinium Salts

A. *Displacement of Ring Substituents in Pyridinium Compounds*

Pyridinium salts of type **III-95**, prepared by the condensation of thiolactam **III-75** with α-bromoacrylic acid derivatives (**III-94**) (93), undergo electrophilic substitution at the unsubstituted position *ortho* to the hydroxyl group (94). Both the oxygen and sulfur substituents are necessary to act as electron donors for, without them, the reaction does not occur. The postulated mechanism is shown in the sequence **III-96**.

Pyridinium compounds having an *N*-tosyl group will undergo methanolysis, yielding the corresponding *N*-methyl derivatives (55). The 2-, 4-, and 6-halogen substituents of some *N*-methylpentahalopyridinium fluorosulfonates are extremely reactive toward nucleophilic reagents (94a).

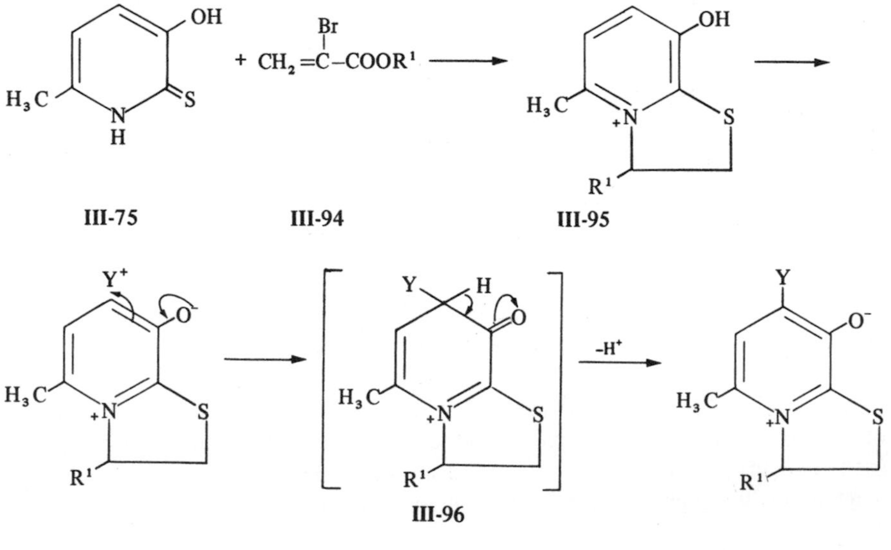

OH, H_3C, N, H, S + $CH_2{=}C(Br){-}COOR^1$ ⟶ OH, H_3C, +N, S, R^1 ⟶

III-75 III-94 III-95

Y^+, O⁻, H_3C, +N, S, R^1 ⟶ [Y, H, O, H_3C, +N, S, R^1] $\xrightarrow{-H^+}$ Y, O⁻, H_3C, +N, S, R^1

III-96

B. *Condensation Reactions Leading to Pyridinium Alcohols and Olefins*

Pyridine-2-aldehyde methiodide (**III-97**) has been condensed with 2-methylquinolinium iodoalkylates and with quaternary salts of type **III-98** to give the ethylene derivatives **III-99** (95), while quinolinium salts of structure **III-100** undergo reactions with diazoamino compounds converting the methyl group to a

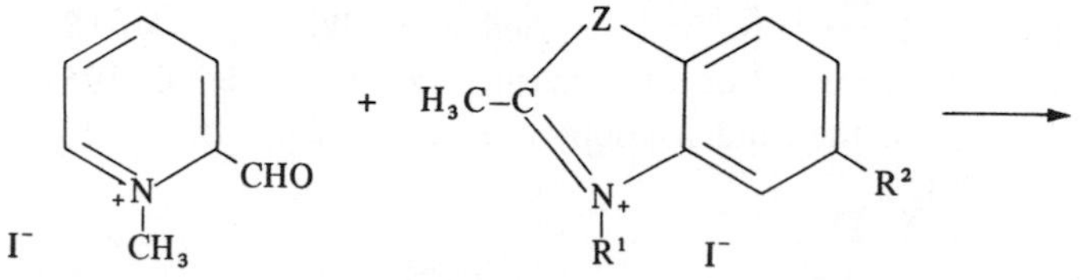

$Z = S, Se$
$R^1 = CH_3, C_2H_5$
$R^2 = H, OCH_3$

III-97 **III-98**

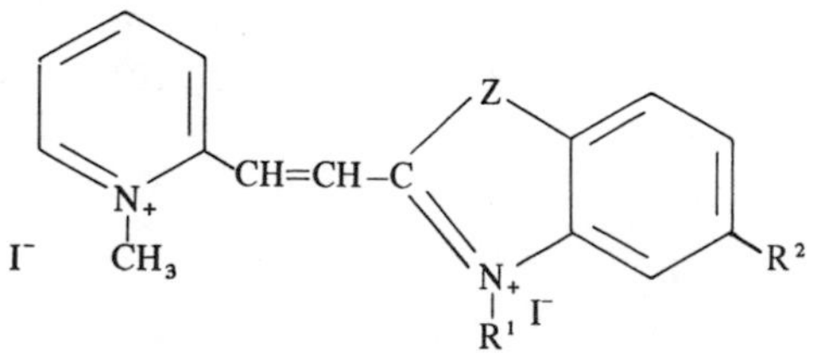

III-99

hydrazone (96). When **III-100** is treated with the salt **III-101**, blue-black methyleneazo dyes are produced (96). The condensation of phosphazoaldehyde **III-102** with a 2-methylpyridinium salt has also been reported (97).

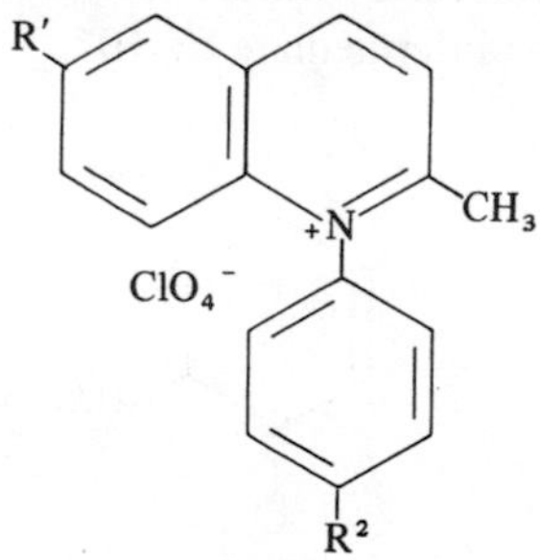

III-100 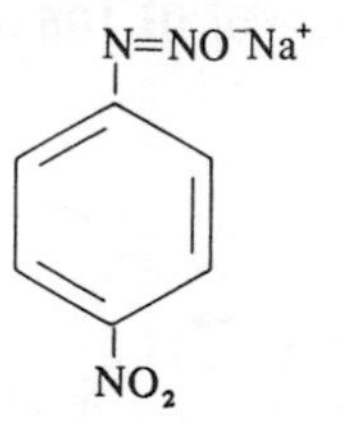 **III-101** 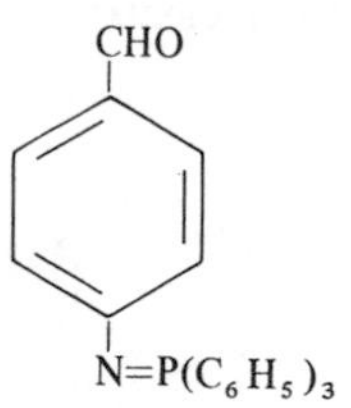 **III-102**

The condensation of 1-methylpyridinium iodide with benzaldehyde has been studied recently in some detail (98, 99). Some years ago, Kröhnke had reported that the piperidine-catalyzed condensation of 1-methylpyridinium bromide with benzaldehyde occurs at the 1-methyl group, *via* the intermediate ylid **III-103** (100). Yet it has been shown by deuterium exchange studies that ylid **III-104** forms considerably faster under the Kröhnke conditions (98). From NMR and kinetic studies Howe and Ratts conclude that ylid **III-103**, although formed more slowly, reacts faster than ylid **III-104**, perhaps at least partially for steric reasons. These authors also find that, under the proper conditions, ylid **III-104** will condense with benzaldehyde. When ylid **III-104** is formed by the thermal decarboxylation of homarine hydrochloride **(III-105)** in excess benzaldehyde, 83% of the condensation occurs at the 2-position and 17% at the 1-methyl group (99).

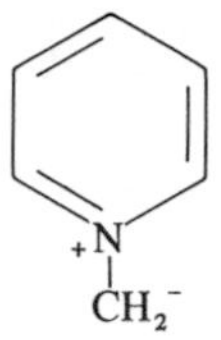

CH_2^- **III-103** CH_3 **III-104** COOH Cl^- CH_3 **III-105**

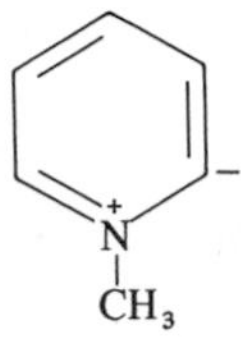

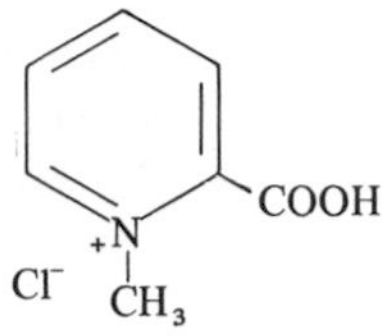

5. By Cyclization Reactions

The mechanism of the ring closure of azatrienes **III-106** to give pyridinium ions, a reaction discovered many years ago by Zincke (101), has now been investigated further (102–104). From a series of careful kinetic studies Marvell and co-workers concluded that the data are in accord with the rate-determining step being an electrocyclization of **III-106** to the intermediate **III-107**.

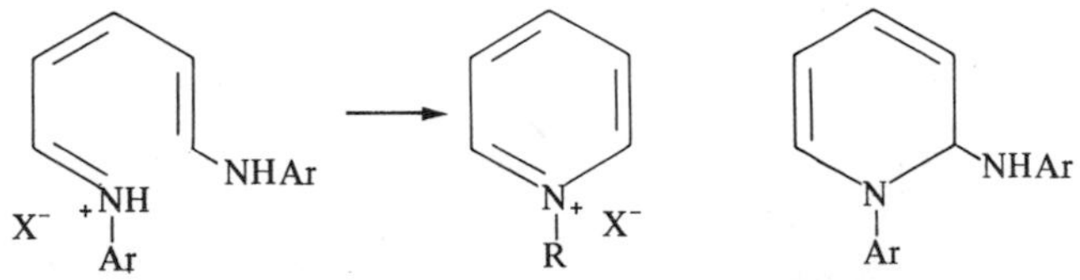

NHAr X^- $^+$NH Ar → N$^+$ X^- R NHAr N Ar

III-106 **III-107**

Pyridinium compounds have been prepared using sugars as starting materials. Thus 1,2-*O*-cyclohexylidine-α-D-xylose was converted to a tosylate and then to **III-108** by heating the tosylate with *p*-toluidine (105). When this substance was heated with methanolic hydrochloric acid, it rearranged quantitatively to the pyridinium salt **III-109**. This preparation is patterned after several earlier reports by Paulsen and co-workers who rearranged such 5-alkylamino sugars to 3-hydroxy-, 2-cyano-5-hydroxy-, and 2-hydroxymethyl-5-hydroxypyridinium salts (106).

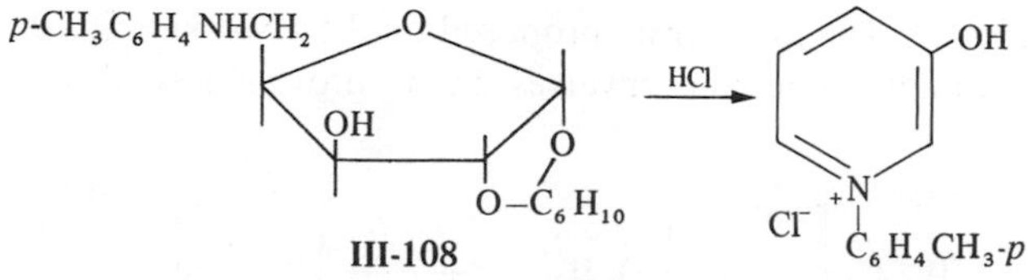

p-$CH_3C_6H_4NHCH_2$ … OH … O—C_6H_{10} $\xrightarrow{HCl}$ OH … Cl^- ^+N … $C_6H_4CH_3$-p

III-108 **III-109**

The preparation of quaternary pyridinium salts by ring closure reactions starting with furans has seen renewed interest. The method for preparing 3-pyridinols by the oxidation of 2-(α-aminoalkyl)-furans with chlorine water (107, 108) has been extended to include *N*-monoalkyl-substituted 2-(α-aminoalkyl)furans (**III-110**), whereby quaternary 3-hydroxypyridinium chlorides (**III-111**) are formed (109).

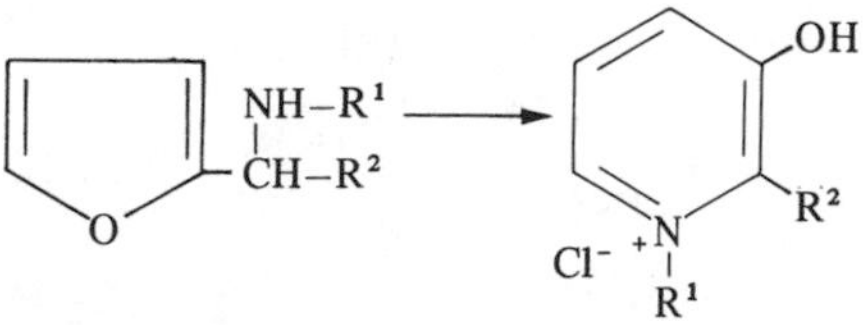

NH—R^1, CH—R^2, O $\longrightarrow$ OH, R^2, Cl^- ^+N, R^1

$R^1 = CH_3, C_2H_5,$
$CH_2C_6H_5, C_6H_5$
R^2 = H, CN, $CONH_2$

III-110 **III-111**

Optically active 3-hydroxy-6-methylpyridinium salts of type **III-113** where R contains an asymmetric center were prepared by the acid-catalyzed ring cleavage and reclosure of the 5-hydroxymethylfurfural derivatives (**III-112**) (110, 111).

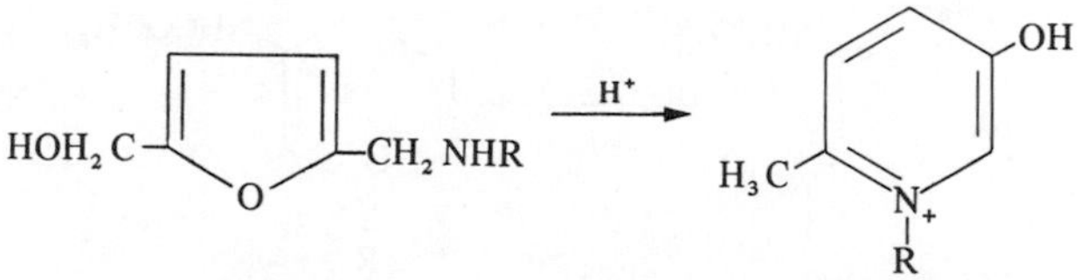

HOH_2C—(O)—CH_2NHR $\xrightarrow{H^+}$ OH, H_3C, N^+, R

III-112 **III-113**

The authors state that the derivatives were prepared to study the stereochemistry of the Menschutkin reaction (Section I.1.A), and the mechanism proposed for the reaction involves initial protonation of the hydroxymethyl group, followed by loss of water and hydrolytic cleavage of the furan ring. Undheim and Gacek (112) report a similar reaction using 2,5-dimethoxy-2,5-dihydrofurans (**III-114**). These compounds likewise rearrange in acid solution to give 3-hydroxypyridinium derivatives (**III-115**) where an asymmetric center in the R group is preserved (111, 112). The dimethoxy derivatives **III-114** are prepared by the electrolytic oxidation of the corresponding furan derivatives in methanol solution. The mechanism proposed by these workers for the formation of the pyridinium compound involves inital protonation of the furan ring

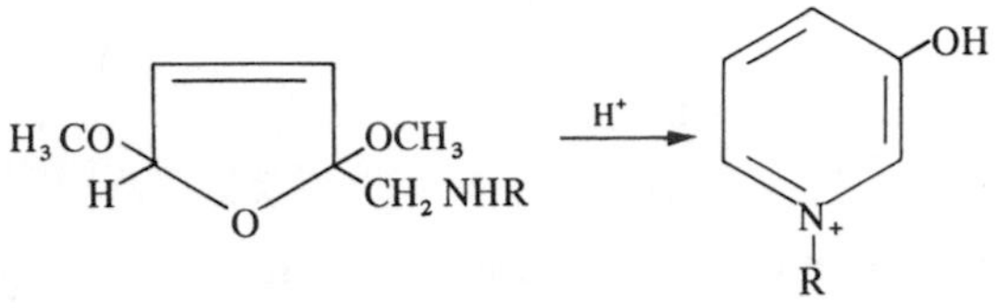

H_3CO, H, O, OCH_3, CH_2NHR —H^+→ OH, N^+, R

III-114 **III-115**

oxygen atom (**III-116**). The nitrogen atom then attacks the electrophilic C-5 position, resulting in the breaking of the carbon–oxygen bond and the formation of **III-117**. Loss of two molecules of methanol then leads to **III-118**.

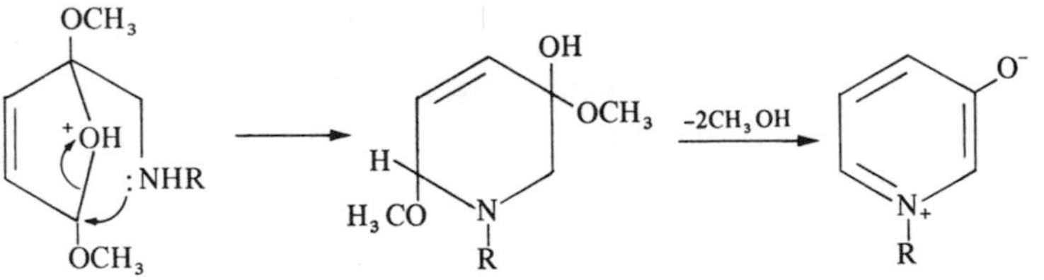

OCH_3, $\overset{+}{O}H$, :NHR, OCH_3 → OH, OCH_3, H, H_3CO, N, R —$-2CH_3OH$→ O^-, N^+, R

III-116 **III-117** **III-118**

A method (113) of preparing pyridinium salts by the double cleavage of the pyridine ring (114) has been further investigated (115). The reaction of primary amines and cyanogen bromide with 2-acetylaminopyridine leads to the pentamethinium salt **III-119**, which recyclizes at elevated temperature to the pyridinium salt **III-120**.

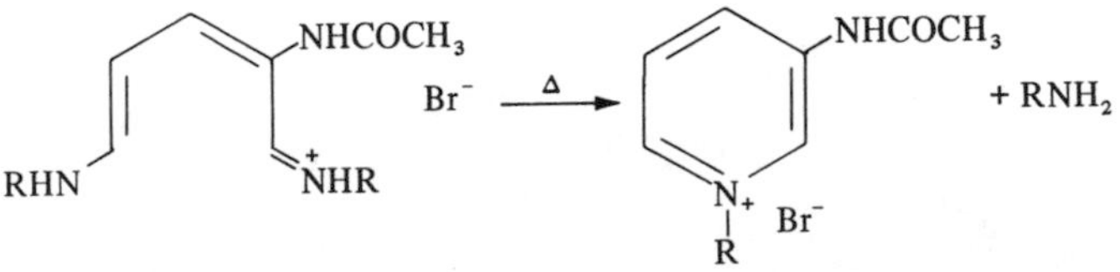

$NHCOCH_3$, RHN, $\overset{+}{N}HR$, Br^- —Δ→ $NHCOCH_3$, N^+, R, Br^- + RNH_2

III-119 **III-120**

Naphthoquinones (**III-121**) react with phosphoranes (**III-122**) to give dibenzoacridinium betaines, which yield the acridinium salts **III-123** on treatment with strong acid (116). A general synthesis of isothiazolo[2,3-*a*]pyridinium salts by the mild bromine oxidation of 2-pyridylthioacetamides, 2-pyridylmethyl thioketones, or 2-benzylthiovinylpyridines has been reported (116a).

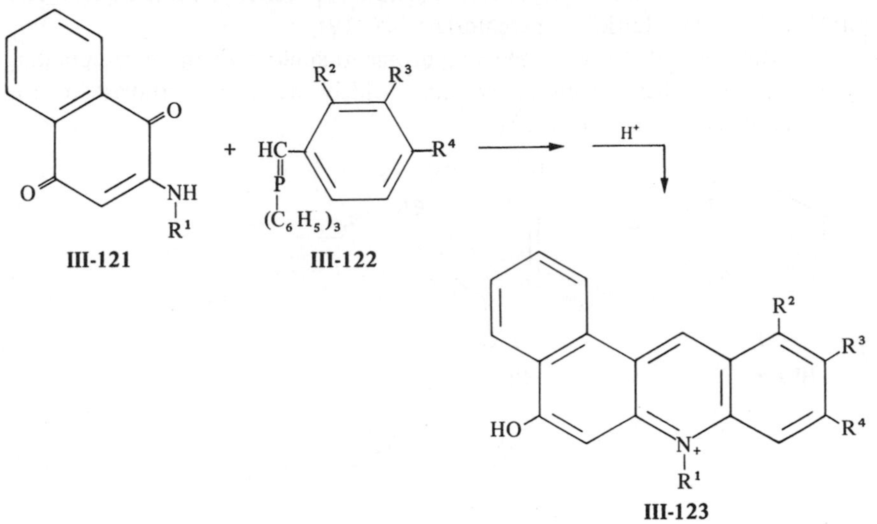

O O NH R^1 + HC P $(C_6H_5)_3$ R^2 R^3 R^4 H^+

III-121 **III-122**

HO N+ R^1 R^2 R^3 R^4

III-123

6. By Oxidation of 1-Substituted Dihydropyridines

The facile nature of the oxidation of 1-substituted reduced pyridines to pyridinium salts is exemplified by the ready air oxidation of compounds of type **III-124** to the corresponding quaternary salts **III-125** in boiling ethanol (117),

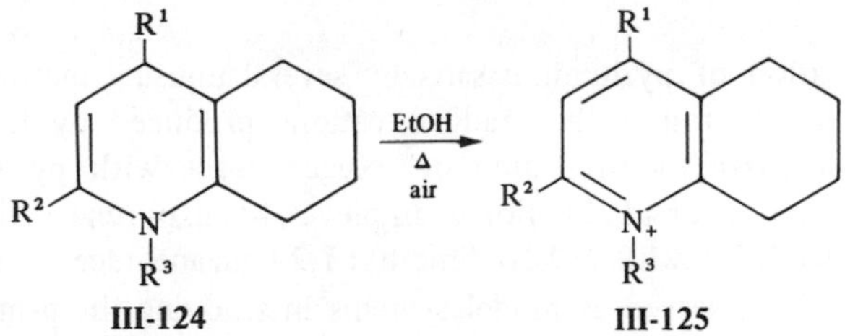

R^1 R^2 N R^3 EtOH Δ air R^1 R^2 N+ R^3

III-124 **III-125**

$R^1 = CH_3, C_6H_5, (CH_3)_3C$
R^2 = H, ring residue
$R^3 = CH_3$, ring residue

whereas chromic acid was used to oxidize 3,5-diacetyl-1,4-dihydro-1,4-diphenylpyridine to the corresponding pyridinium salt (118).

The tricyanodihydropyridine **III-126**, formed by the addition of KCN to 3,5-dicyano-*N*-methylpyridinium *p*-toluenesulfonate, rearranges to the 4-isomer (**III-127**) when heated. The latter substance yields the hydroxy derivative **III-128** when oxidized with dinitrogen tetroxide, and this product yields the quaternary salt **III-129** on the addition of perchloric acid (119).

The oxidations of 1-substituted dihydronicotinamides to the corresponding pyridinium salts have been investigated widely as model studies for the nicotinamide coenzymes (120, 121).

NC, CN, CN, H, N, CH_3 — **III-126** $\xrightarrow{\Delta}$ H, CN, NC, CN, N, CH_3 — **III-127** $\xrightarrow{N_2O_4}$

CN, NC, CN, OH, H, N, CH_3 — **III-128** $\xrightarrow{HClO_4}$ CN, NC, CN, ^+N, ClO_4^-, CH_3 — **III-129**

7. By Other Methods

The preparation of pyridinium salts by several unusual methods have been reported. Rochlitz found that radical cations produced by the reaction of aromatic carcinogens with iodine or oxygen react with pyridine to form pyridinium compounds (122). For example, 3,4-benzpyrene yields the pyridinium salt **III-130**, while 9,10-dimethyl-1,2-benzanthracene gives **III-131**. These studies have served as model systems in studying the primary processes involved in chemically induced carcinogenesis.

Another unusual preparation of a quaternary pyridinium salt involved treating ferrocenylphenylcarbinol with sodium tetraphenylborate, whereby ferrocenylphenylcarbonium tetraphenylborate (**III-132**) was obtained. When this substance

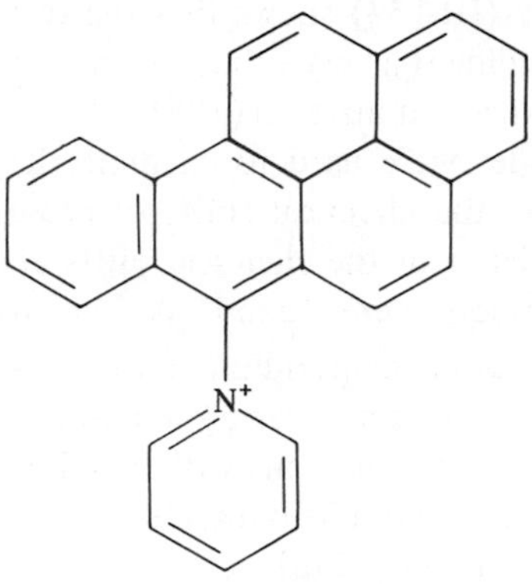

III-130

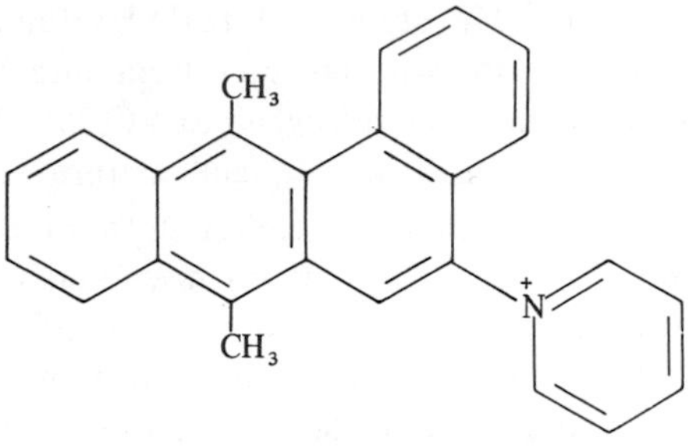

III-131

was treated with pyridine, the pyridinium tetraphenylborate **III-133** was obtained in 88% yield (123).

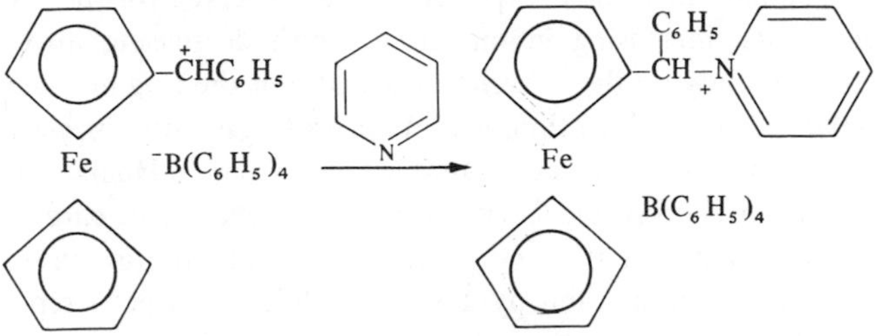

III-132

III-133

Acetophenone reacts with thionyl chloride in pyridine to yield an *N*-α-styrylpyridinium salt. A mechanism involving initial salt formation between thionyl chloride and pyridine, followed by attack by this species at the carbonyl oxygen atom has been proposed (123a).

II. Properties and Reactions

1. General Properties

Pyridinium salts have been used to measure N(sp^3)-C(sp^2) and N(sp^2)-C(sp^2) bond distances by x-ray crystallographic methods to determine the importance of resonance contributions in the amide bond and other such structures (124).

An x-ray study of pyridinium dicyanomethylide (**III-134**) shows that the two cyano groups are inclined to the plane of the pyridine ring, even though a planar structure would be more highly favored by resonance stabilization (125).

In an NMR study, 1,4-diethylpyridinium halides were used to examine the extent of ion-ion and ion-solvent interactions on the chemical shifts of those protons near the nitrogen atom (126). A dependence of the chemical shifts on concentration, solvent, and nature of the anion were found. A similar dependence of the chemical shifts of the ring protons of pyridinium salts was reported recently by Russian workers (127), who found a linear correlation of the chemical shift values of the β- and γ-pyridinium ring protons with σ and σ° values of substituents at the 1-position (N atom). The tetraphenylborate ion has been studied as an NMR shift reagent for pyridinium ions (127a).

The effects of *meta* and *para* substituents on *N*-methyl proton and ^{13}C chemical shifts, as well as the ^{14}N pyridinium nitrogen chemical shift, were reported by Simon and co-workers, who found excellent shift correlations with Hammett substituent constants when σ^{+} values were used for donor substituents (128). A comparable substituent dependence was observed for the ^{13}C-H methyl coupling constants, increasing in magnitude with decreasing nitrogen charge density. A similar study on the ^{14}N chemical shifts of the ring nitrogen atom in a series of *ortho* substituted 1-ethylpyridinium salts gave no correlation of shift with total electron densities as calculated by the extended Hückel theory (129), whereas such a correlation had been found in the *meta* and *para* series (128). The coupling constants for the ring protons in *N*-trideuteriomethylpyridinium, pyridazinium, pyrimidinium, and pyrazinium iodides have been reported (130).

N-Alkylpyridinium salts undergo base-catalyzed hydrogen exchange of the α-protons by a simple deprotonation-protonation process *via* the intermediate ylid **III-135**, and the ability of substituents R to influence the rates of pyridinium ylid formation under buffer conditions has been studied (131). The rates were found to correlate well with the Taft σ_{I} inductive parameter. Such base-catalyzed hydrogen-deuterium exchange in *C*-alkyl side chains of quaternized and unquarternized nitrogen containing heterocycles has been reported widely (132).

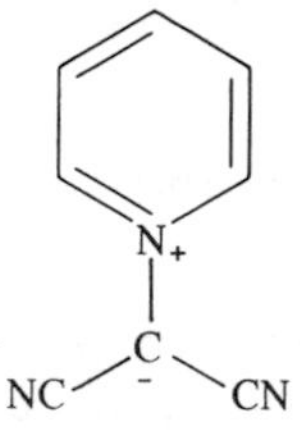

III-134 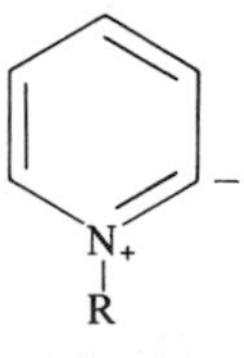 **III-135**

Katcka and Urbanski reported an infrared study of a series of quaternary pyridinium salts wherein they discuss spectral changes in the aromatic C-H, C=C, and C=N vibration regions in terms of structural changes in the pyridine ring brought about by the quaternization process (133). Infrared band assignments have also been recently reported for 2- and 4-pyridone hexachloroantimonate and hexachlorostannate quaternary derivatives (134).

A number of charge transfer studies have involved quaternary pyridinium ions (135). The unusual absorption spectra of alkylpyridinium iodides was first observed by Hantzsch (136) and is now well recognized as a transition involving charge-transfer from the iodide ion to the pyridinium ion (137, 138). In 1960, Kosower and Skorcz reported an extensive study of the charge-transfer bands of 20 alkylpyridinium iodides which established the effects of alkyl substitution on the transition energies of these substances (139).

In a study of coenzyme models, a series of compounds having structure **III-136** was investigated, where R is aryl, imidazolyl, thiomethyl, or a quaternary ammonium group. In these substances, the observed ultraviolet absorption transition intensities are associated with the electron affinity of the pyridinium ion, and it was possible to relate the wave length of the intramolecular charge transfer transition to the electron donor ability of the R group (140).

In a related series, de Boer and co-workers studied compounds of type **III-137** where n = 1, 2, 3, 4, and found that the charge transfer band intensity depends strongly on the methylene chain length (141). To aid in interpreting the results, these workers studied the acceptor properties of the *N*-methylcyanopyridinium ions and found the 4-cyano compounds to be the strongest acceptor.

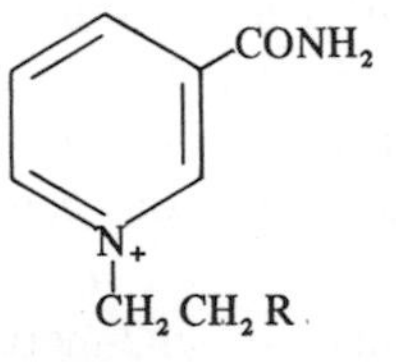

CONH$_2$ N$^+$ CH$_2$CH$_2$R

III-136

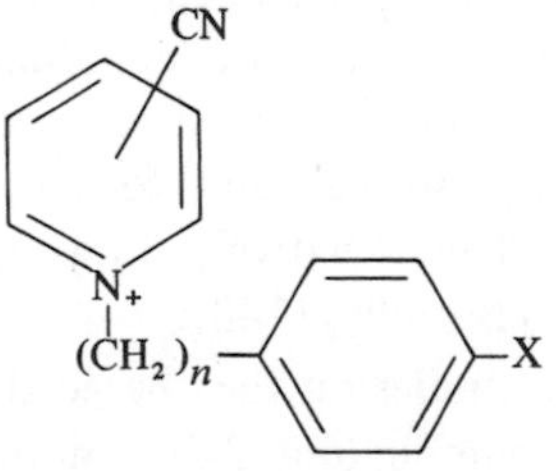

CN N$^+$ (CH$_2$)$_n$ X

X = H, OCH$_3$

III-137

Itoh has studied the intramolecular association of the pyridinyl and pyridinium termini of the cation-radicals **III-138** using e.p.r. and light absorption spectroscopy. The observations indicated that those cation-radicals where n = 3,4 undergo intramolecular association even at room temperture (142). Solvent polarity effects on the spectra were noted, and the investigations were subsequently extended to the diradicals (142a). An intramolecular charge transfer band associated with the betaine **III-139** has also been studied (143).

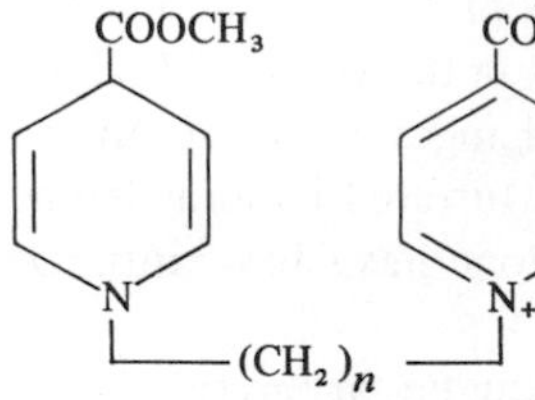

$n = 3, 4, 5$

III-138

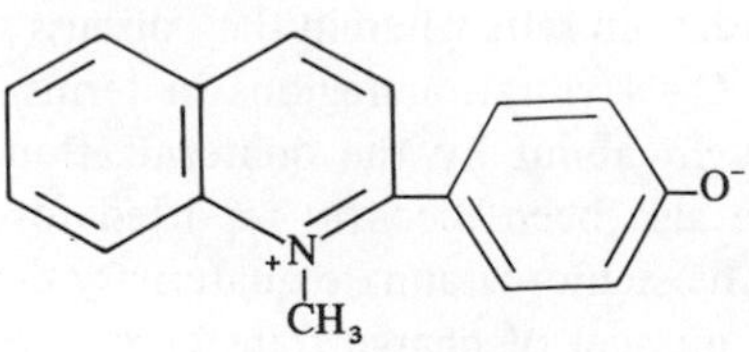

III-139

Farcasin and Farcasin found that triphenylpyrylium perchlorate in pyridine solution gives rise to an ultraviolet absorption band which may be due to compounds containing a quaternary pyridinium linkage, but which the authors feel is more probably due to the formation of a charge transfer complex (144). It was suggested further that such complexes may play an intermediary role in the reaction of pyrylium salts with nucleophiles, such as in the conversion of these substances to quaternary pyridinium salts (Section I.3).

Pyridinium cyclopentadienylides and benzimidazolide (**III-60** and **III-59**; Section I.1.B) show dramatic color variations when dissolved in different solvents, a phenomenon which has been interpreted in the case of the cyclopentadienylides as due to an intramolecular charge-transfer transition (51, 52, 145).

Pyridinium iodide charge transfer complexes were used by Mackay and Poziomek in a study of reactions and solvation properties of nonspherical cation spherical anion ion pairs (146), and a flash photolytic investigation of the 1-methylpyridinium, 1-methylcollidinium, and 1-methylquinolinium iodide charge transfer bands has been reported (147). Intramolecular charge transfer transitions can also be used for rotatory dispersion or circular dichroism measurements. Their relatively long wavelength absorptions can be of distinct advantage in such studies (148).

Recent x-ray studies on the charge transfer salts 1-ethyl-2-methylquinolinium iodide (149), quinolinium 2-dicyanomethylene-1,1,3,3-tetracyanopropanediide (150), and *N*-methylacridinium iodide (151) have been reported. In the acridinium derivatives the iodide ion resides above the C–N bond, a fact which may support Kosower's suggestion that in the 1-alkylpyridinium system, the iodide anion is situated above the plane of the ring (152).

The effect of pressure on the conductance of the pyridinium salt **III-140** in acetone and isobutyl alcohol was determined in connection with a study on the effect of pressure on charge transfer complexes (153). A decrease in conductance with increasing pressure was observed, being more pronounced in isobutyl alcohol than in acetone. Conductivity measurements of polycrystalline *N*-butylpyridinium and *N*-butylquinolinium iodides have also been carried out, and the results compared with the absorption spectra of these compounds (154), while a magnetoelectric effect occurring in molten ethylpyridinium bromide has been observed (155).

Pyridinium salts have been examined polarographically (156), and the technique, together with solubility and infrared spectral evidence, was used to establish that 1-acylpyridinium chlorides are indeed true pyridinium salts with 1-acyl substituents (157). The polarographic reduction of 1,3-bis(*N*-pyridinium-4-aldoxime)propane (TMB-4) (**III-141**), a reactivator for inhibited cholinesterase, occurs in three stages, one of which is reversible, and involves up to 13 electrons (158).

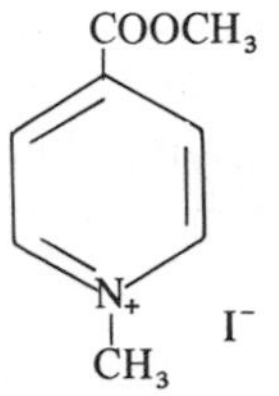

$COOCH_3$ I^- CH_3

III-140

$HON{=}HC{-}$ $N^+{-}(CH_2)_3{-}{}^+N$ $-CH{=}NOH$

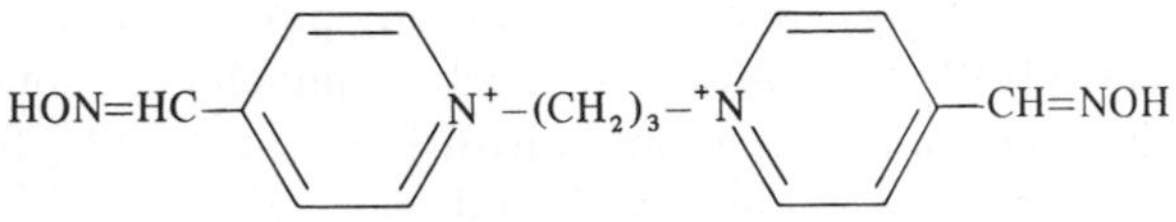

III-141

The dipyridylium dihalides **III-142** and **III-143** are herbicides known commercially as "Diquat" and "Paraquat," respectively. Both produce highly stable free radicals on one-electron chemical (159) or electrochemical (in the case of "Diquat") (160) reduction. The photochemical breakdown of the dipyridylium herbicides has been reviewed (161).

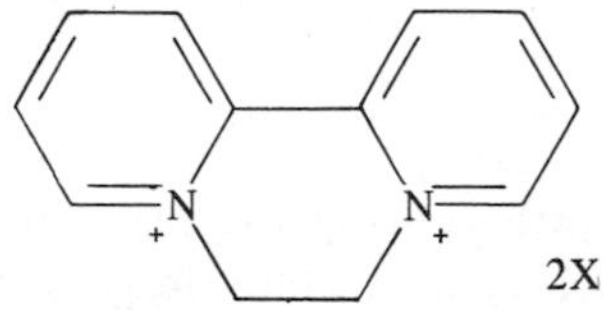

$2X^-$

III-142

$H_3C{-}N^+$ $N^+{-}CH_3$ $2X^-$

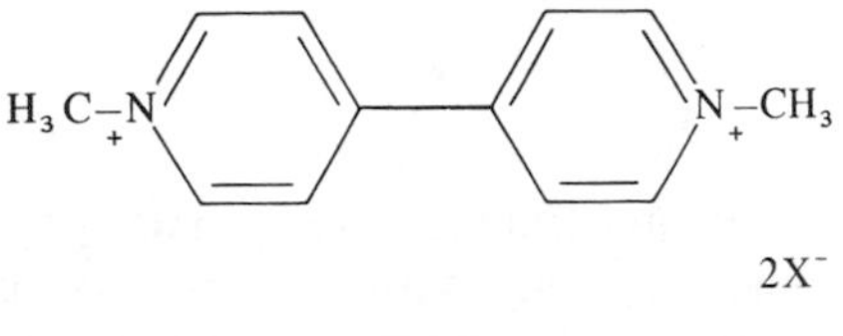

III-143

The use of the quaternary tetrafluoroborate **III-144** as a catalyst in the pyridine-pyridone equilibration of **III-145** and **III-146** has been reported (162, 163). In the methoxypyridine system, the use of a suitably deuterated pyridinium salt enabled the investigators to elucidate a mechanism of methyl transfer involving both oxygen-to-oxygen and oxygen-to-nitrogen transfer but no nitrogen-to-oxygen transfer.

Pyridinium salts having long alkyl chains may act as surfactants, a common example being the bactericide cetylpyridinium chloride. Anacker and Ghose studied the light-scattering properties of cetylpyridinium salt solutions and concluded from their observations that micelle structure depends on the nature of the anion, the aggregating power of an anion roughly paralleling its ability to disrupt the structure of water (164). Cetylpyridinium chloride has been used in microcalorimetric studies on model systems for biological reactions (165), and Kundu investigated its adsorption at the heptane-water interface (166). A review of reaction kinetics in the presence of micelle-forming surfactants has appeared recently (166a).

The pyridinium trifluoromethanesulfonate **III-147** shows antiseptic properties, having about the same bacteriostatic activity as cetylpyridinium chloride but somewhat less fungistatic activity. Solutions of **III-147** show much lower surface tension values than do equivalent concentrations of cetylpyridinium chloride (167). Other surfactant studies have examined the interaction of dodecylpyridinium chloride with congo red (168), have involved the preparation of a series of *N*-(ω-aryloxyalkyl)pyridinium and piperidinium salts for solubilization studies (169), have investigated alkyl quinolinium ions as haemolytic agents (170), and have determined partition coefficients for a large number of pyridinium and quinolinium alkylsulfates (171).

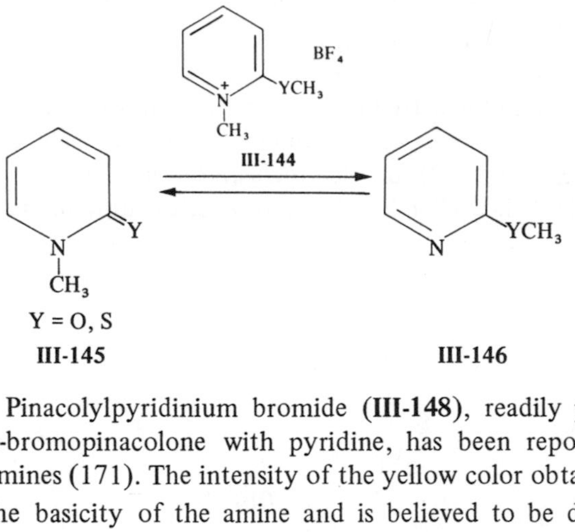

BF_4 YCH_3 CH_3

III-144

Y CH_3 $Y = O, S$ **III-145**

YCH_3 **III-146**

$CF_3SO_3^-$ $CH_2(CF_2)_6CF_3$ **III-147**

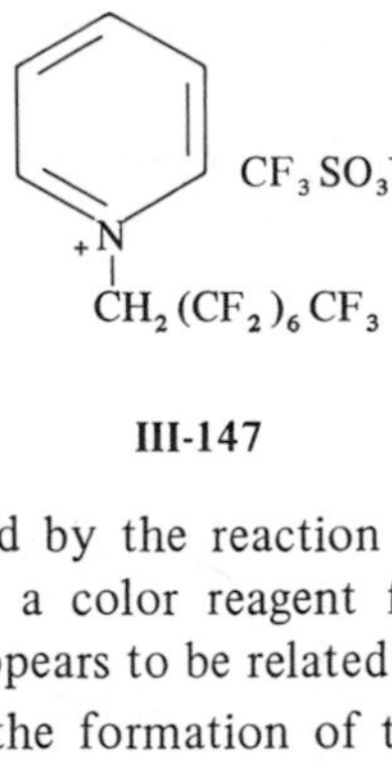

Pinacolylpyridinium bromide (**III-148**), readily prepared by the reaction of α-bromopinacolone with pyridine, has been reported as a color reagent for amines (171). The intensity of the yellow color obtained appears to be related to the basicity of the amine and is believed to be due to the formation of the resonance stabilized ylide **III-149**. Similarly, 4-(4-nitrobenzyl)pyridinium salts have been reported as intermediates in a qualitative and quantitative colorimetric analysis method for diazoalkanes (173) and β-chlorovinyl ketones (174). The salts give colored benzylidene dihydropyridine compounds when treated with

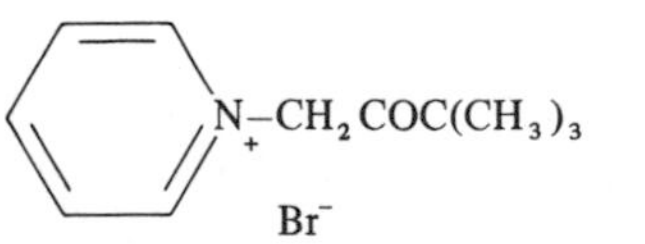

$\overset{+}{N}-CH_2COC(CH_3)_3$ Br^- **III-148**

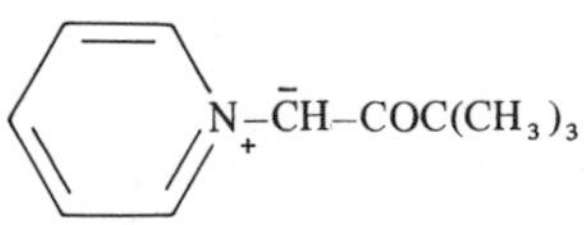

$\overset{+}{N}-\bar{C}H-COC(CH_3)_3$ **III-149**

base (cf. Section II.3.A). Cyanine dyes containing the quinolinium ring system have been prepared (174a).

The use of pyridinium salts as standards to establish a solvent polarity scale has been reported (91). The salts are derivatives of the 1-*p*-hydroxyphenyl-2,4,6-triphenylpyridinium system, and in the presence of base they are converted to betaines which give solvent dependent bands in the 500 to 900 nm region.

An additional study of the thermal rearrangements of 1-alkylpyridinium halides (Ladenburg rearrangement) has been reported (175). The authors confirmed the fact that the rearrangement of the *N*-alkyl group takes place predominantly to the α- and γ-positions, and they propose a mechanism that involves the initial homolytic cleavage of the nitrogen-alkyl bond, followed by recombination of the alkyl radical at the appropriate ring position. An HMO study of this rearrangement supports this mechanism (176).

The distinguishing properties of pyridinium salts have prompted their use in certain kinetic studies. For example, the *N*-methyl-4-oxopyridinium iodide group **(III-150)** was used as a leaving moiety in a solvolytic investigation of the 2-bicyclo[3.1.0]hexyl cation **III-151** to avoid complications due to internal return (177), and *N*-alkylpyridinium cations have been investigated extensively as model compounds in assessing the chemical properties of pyridine coenzymes such as nicotinamide adenine dinucleotide (NAD). For example, in such a study it was shown that complex formation between 3-carbomethoxy-1-methylpyridinium cation **(III-152)** and 8-chlorotheophyllinate anion **(III-153)** decreased the reactivity of the ester toward specific base-catalyzed hydrolysis (178).

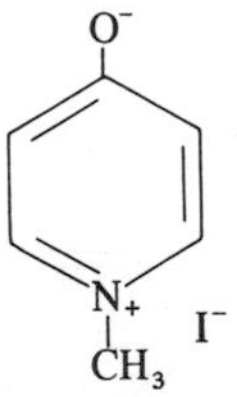

III-150 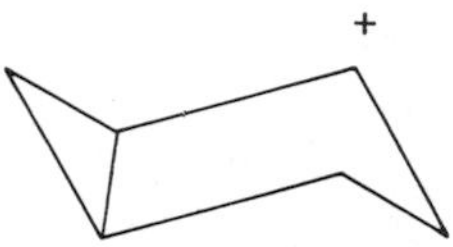 **III-151**

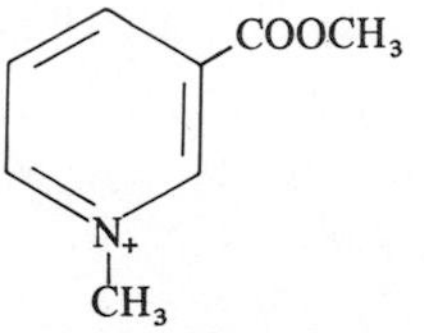

III-152 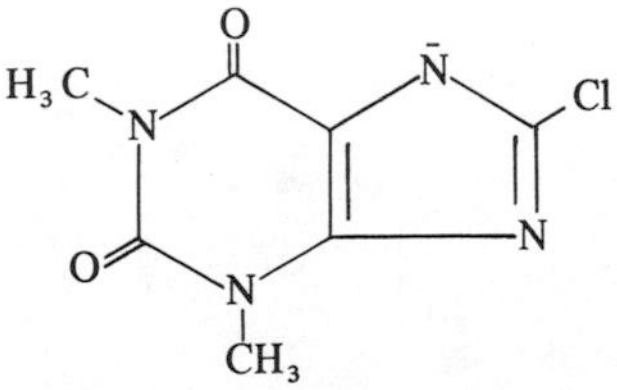 **III-153**

The kinetics of the Zincke ring closure reaction to yield *N*-arylpyridinium salts (Section I.5) in basic media have been investigated (103). The dipole moment of *N*-methyl-3-pyridone has been found to be 7.2 D (178a), and *N*-arylpyridinium-3-oxides have been shown to vaporize as mesoions in mass spectral studies (178b).

2. Pyridone Formation – Anion Attack on the Pyridinium Ring

In 1956, Berson and Cohen reported a new synthesis of 2-pyridones which consisted of treating 2-methylpyridinium compounds **III-154** with iodine and pyridine to give the dipyridinium salts **III-155**; these underwent hydrolysis smoothly in base to give the pyridones **III-156** (179). A similar method was recently reported by Matsumura and coworkers who treated 2-chloromethylpyridine derivatives (prepared from the 1-oxides and tosyl chloride) with pyridine and then dimethyl sulfate which yielded the dipyridinium salts **III-155** (R^2 = CH_3). Aqueous alkali treatment at low temperature gave the corresponding 2-pyridones (180). The structure of the Diels-Alder adduct of maleic anhydride with 1-methyl-2(1*H*)pyridone (**III-156**; R^1 = H, R^2 = CH_3) itself has been confirmed, and the method provides an interesting route to the isoquinuclidine ring-system (181).

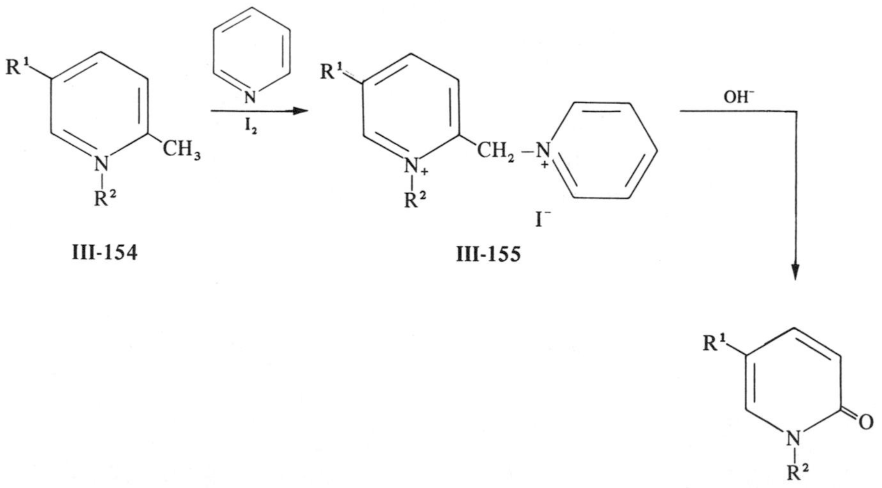

III-154 **III-155** **III-156**

In aqueous base, *N*-alkylpyridinium salts can exist partly as the pseudobase (*cf.*, **III-157** or **III-158**), which can be oxidized by selected oxidizing agents to the corresponding pyridones; in these oxidations the 4-pyridone, expected from the pseudobase **III-159**, has been observed only when a 3-cyano substituent was present and then in very small amount (182, 183). Of those oxidizing agents studied, alkaline potassium ferricyanide, first reported by Decker (184), still appears to be the one of choice (185), and 3-substituted pyridinium salts usually yield a mixture of the 2- and 6-pyridones (186).

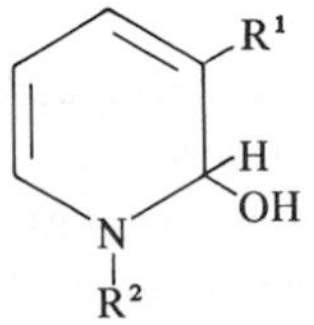

III-157 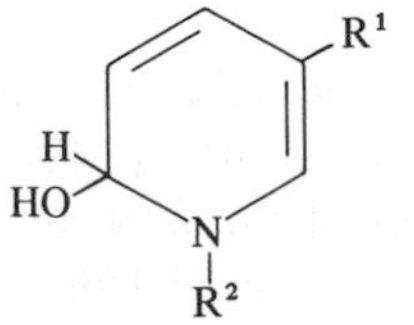 **III-158** 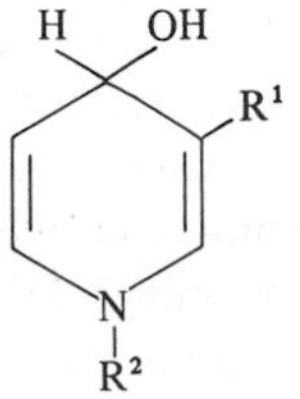 **III-159**

To gain further insight into the mechanism of this oxidation, Abramovitch and Vinutha carried out a quantitative study of the oxidation of several 3-substituted methylpyridinium iodides, including attempts to establish the rate-determining step using the deuterium isotope effect (183). On the basis of their results they concluded that both steric and electronic effects exert an influence on the position of oxidation; for example, 3-methyl and 3-cyano groups were found to activate the ring and direct an oxidative attack at C-2, while a 3-methoxycarbonyl group deactivates and directs exclusively to C-6. It was suggested that the rate determining step is the formation of a pseudobase-ferricyanide complex.

Möhrle and Weber reported similar results in a study on the position of oxidative attack on 3-substituted methylpyridinium salts (186). They further found that a 3-nitro substituent yields only the 6-pyridone, but under their conditions the 3-cyano substituent directed oxidative attack approximately equally at C-2 and C-6. The ferricyanide oxidation of 1-methyl-3-hydroxypyridinium compounds has been reported not to yield pyridones (187).

Nitro groups in the C-3 and C-5 positions of pyridinium salts activate the nucleophilic displacement of a C-2 halogen substituent, and these substances can in fact add a second mole of reagent to give dihydropyridine derivatives. Thus the 2-chloro-5-nitropyridinium salt **III-160** reacts with excess methoxide ion to give the substitution product **III-161** initially, which can then react further to give the addition product **III-162**. The dihydropyridine derivatives lose readily a molecule of dimethyl ether and are converted to the corresponding 2-pyridones; the 3-nitropyridinium salts behave similarly (188).

NO_2, Cl, $\overset{+}{N}$, R BF_4^- — III-160 $\xrightarrow{CH_3O^-}$ NO_2, OCH_3, $\overset{+}{N}$, R BF_4^- — III-161 $\longrightarrow$ NO_2, OCH_3, OCH_3, N, R — III-162

The conversion of 2-halopyridinium salts to pyridones by displacement with hydroxide ion has been used in the synthesis of β-carboline derivatives (Section II.4.A) (189). Investigations on the addition of hydroxide ion to pyridinium salts have been reviewed (305a).

The formation of 2-cyano-1-methylpyridinium iodide in the metabolism of 2-aldoximino-1-methylpyridinium iodide (2-PAM), an antidote for anticholinesterase poisoning, has prompted interest in the further decomposition of cyanopyridinium salts. Kosower and Patton studied the basic hydrolysis of 1-methyl-2-cyano-, -3-cyano-, and -4-cyanopyridinium salts, and found that the 2- and 4-cyano derivatives undergo facile transformation to the corresponding pyridones and carbamidopyridinium ions while the 3-cyano compound **(III-163)** yields the 3-carbamido derivative **(III-164)** and the ring-opened cyanopentadienal **III-165** (190). The product ratios are pH dependent, and the order of

CN, $\overset{+}{N}$, CH_3 ClO_4^- — III-163 $\xrightarrow{OH^-}$ $CONH_2$, $\overset{+}{N}$, CH_3 ClO_4^- — III-164 + CN, OHC, NH, CH_3 — III-165

reaction rates of hydroxide ion with the cyano group was found to be 2- $\gg$ 4 -> 3-.

The castorbean plant (*Ricinus communis,* Euphorbiaceae) contains the alkaloid ricinine **(III-166)** which is biosynthetically derived from nicotinamide (191), while another species of Euphorbiaceae, *Trewia nudiflora*, contains the alkaloid nudiflorine **(III-167)** (192). Recently, it has been shown that at least seven species of Euphorbiaceae contain enzyme systems which can cause *in vitro* oxidation of 3-cyano-1-methylpyridinium salts (193). Three enzyme entities which catalyze the oxidation of 3-cyano-1-methylpyridinium perchlorate **(III-163)** to the corresponding 4- and 6-pyridones have been isolated from *Ricinus communis* (194), and although labeled free 3-cyano-1-methylpyridinium ion appears not to be incorporated into ricinine (195), it is nevertheless felt that

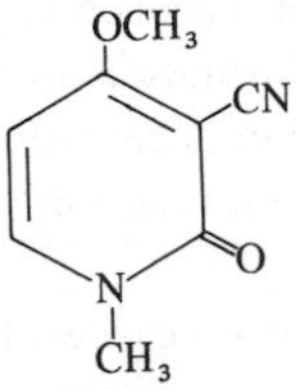

III-166 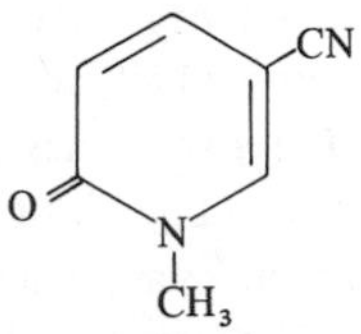 **III-167**

this enzyme system must play an important role in alkaloid biosynthesis in this plant family (193).

Partly because of the biochemical importance of the addition of nucleophiles to pyridinium ions, such reactions have received considerable study (196, 197). The addition is directed to the centers of low electron density, namely the 2-, 4-, and 6-positions, with attack at positions 2- and 6- being the most common due to their proximity to the positive nitrogen atom. Exceptions to this appear to be in thermodynamically controlled cyanide addition reactions (198), hydride ion additions to pyridinium ions containing a sterically large nitrogen substituent (199), and reductions involving hydrosulfite ions (200; *cf.* however ref. 322). Thus Lyle and Gauthier found that in organic solvents, *initial* attack does occur at positions 2- and 6-, but the 1,2- or 1,6-dihydropyridines so formed can rearrange to the thermodynamically more stable 1,4-dihydro derivatives (198). The stabilities of the dihydropyridines, relative to their benzenoid counterparts, make them good model compounds for studying nucleophilic aromatic substitution reactions in general (198).

In contrast to the findings of Lyle and Gauthier with organic solvents, cyanide ion attack on *N*-alkyl-3-carbamoylpyridinium ions in aqueous solution occurs initially at the 4-position (201, 201a), and it was found that the rate and equilibrium constants for this reaction are increased markedly by low concentrations of cationic surfactants (202, 203). Such micelle systems provide an opportunity for investigating reactions of water-insoluble substrates in an essentially aqueous environment (203). In dimethylsulfoxide, cyanide ion addition also occurs mainly at the 4-position (203a). 3,5-Dicyano-1-methylpyridinium tosylate adds cyanide ion mainly at the 2-position in aqueous solution; the product easily rearranges to the 1,4-dihydro derivative (203b), while 3,5-dicyano-1,2,4,6-tetramethylpyridinium tosylate yields the 1,4-dihydro adduct along with pyridine methenes arising from the abstraction of a proton from either the C-2 or the C-4 methyl group (203c). The effect of structure of the pyridinium ion on the affinity of cyanide ion attack has been reported (204), as has the effect of changing the nature of the anion (205).

Aqueous sodium cyanide reacts at room temperature with dodecylpyridinium bromide to give the viologen cation radical **III-168**, while under the same conditions etnylpyridinium bromide fails to react. It has been suggested that this difference in reactivity may be due to micelle formation in the case of the dodecyl derivative, which is known to occur in the concentrations used in these studies (206). Later work has shown that an analogous reaction occurs even with pyridinium salts lacking a long *N*-alkyl chain (349).

In contrast to the cyanide addition products discussed earlier, the 3,5-diacetyl-1,4-diphenylpyridinium salt **III-169** reacted with aqueous potassium cyanide at room temperature to yield the cyano derivative **III-170**, (Z = CN). In like fashion, **III-169** reacts with aqueous sodium carbonate, hydroxide, or acetate to give the hydroxypyridinium salt **III-170**, (Z = OH). Both the cyano and hydroxy compounds are converted back to **III-169** on treatment with hydrochloric acid (118).

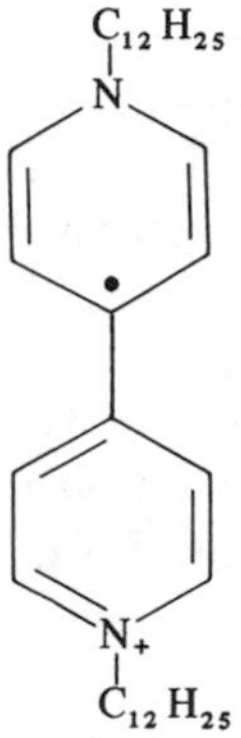

$C_{12}H_{25}$ N ... N+ $C_{12}H_{25}$

III-168

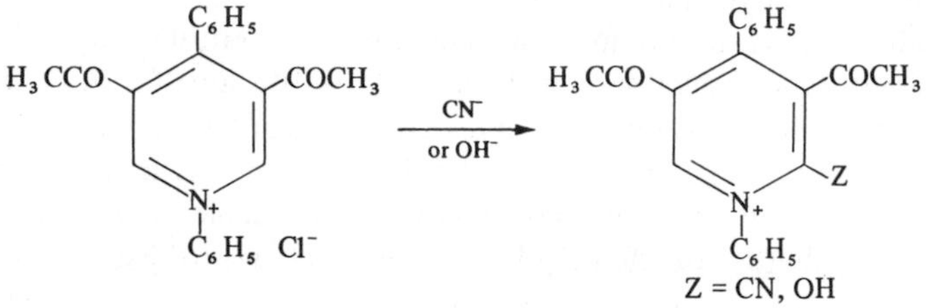

H_3CCO C_6H_5 $COCH_3$ N+ C_6H_5 Cl^-

III-169

$\xrightarrow[\text{or } OH^-]{CN^-}$

H_3CCO C_6H_5 $COCH_3$ Z N+ C_6H_5

Z = CN, OH

III-170

Attempts to cyclize the pyridinium derivatives **III-171** to the spiranes **III-172** failed consistently (207), presumably due to dealkylation and rearomatization of the pyridine ring (208). In the quinoline ring system, however, spirane formation was successful (209).

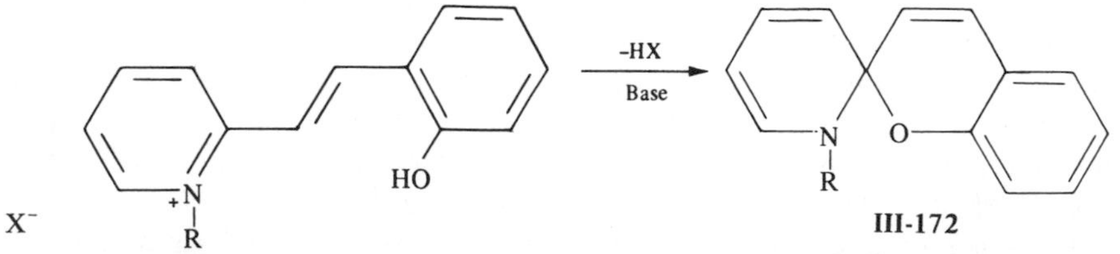

X^- +N R HO

III-171

$\xrightarrow[\text{Base}]{-HX}$

N R O

III-172

The reaction of 3-cyanopyridinium halides **(III-173)** with methylmagnesium bromide gave mixtures of 1,6- and 1,2-dihydropyridines **(III-174, III-175)**,

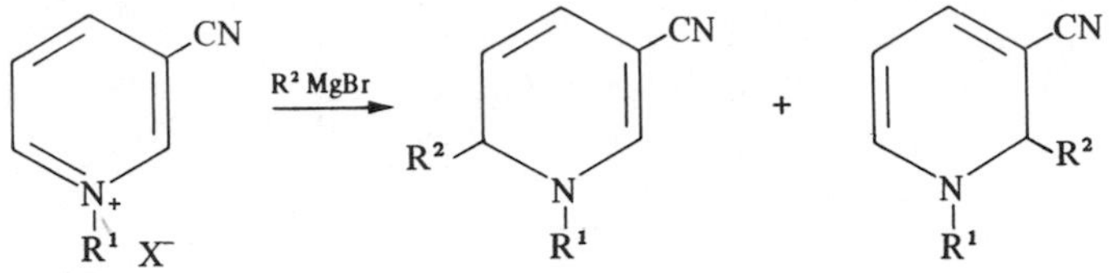

III-173 **III-174** **III-175**

whereas aryl Grignard reagents yielded only the 6-aryl-1,6-dihydropyridines **(III-174**; R^2 = Ar). Methyl- and phenylcadmium gave similar results when treated with 3-methoxycarbonylpyridinium ions, the addition occurring at the 6-position in all cases and also at the 2-position with the methyl reagent and in one case with the phenyl reagent. Phenyllithium reacted vigorously with 1-benzyl-3-cyanopyridinium bromide but no product could be isolated, whereas diphenylmercury was unreactive (210). Grignard reagents add in similar fashion to 1-arylquinolinium salts to yield 2-substituted 1-aryl-1,2-dihydroquinolines (211), while lithium *o*- and *m*-carboranes add in like manner to 1-methylquinolinium and 1-methylpyridinium iodides to give the 1,2- and 1,4- dihydro-derivatives, respectively. The addition of these reagents in a 1,4-manner in the latter case was attributed to their comparatively weak nucleophilic nature (211a). Benzylmagnesium chloride reacts with 1-methyl-2,4,6-triphenylpyridinium perchlorate to yield the 4-benzyl-1,4-dihydropyridine derivative (212), while benzyl Grignard reagents have been added to 4-substituted pyridinium salts to yield C-2 benzyldihydropyridines in the synthesis of morphine related compounds (213).

Quaternary pyridinium salts react with cyclopentadiene, indene, and fluorene anions at the 4-position to give the adducts **III-176 – III-178** (213a–216). These

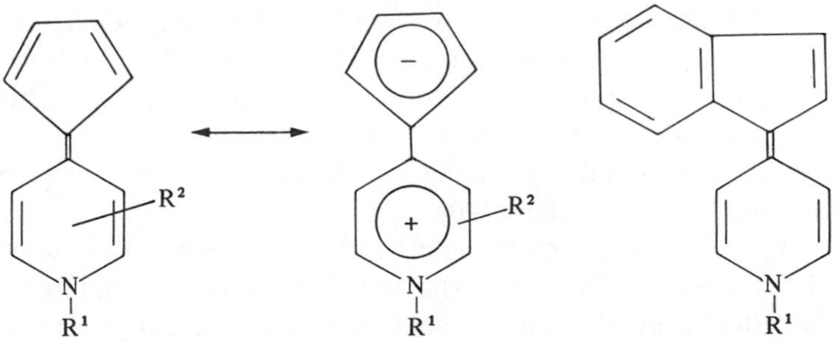

III-176 **III-177**

substances are of interest because they are isoelectronic with the potentially aromatic and as yet unknown sesquifulvalene **(III-179)** (216a). Similar compounds

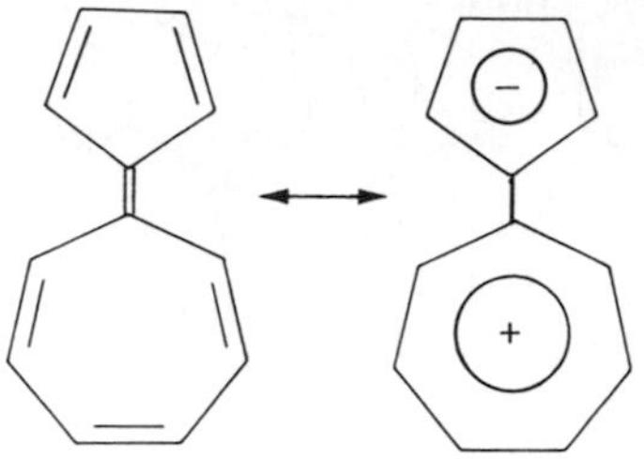

III-179

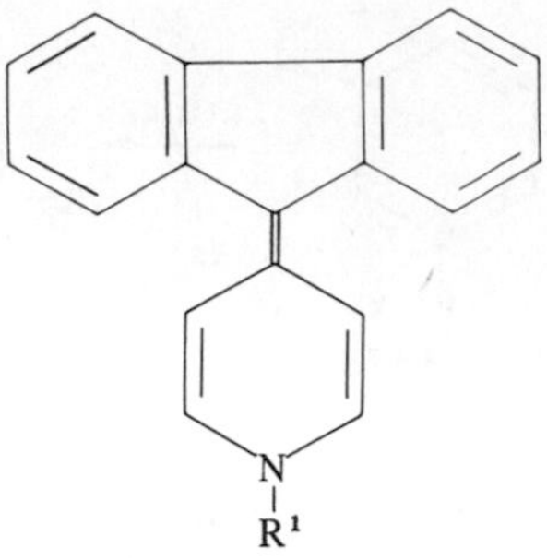

III-178

as well as the C-2 isomers, have also been prepared by treating pyridinium ions having a displaceable substituent such as halogen or aryloxy in the 2- or 4-position with cyclopentadienyl anion (217–219).

Other reactions involving anionic attack of the pyridinium ring system entail the displacement of a C-2 or C-4 sulfonic acid group by methoxyl by passing the pyridinium sulfobetaine (**III-180**) in 70% methanol-water solution through a Dowex-1 ion exchange column (220). Under basic conditions, methyl ketones,

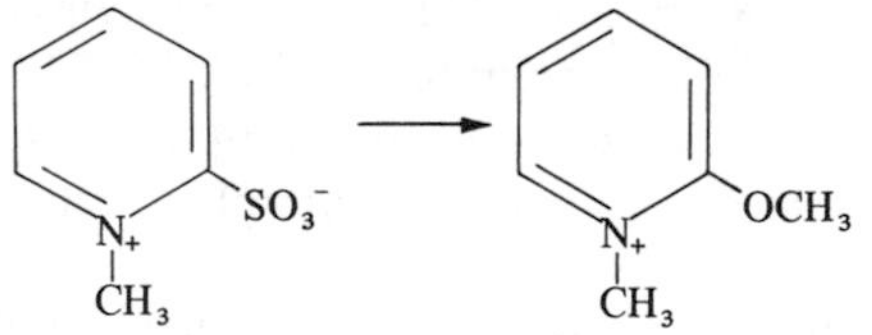

III-180

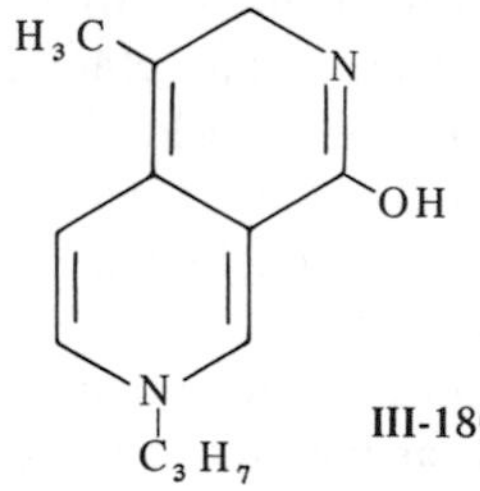

III-180a

nitromethane, diethyl malonate, or desoxybenzoin will add to the pyridinium ring to yield 1,4-dihydropyridine derivatives (221). The base-catalyzed reaction of acetone with 1-propylnicotinamide iodide yields the adduct **III-180a** (222). Klopman (223) has suggested a mechanism for nucleophilic addition to pyridinium ions in which the position of attack is determined by the hardness or softness of the nucleophile, providing an alternative to the charge-transfer mechanism proposed by Kosower (340).

The bisulfite ion adds to the 2-amino-1-methylpyrimidinium ion in water to yield the 1,6-dihydro-6-sulfonate and what is believed to be the 3,4-dihydro-4-sulfonate derivative. Only the former could be crystallized out of concentrated aqueous solutions and its structure was determined by x-ray analysis (222a).

3. Anhydro Bases

A. *Methides*

The preparation of methides of type **III-182** has been reported and involves treatment of the corresponding pyridinium salts **(III-181)** with base. The

$$\text{III-181: 4-}(CH_2C_6H_4NO_2\text{-}p)\text{-pyridinium, } N^+\text{-}R^1\ ClO_4^- \xrightarrow{\text{Base}} \text{III-182: 4-}(=CHC_6H_4NO_2\text{-}p)\text{-1,4-dihydropyridine, } N\text{-}R^1$$

$R^1 = CH{=}CHCOR^2, CH_2R^3$

III-181 **III-182**

products are the basis for quantitative colorimetric analyses for diazoalkanes and β-chlorovinyl ketones (Section II-1) (173, 174).

Methides have been postulated as intermediates in the photoaddition of alcohols, water, and ether to the side-chain olefinic bond of quaternary salts of ethenepyridines (223a) and in the reaction of 4-alkylpyridine-1-oxides **(III-183)** with acid anhydrides (224). The latter reaction yields the products **III-185** and **III-186**, arising from the 1,4-dihydropyridine **III-184**. 1,4-Dihydropyridines are also obtained by the reaction of methyl ketones with pyridinium salts. The products yield anhydro bases on dehydrogenation with nitrosodimethylaniline or chloranil (221).

In 1964, Roth and Möhrle found that triphenyltetrazolium chloride (TTC) would oxidize *N*-alkylpyridinium-2- and 4-carbinols **(III-187)** to the corresponding formyl derivatives, but that the 3-carbinols remained unaffected (225, 392). The rationale for these results involved the postulation of an anhydro base intermediate which then reacts further to form the formyl compounds. Since the 3-carbinols cannot form anhydro bases, they remain unreactive.

When *N*-methylpyridinium-2-carbinol hydroxide was treated with triphenylboron, two boron complexes were obtained. In one, a bimolecular reduction to a glycol **(III-188)** had occurred, while the other consisted of a salt composed of a molecule of the parent pyridinium system and a molecule of its anhydro base (226). The glycol could be prepared in the free state by the air oxidation of a basic solution of *N*-methylpyridinium-2-carbinol; the 4-carbinol derivative reacted similarly (227).

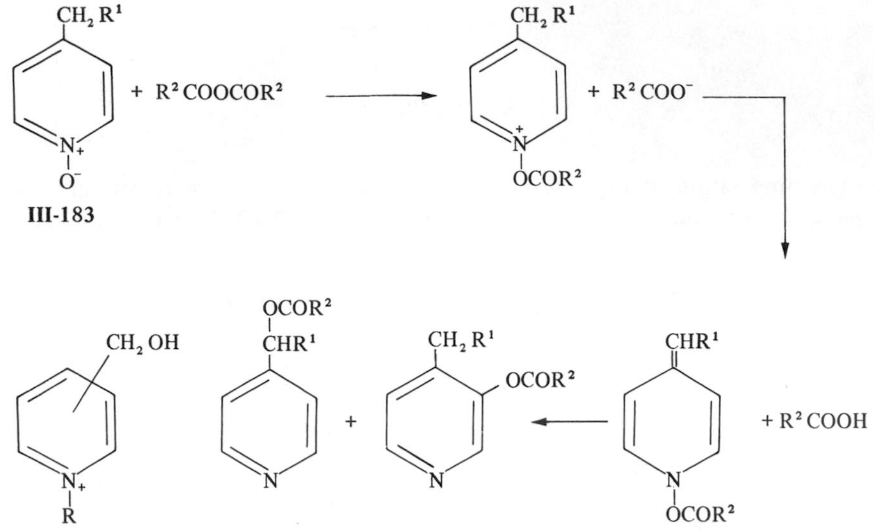

CH$_2$R^1 + R^{2}COOCOR^2 → CH$_2$R^1 + R^{2}COO$^-$

N+ O$^-$ N+ OCOR^2

III-183

CH$_2$OH   OCOR^2 CHR^1 CH$_2$R^1 OCOR^2 CHR^1

N+ R   N + N ← N OCOR^2 + R^{2}COOH

III-187 III-186 III-185 III-184

Triphenyltetrazolium chloride has also been used to convert the steroid side chain **III-189** to the hydroxyacid **III-190** in 65% yield (228).

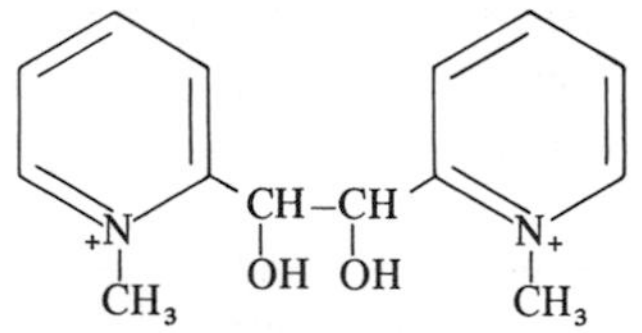

CH—CH OH OH   CH$_2$OH C=O   TTC →   COOH C=O

+N CH$_3$ N+ CH$_3$

III-188 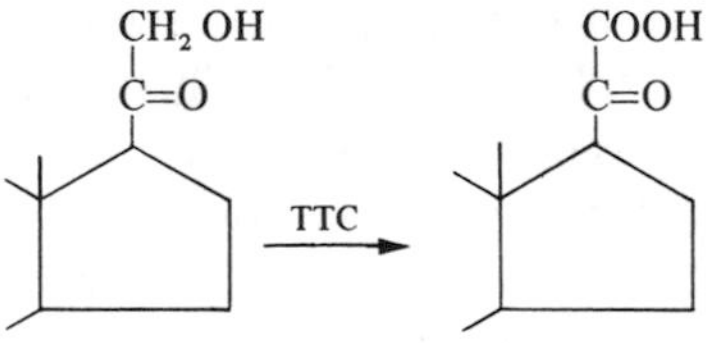 III-189 III-190

A convenient route to indolizines (**III-193**) has been reported by Melton, Taylor, and Wibberley (229, 266), commencing from the quaternary salt **III-191** *via* the intermediate anhydro base **III-192**. The method has been investigated further by Kröck and Krohnke (230), and has been extended by Taylor and Wibberley to 5-azaindolizines (231).

The condensation of *o*-hydroxyaldehydes with heteroaromatic compounds having the *N*-methylated γ- (or α-) picoline ring system (e.g., **III-194**) can yield either **III-195** or **III-196**, depending on the heteroaromatic system involved

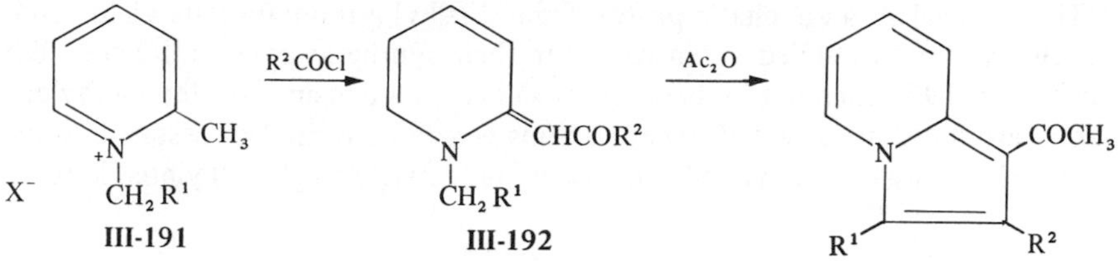

III-191 III-192 III-193

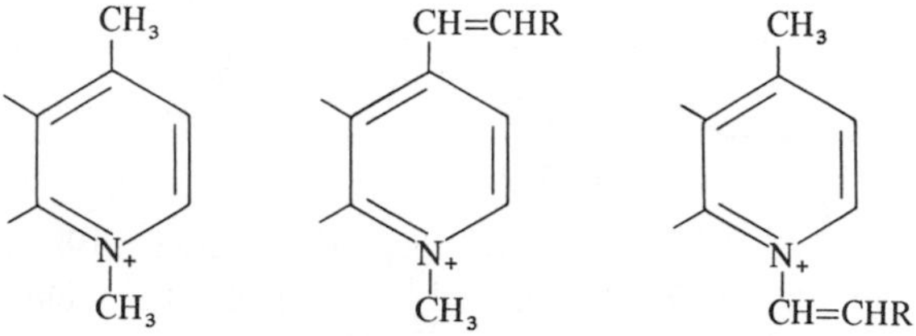

III-194 III-195 III-196

(232). The mechanism of the exchange of hydrogen atoms of *C*-alkyl side chains of quaternary heterocycles has been investigated widely (132).

B. *Ylids–Betaines*

Pyridinium ylids are of particular interest because *d*-orbital stabilization on nitrogen cannot occur, in contrast to those containing phosphorus or sulfur. Two structures are possible for *N*-methylpyridinium ylids (**III-103** and **III-104**)

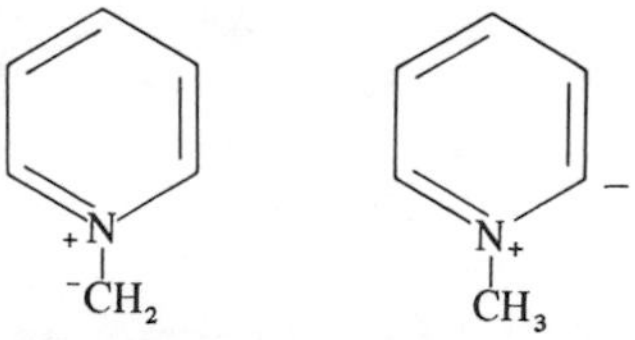

III-103 III-104

(Section I.4.B), and base-catalyzed deuterium exchange studies indicating that, in general, exchange of the C-2 and C-6 protons takes place (98, 99, 233, 234). The effects of different *N*-substituents on this exchange have also been investigated (131).

The removal of an aliphatic proton from *N*-alkylpyridinium salts yields ylids which have been studied extensively for their synthetic potential. Thus ylids such as **III-197** generated by bases from salts containing an activated methylene group are found to be as reactive as anions generated from β-ketoesters in some cases. They may be *C*-alkylated and cleaved reductively to give alkylated ketones

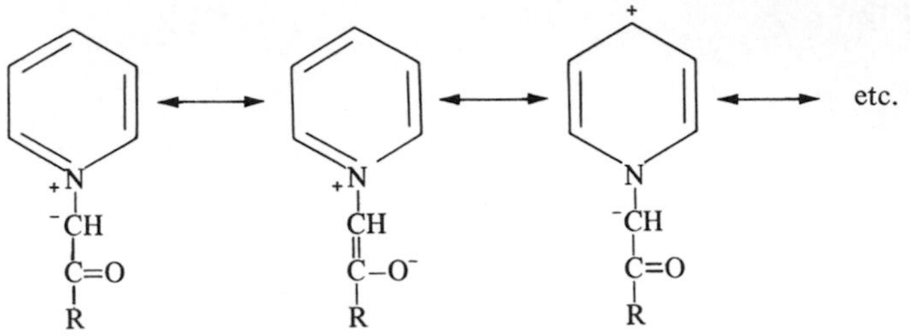

III-197

or esters (235–237). The ylids may also be acylated with acid chlorides or anhydrides and the products cleaved reductively to yield β-diketones or β-ketoesters (238–240).

Ylids of type **III-197** react with cyclopropenones to give α-pyrones (241, 242), whereas the allylpyridinium betaines **III-198** react with diphenylcyclopropenone to give 1,2,6-triphenylphenol (**III-199**) (243). A mechanism involving a ketene intermediate has been proposed (243). The mass spectra of a series of deuterated phenacylpyridinium betaines (**III-197**; R = phenyl) have been determined (244) and have been compared with those reported earlier for *N*-acyliminopyridinium betaines (245).

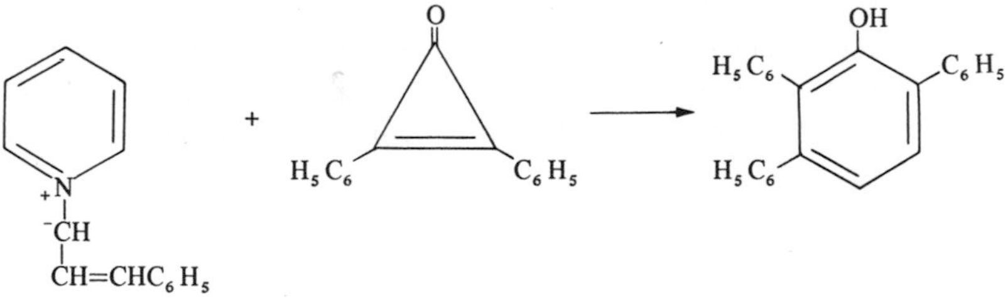

III-198 **III-199**

o-Hydroxyphenacylpyridinium salts (**III-200**) yield chromones (**III-201**) when heated with acid anhydrides (246), and such anhydrides are apparently also necessary to effect a reaction of phenacylpyridinium salts with quinoline-1-oxide (**III-202**) to give the 2-substituted quinoline derivatives **III-203**. The pyridinium group of **III-203** can be cleaved reductively with zinc in acetic acid yielding a new route to quinaldyl ketones (247, 247a). The same reaction has been reported

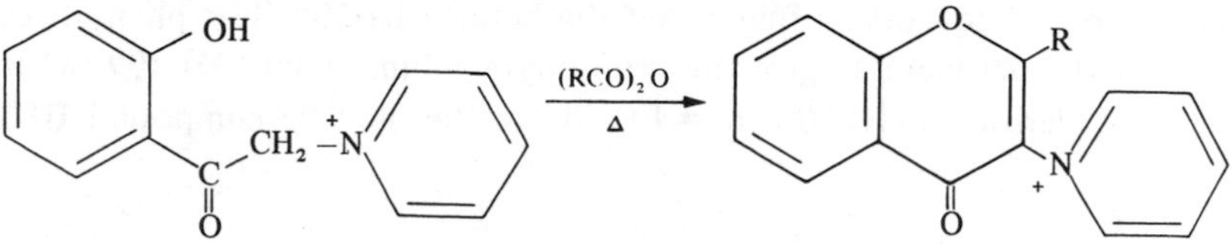

III-200 III-201

for pyridine-1-oxides (247a). Phenylglyoxals react with two moles of phenacylpyridinium bromides to give red tribenzoyldihydroindolizines (247b).

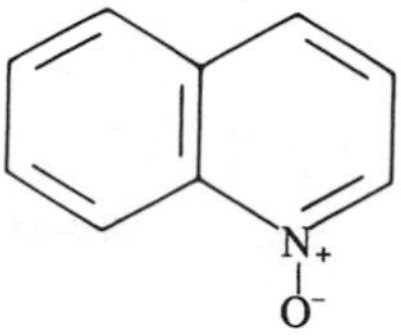

III-202 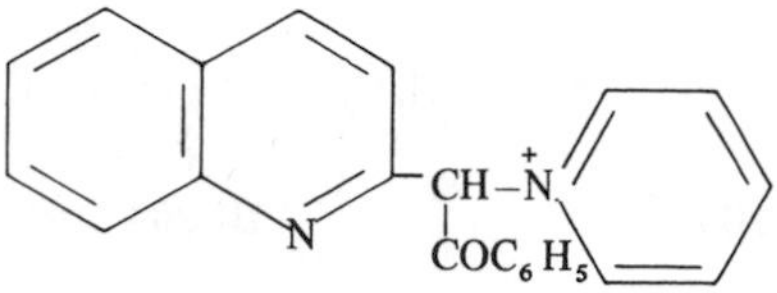 III-203

An interesting 1,3-dipolar addition reaction involving pyridinium carbethoxycyanomethylide (**III-204**) has been reported. When this ylid was reacted with dimethyl acetylenedicarboxylate (**III-205**), the product **III-206** was obtained (248). More commonly, pyridinium and related heteroaromatic ylids react with acetylenedicarboxylic acid derivatives to yield indolizines (**III-207**) (249-255). The sulfonyl ylids give indolizines with loss of the sulfinic acid (16a).

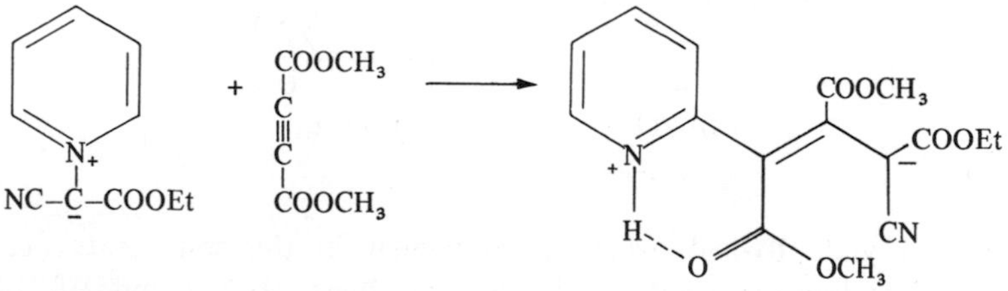

III-204 III-205 III-206

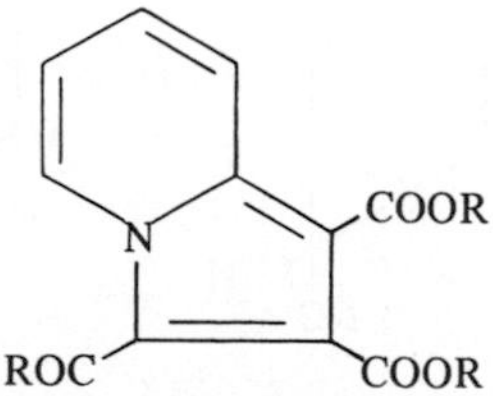

III-207

Wittmann and co-workers found that the betaine **III-208** (R = phenyl) reacts with phenyl isocyanate to give the pyridopyrazinium betaine **III-209**, whereas the diethyl derivative (**III-208**; R = Et) yields the bicyclic compound **III-210**

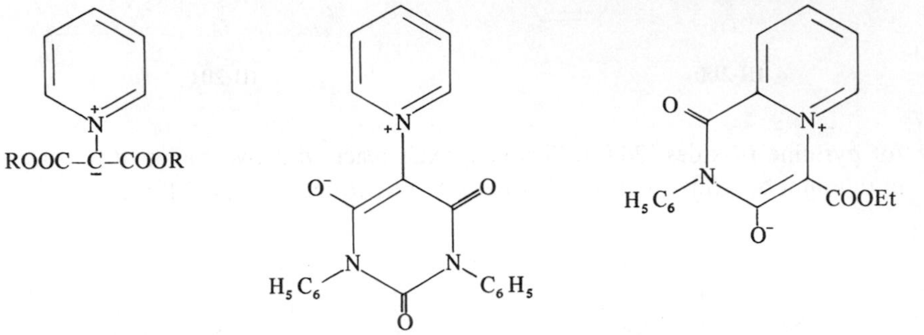

ROOC–C–COOR

III-208 **III-209** **III-210**

(256). Earlier work on the reaction of aryl isocyanates with the betaine **III-211** showed that the imidazole derivatives **III-212** were obtained (257).

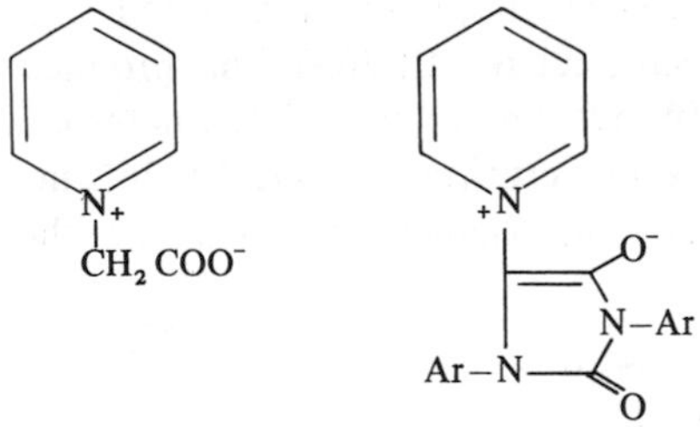

CH_2COO^-

III-211 **III-212**

When the ylide **III-134**, which is nonplanar in the solid state (125), is photolyzed in benzene solution, the two products, **III-213** and **III-214**, are

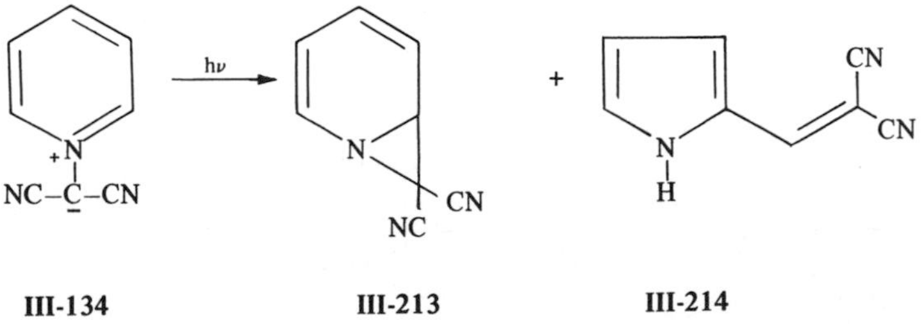

III-134 **III-213** **III-214**

formed (258), while 1-iminopyridinium ylides of type **III-215** yield diazepines (**III-217**) (259, 260), the products being postulated as arising *via* an intermediate diazabicycloheptadiene (**III-216**) (260). It was also found that the pyridooxadiazolone (**III-218**) did not rearrange under the photolysis conditions, a fact interpreted as possibly being due to its inability to form a diaziridine intermediate (260).

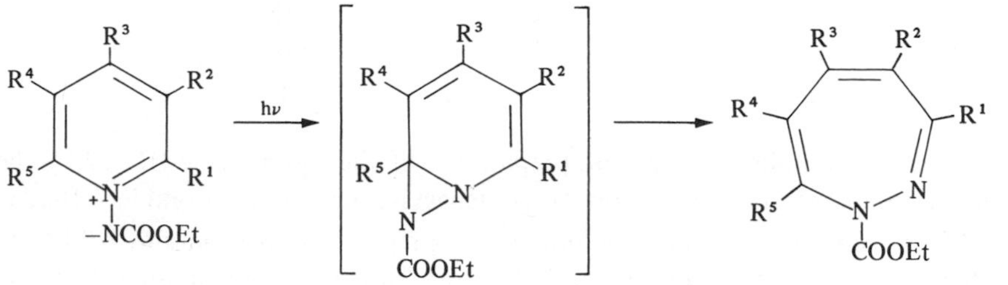

III-215 **III-216** **III-217**

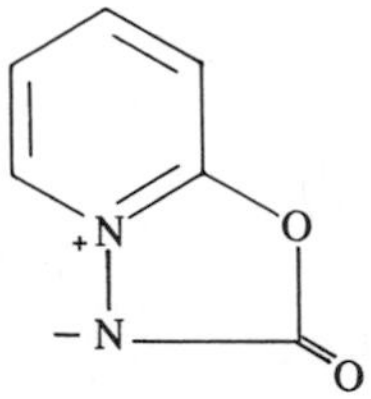

III-218

The desoxybenzoin synthesis developed by Kröhnke and Vogt (261) involving the condensation of benzylpyridinium salts with aryl aldehydes has been extended to the preparation of symmetrically substituted chloro and nitrobenzils (262), and Kröhnke and Ahlbrecht have announced an improved preparation of β-substituted *N*-styrylpyridinium salts from benzylpyridinium salts and aryl aldehydes using a Perkin-type condensation (263). In a similar type of reaction, aldehydo sugars were condensed with phenacylpyridinium iodide in acetic anhydride, pentacetyl derivatives of D-glucose and D-galactose yielding monocondensation products **III-219**, while pentacetyl D-mannose gives the *bis* product **III-220** (264).

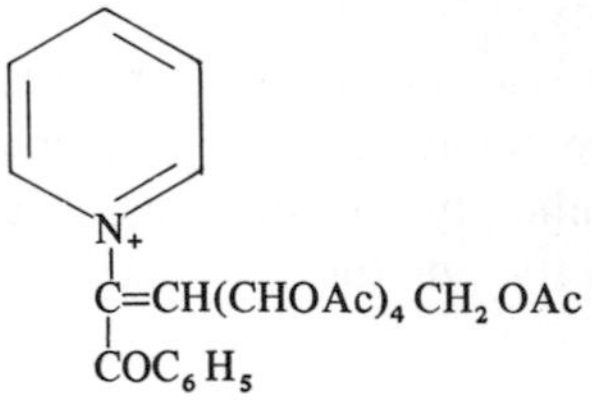

III-219

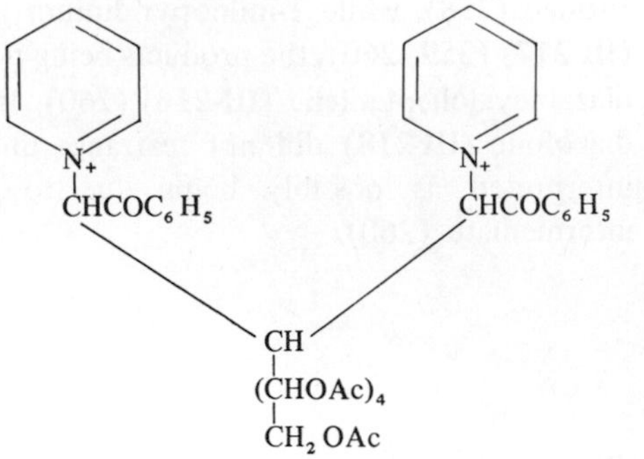

III-220

In 1935, Kröhnke reported the preparation of the enol betaine **III-222** by the Schotten-Baumann benzoylation of 1-phenacyl-2-picolinium bromide **(III-221)** (265). Presumably the same compound was prepared by Melton and Wibberley (266) by the base-catalyzed acylation of 1-phenacyl-2-picolinium bromide **(III-221)** with benzoyl chloride, but these workers assigned the product the structure **III-223** because, on heating with acetic anhydride, a mixture of indolizines was obtained. Later work by Kröck and Kröhnke (231) showed, however, that **III-222** was indeed the true structure, as evidenced from NMR data, as well as an independent synthesis of the substance.

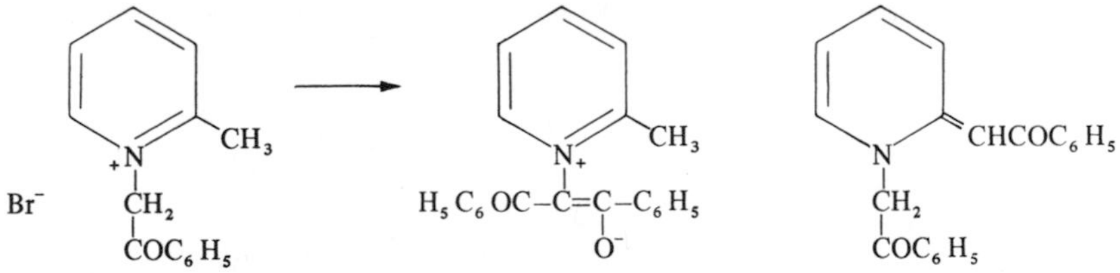

III-221 III-222 III-223

The α-pyridinium acyl group can act as a useful amine protecting system **(III-224)**, which is stable in concentrated or dilute acid solution, but easily removed in dilute base (267). These substances do not form betaines in aqueous base, but rather undergo the cleavage reaction instead, whereas the pyridinium quinolone **III-225** does form a betaine through loss of the N-H proton (267).

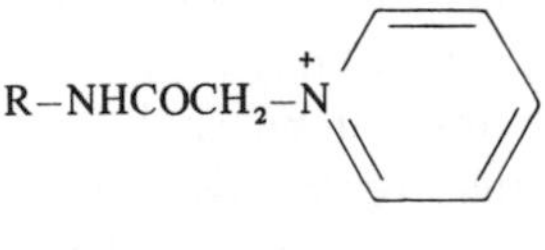

III-224

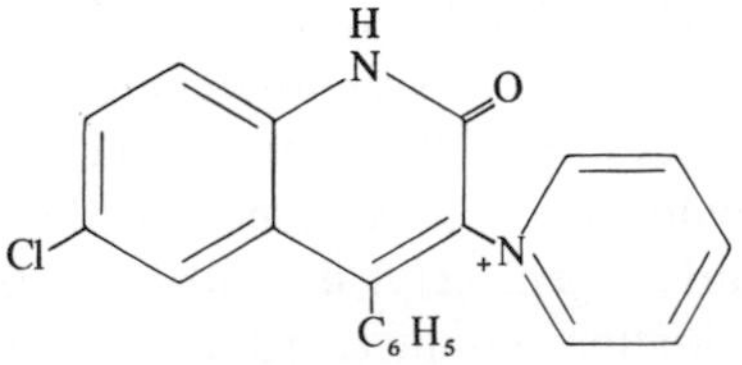

III-225

Kröhnke and Stevernagel have formed betaines by heating *N*-methylene active pyridinium (**III-226**) and isoquinolinium salts with carbon disulfide and base. The products (e.g., **III-227**) undergo *S*-alkylation and dialkylation on treatment

$$\text{Ar–CH}_2\text{–}\overset{+}{\text{N}}\text{(C}_5\text{H}_5\text{)}\ \text{Br}^- + \text{CS}_2 \xrightarrow{\text{OH}^-} \text{Ar–CH(C(=S)S}^-\text{)–}\overset{+}{\text{N}}\text{(C}_5\text{H}_5\text{)}$$

III-226 **III-227**

with alkyl iodides (268). Betaines involving boron have been investigated by Roth and Sarraj (269). The hydroxystyrylpyridinium derivatives **III-228**, readily formed by the aminolysis of styrene oxide with pyridine derivatives, yield the stable betaines **III-229** when treated with triphenylboron. The tetraphenylborate salts of **III-228** which have methyl groups in either positions R^1 or R^3 cleave to give benzaldehyde and the corresponding *N*-methylpyridinium tetraphenylborate when heated in the presence of a trace of acetic acid (270).

$$\text{C}_6\text{H}_5\text{–CH(OH)–CH}_2\text{–}\overset{+}{\text{N}}\text{(pyridine; R}^1\text{, R}^2\text{, R}^3\text{)} \longrightarrow \text{C}_6\text{H}_5\text{–CH(}\overset{-}{\text{O}}\text{B(C}_6\text{H}_5\text{)}_3\text{)–CH}_2\text{–}\overset{+}{\text{N}}\text{(pyridine; R}^1\text{, R}^2\text{, R}^3\text{)}$$

III-228 **III-229**

The zwitterionic species **III-104**, reported to be the type of intermediate involved when certain pyridinium derivatives undergo α-hydrogen exchange (271), has been prepared by the base-catalyzed cleavage of the 2-hydroxymethyl-, 2-acetoxymethyl-, or the 2-formylmethyl hemi-acetal derivatives (**III-230**) (272).

The effect of a γ-substituent on the stability of pyridinium ylids is not apparent from resonance theory because an electron-donating group can delocalize the positive charge while an electron-withdrawing group can delocalize the negative charge. Two studies have appeared recently that settle this question. Phillips and Ratts compared the pK_a's of pyridinium phenacylides of type **III-231** (273), while Douglass, Tabor, and Spradling measured the basicities of 4-substituted pyridinium dicarbethoxymethylides (**III-232**) (274). The studies show that electron-withdrawing groups are stabilizing and electron-donating groups destabilizing, the same result as that found for pyridine-1-oxides (275). In general, it was found (273) that nitrogen ylids are more strongly basic than their sulfur, phosphorus, or arsenic counterparts, which is at least in part

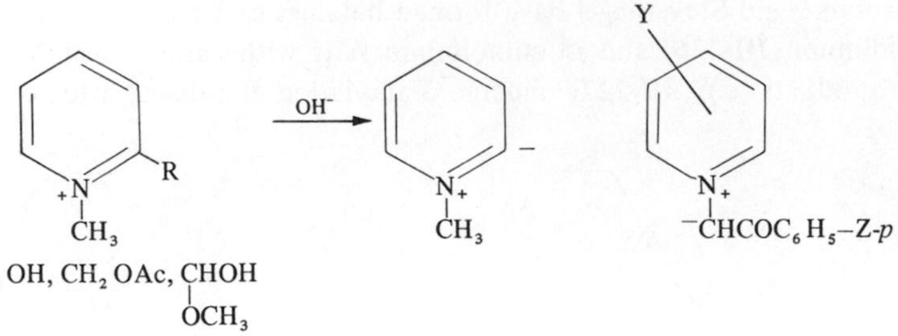

$R = CH_2OH, CH_2OAc, CH(OCH_3)OH$

III-230 **III-104** **III-231**

ascribed to the fact that *N*-ylids can have no back donation of *d*-orbital electrons.

Undheim and Greibrokk have studied the base-catalyzed racemization of pyridinium derivatives of optically active acids **(III-233)**, involving the abstraction of the methine proton (276). When R = H, reprotonation leads to racemization, but when R = CH_3 configuration is retained. The latter result has been attributed to steric interaction between the methyl and carboxyl groups preventing the inversion of the pyramidal carbanion.

The pyridinium betaine **III-234** has been prepared by passing an aqueous solution of 1-methyl-3-hydroxypyridinium iodide through an Amberlite IRA-401 column (277). The substance undergoes 1,3-dipolar additions with acrylonitrile, methyl acrylate, maleic anhydride, and other dienophiles to give structures such as **III-235**, which, in turn, are converted readily by Hofmann elimination to tropones and tropolones (278, 278a).

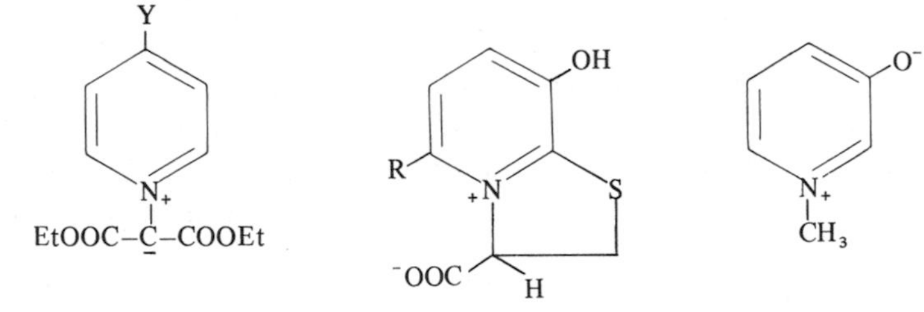

III-232 **III-233** **III-234**

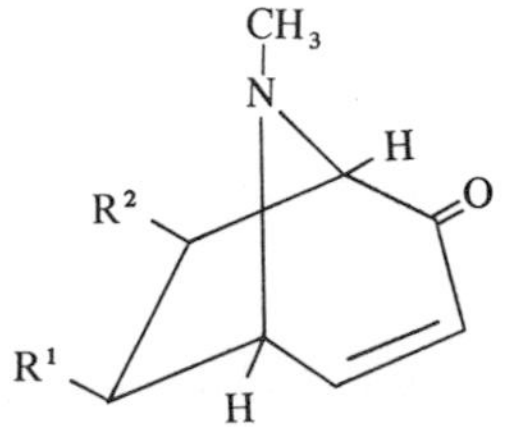

III-235

4. Reduction

A. *Reduction to Piperidines, Dihydro, and Tetrahydro Pyridines*

A systematic study of the *cis-trans* isomer ratios of dimethylpiperidines (lupetidines) produced by the catalytic hydrogenation of lutidine methiodides using either Adams' catalyst or Raney nickel (W-2) has been reported (279). In general, the *ortho*-dimethylpyridinium salts gave almost exclusively the *cis* isomers, the *meta*-dimethyl derivatives gave predominantly the *cis* isomers, while the *para* derivative yielded greater mixtures of both isomers. Furthermore, cyanopyridinium salts have been reduced to the corresponding piperidines using a platinum oxide or rhodium on carbon catalyst under conditions where the cyano group is not reduced (280).

Although catalytic hydrogenation of pyridinium salts usually gives the corresponding piperidines (281), such reduction of 1-alkyl-3-acylpyridinium salts of type **III-236**, especially under basic conditions, generally yields 1-alkyl-3-acyl-2-piperideines (**III-237**) [platinum oxide, sodium bicarbonate (282); Raney nickel W-5, triethylamine (283); palladium, triethylamine, (284, 285); palladium, no base (286)]. It has been suggested that the formation of a vinylogous amide system in **III-237** causes the arrest of the reduction at the tetrahydro stage (285). Yet it was found that 2-alkyl-3-acylpyridinium salts are reduced completely to piperidines in basic solution. Presumably in this case the species reduced is now **III-238** which would be expected to be in equilibrium with the quaternary pyridinium system (285).

Ferles and co-workers have studied the reduction of quaternary pyridinium salts using lithium aluminum hydride (287), sodium borohydride (288, 289), and aluminum hydride (290). All of these reagents give predominantly 3-piperideines, with small amounts of hexahydro derivatives produced occasionally. In general, where several tetrahydro products are possible, one largely predominates. A useful comparison of the product ratios obtained from picolinium and

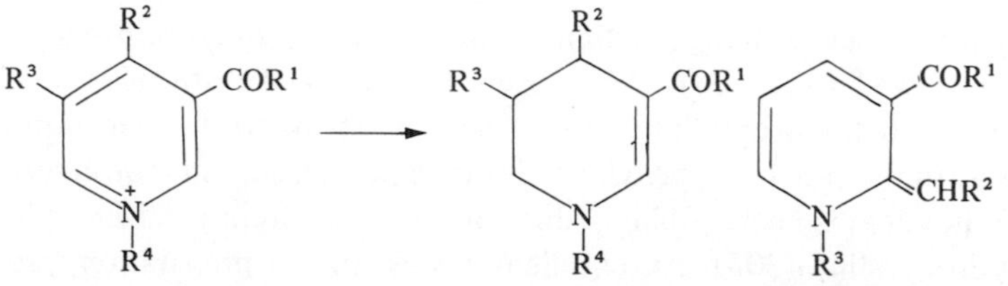

III-236 **III-237** **III-238**

lutidinium salts using the three metal hydride reagents has been made (290). 1-Alkoxypyridinium salts yield pyridines, 3-piperideines, and piperidines on reduction with sodium in ethanol, metal hydrides, or by electrolysis (291).

Sodium borohydride reduction of pyridinium salts to tetrahydropyridines occurs *via* initial formation of the 1,2-dihydro intermediate, followed by protonation at the central position of the resulting dienamine and further reduction of the immonium system to the tetrahydro product **(III-239)** (292–294). In the case of the 3-cyano-1-methylpyridinium system it has been shown that the water solvent cannot be the protonating species by itself in the formation of the immonium salt intermediate (295, 296).

BH_4^- → H^+ → BH_4^- →

III-239

When the R group in **III-239** is larger than methyl, some reduction to the piperidine system also occurs, the amount increasing with increasing size of R (297). The piperidine ring is formed by initial borohydride attack at the 4-position and consequently such an attack becomes more favorable as the R substituent increases in size (297). A new synthetic route to 1-unsubstituted 1,2,5,6-tetrahydropyridines commences with the sodium borohydride reduction of 1-benzylpyridinium salts (297a). Products formed by the Diels-Alder dimerization of the 1,2-dihydropyridine intermediates have been isolated (297b).

If the sodium borohydride reduction is carried out in the presence of cyanide ion, the latter becomes an effective competitor with hydride ion in the second stage of the reduction and 6-cyano-1,2,5,6-tetrahydropyridines **(III-240)** result (298). These substances can lose HCN to yield the dihydro products (298, 299), or they can undergo nucleophilic displacement of the cyano group by Grignard reagents to give the 6-alkyl derivatives **(III-241)** (299a).

Although, as a general rule, sodium borohydride reduction of pyridinium salts yields the tetrahydro derivatives, occasionally a dihydro derivative is obtained (292, 300). In the case of 3-substituted pyridinium salts, these are usually the 1,6-dihydro isomers although at times some of the 1,2-dihydro isomer has been obtained as well (301–305). It has been shown that substituted 1,4-dihydro- (292) and 1,6-dihydropyridines (306) are resistant to further borohydride reduction. In the case of 1-phenylpyridinium chloride, sodium borohydride gives the 1,2-dihydro isomer, while reduction with sodium amalgam yields the 1,4-dihydropyridine (305). An excellent review of the preparation, properties, and reactions of dihydropyridines has appeared recently (305a).

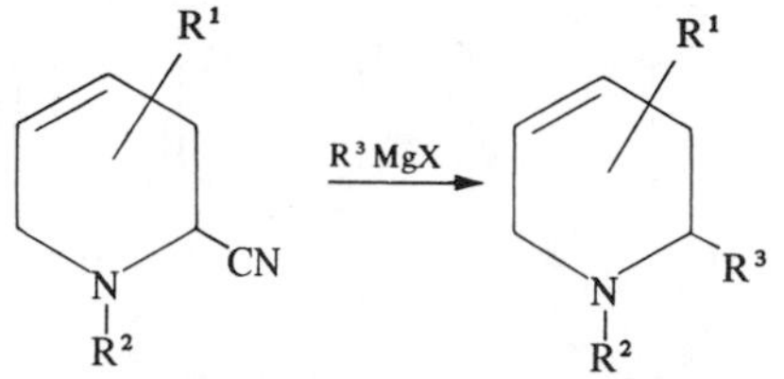

III-240 $\xrightarrow{R^3MgX}$ III-241

A tempting route to indole alkaloids containing the quinolizine ring system (**III-243**) is the reductive cyclization of the pyridinium salts **III-242**. Both Wenkert and Potts have developed methods whereby this can be accomplished with metal hydrides, employing both lithium aluminum hydride and lithium

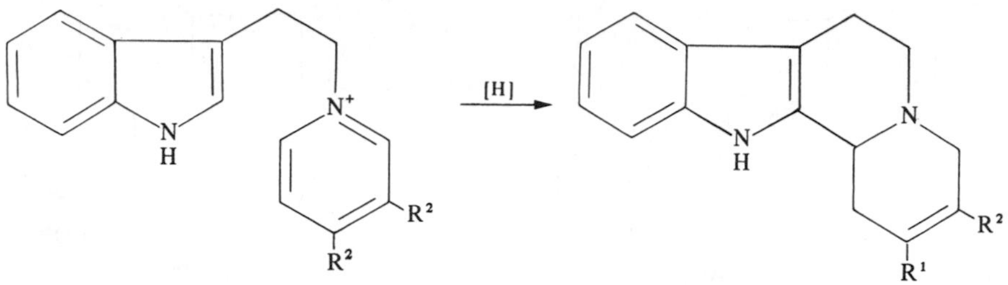

III-242 $\xrightarrow{[H]}$ III-243

tri-*t*-butoxyaluminum hydride (307, 308). Initial attempts to carry out this cyclization using sodium borohydride were unsuccessful, always leading to the uncyclized tetrahydropyridine system (307). However, Fry and Beisler recently reported conditions whereby it can be achieved (299), particularly if Fry's previously discovered cyanide procedure is used (298). Both of the methods they reported lead to **III-243** *via* a dihydropyridine intermediate. The method has also been applied to indolylethylisoquinolinium salts in the synthesis of benzindoloquinolizidine alkaloids (308a), as has a metal-catalyzed partial hydrogenation method (309) developed earlier by the Wenkert group for the tetrahydrocarboline system (309a).

Joule and co-workers also found recently that if the tetrahydro derivative obtained from the sodium borohydride reduction contains a 4-acyl group (**III-244**), then cyclization to **III-247** can be effected using sodium methoxide or sodium hydride-DMF, followed by treatment with HOAc-MeOH (310). An unstable substance was isolated from the base treatment of **III-244** and is believed to be the enamine intermediate **III-245**. The tetrahydro derivatives **III-244** could also be cyclized in an alternative manner to give compounds **III-246**, which contain four of five rings of the Iboga alkaloid skeleton. Other

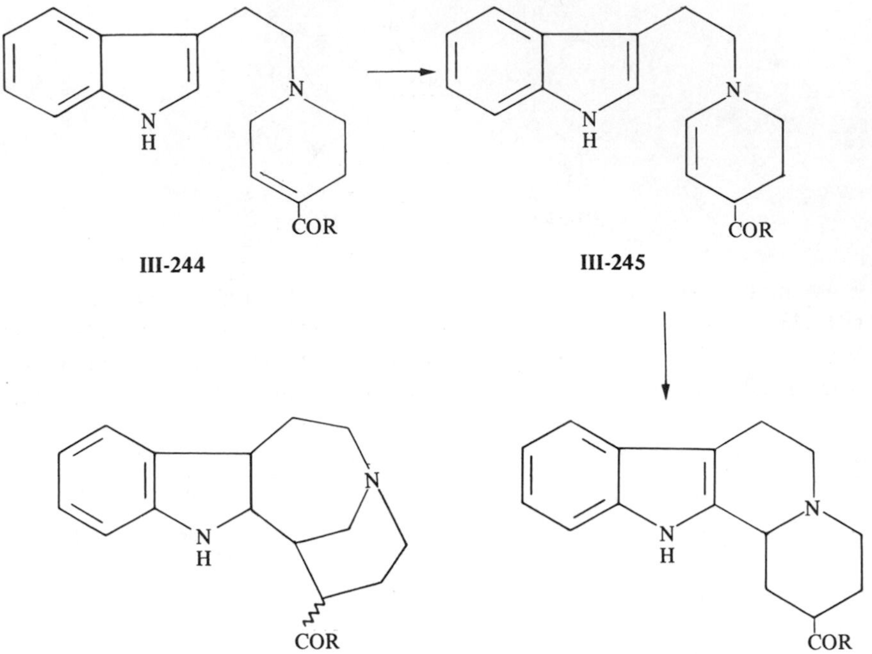

III-244 III-245

III-246 III-247

metal hydride reductions and reductive cyclizations of the **III-242** system have been studied (311, 312).

A potential method of converting a quaternary pyridinium salt to a dihydro derivative was employed by Saunders and Gold (313). The tetrahydropyridinium iodide **III-248**, prepared by the sodium borohydride reduction of 1-methylpyridinium iodide followed by *N*-methylation, was converted to the dibromide (**III-249**) on photolytic bromination. The normal ionic addition of bromine consistently failed, apparently being inhibited by the positive charge. Dehydrobromination of **III-249** with base led to the dihydropyridinium salt **III-250**,

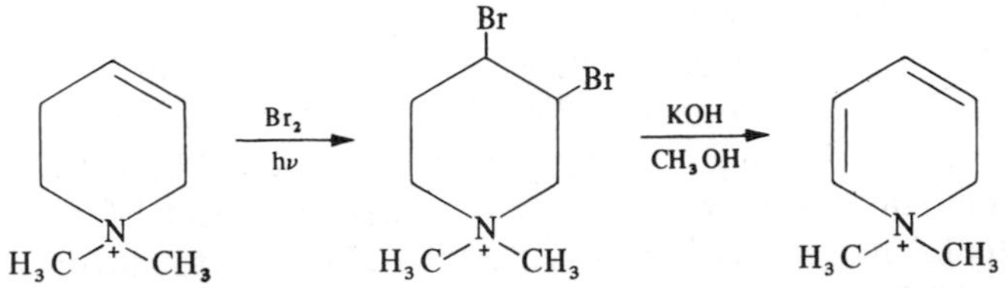

Br_2 / $h\nu$; KOH / CH_3OH

III-248 III-249 III-250

which (along with 1,2-dihydroquinolinium salts) was used in studying the kinetics of aqueous base-catalyzed deuterium exchange reactions (313).

The sodium borohydride reduction of 3-cyanopyridinium salts yields the expected tetra- and dihydro-derivatives (304, 306). The 4-cyano isomer (**III-251**) also gives the expected tetrahydro product when reduced with this reagent in methanol/water. If, however, the reaction is carried out in methanolic sodium hydroxide at 0°, the dimer **III-252** is formed (314), while at temperatures below -20° increasing amounts of **III-253** are formed as the temperature is lowered (315). The dimer **III-253** is very labile and rapidly converts to **III-252** in solution above -20°, and even in the solid state at 15 to 20°. The mechanisms for the formation of these products have been discussed (314, 315). The complex metal hydride reduction of nitrogen heterocycles,

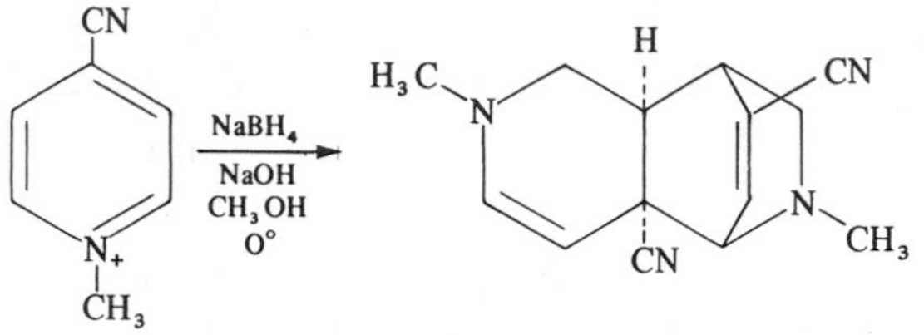

$NaBH_4$, NaOH, CH_3OH, 0°

III-251 **III-252**

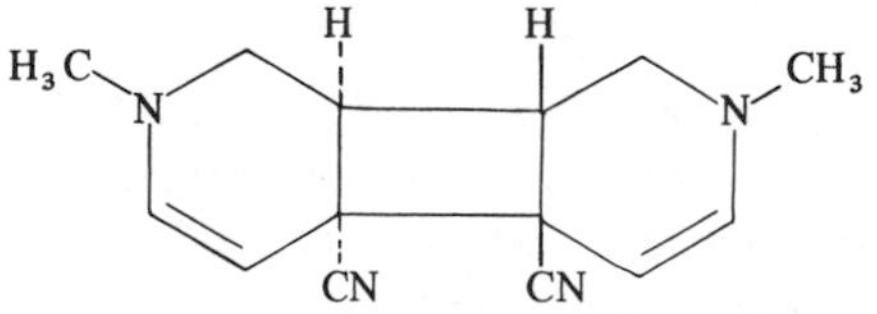

III-253

including pyrimidine salts, has been reviewed (294), and the orientation of the hydroboration of tetrahydropyridines has been studied (316). Isoquinolinium salts undergo reductive alkylation in the presence of aldehydes and sodium borohydride, yielding 4-alkylated tetrahydroisoquinolines (316a).

Quaternary pyridinium salts are reduced by formic acid giving a readily separable mixture of *N*-alkylpiperidines and 1,2,5,6-tetrahydropyridines (317). The ratios of the products formed are substituent dependent, and the mechanism of the reduction is similar to that observed for the sodium borohydride reduction with the decomposition of the formate ion as the source of the hydride ion (318).

The reduction of quaternary pyridinium salts to 1,4-dihydropyridines using sodium dithionite (319, 320) continues to be used in diphosphopyridine nucleotide model investigations (321, 322), discussed more fully in Section II.4.B. The reaction appears to proceed satisfactorily with pyridinium salts having C-4 (118, 322) and C-3, C-5 (300) substituents, and has been employed in constructing the indoloquinolizine ring system (322a).

The herbicidal activity of Diquat (**III-142**) and Paraquat (**III-143**) is believed to be due to their abilities to undergo facile one-electron reduction in aqueous solution to give stable cation radicals, which in turn are rapidly reoxidized quantitatively by oxygen to yield peroxide radicals (323–326). In addition to chemical (159) and electrochemical (160) reductions of these substances, one-electron reduction studies of a number of analogs of these herbicides have

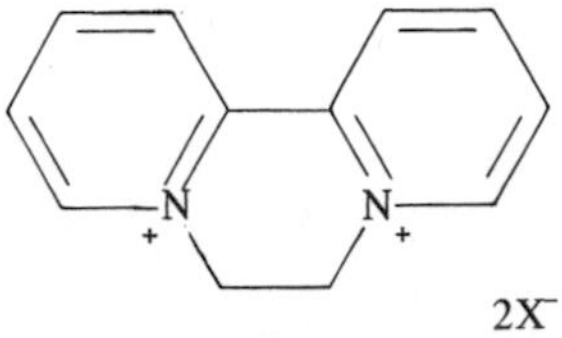

$2X^-$

III-142

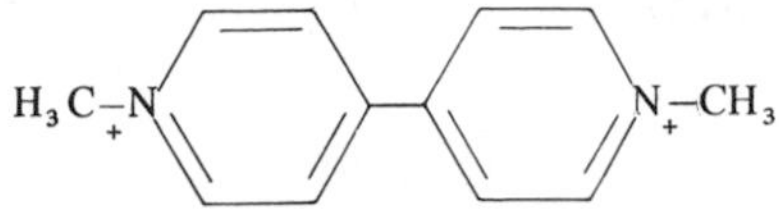

$H_3C-\overset{+}{N}$ $\overset{+}{N}-CH_3$

$2X^-$

III-143

been reported, among them being diquaternary salts having the phenanthroline (324, 327), dipyridylsulfide (328, 329), dipyridylamine (327), dipyridylketone (330), biquinoline (331), terpyridine (332), and dipyridylethylene (333) systems. The polarographic reduction of pyridinium salts has been discussed in Section II.1.

In the mass spectral fragmentation of the *N*-alkyl-1,4-dihydropyridines **III-254** and **III-255**, a C-4 hydrogen atom or methyl group is lost, yielding the corresponding pyridinium ions, which then undergo further fragmentation characteristic of the parent pyridinium ions themselves (334, *cf.* also ref. 335). A similar situation was observed with the 1,2-dihydropyridine derivatives **III-256**, where the preferred initial fragmentation always led to the pyridinium ion **III-257** (335a). Since it appears that initial fragmentation involves the preferred loss of a group from the sp^3 hybridized carbon of such

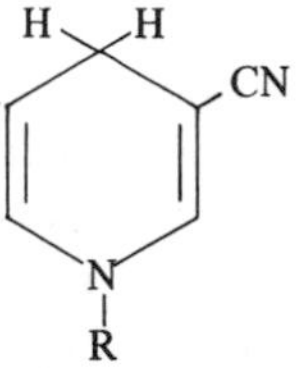

III-254

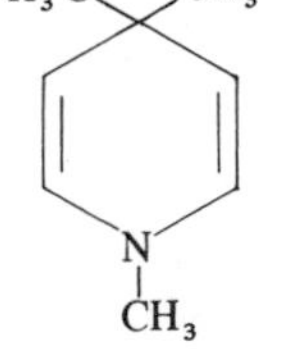

III-255

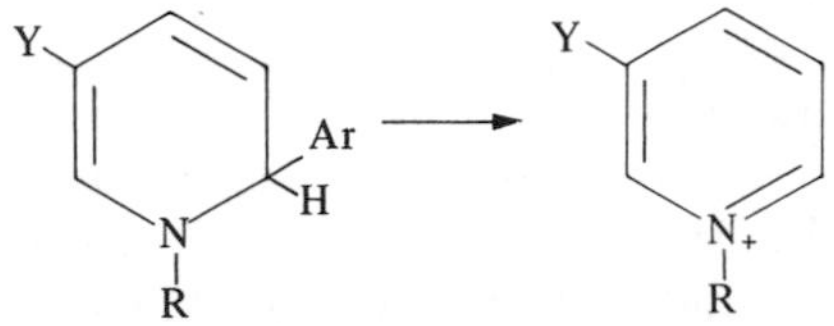

III-256

Y = CN, $COOCH_3$

R = CH_3, $C_6H_5CH_2$

III-257

dihydropyridines, Lyle and White have suggested that mass spectrometry provides a powerful tool for studying the orientation of nucleophilic addition to pyridinium ions (335a).

B. Nicotinamide Coenzymes

The reduction of nicotinamide quaternary salts and their analogs continues to harbor interest due to the great importance of this process in biological systems. Nicotinamide adenine dinucleotide (NAD^+) is the pyridinium salt **(III-258a)** which acts as a coenzyme in biological hydrogen transfer reactions. When NAD^+ is reduced with sodium dithionite a yellow intermediate is produced which decomposes readily to NADH **(III-260a)** and sulfite ion (336). A π-complex structure, similar to the charge transfer complexes of the pyridinium iodides, had been suggested for this intermediate (337), but Caughey and Schellenberg (321) and Biellmann and Callot (322), working with the model systems **III-258b**, showed the correct structure to be **III-259**, one suggested some years earlier by Yarmolinski and Colowick (336). The French workers also studied steric effects in this reduction by investigating various ring-substituted methyl analogs, and compared the dithionite reduction results with those obtained using sodium borohydride (322, *cf.* also ref. 336a). Nicotinamide salts undergo a reaction with sulfite ion to yield an adduct analogous to **III-259** (332b).

$Na_2S_2O_4$ → $+ SO_3^{2-}$

III-259 **III-260**

III-258 a R = ribose-P-P-ribose-adenine

III-258 b R = $C_6H_5CH_2$, *p*-$ClC_6H_5CH_2$, *o*-$ClC_6H_5CH_2$

The oxidation of 1,4-dihydropyridines related to NADH by various agents appear to indicate that the hydrogen transfer process may proceed by either a hydride or a free radical mechanism (338, 339). The suggestion has been made (340) and evidence cited (341) that the hydride transfer process may be a two step electron-hydrogen transfer reaction.

The electrochemical (polarographic) reductions of NAD^+ (342, 343) and NADP (344) have been investigated, as have similar studies with model pyridine nucleotide systems (160, 343, 345, 346, 346a). In apparent contrast with previously published data, the 2-aldoxime analog of NAD^+ has been reported to show no coenzyme activity with horse liver and yeast alcohol dehydrogenases (346b), and an attempt to effect intramolecular hydride transfer to the pyridinium ring in a dehydrogenase enzyme model system was unsuccessful (346c). The properties of the nicotinamide coenzymes have been reviewed (347).

C. *Biomolecular Reduction Products*

In 1964, Kosower and Cotter found that 4-cyano-1-methylpyridinium iodide (**III-261**) reacts with sodium dithionite in aqueous solution to yield the stable cation radical methylviologen **III-262**, which, in turn, undergoes oxidation in the presence of oxygen or iodine to Paraquot (**III-143**) (**348**). The mechanism proposed for this dimerization involves the initial free radical coupling of two methyl cyanopyridinium molecules to give **III-263**. The expulsion of cyanide ion effects

CN

CH_3 I^-

III-261

H_3C-N $N-CH_3$

III-262

O_2 or I_2

$Na_2S_2O_4$

$Na_2S_2O_4$

CN

H_3C-N $N-CH_3$ $\xrightarrow{-2CN^-}$ H_3C-N $N-CH_3$

NC

III-263

III-143

rearomatization of the pyridinium rings, yielding **III-143**, which is in turn reduced to the radical cation **III-262** (348). This radical cation is also formed in an unusual reaction when the dimethiodide of di-(4-pyridyl) ketone is treated with concentrated aqueous alkali (348a). Later work showed that 4-acylpyridinium salts yield nonviologenic stable long-lived free radicals when treated with concentrated aqueous base (348b).

Dodecylpyridinium bromide forms the viologen cation radical **III-168** directly when treated with aqueous sodium cyanide at room temperature, possibly due to micelle formation (206) (see, however, ref. 349), while under air-free conditions cyanide in aqueous methanol converts the 1-phenyl- and 1-benzylpyridinium salts to the $\Delta^{4,4'(1H,1'H)}$-bipyridines **(III-264)** (349). The 1,2-dimethylpyridinium salt behaves similarly. The bipyridylidenes **III-264** can also be

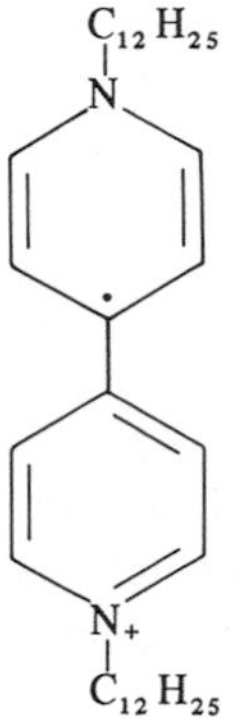

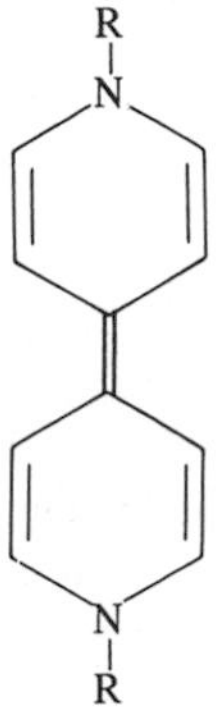

$R = C_6H_5, C_6H_5CH_2$

III-168 **III-264**

obtained by active metal reduction of the corresponding 4,4′-bipyridylium salts but the yields are reported to be poor and the products are contaminated frequently with the radical cation (349). A superior method was advanced by Colchester and co-workers, who were able to prepare both the radical cation **III-262** and the dihydrobipyridylidenes **III-264** ($R = CH_3$) quantitatively by the alkaline dithionite reduction of **III-143** (350). The cation radical **III-262** has also been prepared by the reduction of **III-143** (paraquat ion) with methoxide ion in methanol. It has been suggested that the reducing agent in this case is formaldehyde, formed from the methoxide ion (350a). Under more forcing conditions, reduction to the dihydrobipyridylidenes **III-264** ($R = CH_3$) can occur (350). The *N,N′*-diacetyl derivative of **III-264** ($R = COCH_3$) has been oxidized by air or iodine to the corresponding dipyridinium cation *via* the intermediate cation radical (351).

The electrochemical reduction of the pyridinium bromides **III-265** was investigated by Underwood and co-workers, who postulated from polarographic evidence that coupling occurs between the C-2 positions of the pyridine rings, leading to **III-266** in the presence of oxygen (160). Wallenfels and Gellrich obtained similar tetrahydrobipyridyl compounds by the reduction of 1-substituted nicotinamides with chromous ion, magnesium, or zinc-copper, the last

H_2NOC $CONH_2$ $\overset{e^-}{\underset{O_2}{\longrightarrow}}$

$CH_2-(CH_2)_n-CH_2$

$n = 0–3$

III-265

H_2NOC $CONH_2$

$CH_2-(CH_2)_n-CH_2$

III-266

giving the best yields (352). These workers assigned the dimers the tetrahydro-6,6′-bipyridyl structures on the basis of ultraviolet absorption data, but later evidence from proton magnetic resonance spectroscopy does not unambiguously rule out a 4,4′-structure (353).

5. Ring Opening Reactions

In 1904, Zincke and König independently discovered that the pyridine ring of certain pyridinium salts could be cleaved; Zincke (101, 354) used 2,4-dinitrochlorobenzene and König (113, 355) employed cyanogen bromide to prepare the salts. A number of other investigators have since studied these reactions (392) and König extended the original reaction to a preparation of certain 1,2,3-triazoles by using a double pyridine ring opening sequence (115). In this scheme, 3-acetylaminopyridine (**III-267**) is treated with cyanogen bromide in the presence of primary aromatic amines to yield the pentamethinium salts **III-268**, which in turn undergo facile ring closure to the 3-acetylamino-*N*-arylpyridinium salts **III-269**. Acid hydrolysis of **III-269** cleaves the acetyl group and diazotization of the amino group in the product (**III-270**) leads readily to the triazole acrolein derivatives **III-271**. These compounds could be converted to other products, such as the *trans* isomer **III-272** and the carboxylic acid **III-273**.

Electron-withdrawing groups attached to the pyridinium ring aid in its cleavage and a systematic study of the effect of ring substituents on ring cleavage has been reported for the reaction of dinitrophenyl chloride derivatives of 3- and 4-substituted pyridines with aromatic amines (356). The reaction was found to

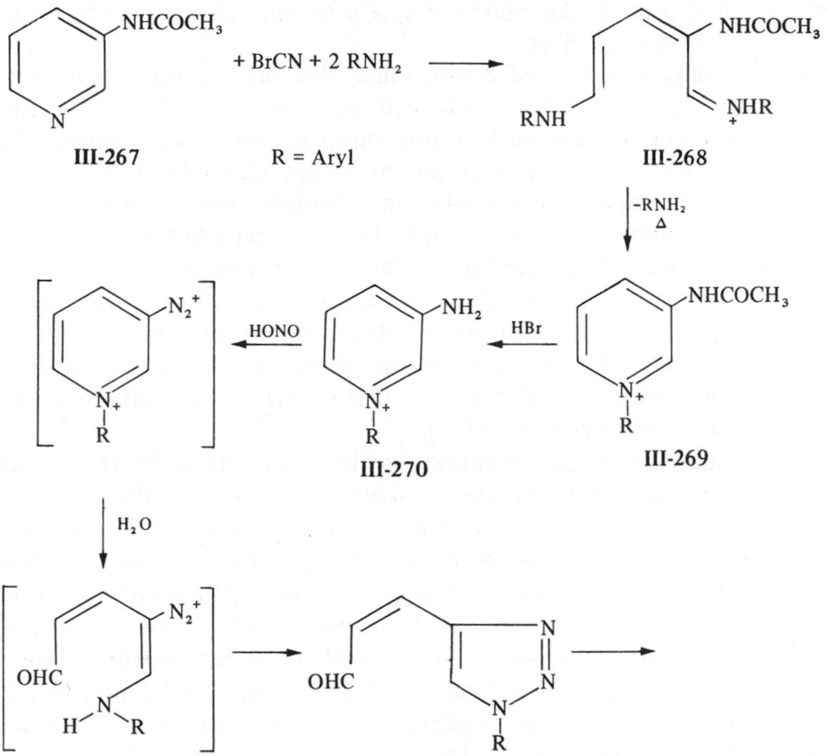

III-267 R = Aryl III-268 III-269 III-270 III-271 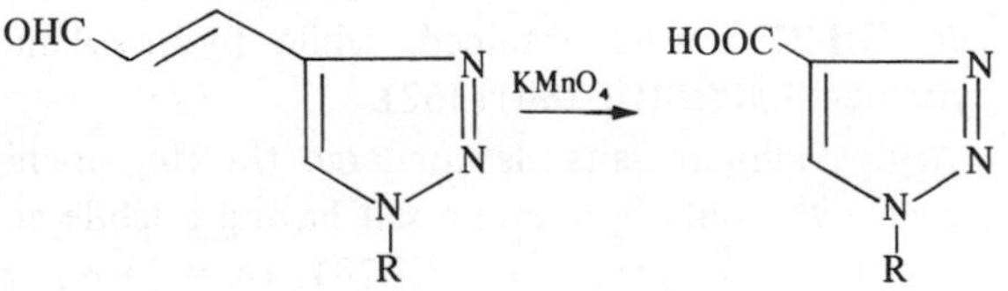 III-272 III-273

depend markedly on the nature of the substituent and on its position in the pyridine ring. Similar results were observed when the pyridine ring cleavage was carried out by the cyanogen bromide method (357). The importance of the position of the substituent was demonstrated with the acetamido group where a shift from the β- to the α-position led to considerable resistance to ring cleavage in both the dinitrophenyl chloride derivatives and in the cyanogen bromide reactions. Substituents in the α-position appear to add additional hindrance to cleavage because of steric effects.

Correspondingly, Kosower and Patton found that the 1-methyl-3-cyanopyridinium system is cleaved substantially with base, while the 2- and 4-cyano derivatives did not undergo such a ring opening, but yielded instead the 1-methyl-2- and 4-pyridones with the expulsion of cyanide ion (190).

The electron withdrawal can also take place through a group attached at the pyridinium nitrogen atom, and a study has been reported where such a substituent is a benzene ring containing electron-withdrawing groups (358). A heterocyclic ring so attached can also affect the necessary electron withdrawal, and in this regard, Lira found that the salt **III-274** undergoes facile ring opening despite the presence of the amino group, the electron-donating ability of which is apparently not sufficient in this case to offset the electron attracting effect of the pyrimidine ring nitrogen atoms (359).

Zincke found that phenylhydrazine would open the pyridine ring of 2,4-dinitrophenylpyridinium chloride to yield the hydrazone **III-275**, (R = C_6H_5) (101). This reaction was further investigated by Beyer and Thieme, who used a number of heterocyclic hydrazines (360). When the corresponding hydrazones (**III-275**; R = 2-pyridyl, 2-quinolyl, 2-(4-methyl)thiazolyl, etc.) were heated in glacial acetic acid or alcoholic hydrochloric acid, the betaines **III-276** were formed. The latter underwent 1,3-dipolar addition reactions with acrylonitrile to give tetrahydropyrazolopyridines **III-277**. The use of 2,4-dinitrophenylisoquinolinium chloride and acrylonitrile, methacrylic acid, or allyl isothiocyanate yielded the corresponding tetrahydropyrazoloisoquinolines **III-278**.

Recently, Tamura and co-workers prepared a large number of *N*-substituted iminopyridinium and iminoisoquinolinium betaines using this ring opening ring cyclization procedure (361, 362), including some containing deuterium for mass spectrometric studies (245). If hydrazine itself was used, the *N*-aminopyridinium salts (**III-279**) were obtained, while hydroxylamine gave the corresponding pyridine-1-oxides (**III-280**) (362).

Acylpyridinium salts also undergo the ring opening reaction (363), but until recently the only pyridinium salt having a labile acyl group known to undergo successful ring opening was **III-281**, (n = 2) reported by Schwarzenbach and Weber (364). Further examples of compounds having structure **III-281**, (n = 1) were reported by Fischer (365) who prepared the acylvinylpyridinium salts by the reaction of pyridine with β-chlorovinyl ketones or with propargyl aldehyde (75). When treated with sodium hydroxide these salts yielded the ring-cleaved products **III-282**, which in turn, gave (reversibly) the aldehydes **III-283** on acidification.

A spectroscopic method for determining *gem* polyhalogen derivatives has been reported by Bartos and is based on the pyridinium ring opening reaction (366).

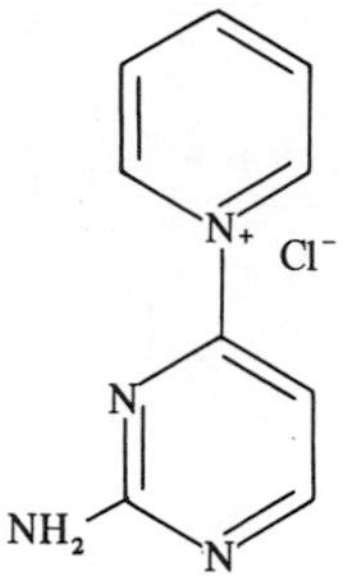

N^+ Cl^-

N

NH_2 N

N–NHR

NH

NO_2

NO_2

HOAc

Δ

N^+

N^-

R

III-274

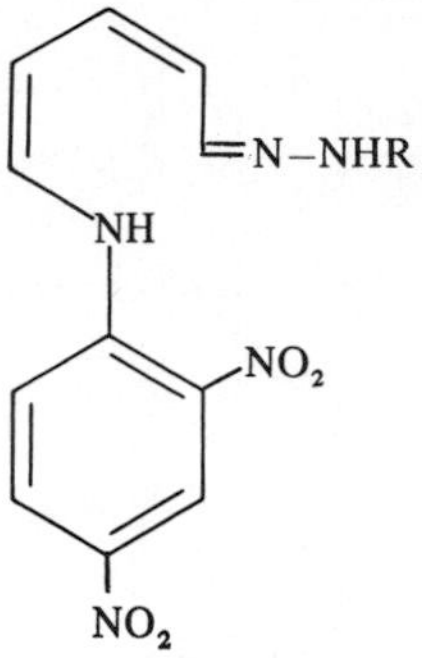

III-275

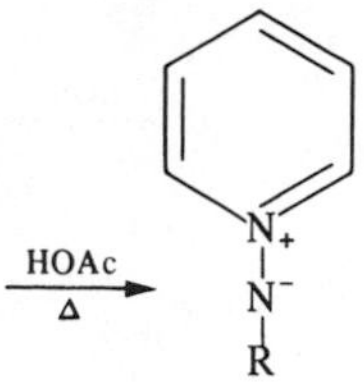

III-276

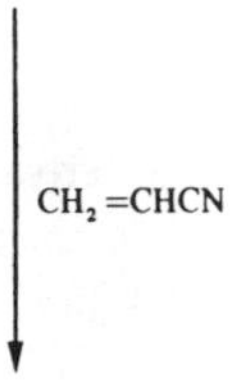

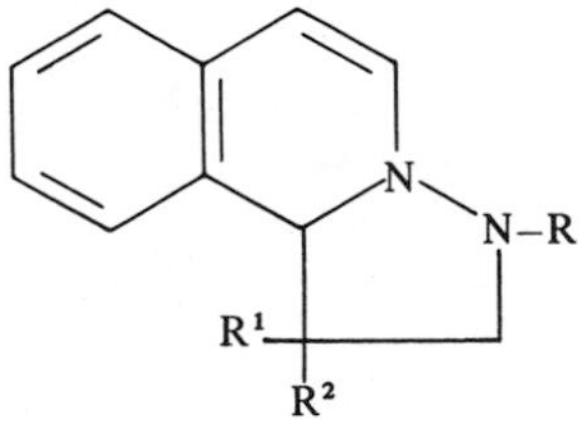

CH_2 =CHCN

H

CN

N

N

R

N

N–R

R^1

R^2

R^1 = H, CH_3

R^2 = CN, COOH, NCS

III-278

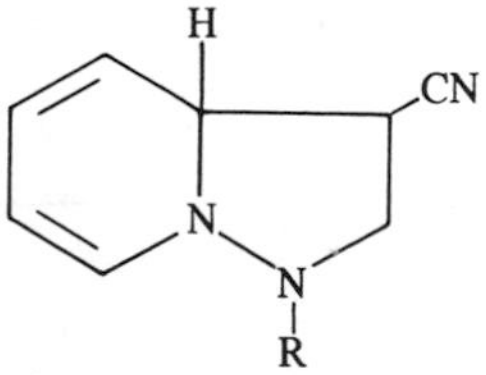

III-277

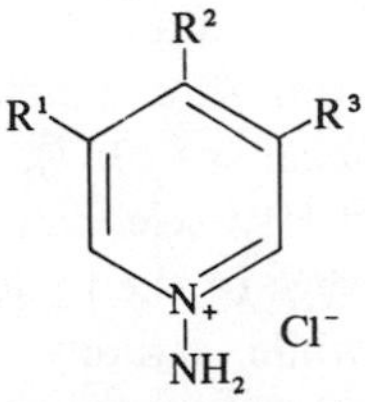

R^2

R^1 R^3

N^+ Cl^-

NH_2

III-279

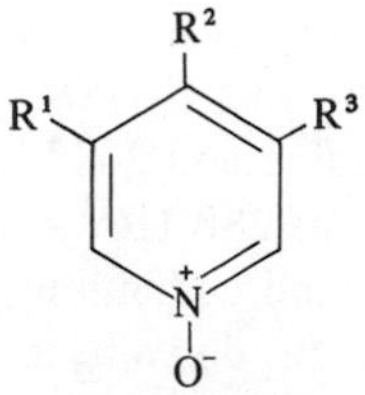

R^2

R^1 R^3

N^+

O^-

III-280

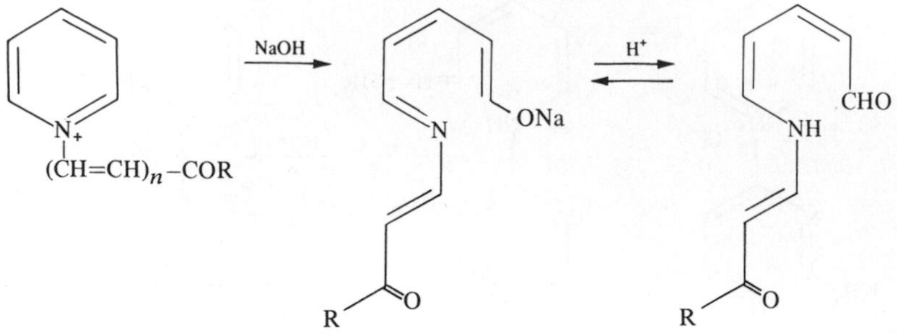

NaOH →; H⁺ ⇌

N^+ — $(CH{=}CH)_n{-}COR$; N … ONa, R … O ; NH … CHO, R … O

III-281 **III-282** **III-283**

The halogen derivative (e.g., **III-284**) is treated with pyridine and the pyridinium salt **III-285** is then ring cleaved with base to yield the glutaconic dialdehyde derivative **III-286**. When reacted with *p*-aminobenzoic acid **III-286** gives the fluorescent Schiff's base **III-287**.

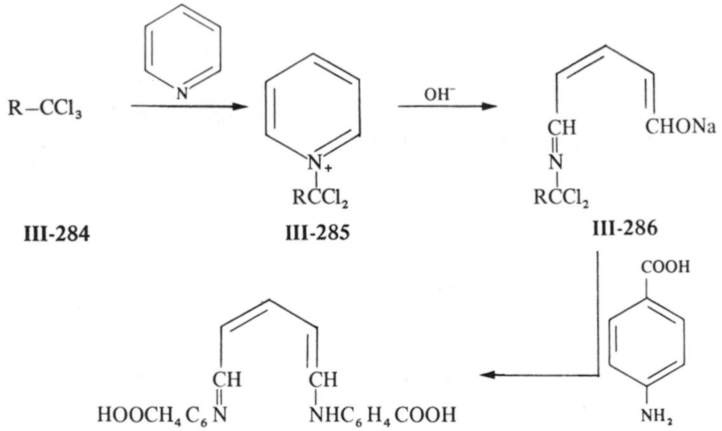

$R{-}CCl_3$ → (pyridine) → N^+ — $RCCl_2$ → OH^- → CH=N—$RCCl_2$, CHONa

III-284 **III-285** **III-286**

COOH, NH_2

CH=N—C_6H_4COOH (HOOCH$_4$C$_6$N), CH—NHC_6H_4COOH

III-287

Tamura and co-workers have investigated the reaction of various nucleophiles with the pyridinium salt **III-288**, prepared from an equimolar mixture of the corresponding pyridine and 3-chloro-2-cyclohexenone (367). In the case of the parent compound **III-288** (R^1 = R^2 = R^3 = H) nucleophiles such as sodium hydroxide, aniline, and carbanions cause cleavage of the pyridinium ring, giving the glutaconaldehyde derivatives **III-289**, while carbonyl reagents such as semicarbazide, phenylhydrazine, and hydroxylamine attacked preferentially the keto group, yielding the corresponding carbonyl derivative.

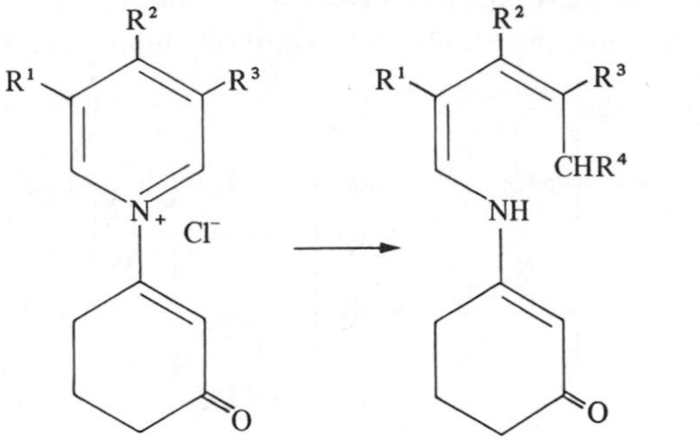

R^4 = O, NC_6H_5, C(CN)COOEt

III-288 III-289

A new quinoxaline synthesis has been reported, based on the pyridinium ring opening of compounds of structure **III-290** (398). When these salts were treated with sodium hydroxide, the quinoxaline derivatives **III-291** were obtained, which in turn yielded the cleaved products **III-292** and **III-293** on reduction with sodium borohydride followed by acid hydrolysis.

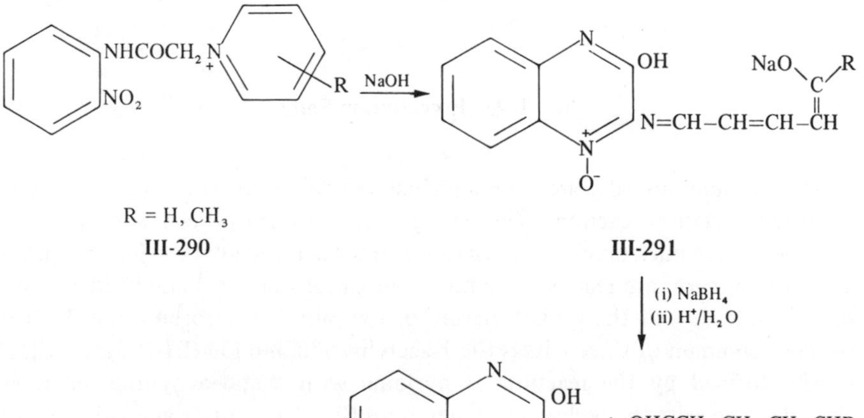

R = H, CH_3

III-290 III-291

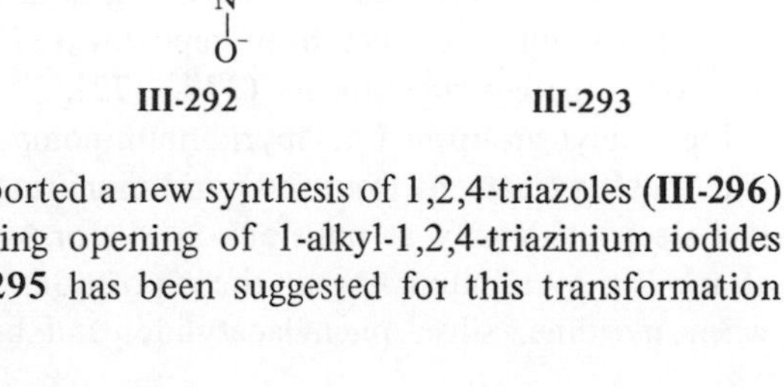

(i) $NaBH_4$
(ii) H^+/H_2O

+ OHCCH=CH–CH=CHR

III-292 III-293

Recently, Lee and Paudler reported a new synthesis of 1,2,4-triazoles **(III-296)** based on the base catalyzed ring opening of 1-alkyl-1,2,4-triazinium iodides **(III-294)**. The mechanism **III-295** has been suggested for this transformation (369).

Quinolizinium salts have been converted to indolizines by a reaction sequence involving the opening of one of the quinolizinium rings (458).

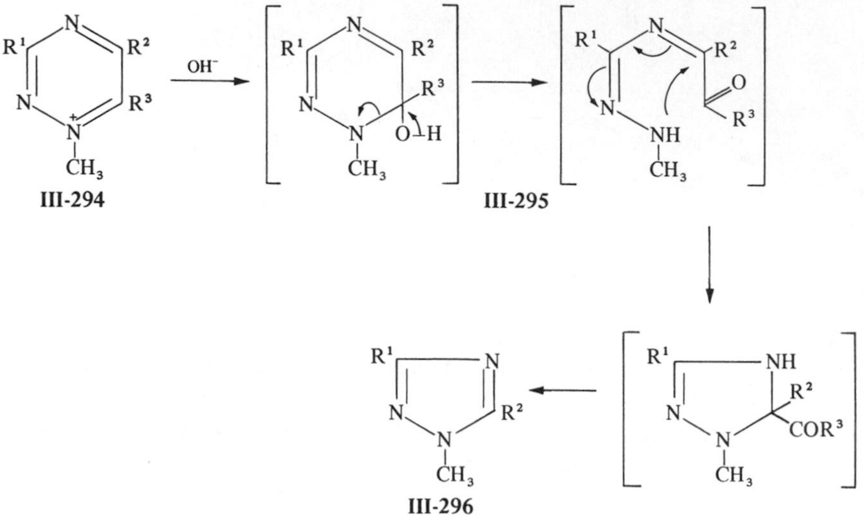

III-294

III-295

III-296

6. 1-Acylpyridinium Salts

1-Acylpyridinium salts are generally believed to be intermediates in pyridine catalyzed acylation reactions. The salts are readily formed because pyridine is a highly effective nucleophile for acylating agents having a good leaving group, and the salts, in turn, are reactive because a resonance stabilized amide structure is not possible due to the tertiary nature of the pyridine nitrogen atom. Perhaps the most common of these salts is the 1-acetylpyridinium ion (**III-297**; R = CH_3), usually formed by the reaction of pyridine with acetic anhydride or acetyl chloride. The salts themselves are generally difficult to isolate and evidence for their existence in transacylation reactions is largely kinetic (370). A structure-reactivity study involving the acylating abilities of a number of substituted 1-acylpyridinium ions has been reported (371), the ions being stabilized by electron donating substituents (370, 372).

The *N*-acyl group in 1-acylpyridinium compounds has the expected effect of withdrawing electrons from the pyridinium ring and effectively places centers of low electron density at positions 2-, 4-, or 6-. Consequently, nucleophiles can attack the pyridinium ring at these positions to yield dihydropyridines. Thus when pyridine, silver phenylacetylide, and benzoyl chloride were heated in carbon tetrachloride solution, approximately 70% of the addition product **III-298** was obtained, which showed the properties of a diene, an amide, an alkyne, and an active hydrogen compound (373). Grignard reagents have been reported to react readily with *N*-ethoxycarbonylpyridinium salts to give the C-2 addition products **III-299** (374), and 1-acylated lepidine and quinaldine can undergo self-alkylation at the 2-position to form dimeric products (374a).

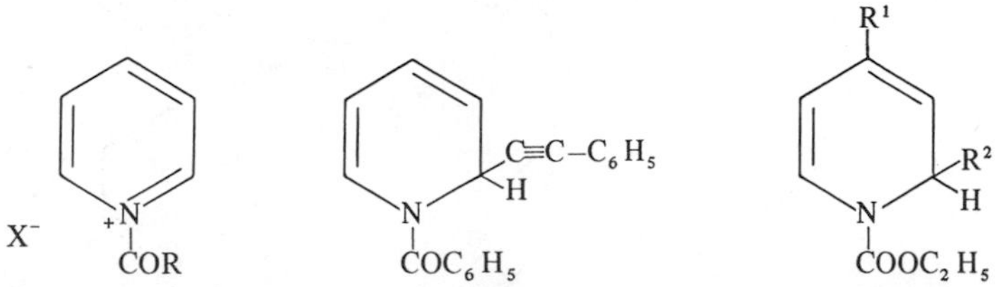

X^- $\overset{+}{N}$–COR

III-297

N–COC_6H_5, C≡C–C_6H_5, H

III-298

R^1, R^2, H, N–$COOC_2H_5$

III-299

N-Acylpyridinium salts have also been condensed with indoles to yield either the 1,4-dihydroacylpyridyl derivative **III-300** (375) or the pyridinium salt **III-301**, the product obtained depending on the solvent and the reactant ratio

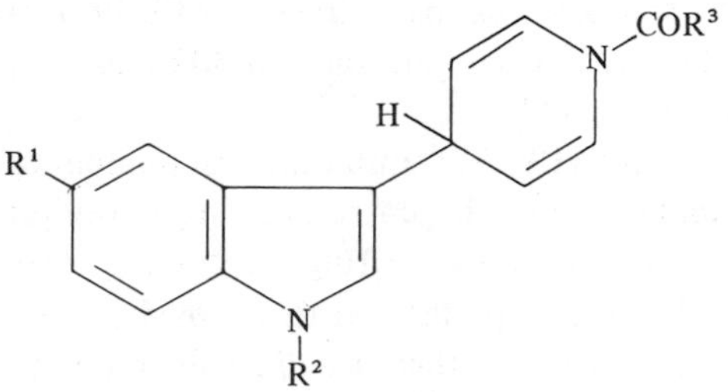

COR^3, H, R^1, N–R^2

III-300

$\overset{+}{N}H$ X^-, R^1, N–R^2

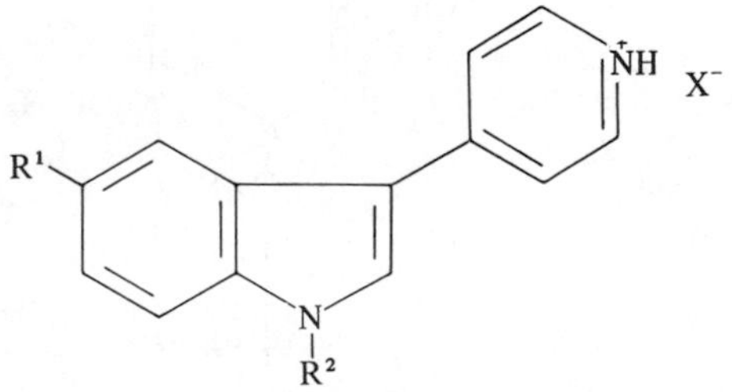

III-301

employed (376). It was shown that **III-300** is an intermediate in the formation of **III-301** and a hydrogen transfer mechanism has been proposed as a key step (376).

N-Acylpyridinium salts (**III-297**) may also act as electrophiles in Friedel-Crafts type reactions, condensing with *N,N*-dimethylaniline in the presence of aluminum chloride to yield the 4-pyridyl derivative **III-302** (377). For this

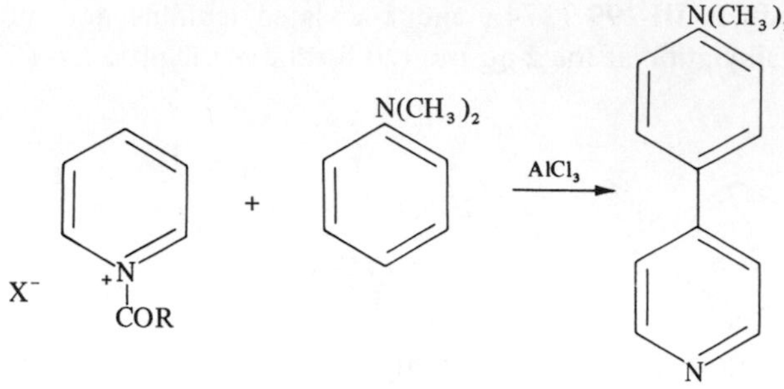

$N(CH_3)_2$

$N(CH_3)_2$

$AlCl_3$

X^- ^+N COR

N

III-297 **III-302**

reaction, it was found that, in general, aroylpyridinium salts gave higher yields than did the aliphatic derivatives, presumably because the aromatic substituents have the stronger electron-withdrawing power. The yields also were dependent on the nature of the anion, increasing in the order Cl, Br, I, and on the length of the aliphatic acyl chain (377). The product **III-302** and its quaternary salts are of interest as bactericides (378).

N-Acyl salts of pyridine (**III-297**), quinoline, isoquinoline, and acridine react with triethyl phosphite to give the corresponding phosphonic acids (**III-303**). The acids are obtained readily by heating mixtures of the heterocycles, acid chlorides, and the trialkyl phosphite, and this provides a method for preparing some phosphonic acids which are otherwise difficult to obtain (379).

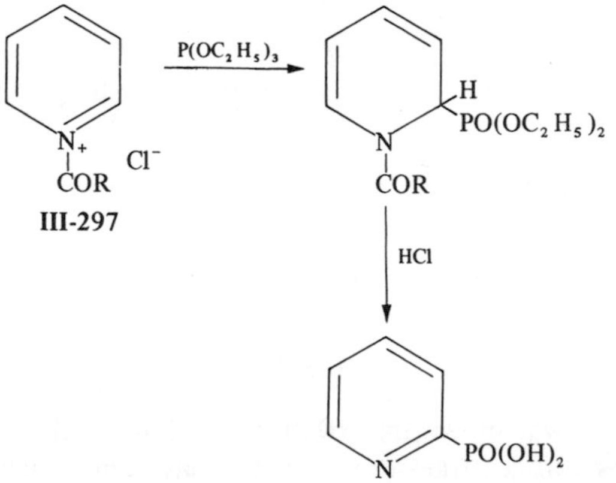

$P(OC_2H_5)_3$

N^+ Cl^- COR

III-297

H $PO(OC_2H_5)_2$

N COR

HCl

N $PO(OH)_2$

III-303

A method which may be useful for protecting free amino groups of peptides has been reported recently, using the stable 1-*t*-butoxycarbonylpyridinium salt **III-304**. The salt reacts with amines in aqueous solution, yielding the *t*-butoxycarbonyl derivative (380).

Johnson and Rumon studied the chemical nature of the carbamoylating agent, carbamoylpyridinium chloride (**III-305**; $R^1 = R^2 = H$), and the methyl derivatives (**III-305**; $R^1 = CH_3$, $R^2 = H$ and **III-305**; $R^1 = H$, $R^2 = CH_3$) (381). In contrast to acylpyridinium chlorides which do not decompose in nonhydroxylic solvents, the carbamoylpyridinium chlorides yield dimethylcarbamoyl chloride (**III-307**) and the corresponding pyridine derivative (**III-306**). It was also found the 2,6-lutidine and 2,4,6-collidine (but not 2,6-dimethoxypyridine) will form acylpyridinium compounds with acetyl chloride while none of these bases will react with dimethylcarbamoyl chloride (381).

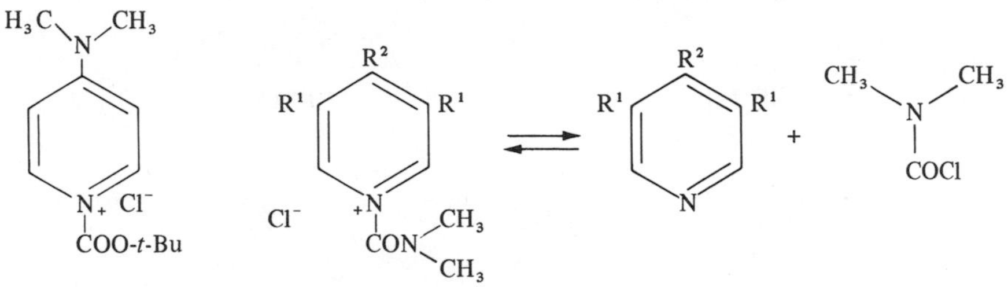

III-304 **III-305** **III-306** **III-307**

N-(*p*-Nitrophenoxycarbonyl) derivatives **III-308** of glycine (R = H), alanine (R = CH_3), and leucine [R = $CH_2CH(CH_3)_2$] react with pyridine to give the isoxazoline-2,5-diones (**III-310**), the cyclizations probably occurring *via* the intermediate *N*-carbamoylpyridinium dipolar ion **III-309** (382). Furthermore, the isoxazoline-2,5-diones can polycondense, yielding the corresponding polyamino acids (**III-311**) and provide a potential synthetic route to peptides and proteins.

In a study of the acetylation of 4-aminopyridine (**III-312**), Wakselman and Guibé-Jampel found that acetylation can occur at either nitrogen atom depending on the reaction conditions, (383). It had been previously observed that 4-acetamidopyridine (**III-313**) is obtained with acetic anhydride, either neat or in acetic acid (384), but the French workers found that 1-acetyl-4-aminopyridine (**III-314**) is obtained with acetyl chloride. Their studies further suggest that **III-314** is probably an intermediate in the formation of **III-313** (383).

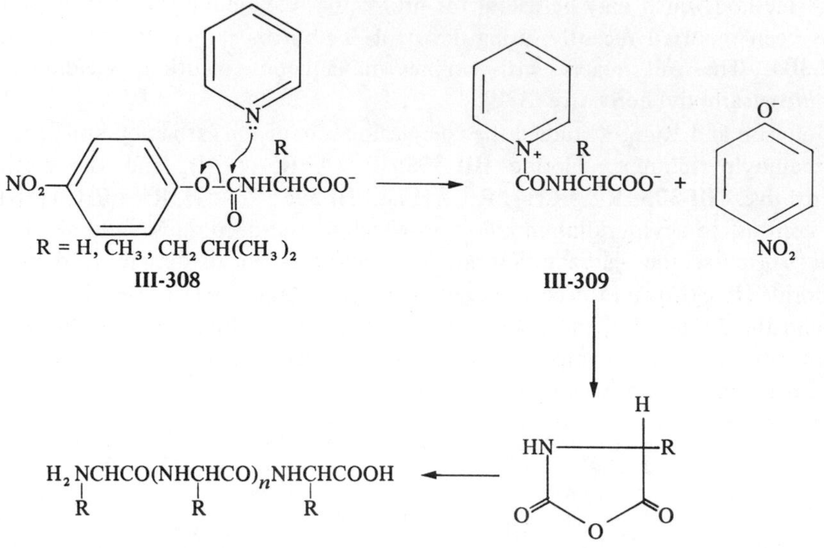

NO_2–C$_6$H$_4$–O–C(=O)NHCH(R)COO$^-$ + pyridine → pyridinium–CONHCH(R)COO$^-$ + $^-$O–C$_6$H$_4$–NO_2

R = H, CH_3, $CH_2CH(CH_3)_2$

III-308 III-309

III-310 → $H_2NCHCO(NHCHCO)_nNHCHCOOH$ (each CH bearing R)

III-311 III-310

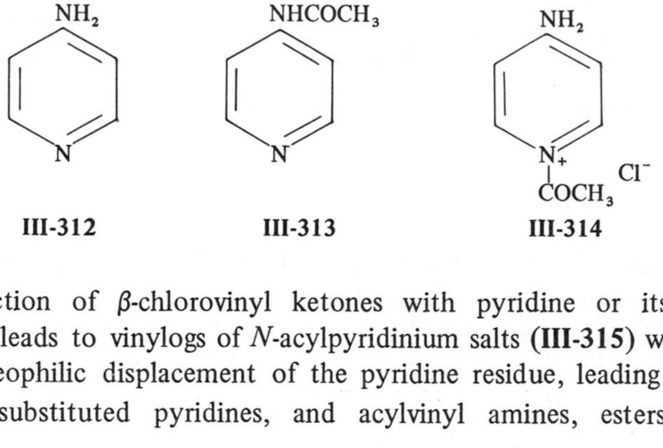

III-312 III-313 III-314

The reaction of *β*-chlorovinyl ketones with pyridine or its substituted derivatives leads to vinylogs of *N*-acylpyridinium salts **(III-315)** which undergo facile nucleophilic displacement of the pyridine residue, leading to pyrazoles **(III-316)**, substituted pyridines, and acylvinyl amines, esters, or azides,

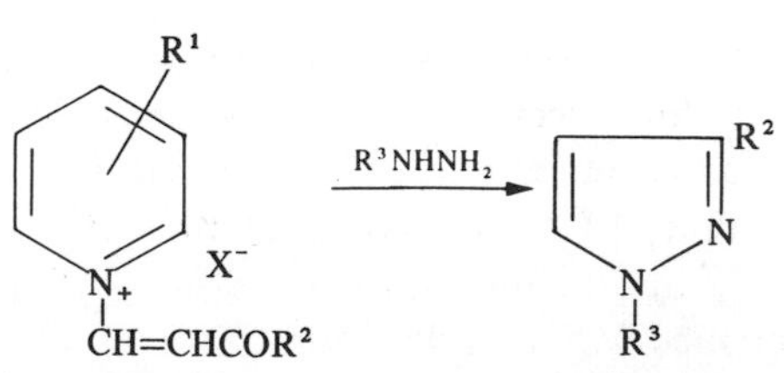

R^3NHNH_2

X^- $CH{=}CHCOR^2$

III-315 III-316

depending on the nucleophile (385). The salts also undergo ring opening when treated with alkali, leading to azaoxonol dyes (386).

Undheim and Tveita have prepared and studied the thiazolopyridinium oxide system **III-317** which exhibits a pH dependent tautomerism with enolic form **III-318** (387). The system was found to be labile, readily undergoing hydrolysis and aminolysis (**III-319**).

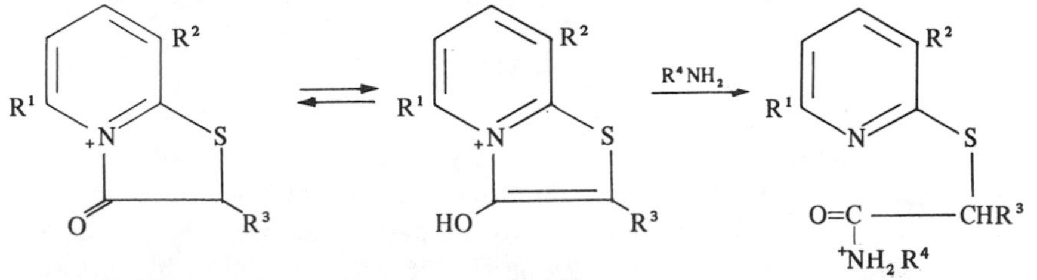

III-317 ⇌ **III-318** $\xrightarrow{R^4NH_2}$ **III-319**

The preparation of *N*-acyldihydropyridines (**III-321**) has been accomplished by treating 1-lithio-2-phenyl-1,2-dihydropyridine (**III-320**) with an acid chloride or an ester (388). The NMR spectra of *N*-acylpyridinium salts have been studied (389). The adduct **III-322** has been isolated in an attempted acetylation using pyridine and acetic anhydride (390).

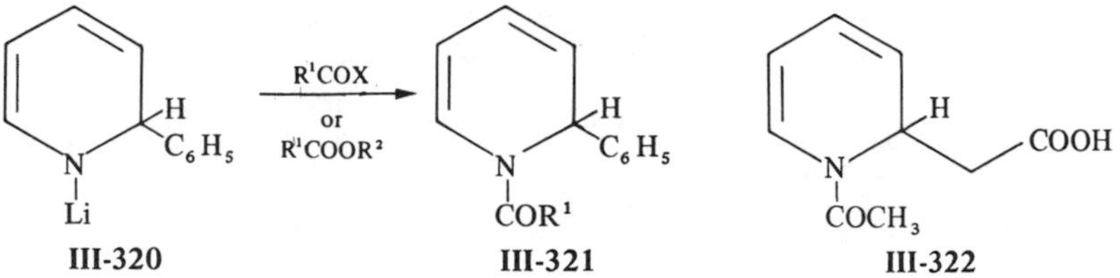

III-320 $\xrightarrow{R^1COX \text{ or } R^1COOR^2}$ **III-321**; **III-322**

7. Pyridinium Salts of Value in General Synthetic Work

Because of the variety of nucleophilic reactions which pyridinium ions undergo (391), and certain other properties characteristic of their structure, some have high potential synthetic utility (235, 393). In this section reactions are covered which in the opinion of the author have use in synthesis and which have not been discussed in other parts of this chapter.

Alkylating agents capable of functioning in biological media have been developed, consisting of an alkylsulfonate function as the alkylating moiety and a quaternary ammonium group to provide water solubility. The most active agent of those prepared was the pyridinium perchlorate **III-323** (220, 393).

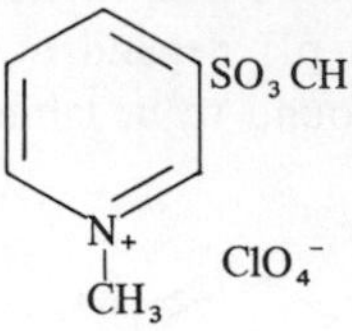

SO_3CH_3

N^+ ClO_4^-

CH_3

III-323

Nace and Nelander prepared a steroid ring A diosphenol (2,3-diketone) by a method reported earlier by Ruzicka, Plattner, and Furrer (394). The bromopregnane derivative **III-324** was converted to the pyridinium bromide **III-325** with pyridine, which then yielded the nitrone **III-326** when treated with *p*-nitrosodiethylaniline. Hydrolysis of the nitrone with dilute hydrochloric acid gave the diosphenol **III-327** in high yield (395). The sequence was also applied to prepare the diosphenol system in the *D*-homo steroid **III-329** from the ketone **III-328** (396).

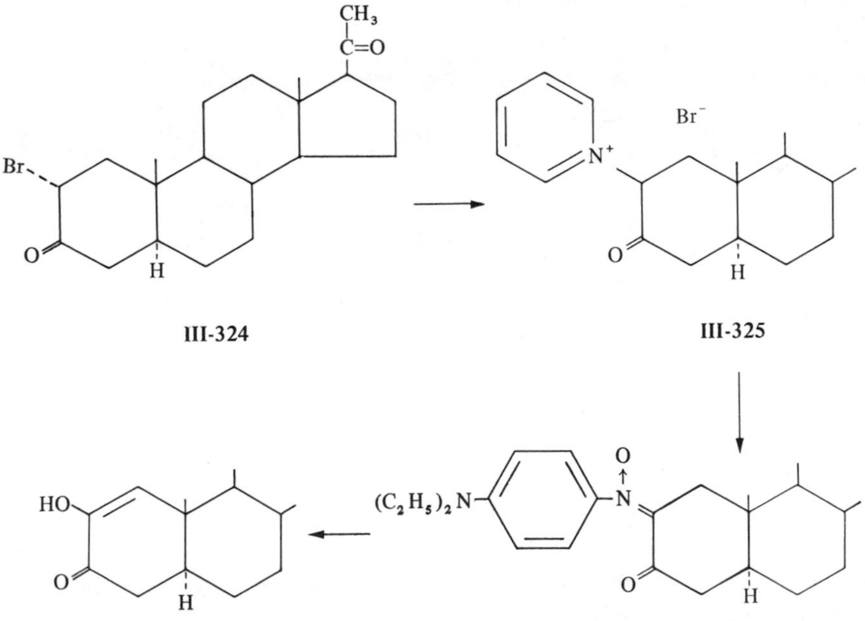

CH_3 C=O Br O H

III-324

Br^- N^+ O H

III-325

HO O H

III-327

$(C_2H_5)_2N$ O ↑ N O H

III-326

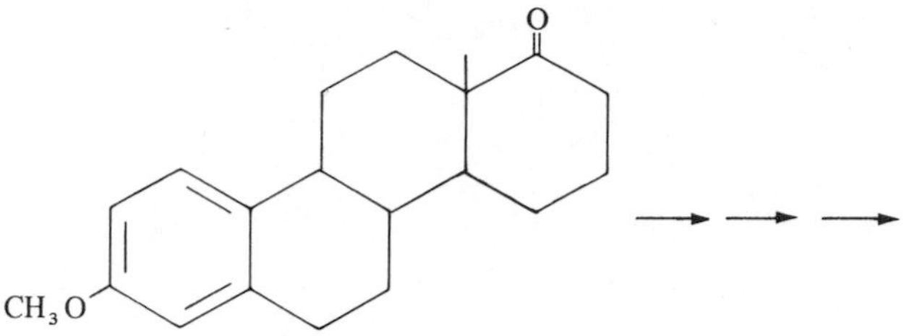

CH$_3$O

III-328

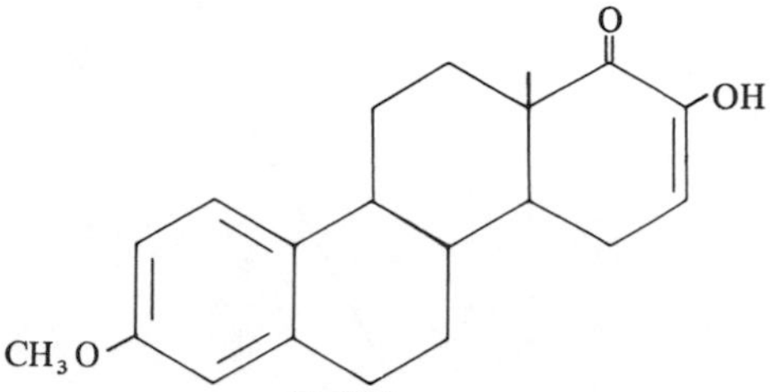

→ → →

CH$_3$O

OH

III-329

The possibility that the pyridinium salts of structure **III-325** reported by Ruzicka and co-workers may actually be 2-keto-3-pyridinium compounds, arising *via* a rearrangement similar to that observed with the acetoxycholestanones (397), was ruled out by the reduction of the pyridinium salts to the 3-keto steroids (398). In this connection, Warnhoff has reported the oxidation of the γ-picolinium salt **III-330** to the *seco* diacid **III-332** both directly by permanganate in basic media or by the base-catalyzed reaction with ethanolic isoamyl nitrite, followed by acid hydrolysis of the pyridinium oxime ester **III-331** (398).

The use of the nitrone procedure to convert 3,4-bis(chloromethyl)furan (**III-333**; Y = Cl) to 3,4-furandicarboxaldehyde (**III-335**) was employed by Novitskii, Khachaturova, and Yur'ev, by converting the dichloro compound to the dipyridinium salt (**III-333**; Y = $\overset{+}{N}C_5H_5$), and then with *p*-nitrosodimethylaniline to the bisnitrone **III-334**. Acid hydrolysis of the nitrone yielded the dialdehyde **III-335** (399).

Other conversions of pyridinium salts bearing an active methylene group to nitrones have been reported (400–402), in several cases yielding unusual products (401, 402).

Lipke has investigated some new potential chlorinating agents (**III-337**), prepared by treating the pyridinium salts **III-336** with chlorine in chloroform at 0° (403). The following chlorinations could be carried out: acetanilide to 4-chloroacetanilide (60%); β-naphthol to 1-chloro-2-hydroxynaphthalene (60%); and salicylic acid to 5-chloro-2-hydroxybenzoic acid (65%). In the reported

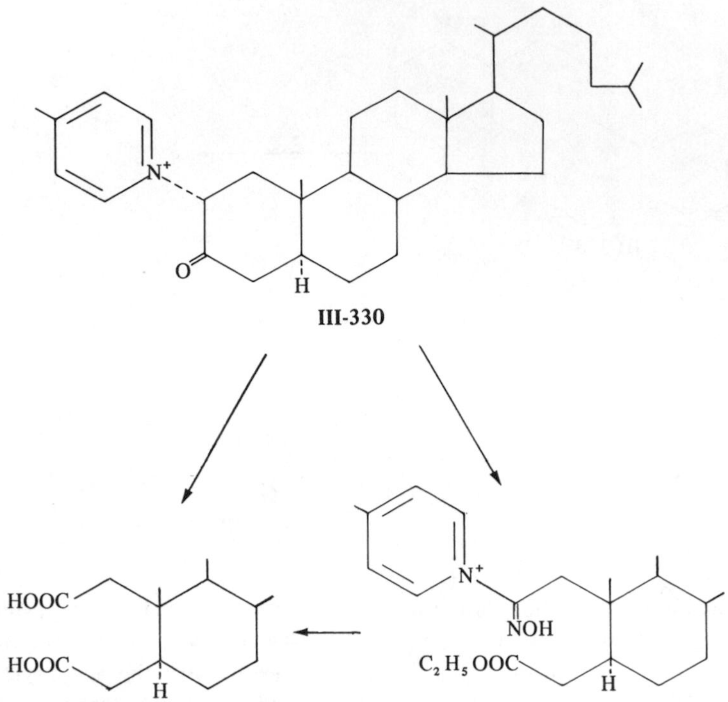

N⁺ O H

III-330

HOOC HOOC H

III-332

N⁺ NOH C_2H_5OOC H

III-331

procedure, the chlorinated products are separated readily from the pyridinium salts because the latter are ether insoluble (403).

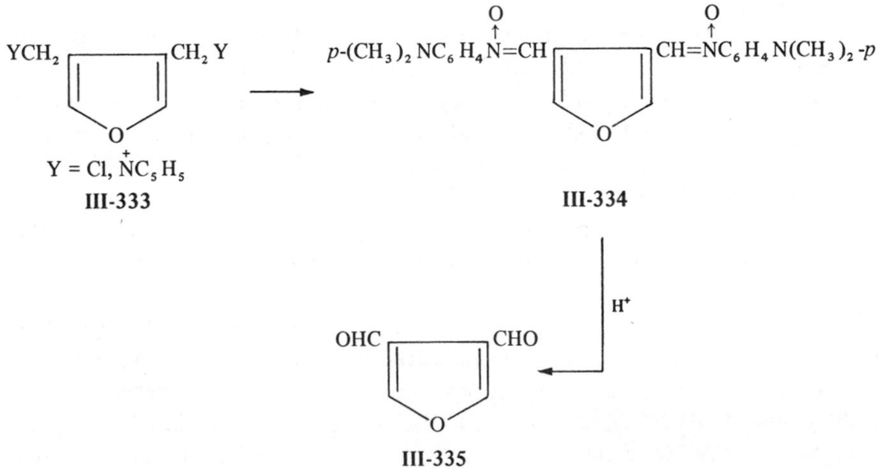

YCH_2 CH_2Y O

Y = Cl, $\overset{+}{N}C_5H_5$

III-333

p-$(CH_3)_2NC_6H_4\overset{O \atop \uparrow}{N}$=CH CH=$\overset{O \atop \uparrow}{N}C_6H_4N(CH_3)_2$-$p$ O

III-334

H^+

OHC CHO O

III-335

N-Nitropyridinium tetrafluoroborates (**III-338**) have the ability to nitrate aromatic substrates in organic solvents at room temperature (404). Interestingly, an alkyl group in the α-position appears to be necessary for nitration. Thus in **III-338** if R = H, no nitration of toluene takes place at 25°, but if R = CH_3, the

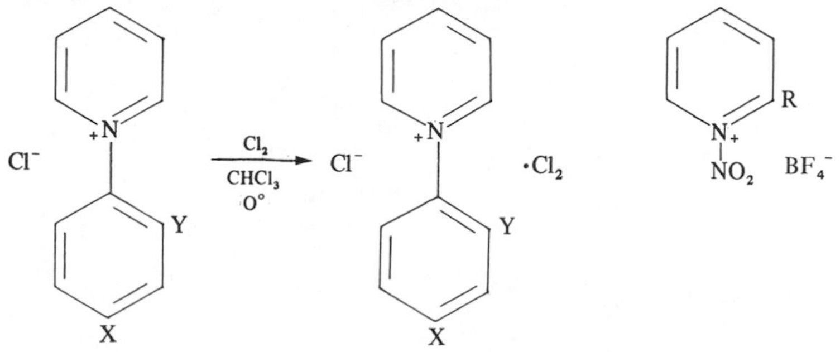

Cl^- +N ... Y, X; $\xrightarrow[\text{CHCl}_3,\ 0^\circ]{\text{Cl}_2}$ Cl^- +N ... Y, X $\cdot Cl_2$; N+ R, NO_2 BF_4^-

X = Cl, Br, I, H
Y = H, I

III-336 **III-337** **III-338**

reaction proceeds quantitatively. The result is interpreted as the α-methyl group preventing the nitro group from achieving coplanarity with the pyridine ring, thus allowing relatively little double bond character between the two nitrogen atoms (404).

8. Formation of Condensed Heterocyclic Systems

Discussion in this section is confined to those structures having a nitrogen atom at a bridgehead position.

A. *Intramolecular Quaternization*

A number of procedures for preparing ring systems containing a quaternary bridgehead nitrogen atom involve a cycloquaternization as a key step in synthesis. The synthesis of the quinolizinium ion (**III-339**) involved such a procedure (405), as did the preparation of the diquaternary salt **III-341**. The latter was obtained on heating the 1,3-dioxalane derivative **III-340** with hydrobromic acid (406). A similar procedure, but using the oxime **III-342** (R^1 = CH=NOH) (407), the 2,4-dinitrophenylhydrazone **III-342** (R^1 = $CH{=}NNHC_6H_3(NO_2)_2$) (408, 409) or the methyl ketone **III-342** ($R^1 = COR^2$) (409, 410) yielded the diquaternary salts **III-343**.

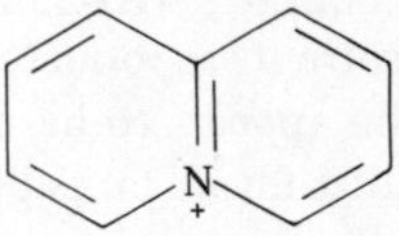

III-339

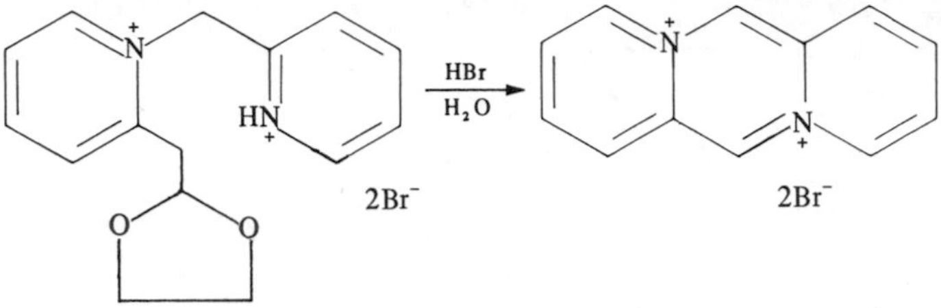

HBr
H_2O

2Br⁻

2Br⁻

III-341

III-340

(i) HBr
(ii) PBr_3

Br⁻ CH_2R^1

R^1

2Br⁻

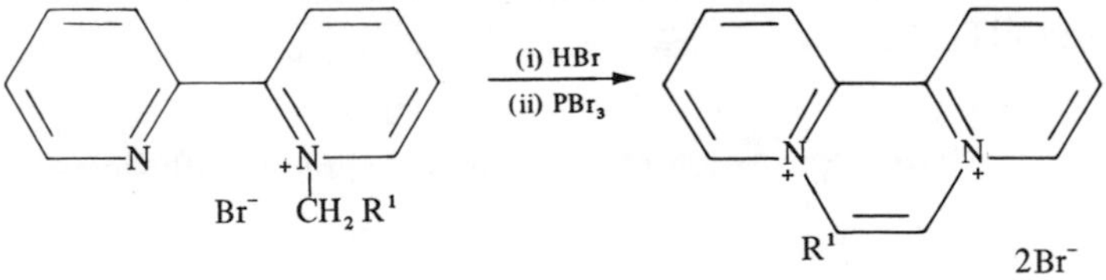

$R^1 = CH{=}NOH, COR^2, CH{=}NNHC_6H_3(NO_2)_2$

$R^1 = H, R^2$

III-342

III-343

The benzo[*c*]quinolizinium ring system (**III-345**) was prepared by cycloquaternization starting with the α-ethoxyketone **III-344** (411), or the stilbazoles **III-346** (412). It has been suggested that the aromatic halogen in the stilbazoles is activated sufficiently to undergo quaternization because the electron-withdrawing power of the pyridine ring is transmitted to the benzene ring through the conjugated system (413). A study of some reactions of this benzoquinolizinium ring system has been reported (414), as has the extension of the **III-346**→**III-345** synthetic sequence to tetracyclic systems of potential interest in the preparation of azasteroids (413).

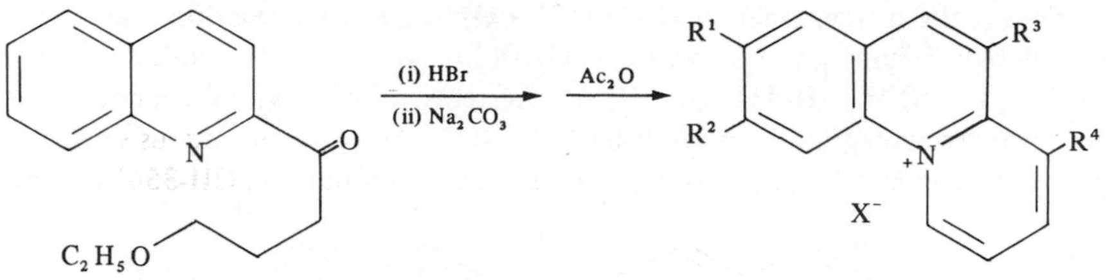

(i) HBr (ii) Na_2CO_3 → Ac_2O →

III-344 III-345

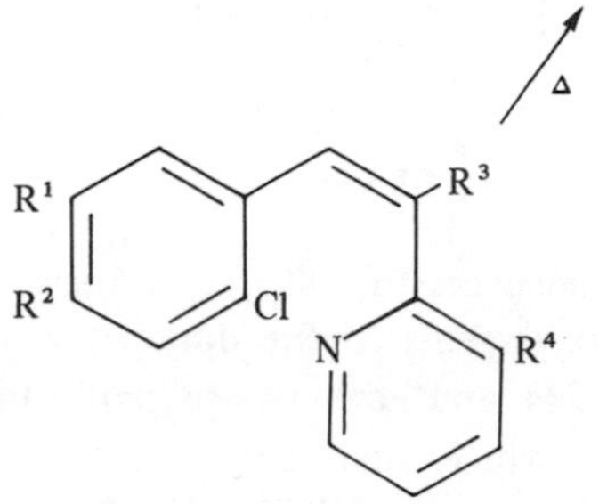

Δ

III-346

Undheim and co-workers have prepared the thiazolopyridinium ring system **III-348** by the intramolecular quaternization of the bromide **III-347** (68); the thiazinopyridinium system **III-349** was made similarly (415).

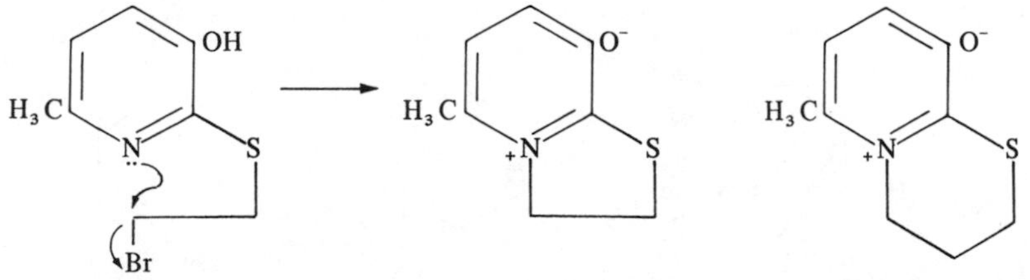

III-347 III-348 III-349

The thiazolopyridinium cation **III-351** (Y = S) has been prepared by the cyclization of the pyridylthio derivatives **III-350** (416–418), and by the ring closure of the 2-pyridinethione compounds **III-352**, (Y = S) (419), while

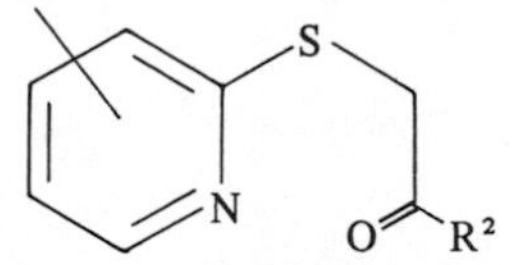

R^1 = H, OH
R^2 = H, alkyl

III-350

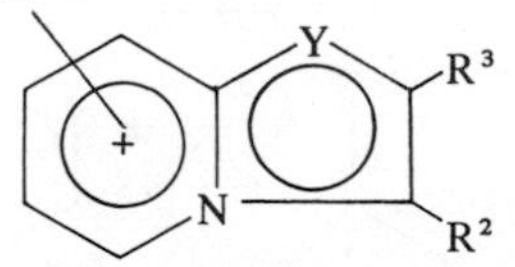

Y = S, O

III-351

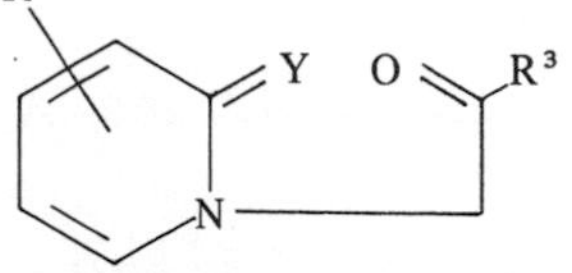

Y = S, O
R^1 = H, CH_3
R^3 = aryl, alkyl

III-352

employing the oxygen analog (**III-352**; Y = O) in the latter reaction leads to the oxazolopyridinium salts **III-351** (Y = O) (420). An alternate synthesis of the **III-351** (Y = S) and **III-351** (Y = O) ring systems, starting with 2-cyanothiazole or 2-cyanooxazole, has been reported (421, 422); however, analogous syntheses of the thiazolopyridinium (**III-353**) and isothiazolopyridinium (**III-354**) systems

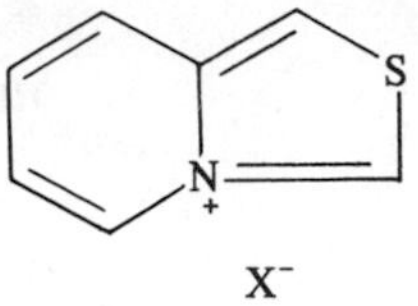

X^-

III-353

X^-

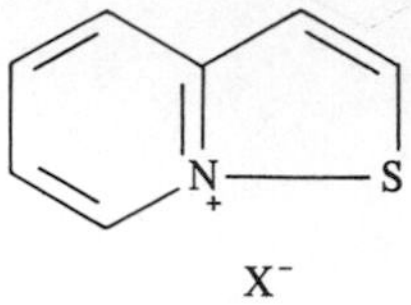

III-354

proved to be unsuccessful (423). Undheim and co-workers have investigated some of the conversions of the dihydro system **III-348** to the cation (**III-351**; Y = S) (418, 424), and have also prepared the homologous thiazinopyridinium system **III-355**, which undergoes facile acid-catalyzed rearrangement to the thiazolopyridinium system **III-351** (Y = S) (415).

Some years ago, Boekelheide and Feeley reported the preparation of the diquaternary system **III-356** by the dimerization of 2-(β-bromoethyl)pyridine (425). A recent attempt, however, to prepare the diketone analog **III-357** by an analogous dimerization using 2-(β-bromoacetyl)pyridine was unsuccessful (426).

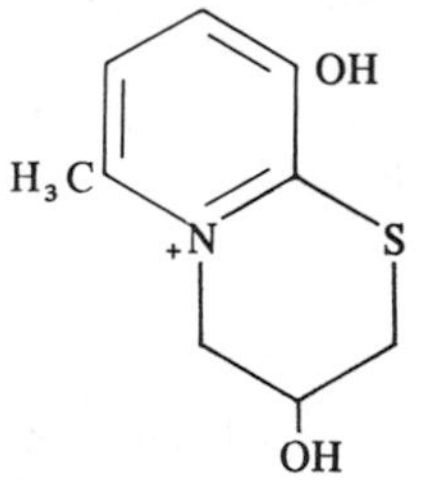

III-355

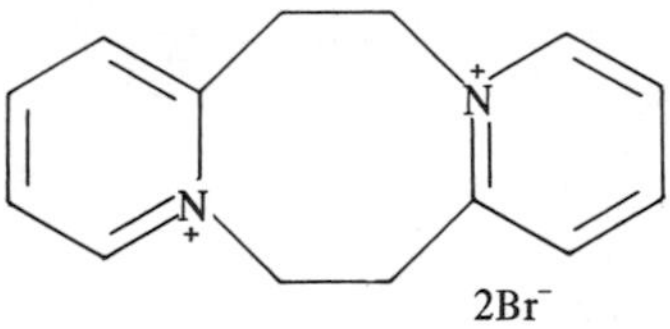

$2Br^-$

III-356

$2Br^-$

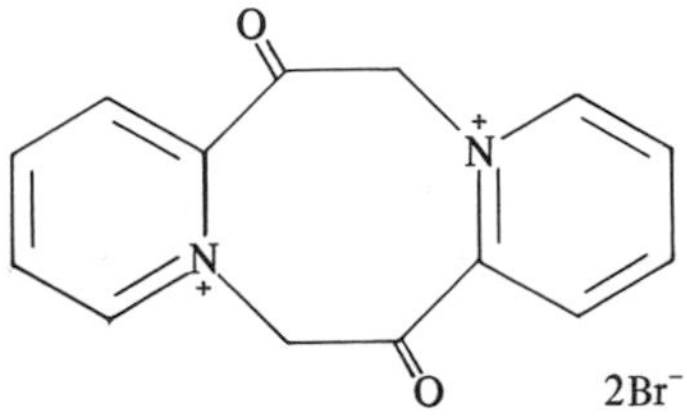

III-357

The dipyridoimidazolium salt **III-360**, of interest in aromaticity studies, could be obtained by the reaction of the 2,2′-bipyridyl **III-358** with methylene iodide or benzylidene chloride, but in general the yields were low and the reaction failed with other *gem*-dihalides (460). The diquaternary salt **III-359** is presumed to be an intermediate (460). The ring system could be prepared in higher yields by first treating the bipyridyls with an α-haloketone to form the monoquaternary salt, followed by cyclization by treatment with bromine or *p*-toluenesulfonyl chloride in pyridine (17) (Section I.1.A.). The diquaternary salts **III-361** (*n* = 2,3,4) had been prepared earlier by Homer and Tomlinson (461).

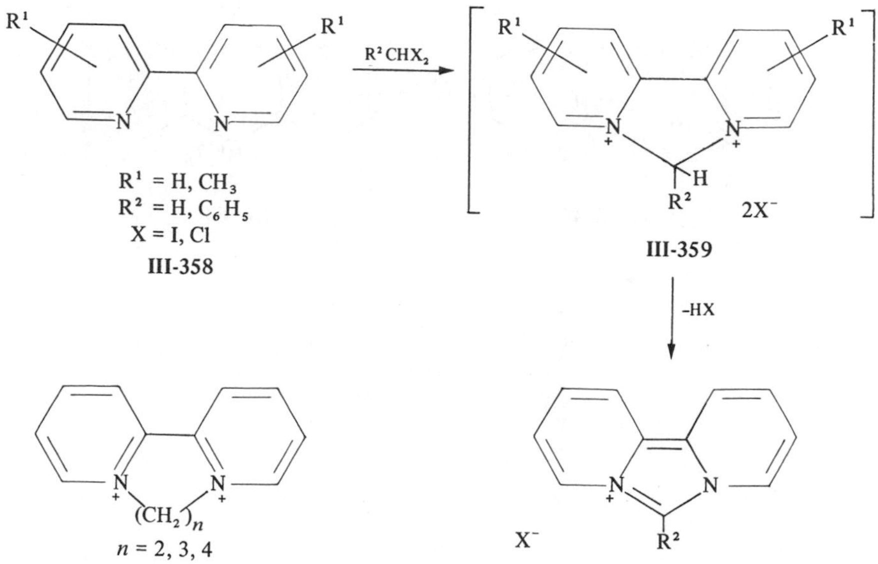

R^1 = H, CH_3
R^2 = H, C_6H_5
X = I, Cl
III-358

R^2CHX_2

$2X^-$
III-359

-HX

X^-
III-360

$(CH_2)_n$
n = 2, 3, 4
III-361

B. *Intramolecular Cyclizations of Pyridinium Salts*

a. CYCLODEHYDRATION. A number of cyclodehydration methods have been employed to prepare ring systems containing a pyridinium nitrogen atom at a bridgehead position and have used pyridinium salts as starting materials. Schraufstätter found that the acetal **III-362** readily underwent hydrolysis and ring closure in boiling 48% hydrobromic acid to give the 3-hydroxyquinolizinium salt **III-363** (427). The acridizinium (**III-364**) and phenanthridizinium

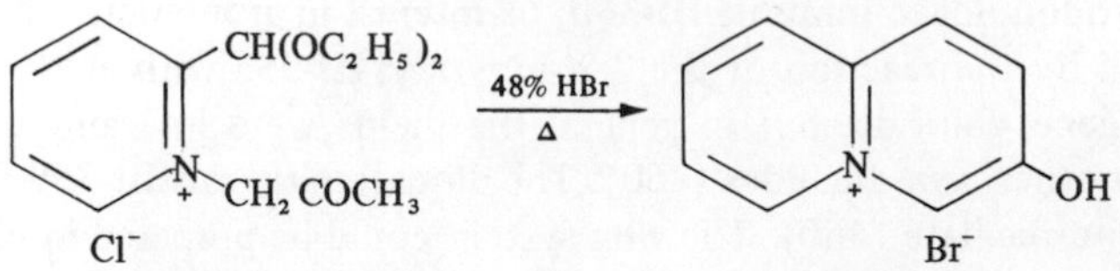

48% HBr, Δ

III-362 III-363

(**III-365**) systems have been prepared in similar fashion and their syntheses and reactions, as well as those of the benzo[*c*]quinolizinium system (**III-345**), have been reviewed (428).

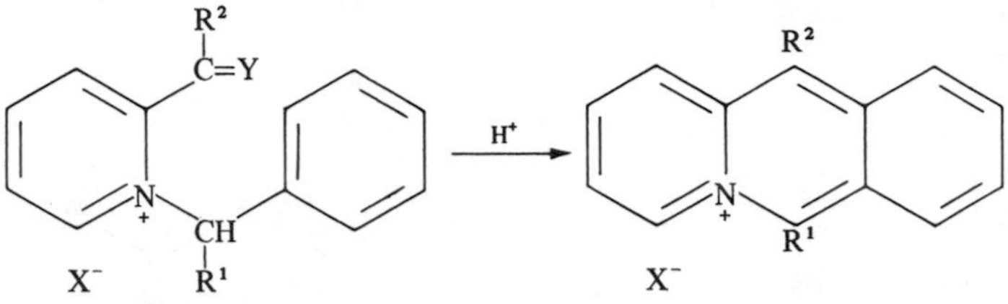

$Y = O, (OCH_3)_2$

III-364

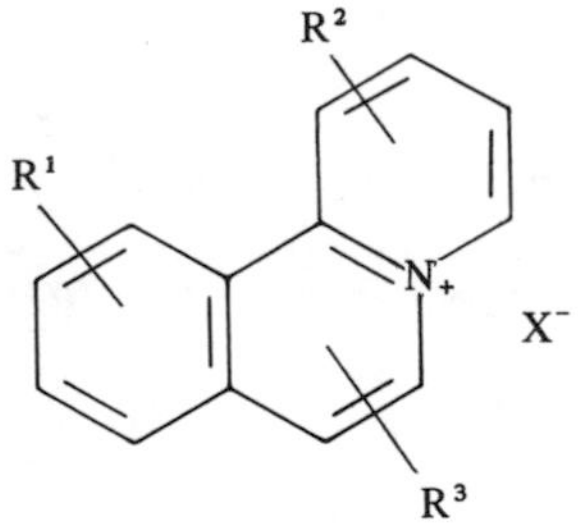

III-365

When the oxime **III-367** was treated with concentrated sulfuric acid, the bromolactam **III-366** was obtained, arising by a Beckmann rearrangement, cyclization, and bromination. The unbrominated compound was formed if the perchlorate salt of **III-367** was used instead of the bromide (429). When **III-367** was treated with 48% hydrobromic acid, the 2-azaquinolizinium salt **III-368** was obtained (429).

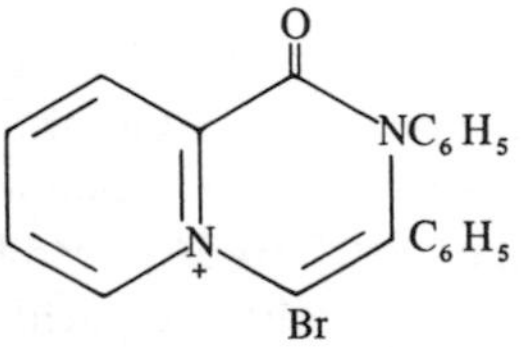

III-366

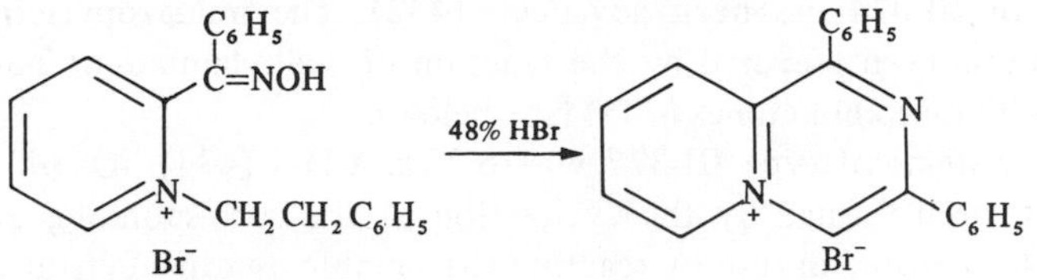

48% HBr

III-367 III-368

The acid catalyzed cyclization of oximes **III-369** leads to the 2-azaquinolizinium oxides **III-370** (429–431), which can be deoxygenated by heating with phosphorous tribromide (429, 431).

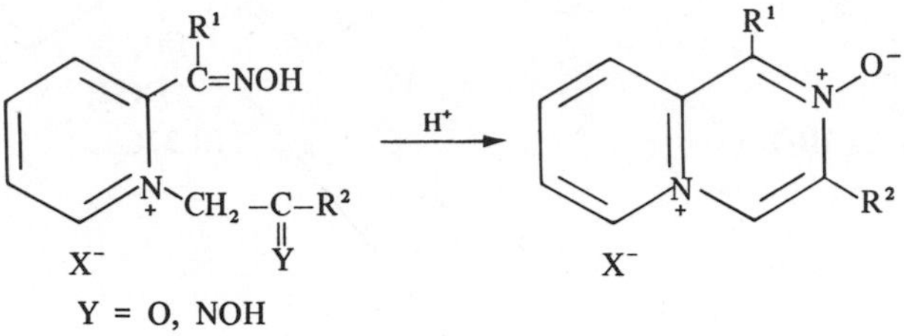

Y = O, NOH

III-369 III-370

Bradsher and co-workers have reported the synthesis of the imidazopyridinium salts **III-372** by the reaction of 2-bromo-1-phenacylpyridinium bromide (**III-371**) with amines (432). On interrupting the reaction with butylamine two intermediates, **III-373** and **III-374**, could be isolated, and a mechanism for the

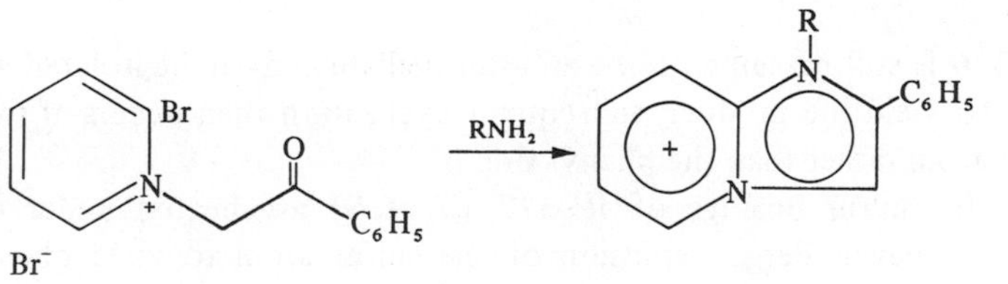

RNH_2

III-371 III-372

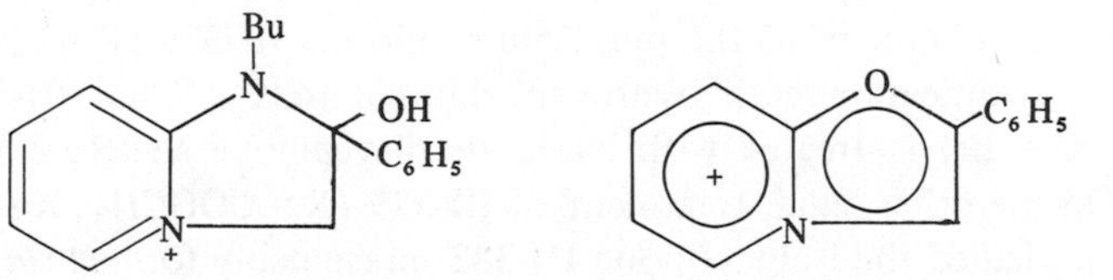

III-373 III-374

formation of **III-374** has been advanced (432). The imidazopyridinium ring system has also been prepared by the reaction of 2-alkylamino- or 2-arylamino-pyridines with α-bromoketones (433) (see below).

Tricyclic systems of type **III-377** where Y is -CH_2- (434), -O- (435), or -S- (436), have been formed by the cyclization of the corresponding quaternary salts **III-376**, prepared in turn by reacting the suitable pyridine derivative **III-375** with bromoacetone. When the latter reaction was attempted with 2-anilinopyridine (Y = NH), the product **III-378** was isolated instead (433). The quaternary

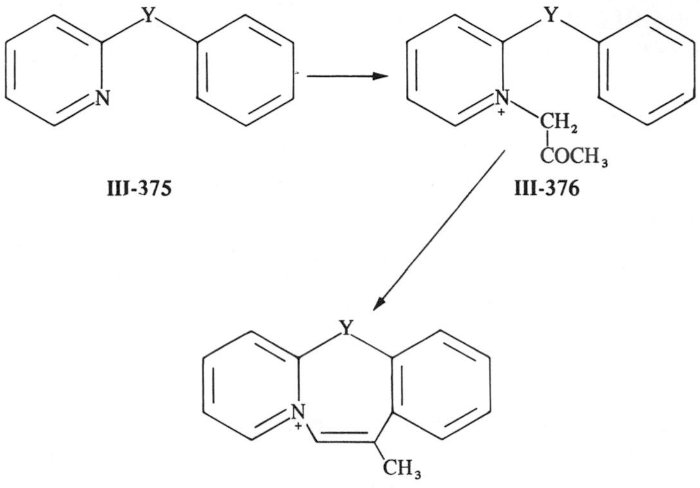

Y N Y N+ CH_2 $COCH_3$

III-375 **III-376**

Y N+ CH_3

III-377

salt **III-376** is still presumed to be an intermediate and can be isolated when Y = N-CH_3. To yield the product, subsequent cyclization then occurs at the anilino nitrogen atom rather than the phenyl ring.

When the sulfur analogs of **III-377** (Y = S) are heated under oxidizing conditions, they undergo expulsion of the sulfur atom to yield phenanthridizinium salts (**III-365**) (437). The reaction appears to proceed *via* a sulfoxide, rather than a sulfone, intermediate (438).

The pyridobenzodiazapine salt **III-380** was obtained by Knowles and coworkers by base treatment of the pyridinium chloride **III-379** (R = CN, X = Cl) followed by acidification with hydrogen chloride (439). When **III-379** (R = $COOCH_3$, X = Br) is treated with base, the diazepinone structure **III-381** is obtained. On the other hand, treatment of **III-379** (R = $COOCH_3$, X = Br) with benzylamine yielded the isoindolinone **III-383**, presumably formed *via* an initial Dimroth rearrangement (440) to the benzylaminopyridine **III-382** (441).

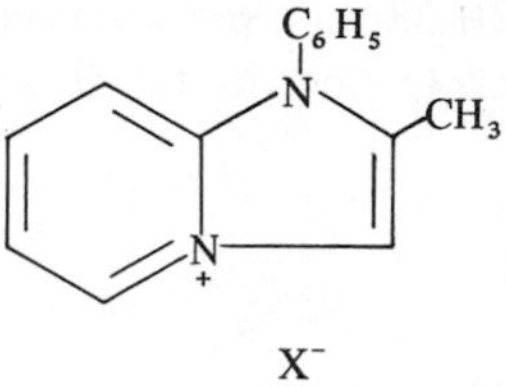

C_6H_5 N CH_3 N X^-

III-378

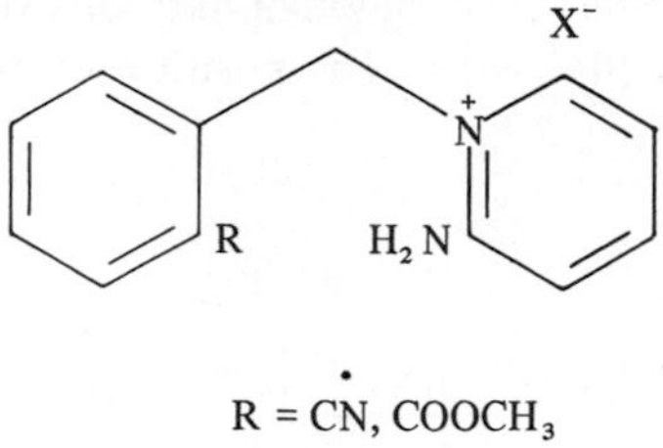

X^- N R H_2N

R = CN, $COOCH_3$

III-379

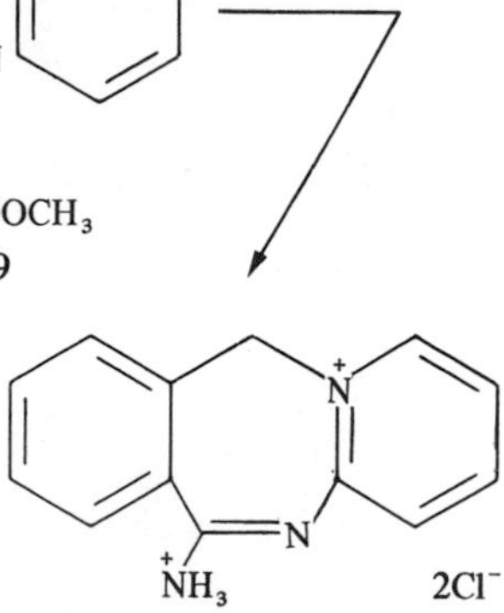

N N $\overset{+}{N}H_3$ $2Cl^-$

III-380

N N O

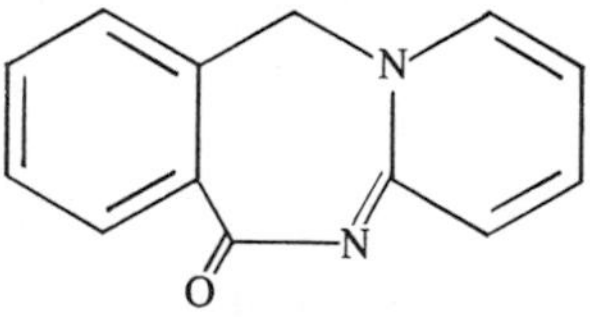

III-381

III-379

R = $COOCH_3$

X = Br

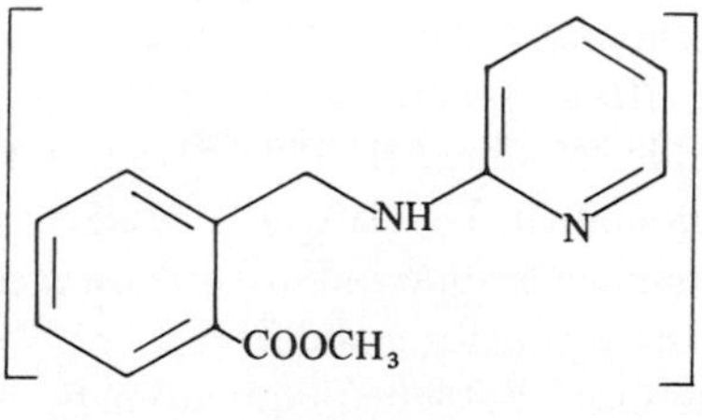

NH N $COOCH_3$

III-382

N N O

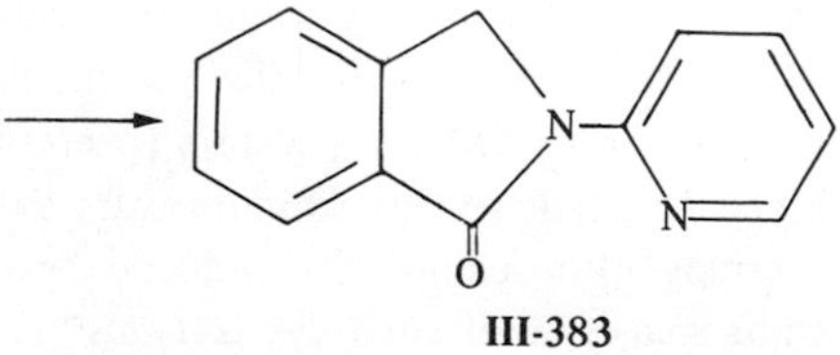

III-383

Indolizines (**III-384**) form indolizinium salts (**III-385**, **III-386**) on protonation with strong acids (441–445). The position of protonation (giving rise to either

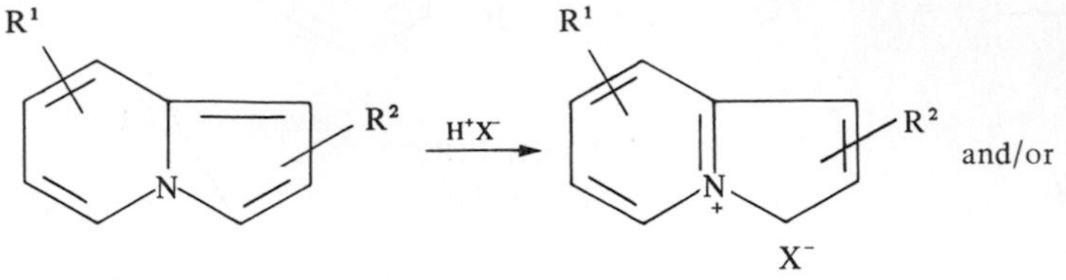

R^1 R^2 $\xrightarrow{H^+X^-}$ R^1 R^2 X^- and/or

III-384 **III-385**

R^1 R^2 X^-

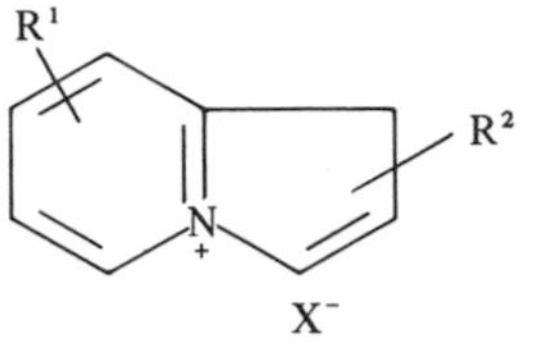

III-386

III-385 or **III-386**) depends on the nature and position of substituents (444, 445), and on the strength of the proton donor (446). Several convenient syntheses of the indolizine ring system involving the cyclization of pyridinium salts are available (444, 447–451), as are other methods (447, 448, 452). The reduction of indolizines has been investigated (453).

b. PHOTOCHEMICALLY INDUCED CYCLIZATION. A procedure for the photocyclization of stilbene derivatives to phenanthrenes (454) has been adapted by Doolittle and Bradsher to the preparation of phenanthridizinium compounds (**III-365**) (455). The method provides a practical synthetic route to these substances having substituents in either of the terminal rings (**III-365**; $R^3 = H$). Along similar lines, Fozard and Bradsher were able to prepare pyridoisoindoles of type **III-388** by the photochemical cyclizations of the pyridinium salts **III-387** (456). When the pyridoisoindolium salts (**III-388**) were treated with aqueous sodium carbonate, they yielded the pyridoisoindoles **III-389**, having a peripherally conjugated ring system (pseudoaromaticity). Lyle and co-workers employed 1-(2-haloaryl)pyridinium salts to study electronic and orientation factors governing photochemically induced cyclodehydrohalogenation reactions. Their results suggest that such cyclizations are unsuccessful when both aromatic rings are electron deficient heterocycles (456a).

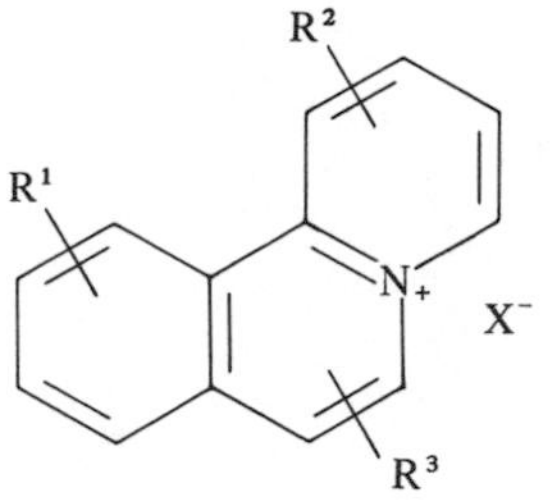

III-365

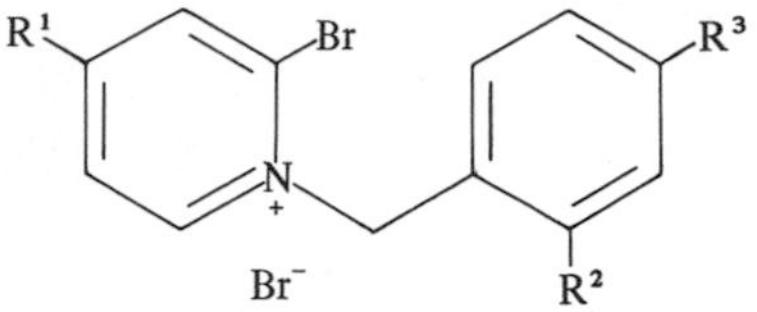

III-387

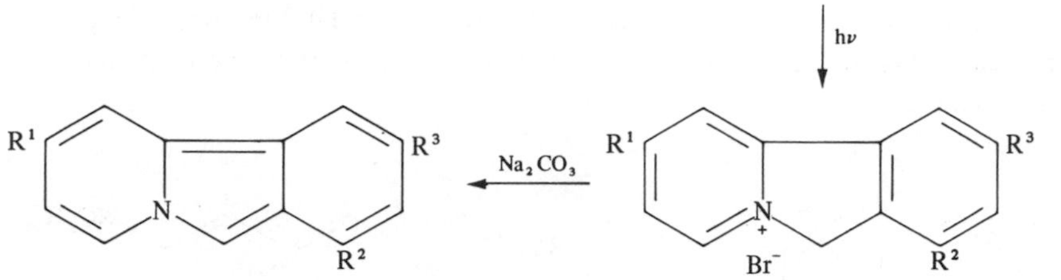

III-389

III-388

c. CYCLOADDITIONS. It has been noted earlier that anionic attack on the pyridinium ring can occur yielding addition products (Section II.2). If this addition occurs intramolecularly, polycyclic ring systems containing a bridgehead nitrogen atom can result. Thus Wilson and DiNinno found that aqueous sodium bicarbonate cyclized the pyridinium salt **III-390** to the 1,6-dihydropyridine **III-391**, strong acid reversing the reaction (457). The isoquinolinium system underwent a similar reaction, cyclization to the 1-position of the isoquinoline ring occurring in this case **(III-392)**.

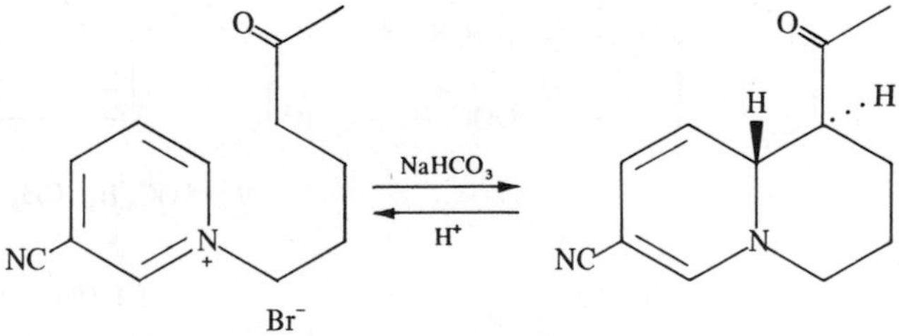

III-390

III-391

The pyridinium salts **III-394** ($R^2 = OC_2H_5$) obtained in two steps from quinolizinium salts **(III-393)** undergo base-catalyzed ring closure to give the indolizines **III-395**, and **III-396** ($R^2 = OC_2H_5$) (458). In the case of alkyl or aryl ketones (**III-394**; $R^2 = CH_3$, aryl), cyclization occurs readily on an alumina

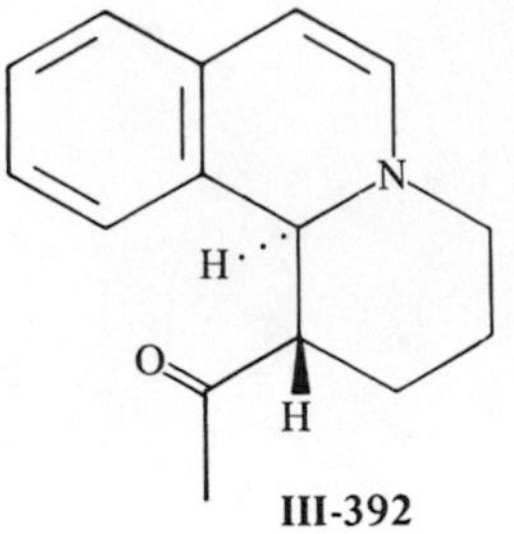

III-392

chromatographic column, from which the indolizines **III-396** ($R^2 = CH_3$, aryl) are easily obtained. A mechanism for the formation of these products has been advanced (458).

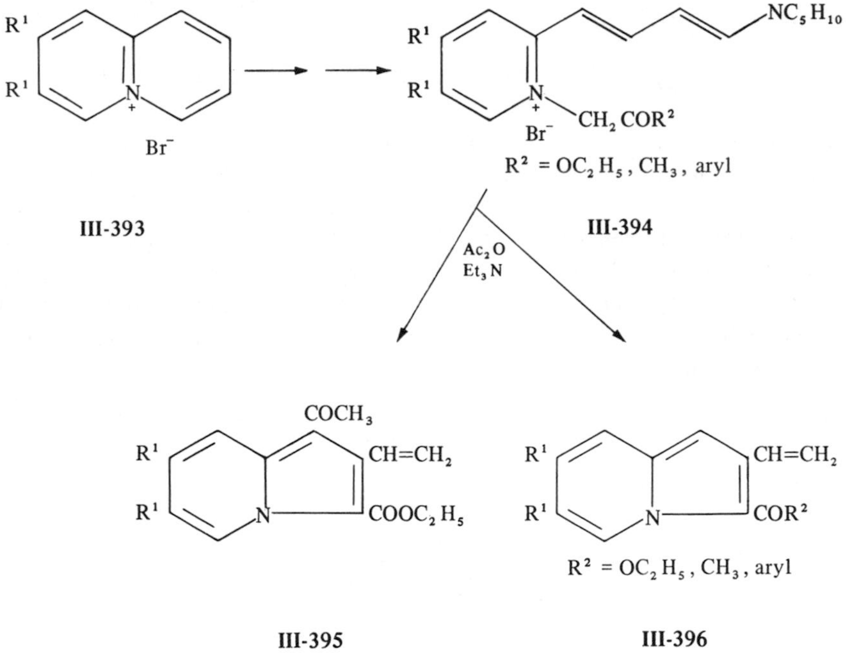

III-393

$R^2 = OC_2H_5, CH_3$, aryl

III-394

Ac_2O
Et_3N

III-395

$R^2 = OC_2H_5, CH_3$, aryl

III-396

Another synthesis of the indolizine system has been reported by Augstein and Kröhnke who found that pyridinium picryl betaines of type **III-397** will undergo cycloaddition and subsequent loss of nitrous acid to give the indolizine **III-398** in very high yield (459). Again, the isoquinolinium system reacts similarly, with ring closure occurring in the isoquinolinium 1-position.

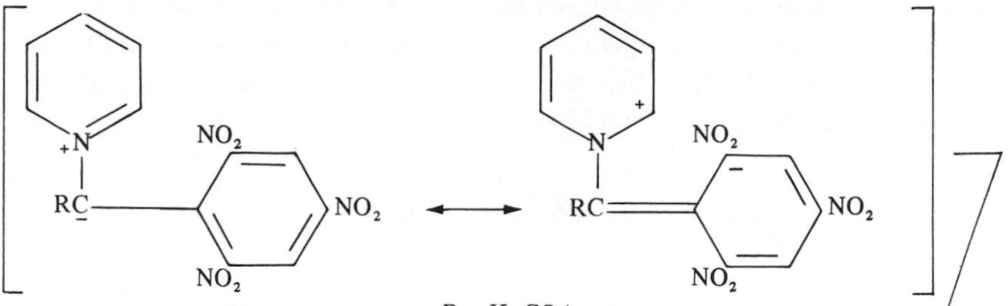

NO_2 NO_2 NO_2 RC ↔ RC NO_2 NO_2 NO_2

R = H, COAr
III-397

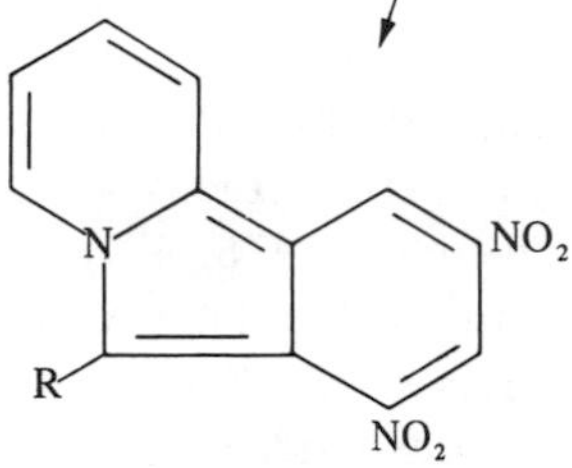

NO_2 NO_2 R

III-398

Tamura and co-workers (451) have found that a similar ring closure occurs with the pyridinium salts **III-399**, leading to the indolizines **III-401** *via* the cycloaddition products **III-400**.

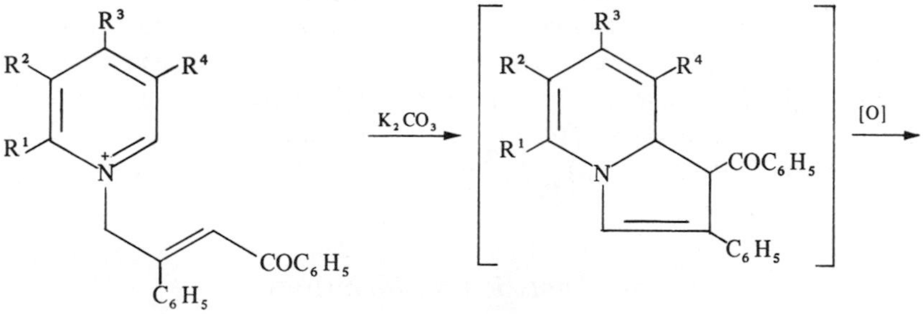

R^3 R^2 R^4 R^1 COC_6H_5 C_6H_5
III-399

$\xrightarrow{K_2CO_3}$

R^3 R^2 R^4 R^1 COC_6H_5 C_6H_5
III-400

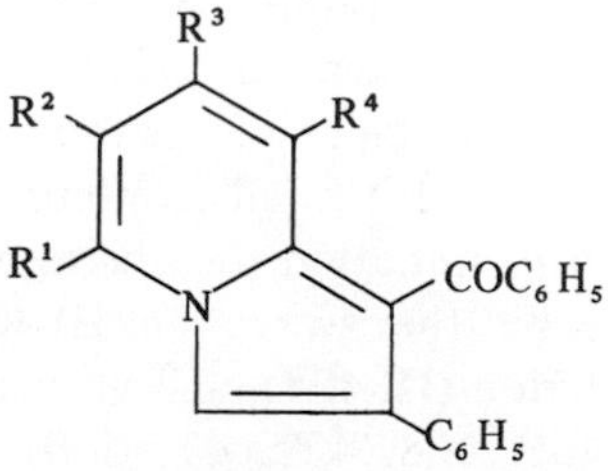

$\xrightarrow{[O]}$

R^3 R^2 R^4 R^1 COC_6H_5 C_6H_5
III-401

Indolizines have also been prepared by reacting pyridinium and other heteroaromatic ylids with acetylenedicarboxylic acid derivatives (16a, 249–255), and the reductive cyclization of pyridinium salts to yield the quinolizine ring system has been discussed in Section II.4.A.

9. Polymerization

Attempts to prepare quaternary salts of vinylpyridines using alkyl halides or protic reagents usually result in spontaneous polymerization (462–474), although in a few cases the monomeric salts have been isolated. Thus Reynolds and Laakso (475) reported in 1955 that the *p*-toluenesulfonate of 4-vinylpyridine (**III-402**; R = H, X = *p*-$CH_3C_6H_4SO_3$) was stable at $-15°$, although it polymerizes readily at higher temperature, and only recently were Salamone and coworkers (476, 477) successful in isolating the trifluoroacetic, hydriodic, and nitric acid salts of 4-vinylpyridine,and the *N*-methylpyridinium salts **III-402** (R = CH_3, X = *p*-$CH_3C_6H_4SO_3$, SO_3OCH_3, I). Generally, however, quaternary salts of vinylpyridines undergo facile polymerization, and such pyridinium polymers have been investigated for use as bactericides, fungicides, insecticides, surface-active agents (478), polyelectrolytes (479–481), esterases (482–486), precipitants and mordants of dyes (487), polysoaps (480, 488), rubbers (489), soluble anion exchange resins (490), and as electrical conductors (491, 492).

Mercier and Dubosc studied the photochromic properties of poly(nitrobenzyl)vinylpyridines (493), and conformational changes in poly(4-vinyl-*N*-benzylpyridinium chloride) have been measured by the paramagnetic tracer technique (494). The quaternization of poly(4-vinylpyridine) by butyl bromide has been investigated kinetically (495), and the adiabatic compressibility of the product has been examined (496).

A. *Mechanistic Considerations*

The polymerization of 4-vinylpyridine has been studied the most extensively, and Kargin, Kabanov, and co-workers suggested a mechanism for the spontaneous polymerization of this substance on quaternization which involves attack of the counterion X^- at the unsubstituted terminus of the double bond to give the zwitterion **III-403**. Chain initiation then takes place by attack of a second vinylpyridinium molecule by the zwitterion **III-403**, forming a new highly resonance stabilized zwitterion (**III-404**), and ultimately leading to the straight chain polymer **III-405** (465, 468, 472, 497–500). It was further found that unquaternized vinylpyridine molecules or other monomers such as acrylonitrile

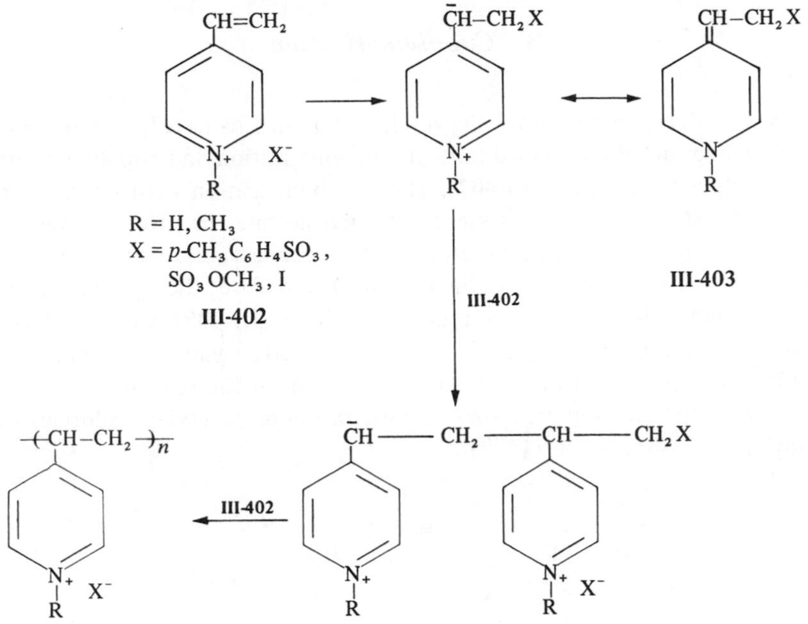

$R = H, CH_3$
$X = p\text{-}CH_3C_6H_4SO_3, SO_3OCH_3, I$

III-402

III-403

III-404

III-405

and styrene normally cannot compete with vinylpyridinium ions for the active zwitterionic center, presumably because the double bond electron density is greater in these monomers, and furthermore, their addition would lead to an intermediate lacking the high resonance stabilization provided by the quaternary pyridinium system (465, 468). However, pyridinium salts have been used as modifiers in methacrylate polymerizations (501, 502), and the vinylpyridinium polymer itself can be converted to polyvinylpyridine by removal of the quaternary group upon treatment with alkali (498).

Recently, evidence has been forthcoming that the polymerization of 4-vinylpyridinium salts may, in fact, be initiated by an unquaternized pyridine molecule, rather than the counterion (476, 477, 503, 504). Furthermore, under certain conditions, acid salts of 4-vinylpyridine yield polymers having pyridinium units in the main chain (**III-406**) (477, 503, 504).

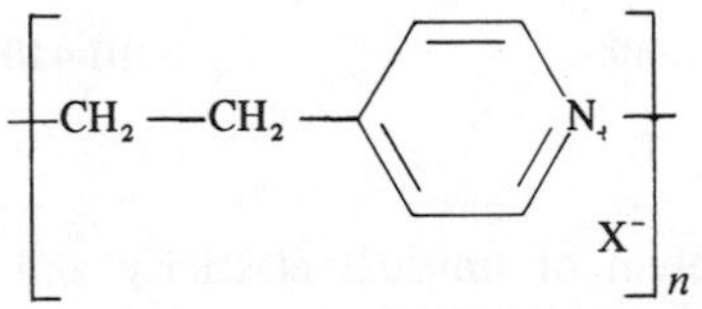

III-406

B. *Copolymerization*

Some copolymerizations involving vinylpyridinium salts have been reported. In 1962, Duling and Price studied both the polymerization and copolymerization of *N*-vinylpyridinium salts (**III-407**). The copolymerization went poorly or not at all with styrene, α-methylstyrene, or vinyl acetate (negative *e* monomers) (505) but did occur with acrylonitrile, methyl acrylate, and methyl methacrylate (positive *e* monomers) (506). Shyluk (462) carried out the copolymerization of 1,2-dimethyl-5-vinylpyridinium methyl sulfate (**III-408**) with acrylamide, methacrylamide, methacrylic acid, and methyl methacrylate, the reactivity of **III-408** being dependent on the dielectric constant of the solvent (507). Berlin and co-workers have reported the coplasticization of poly(vinyl chloride) with methylvinylpyridine rubbers (489).

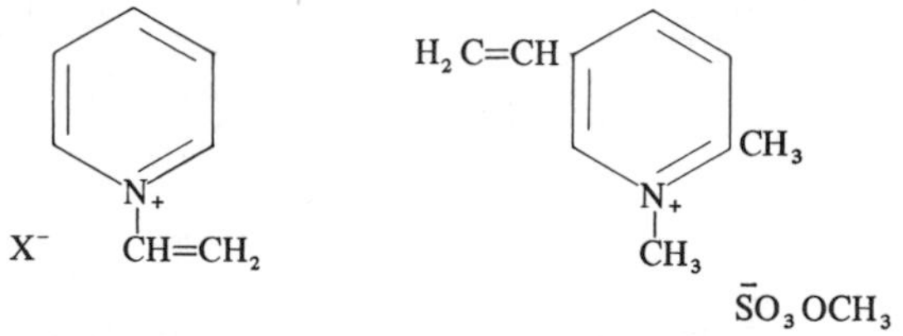

X^- $CH{=}CH_2$ (on N^+)

III-407

$H_2C{=}CH$, CH_3, N^+–CH_3, $\bar{S}O_3OCH_3$

III-408

Two unique quaternary pyridinium salts, desmosine (**III-409**) and isodesmosine (**III-410**), are believed to be the main crosslinks for the polypeptide

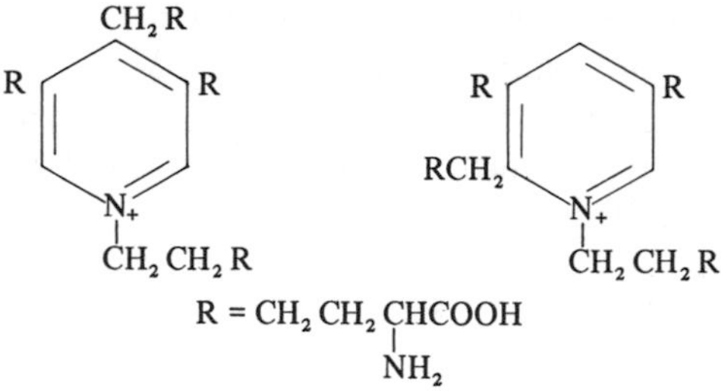

CH_2R, R, R, N^+–CH_2CH_2R

R, R, RCH_2, N^+–CH_2CH_2R

$R = CH_2CH_2CH(NH_2)COOH$

III-409 **III-410**

chains of elastin, a protein of unusual elasticity and tensile strength found mainly in connective tissues (508). A mechanism for the formation of these crosslinking substances in elastin has been proposed (509).

C. *Matrix Polymerization and Enzyme Models*

The polymerization of 4-vinylpyridine on a polymer matrix has been studied as a potential route to synthetic enzymes (465, 469, 470, 472, 498, 510–512). Kabanov and co-workers polymerized 4-vinylpyridine using polystyrenesulfonic acid as an activating matrix, where the vinylpyridine molecules would react with the sulfonic acid groups to form pyridinium salts with polymerization proceeding along the polystyrenesulfonic acid chains (**III-411**) (465, 469, 513).

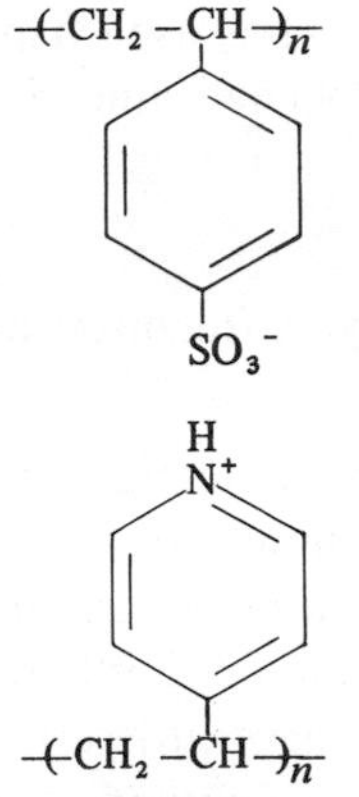

$$-\!\!(CH_2-CH)\!\!-_n$$

SO_3^-

H

N^+

$$-\!\!(CH_2-CH)\!\!-_n$$

III-411

In other investigations, these workers used poly(ethylenesulfonic) (469), poly(acrylic) (470, 477, 511), poly(methacrylic) (502), poly(*L*-glutamic) (470), and polyphosphoric acid (472, 477) matrices. It has been suggested that such polyacid matrix polymerizations are initiated by vinylpyridine, as in the case of vinylpyridinium salts (477, 503, 504).

Okawara and co-workers have incorporated a nicotinamide quaternary salt into a polymer matrix by reacting nicotinyl chloride with polyvinyl alcohol, polyaminostyrene, and an aniline-formaldehyde polymer, followed by quaternization with methyl iodide or methyl sulfate (514). These workers also reacted nicotinamide with chloromethylated polystyrene to give polymers containing nicotinamide moieties in side chains. Reduction of these polymers with sodium dithionite or sodium borohydride yielded polymeric dihydronicotinamides which could function as hydrogen transfer agents (514, 515).

Enzyme models have also been prepared by the partial quaternization of poly(4-vinylpyridine) (482–486, 516–518). Kabanov, Kirsh, and co-workers partially alkylated poly(4-vinylpyridine) with benzyl chloride (482, 484, 486)

2-(2-chloroethyl)pyridine (483), a series of alkyl bromides (484, 486, 516) and 4(5)-(hydroxymethyl)imidazole hydrochloride (485) in the preparation of synthetic esterase enzymes, and with bromoacetic acid in the synthesis of a polymer having ascorbate oxidase activity (517). The kinetic behavior of some of the alkylated polymers was similar to that of α-chymotrypsin (486, 518).

III. Pyridinium Compounds of Biological Interest

Pyridinium compounds have been investigated for a wide variety of biological activities and in this section a brief summary of some of the more recent investigations in this area is given (see also Chap. XVI). The role of quaternary pyridinium salts in nicotinamide coenzyme chemistry has already been mentioned (Section II.4.B.).

1. Specific Enzyme Inhibitors and Reactivators

Cavallito and co-workers have found that certain styrylpyridinium salts are potent choline acetyltransferase inhibitors (519–522). The inhibitory potency was diminished by highly electronegative substituents on the styryl phenyl ring, but enhanced by chlorine and bromine. Furthermore, the nature of the pyridine quaternizing group was found to have little effect and a hydrophilic substituent could be used to increase water solubility (520, 522). The pyridinium portion is presumed to play an important role in the binding of these substances to the enzyme (521).

As part of a very extensive study of irreversible enzyme inhibitors, B. R. Baker and co-workers have found that pyridinium salts of type **III-412** act as inhibitors of serum complement, a result of potential interest in the problem of mammalian organ and tissue transplant rejection (523–526). A number of the salts were also excellent irreversible α-chymotrypsin inhibitors (424, 527), while similar quaternary salts lacking the sulfonyl fluoride group acted as reversible inhibitors (527).

The 3-bromoacetylpyridinium bromides **III-413** slowly inactivate lactate dehydrogenase (528), and a series of *N*-benzylpyridinium chlorides are inhibitors of the yeast alcohol dehydrogenase-catalyzed oxidation of ethanol (529). A series of *N*-methylpyridinium compounds have been studied as substrates and inhibitors of pig kidney diamine oxidase (530).

Substances containing the quaternary pyridinium system have been studied as reactivators of the enzyme acetylcholinesterase inhibited by phosphorylation. Organophosphorus compounds which have the ability to rapidly and irreversibly inhibited this enzyme have gained importance as insecticides and nerve agents.

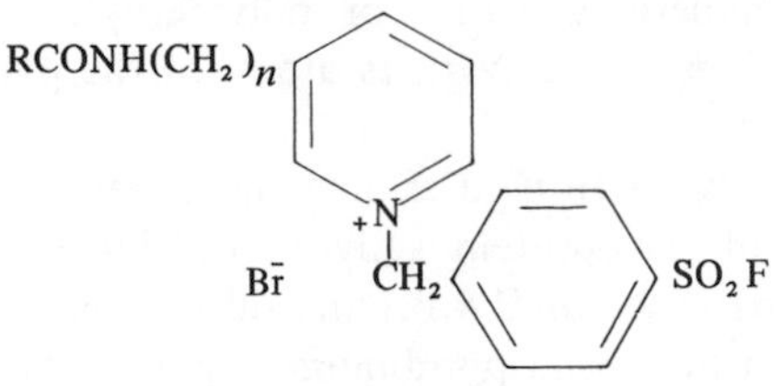

RCONH(CH_2)$_n$ … $^+$N … Br^- CH_2 … SO_2F

n = 0, 1

III-412

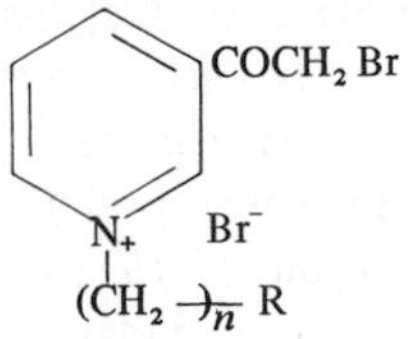

$COCH_2Br$ … N^+ Br^- (CH_2)$_n$R

n = 2-5

R = CH_3, COOH

III-413

These substances either phosphorylate or phosphonylate the enzyme, thus deactivating it (531, 532a), and reactivation can be effected by removing the phosphorus containing group with nucleophilic reagents such as oximes or hydroxamic acids (532b–535). Three such compounds of high effectiveness are *N*-methylpyridinium-2-aldoxime chloride or iodide (2-PAM) (**III-414**), the trimethylenebispyridinium oxime **III-415** (TMB-4), and the ether analog **III-416** (LüH6, toxogenin) (536). However, many others have been studied as well (13, 40, 537–547a). This topic is discussed in detail in Ch. XVI.

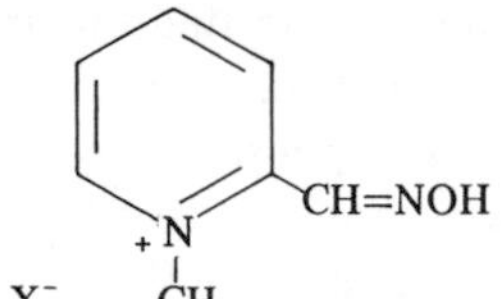

CH=NOH $^+$N X^- CH_3

III-414

HON=CH … CH=NOH … N^+ … N^+ … CH_2—Y—CH_2 $2Br^-$

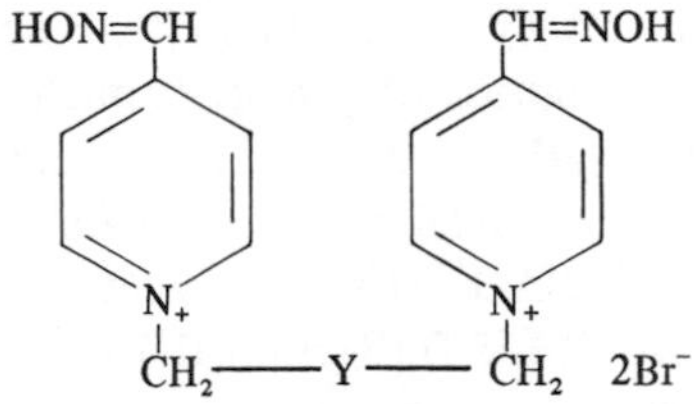

Y = CH_2

III-415

Y = O

III-416

In attempts to elucidate the nature of the anionic site in human blood serum butyrylcholinesterase, Augustinsson used a series of carbinol acetates of pyridine and *N*-methylpyridinium salts as substrates (548). Comparing with acetylcholinesterase, he found evidence suggesting that esteratic sites of the two esterases are not the same, and that butyrylcholinesterase contains a second nonesteratic site differing from the anionic site of acetylcholinesterase (549). The rates of hydrolysis of *N*-methylpyridinium *O*-acetylaldoxime (550) and *O*-acetylketoxime (551) iodides have been investigated in studies designed to elucidate further the mechanism of phosphorylated cholinesterase reactivation. An attempt to correlate the relative curaremimetic activities of polymethylene-bis-pyridinium salts with π-electron densities on the various atoms of the pyridinium rings has been reported (552).

In some cases, pyridinium salts have been used as enzyme models. Thus in an investigation of the mechanism of flavoprotein activity, the kinetics of the nonenzymatic oxidation of NADH (Section II.4.B.) and dihydrolipoic acid by synthetic substances having both a flavin and pyridinium ring were determined (553). Studies designed to elucidate the mechanism of the enzymatic hydrolysis of a nitrile function have also been reported (554). While the nitrile group is normally relatively resistant to chemical hydrolysis, several enzymatic systems are known which will effect this reaction with relative ease (555, 556). Zervos and Cordes studied the hydrolysis of *N*-benzyl-3-cyanopyridinium bromide by mercaptoethanol, and their results support the supposition that nitrilases carry out their hydrolysis functions by the participation of thiol groups (554).

2. Antitumor and Antileukemic Agents

A number of pyridinium salt derivatives have been investigated for carcinostatic activity. Following the premise that certain neoplastic cells show a greater dependence on glycolytic energy than normal cells, Ross and Lovesey surmised that the interruption of glycolysis might be a means of arresting the growth of tumor cells (201a, 336a, 557). In the glycolytic pathway of carbohydrate metabolism, an oxidative step involving glyceraldehyde phosphate requires that the cofactor NAD^+ (**III-258a**) accept a hydrogen atom in the β-orientation yielding NADH (558) (Section II.4.B). Since cancer cells are deficient in NAD^+, this cofactor must be regenerated from NADH in these cells to provide continuing energy production, a process which requires the loss of the hydrogen atom of α-configuration (559). Since it was known that *in vivo* incorporation of substituted nicotinic acids into the NAD^+ molecule can occur (560), Ross theorized that the 4-substituted nicotinic acid moiety should have an inhibitory effect on the proliferation of cancer cells since no α-hydrogen would now be available to regenerate NAD^+ (557). In agreement with this, it was indeed found that a 4-methyl group increases the tumor inhibition properties of such substances (201a).

A slightly different approach was taken by Friedman, Pollak, and Khedouri (561). Based on the fact that certain alkylating agents possess antitumor activity (562), these investigators prepared the compounds **III-417** and **III-418**, the latter being an active alkylating agent of the β-chloroethylamino (nitrogen mustard) type. Thus it was reasoned that for neoplastic cells having a high capacity to reduce **III-417** to **III-418** or a low capacity to carry out the reverse oxidation

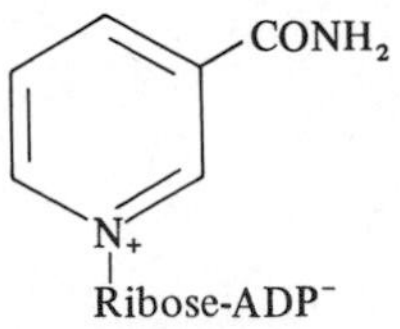

CONH₂ structures:

N+ Ribose-ADP⁻ **III-258a**

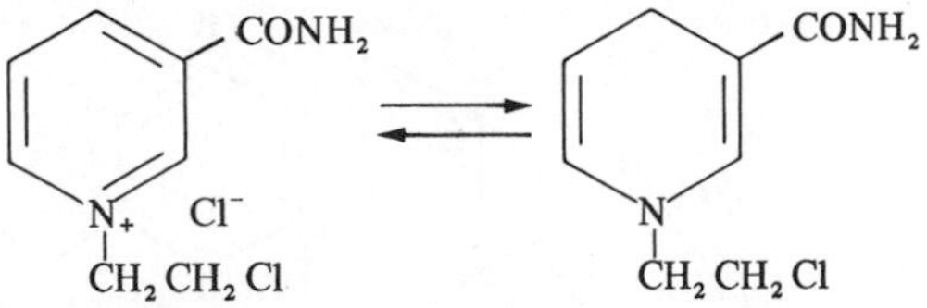

N+ Cl⁻ CH_2CH_2Cl **III-417**

N CH_2CH_2Cl **III-418**

relative to normal cells, these substances constitute potential antitumor agents which might localize themselves preferentially in the tumor (561).

The pyridinium ring has also been incorporated in other structures containing the nitrogen mustard function. Thus Schulze and co-workers studied azomethines of the type **III-419**, which were found to be active against Ehrlich ascites carcinoma (563).

Gruszecki and Borowski reported that the same tumor responds to *in vitro* treatment with the pyridinium acridine derivatives **III-420** (564), and prepared pyridinium salt derivatives of the 4-acridinylmethyl system (**III-421**) for similar studies (565).

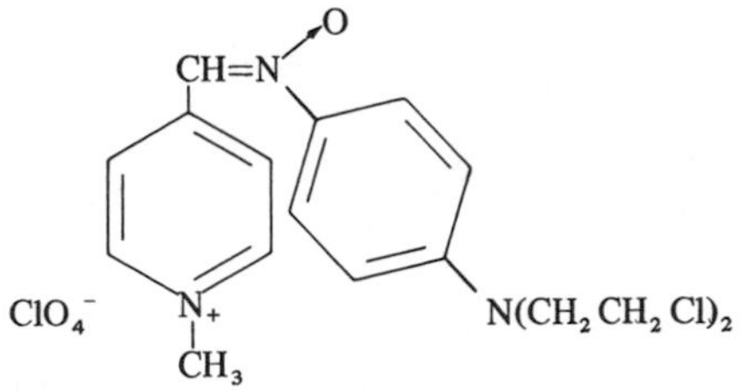

ClO_4^- CH=N→O, N+ CH_3, $N(CH_2CH_2Cl)_2$

III-419

Cl⁻ N+ R^1 N R^2

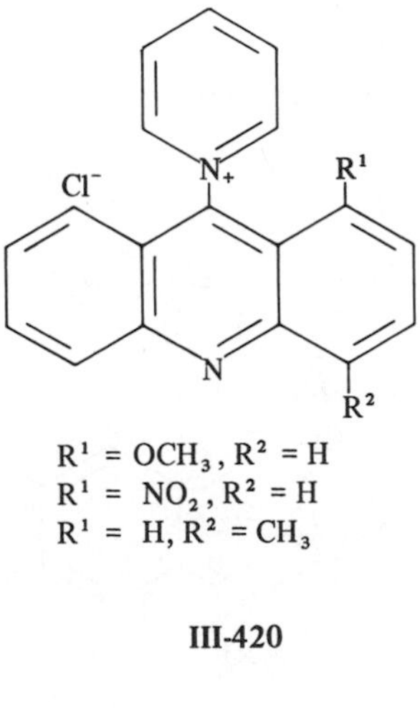

$R^1 = OCH_3, R^2 = H$
$R^1 = NO_2, R^2 = H$
$R^1 = H, R^2 = CH_3$

III-420

Certain derivatives of aminoacridines have been found active against leukemia (566, 567), as have other pyridinium salts. Cain and co-workers prepared numerous variations of compounds of type **III-422** and found that certain ones showed antileukemic activity. The large diversity of structures studied by these workers allowed the determination of some structure-activity relationships (566, 568–570).

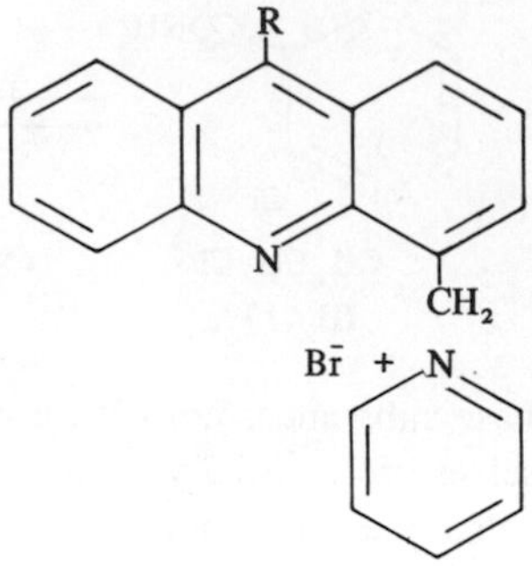

R = H, Cl

III-421

In 1957, Lund found that when anthracene is subjected to electrolytic oxidation in the presence of pyridine and sodium perchlorate, the pyridine can trap the intermediate, yielding the 9,10-dihydrodipyridinium salt **III-423** (571).

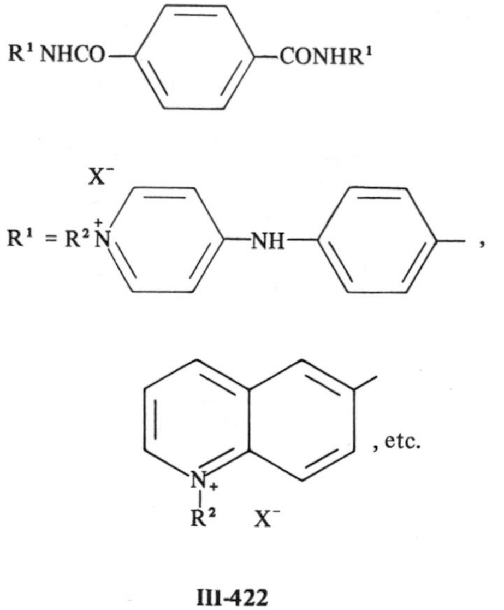

$R^1NHCO-C_6H_4-CONHR^1$

$R^1 = R^2\overset{+}{N}C_5H_4-NH-C_6H_4-$, etc. X^-

$2ClO_4^-$

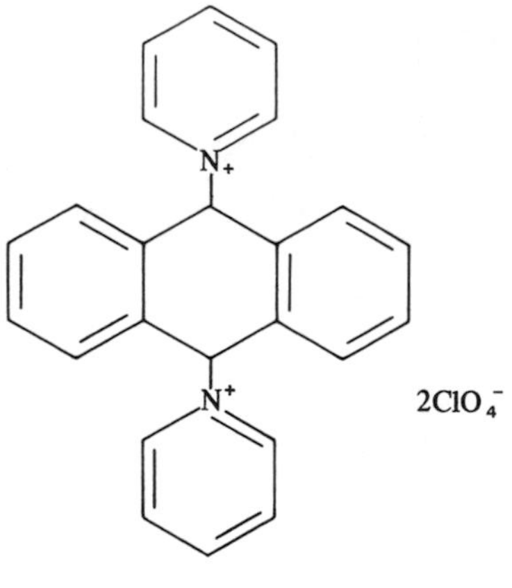

III-423

III-422

Along similar lines, Rochlitz found that pyridine can capture radical cations of aromatic carcinogens such as 3,4-benzpyrene and 9,10-dimethyl-1,2-benzanthracene. Since such radical ion intermediates may play a primary role in chemical carcinogenesis, their entrapment as pyridinium salts may provide a means of inactivating the carcinogenicity of these substances (122).

3. Hypoglycemic and Hypotensive Agents

Researchers at the Lederle Laboratories have studied a large number of quaternary pyridinium salts having a heterocyclic ring system attached at C-4 (**III-424**) as potential oral hypoglycemic agents. Hypoglycemic activity (as measured in mice, and in one case also chicks, 580) was found with pyrazolyl (572), thiazolyl (573), oxazolyl (574), isoxazolyl (575–577), and 1,2,4-oxadiazolyl (578) derivatives, as well as those containing a furan, thiophene, pyrrole (579), and 3-indolyl (580) ring. No activity was observed with compounds tested having a 1,2,4-triazolyl, 1,3,4-thiadiazolyl, tetrazole, imidazole (581), 2-indolyl (580), or 4-pyrimidinyl (582) ring system, while the 1,3,4-oxadiazolyl derivative showed a slight nondose related hypoglycemic response (581). It was also found that for those substances showing hypoglycemic properties, quaternization was necessary for biological activity, and it was thus concluded that a positive charge on the pyridine nitrogen atom may be a requirement for such activity. The 1-oxides, however, which have at least a partial positive charge on the nitrogen atom, were found to be inactive (583).

Several studies have been reported on the use of quaternary pyridinium salts as hypotensive agents. Karten and Schwinn found that maleimide derivatives of type **III-425** showed a short hypotensive effect, but possessed a marked topical anesthetic activity (26). The indole pyridinium salts **III-426** were found to be

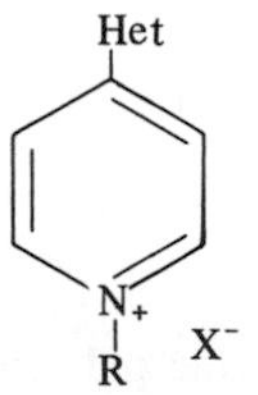

III-424

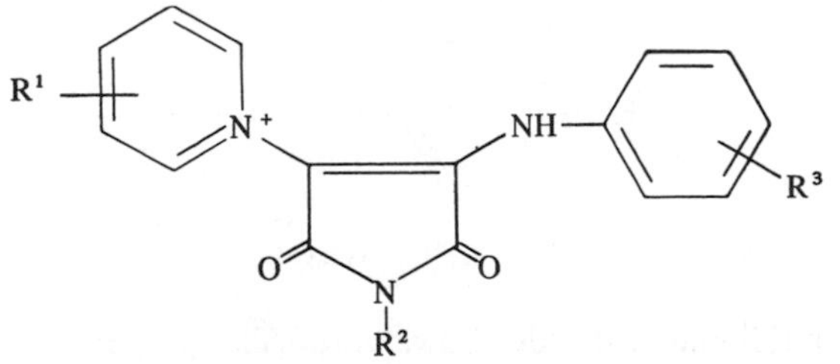

III-425

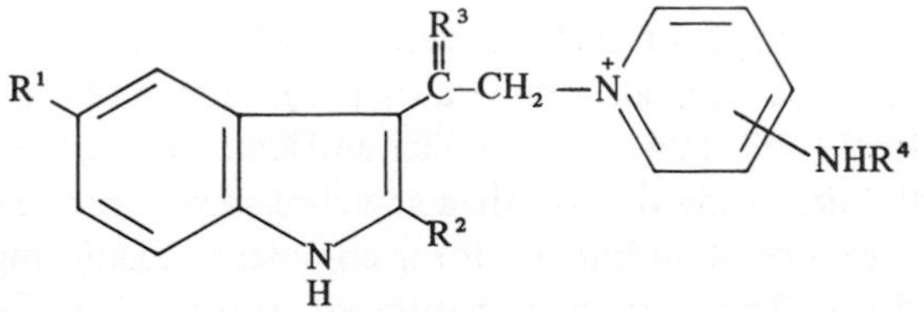

R^1 = H, OCH_3 R^3 = H_2, O

R^2 = H, CH_3 R^4 = H, Acyl **III-426**

devoid of hypotensive activity except at high doses. However, the reduction of the pyridinium system to a piperidine ring yielded some highly active hypotensive agents (584). It has also been observed that 1-amino-4-phenylpyridinium chloride causes a decrease in systolic blood pressure without any change in heart rate, and produces a long lasting antihypertensive effect in renal hypertensive dogs (585, 586). Several substances containing the pyridinium ion as one part of a bisquaternary system show some hypotensive activity (587).

4. Antimicrobials, Anthelmintics, and Herbicides

Sheinkman, Kost, and co-workers prepared the quaternary pyridinium salts **III-427** and found them to be active bactericides, fungicides, and herbicides (588), while Pedrazzoli and co-workers reported that the β-alkoxyphenethylpyridinium bromides **III-428** possess antibacterial, antifungal, and antitrichomonas activity (589).

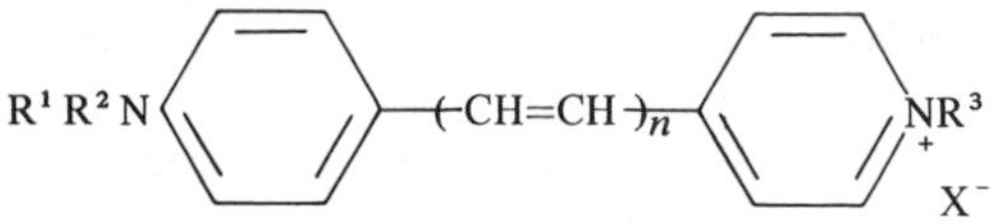

$R^1R^2N-C_6H_4-(CH=CH)_n-C_5H_4\overset{+}{N}R^3\ X^-$

$n = 0, 1$

III-427

$R^1,R^2-C_6H_3-CH(OR^3)-CH_2-\overset{+}{N}C_5H_5\ Br^-$

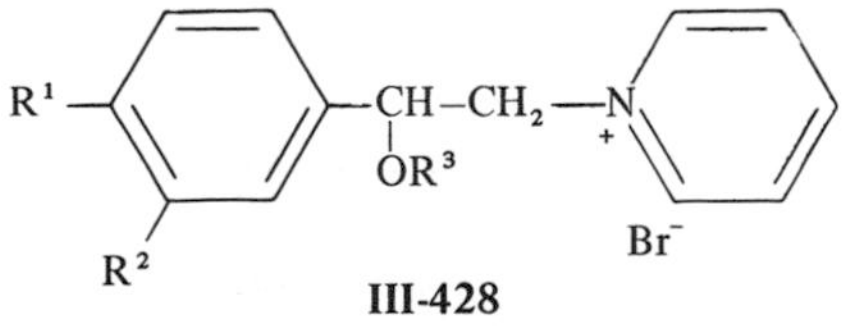

III-428

Acetylpyridinium chloride shows germicidal properties over a wide pH range (590), but more commonly germicidal pyridinium salts belong to the so-called surface-active agents, a series of compounds having both hydrophobic and hydrophilic moieties and capable of reducing the surface tension of solutions (591, 592). Of these, cetylpyridinium chloride (**III-429**) is one of the oldest and most active currently in use, although laurylpyridinium chloride and other pyridinium derivatives (593, 594) show similar antiseptic properties. Pyridinium salts of the type **III-430**, having the so-called acylcholaminoformylmethyl group, can combine with elemental iodine to form complexes exhibiting germicidal properties (591, 592). The antiseptic nature of these substances has been attributed to the slow release of iodine on contact with skin or mucous membrane (592, 594a). The highly fluorinated pyridinium salt **III-147** has been

N^+ Cl^- $(CH_2)_{15}CH_3$

III-429

N^+ $CF_3SO_3^-$ $CH_2-(CF_2)_6CF_3$

III-147

Cl^- ^+N $CH_2CONHCH_2CH_2OC(=O)(CH_2)_nCH_3$

$n = 6\text{-}12$

III-430

reported to show approximately equivalent bacteriostatic activity to cetylpyridinium chloride, but somewhat less fungistatic activity. Nevertheless, its relative lack of toxicity and of irritation to mucous membranes as well as other desirable properties are said to be of advantage in its use as a skin, surgical wound, and possibly oral antiseptic (167). Quaternary pyridinium salts having the pyridinium system appended to fatty ester chains show biocidal activity against *Escherichia coli* and *Staphylococcus aureus* (*Micrococcus pyogenes*) (36).

Coccidiosis is an intestinal infection caused by microscopic protozoan parasites, and afflicts poultry and other animals. A commercial drug to combat this disease is amprolium, the quaternary pyridinium salt **III-431** (R = *n*-propyl) which acts as a thiamine antagonist (595). A number of variations of this structure were also found to possess activity (595, 596), including compounds having R = cyclopropyl, cyclopropylmethyl, and cyclobutylmethyl (597). The tetrahydropyridine derivatives **III-432**, prepared by the borohydride reduction of the corresponding pyridinium salts, showed activity against the rodent malarial parasite *Plasmodium berghei* (598).

McFarland and Howes have reported a series of anthelmintic agents having the 1-arylvinylpyridinium system **III-433** (599), which apparently behave biologically differently from anthelmintics of structure **III-434** studied earlier by Wood and associates (600, 601). The corresponding 2- and 4-styrylquinolinium salts are also active anthelmintics (602), as are the 2-(*p*-dialkylaminophenyl)-1-methylquinolinium salts (603). However, in the latter case, the pyridinium analogs are inactive (604).

The herbicidal activity of the quaternary pyridinium salts Diquat and Paraquat as well as some of their analogs has been mentioned in Section II.4.A. In addition to these compounds, Black and Summers investigated the herbicidal

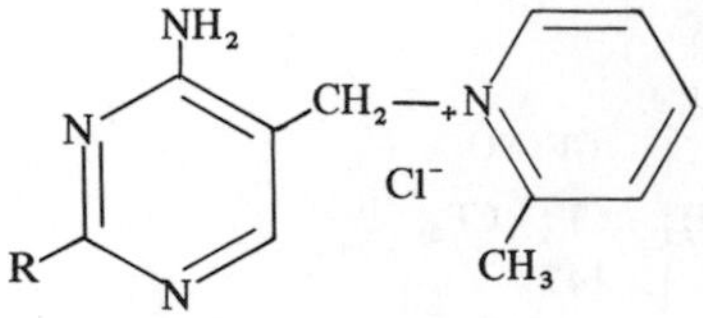

NH$_2$ N R N CH$_2$—$^+$N Cl$^-$ CH$_3$

III-431

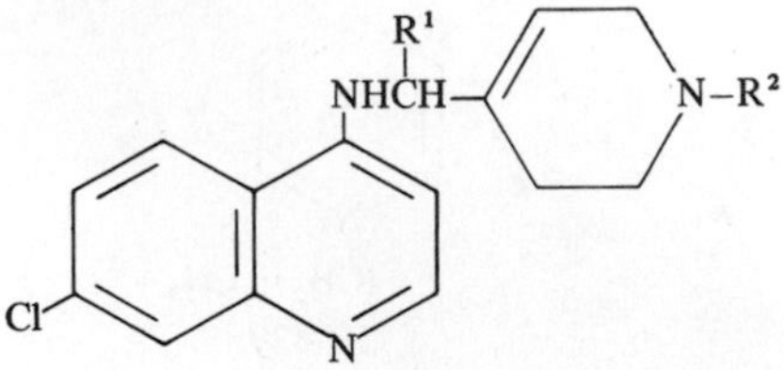

R^1 NHCH N–R^2 Cl N

III-432

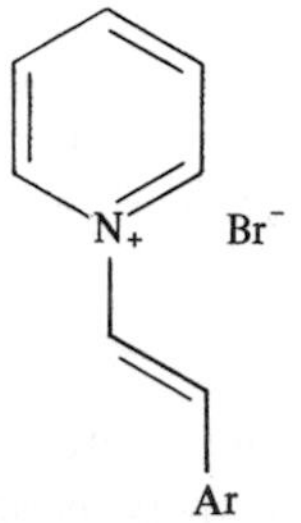

N$_+$ Br$^-$ Ar

III-433

Ar N$_+$ CH$_3$ Cl$^-$

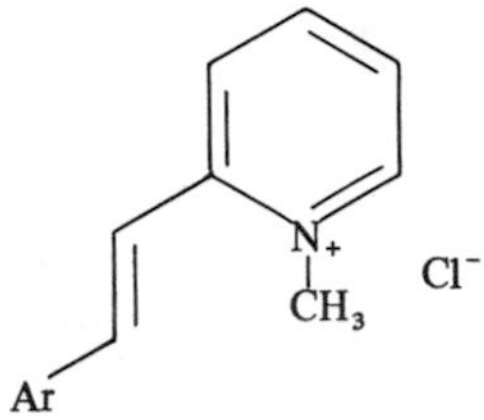

III-434

activities of the dipyridopyrazinium compounds **III-435** and **III-436**, and found the former to be highly active against several plant species in postemergent herbicide tests (605). The latter showed slight activity (605), while the pyridylpyridinium analogs **III-67** were inactive (62). The activities of other analogs are reported in references cited in Section II.4.A. Pyridinium compounds containing the arylthiophosphate or dithiophosphate group in the anionic portion of the salt have been reported as being active herbicides (606).

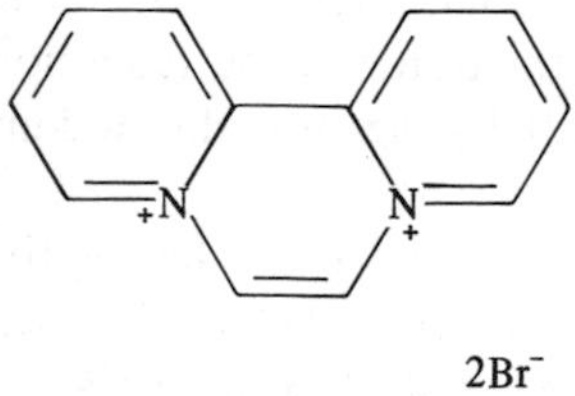

+N N+ 2Br$^-$

III-435

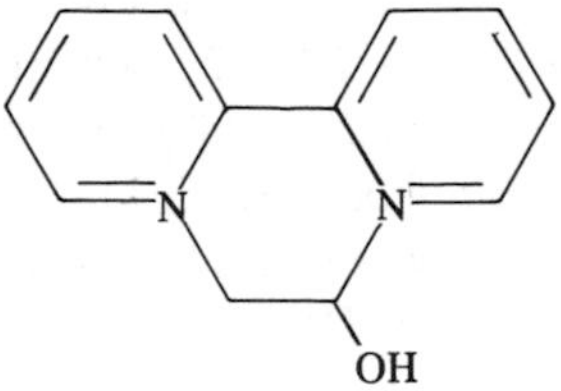

N N OH

III-436

N$_+$ N$_+$ R 2Br$^-$

R = CH$_3$, C$_2$H$_5$

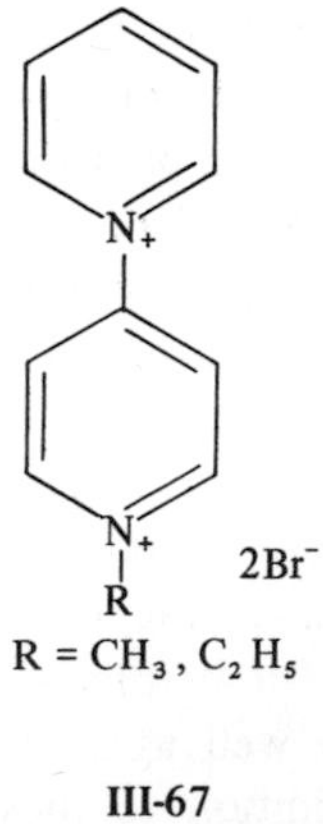

III-67

5. Miscellany

The pyridinium salts **III-437** and **III-438** have been found to harbor antiradiation activity (607), and pyridinium salts of type **III-439** have been investigated as cholesterol-solubilizing agents for the treatment of gallstones (608). A mechanism has been proposed for the formation of the pyridinium

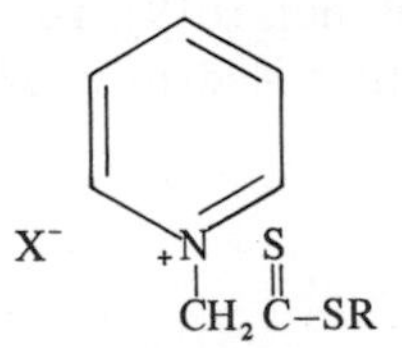

X^- $\overset{+}{N}$–$CH_2C(=S)$–SR

$R = p\text{-}NO_2-C_6H_4-CH_2-, CH_2=CH-CH_2-$

III-437

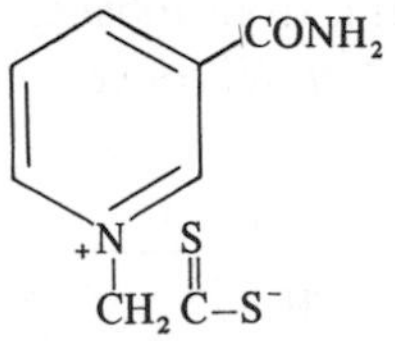

3-$CONH_2$; $\overset{+}{N}$–$CH_2C(=S)$–S^-

III-438

Y; $\overset{+}{N}$–R X^-

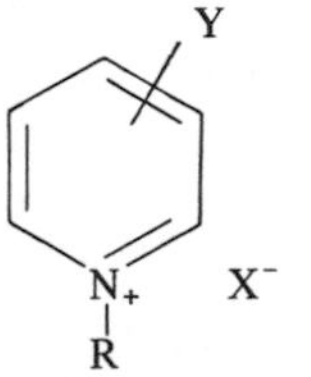

$Y = 3\text{-COOH}, 3\text{-}SO_3^-, 4\text{-COOH}$

$R = C_{10}H_{21}, C_{12}H_{25}, C_{14}H_{29}, C_{16}H_{33}$

III-439

amino acids, desmosine, and isodesmosine, believed to be crosslinks between the polypeptide chains of elastin (509). Maley and Bruice have used quaternary pyridinium salts as models for pyridoxal in studies on the transamination of amino acids (609).

Quaternary pyridinium compounds have been used in attempts to map active sites of enzymes. Westheimer amd co-workers prepared the 3-diazoacetoxymethyl analog of DPN^+ containing a radioactive label and studied its use in identifying the active sites in various dehydrogenases (610). Deranleau investigated use of *N*-methylnicotinamide chloride and *N*-methylisonicotinamide chloride as charge transfer probes in biological systems (611, 612). *N*-Alkylpyridinium halides have been used to study the bonding in hemin complexes (613), and cetylpyridinium bromide has been employed to isolate 2-amino-1-naphthyl hydrogen sulfate from the urine of dogs given the bladder carcinogen 2-naphthylamine (614). Pyridinium and quinolinium alkylsulfates have been utilized in studying the intestinal absorption of organic ions (615), while phenanthridinium salts have been useful in investigating the biliary excretion of organic cations, as well as in a number of other biological studies (616).

In addition to NAD^+, desmosin, and isodesmosin already discussed, several other quaternary pyridinium salts have been found to occur naturally (617). Two of the more recent ones isolated include the terpene alkaloids **III-440** from the medicinal plant *Valeriana officinialis*, **III-440** (R = H), showing inhibition of cholinesterase activity (618), and the pyridinium amino acid, nicotianine (**III-441**), from tobacco leaves, stems, and roots (619). Tobacco plants also have the ability to convert administered nicotinic acid to its *N*-glucoside derivative (620).

CH_3, H, CH_2R, N^+, $CH_2CH_2C_6H_4OH$-*p*

R = H, OH

III-440

COO^-, N^+, NH_2, $CH_2CH_2CHCOOH$

III-441

IV. References

1. J. Menschutkin, *J. Russ. Phys. Chem. Soc.*, **34**, 411 (1902); *Chem. Zentr.*, 73 (II), 86 (1902).
2. A. Fischer, W. H. J. Galloway, and J. Vaughn, *J. Chem. Soc.*, 3591 (1964).
2a. L. W. Deady and J. A. Zoltewicz, *J. Org. Chem.*, **37**, 603 (1972).
3. H. C. Brown and B. Kanner, *J. Amer. Chem. Soc.*, **75**, 3865 (1953).
4. Y. Okamoto and Y. Shimagawa, *Tetrahedron Lett.*, **3**, 317 (1966).
5. W. J. le Noble and Y. Ogo, *Tetrahedron*, **26**, 4119 (1970).
6. Y. Okamoto and Y. Shimakawa, *J. Org. Chem.*, **35**, 3752 (1970).
7. R. Damico and C. D. Broadlus, *J. Org. Chem.*, **31**, 1607 (1966).
8. G. Briegleb, *Angew. Chem. Int. Ed.*, **2**, 545 (1963).
9. J. D. Reinheimer, J. D. Harley, and W. W. Meyers, *J. Org. Chem.*, **28**, 1575 (1963).
10. E. M. Kosower, *J. Amer. Chem. Soc.*, **61**, 488 (1939).
11. K. R. Brown, *J. Amer. Chem. Soc.*, **85**, 1401 (1963).
12. P. Haberfield, A. Nudelman, A. Bloom, R. Romm, and H. Ginsberg, *J. Org. Chem.*, **36**, 1792 (1971).
13. C. N. Corder and J. L. Way, *J. Med. Chem.*, **9**, 638 (1966).
14. R. Royer, J. P. Bachelet, and P. Demerseman, *Bull. Soc. Chim. Fr.*, 878 (1969).
15. Bamberger, *Ber.*, **20**, 3344 (1887); R. Royer, E. Bisagni, and C. Hudry, *Bull. Soc. Chim. Fr.*, 933 (1961).
16. N. Saldabols, L. N. Alekseeva, B. Brizga, L. Kruzmetra, and S. Hillers, *Khim-Farm. Zh.*, **4**, 20 (1970); *Chem. Abstr.*, **73**, 77136p (1970).

16a. R. A. Abramovitch and V. Alexanian, unpublished results (1973).
17. I. C. Calder and W. H. F. Sasse, *Aust. J. Chem.*, **21**, 1023 (1968).
18. F. Kröhnke, *Chem. Ber.*, **66**, 1386 (1933).
19. L. C. King, *J. Amer. Chem. Soc.*, **70**, 242 (1948).
20. L. M. Litvinenko and L. A. Perelman, *J. Org. Chem. U.S.S.R.*, **3**, 900 (1967).
21. K. Undheim and T. Gronneberg, *Acta Chem. Scand.*, **25**, 18 (1971).
22. T. Cohen and I. H. Song, *J. Amer. Chem. Soc.*, **87**, 3780 (1965).
23. W. G. Phillips and K. W. Ratts, *Tetrahedron Lett.*, 1383 (1969).
24. J. A. VanAllan and G. A. Reynolds, *J. Org. Chem.*, **28**, 1019, 1022 (1963).
25. G. A. Reynolds, R. E. Adel, and J. A. VanAllen, *J. Org. Chem.*, **28**, 2683 (1963).
26. M. J. Karten and A. Schwinn, *J. Med. Chem.*, **9**, 702 (1966).
27. O. Neilands and B. Karele, *J. Org. Chem. U.S.S.R.*, **1**, 1884 (1965).
28. O. Neilands, *J. Org. Chem. U.S.S.R.*, **1**, 1888 (1965).
28a. R. A. Abramovitch and I. Shimkai, *J. C. S. Chem. Comm.*, 569 (1973).
29. J. Jonas, M. Kratochvil, J. Mikula, and J. Pichler, *Collect. Czech. Chem. Commun.*, **36**, 202 (1971).
30. A. Piskorska-Chlebowska, *Rocz. Chem.*, **40**, 1207 (1966); *Chem. Abstr.*, **66**, 29011c (1967).
31. A. Skrobacz, *Acta Pol. Pharm.*, **28**, 11 (1971).
32. A. A. Stepanyan, S. G. Agbalyan, and G. T. Esayan, *Arm. Khim. Zh.*, **22**, 688 (1969); *Chem. Abstr.*, **71**, 123435u (1969).
33. B. A. Porai-Koshits, I. Y. Kvitko, I. V. Frankovskaya, and O. V. Favorskii, *Zh. Prikl. Khim.*, **37**, 1081 (1964); *Chem. Abstr.*, **61**, 5603e (1964).
34. G. Ferre and A. L. Palomo, *Anales Real Soc. Espan. Fis. Quim.* (Madrid), **65**, 163 1969).
35. G. Ferre and A. L. Palomo, *Tetrahedron Lett.*, **26**, 2161 (1969).
35a. F. Ramirez, S. Glaser, P. Stern, P. D. Gillespie, and I. Ugi, *Angew. Chem. Int. Ed.*, **12**, 66 (1973).
36. D. H. Wheeler and J. Gross, *J. Amer. Oil Chem. Soc.*, **42**, 924 (1965).
37. Y. A. Serguchev and E. A. Shilov, *Ukr. Khim. Zh.*, **34**, 969 (1968); *Chem. Abstr.*, **70**, 28320y (1969).
38. U. E. Diner and J. W. Lown, *Chem. Commun.*, 333 (1970); *Can. J. Chem.*, **49**, 403 (1971).
39. V. A. Nefedov, *J. Gen. Chem. U.S.S.R.*, **36**, 1513 (1966).
40. Y. Ashani, H. Edery, J. Zahavy, W. Künberg, and S. Cohen, *Israel J. Chem.*, **3**, 133 (1965).
41. A. Giner-Sorolla, I. Zimmerman, and A. Bendich, *J. Amer. Chem. Soc.*, **81**, 2515 (1959).
42. P. Hrnciar, *Chem. Zvesti*, **19**, 360 (1965); *Chem. Abstr.*, **63**, 5616d (1965).
43. K. L. Nagpal, P. C. Jain, P. C. Srivastava, M. M. Dhar, and N. Anand, *Indian J. Chem.*, **6**, 765 (1968).
44. W. A. Szarek, B. T. Lawton, and J. K. N. Jones, *Tetrahedron Lett.*, 4867 (1969).
45. C. K. Bradsher, J. C. Parham, and J. D. Turner, *J. Heterocycl. Chem.*, **2**, 228 (1965).
45a. J. A. Campbell, J. C. Babcock, and J. A. Hogg, *J. Amer. Chem. Soc.*, **80**, 4717 (1958).
46. N. Kornblum and G. P. Coffey, *J. Org. Chem.*, **31**, 3449 (1966).
47. C. Reichardt, *Chem. Ber.*, 99, 1769 (1966).
48. Q. Stahl, F. Lehmkuhl, and B. E. Christensen, *J. Org. Chem.*, **36**, 2462 (1971).
49. T. Kobayashi and N. Inokuchi, *Tetrahedron*, **20** 2055 (1964).
50. T. Hino, M. Nakagawa, T. Wakatsuki, K. Ogawa, and S. Yamada, *Tetrahedron*, **23**, 1441 (1967).
51. G. V. Boyd, *Tetrahedron Lett.*, 3369 (1966).

52. D. Lloyd and J. S. Sneezum, *Tetrahedron,* **3**, 334 (1958).
53. T. Nozoe, K. Takase, and N. Shimazaki, *Bull. Chem. Soc. Jap.*, **37**, 1644 (1964).
54. J. W. Grochowski and K. Okon, *Rocz. Chem.*, **37**, 1437 (1963); *Chem. Abstr.*, **60**, 9242h (1964).
55. A. Kirkien-Konasiewicz, G. M. Sammy, and A. Maccoll, *J. Chem. Soc. B,* 1364 (1968).
56. H. L. Sharma, V. N. Sharma, and R. L. Mital, *Can. J. Chem.*, **44**, 1327 (1966).
57. R. L. Letsinger and O. B. Ramsay, *J. Amer. Chem. Soc.*, **86**, 1447 (1964).
58. R. L. Letsinger, O. B. Ramsay, and J. H. McCain, *J. Amer. Chem. Soc.*, 87, 2945 (1965).
59. K. E. Steller and R. L. Letsinger, *J. Org. Chem.*, **35**, 308 (1970).
60. E. Koenigs and H. Greiner, *Chem. Ber.*, **64**, 1049 (1931).
61. R. F. Evans, H. C. Brown, and H. C. van der Plas, *Org. Syn.*, **43**, 97 (1963).
62. A. L. Black and L. A. Summers, *Aust. J. Chem.*, **23**, 1495 (1970).
63. D. Jerchel, H. Fischer, and K. Thomas, *Chem. Ber.*, **89**, 2921 (1956).
64. D. L. Garmaise and G. Y. Paris, *Chem. Ind.* (London), 1645 (1967).
64a. R. N. Haszeldine, R. B. Rigby, and A. E. Tipping, *J. Chem. Soc. Perkin I*, 676 (1973).
65. R. Frampton, C. D. Johnson, and A. R. Katritzky, *Ann. Chem.*, **749**, 12 (1971).
66. C. H. Jarboe and C. M. Schmidt, *J. Med. Chem.*, **13**, 333 (1970).
67. C. H. Jarboe and J. A. Schaefer, *J. Med. Chem.*, **13**, 1026 (1970).
68. K. Undheim, P. O. Tveita, L. Borka, and V. Nordal, *Acta Chem. Scand.*, **23**, 2065 (1969).
69. J. W. Bunting and W. G. Meathrel, *Can. J. Chem.*, **48**, 3449 (1970).
69a. R. M. Bystrova and Y. M. Yutilov, *Khim. Geterotsikl. Soedin.*, 570 (1973).
70. G. B. Kauffman and K. L. Stevens, *Inorg. Syn.*, **7**, 173 (1963).
71. C. F. Hammer and C. E. Costello, *Tetrahedron Lett.*, 1743 (1971).
72. G. S. Fonken and F. A. Mackellar, *J. Med. Chem.*, **13**, 1246 (1970).
73. L. F. Fieser and M. Fieser, "Reagents for Organic Synthesis," Vol. 1, Wiley, New York, 1967, p. 75.
74. V. Dressler and K. Bodendorf, *Arch. Pharm.* (Weinheim), **303**, 481 (1970).
75. G. W. Fischer, *Z. Chem.*, **8**, 269 (1968).
76. A. E. Pohland and W. R. Benson, *Chem. Rev.*, **66**, 161 (1966).
77. J. J. O'Connor and I. A. Pearl, *J. Electrochem. Soc.*, **111**, 335 (1964).
78. C. Toma and A. T. Balaban, *Tetrahedron,* Suppl. 7, 9 (1966).
79. A. T. Balaban and C. Toma, *Tetrahedron,* Suppl. 7, 1 (1966).
80. A. B. Susan and A. T. Balaban, *Rev. Roum. Chim.*, **14**, 111 (1969); *Chem. Abstr.*, **71**, 61150m (1969).
81. A. N. Narkevich, G. N. Dorofeenko, and Y. A. Zhdanov, *Dokl. Akad. Nauk S.S.S.R.*, **176**, 103 (1967); *Chem. Abstr.*, **68**, 78605z (1968).
82. C. Toma and A. T. Balaban, *Tetrahedron,* Suppl. 7, 27 (1966).
83. Y. A. Zhdanov, G. N. Dorofeenko, and A. N. Narkevich, *J. Gen. Chem. U.S.S.R.*, **33**, 2357 (1963).
84. A. N. Narkevich, G. N. Dorofeenko, and Y. A. Zhdanov, *J. Gen. Chem. U.S.S.R.*, **36**, 838 (1966).
84a. J. A. VanAllan, G. A. Reynolds, and C. C. Petropoulos, *J. Heterocycl. Chem.*, **9**, 783 (1972).
85. M. H. O'Leary and G. A. Samberg, *J. Amer. Chem. Soc.*, **93**, 3530 (1971).
86. K. Dimroth, *Angew Chem.*, 72, 331 (1960).
87. K. Dimroth and K. H. Wolf, "Newer Methods in Preparative Organic Chemistry," W. Foerst (Ed), Vol. 3, Academic Press, New York, 1964, p. 357.

88. K. Dimroth, C. Reichardt, T. Siepmann, and F. Bohlmann, *Ann. Chem.*, **661**, 1 (1963).
89. K. Dimroth, C. Reichardt, and A. Schweig, *Ann. Chem.*, **669**, 95 (1963).
90. C. Reichardt and K. Dimroth, *Fortsch. Chem. Forsch.*, **11**, 1 (1968).
91. K. Dimroth and C. Reichardt, *Ann. Chem.*, **727**, 93 (1969).
92. V. Snieckus and G. Kan, *Chem. Commun.*, 1208 (1970) and references cited therein.
93. K. Undheim and V. Nordal, *Acta Chem. Scand.*, **23**, 1975 (1969).
94. K. Undheim, T. Wiik, L. Borka, and V. Nordal, *Acta Chem. Scand.*, **23**, 2509 (1969).
94a. E. Ager and H. Suschitzky, *J. Chem. Soc. Perkin I*, 2839 (1973).
95. B. P. Lugovkin, *Khim. Geterotsikl. Soedin.*, 1071 (1968); *Chem. Abstr.*, **70**, 87761r (1969).
96. G. T. Pilyngin, S. V. Shinkorenko, V. V. Stashkevich, and O. M. Stashkevich, *Khim. Geterotsikl. Soedin.*, 1075 (1968); *Chem. Abstr.*, **70**, 116214x (1969).
97. I. N. Zhmurova, A. A. Tukhar, and R. I. Yurchenko, *J. Org. Chem. U.S.S.R.*, **39**, 2150 (1969).
98. R. K. Howe and K. W. Ratts, *Tetrahedron Lett.*, 4743 (1967).
99. K. W. Ratts, R. K. Howe, and W. G. Phillips, *J. Amer. Chem. Soc.*, **91**, 6115 (1969).
100. F. Kröhnke, *Angew. Chem.*, **65**, 605 (1953).
101. T. Zincke, *Ann. Chem.*, **330**, 361 (1903); **333**, 296 (1904).
102. E. N. Marvell, G. Caple, and I. Shahidi, *Tetrahedron Lett.*, 277 (1967).
103. E. N. Marvell, G. Caple, and I. Shahidi, *J. Amer. Chem. Soc.*, **92**, 5641 (1970).
104. E. N. Marvell and I. Shahidi, *J. Amer. Chem. Soc.*, **92**, 5646 (1970).
105. V. I. Veksler, *J. Org. Chem. U.S.S.R.*, **38**, 1599 (1968).
106. H. Paulsen, K. Todt, and K. Heyns, *Ann. Chem.*, **679**, 168 (1964).
107. N. Elming, S. V. Carlsten, B. Lennart, and I. Ohlsson, British Patent 862,581 (1957); *Chem. Abstr.*, **56**, 11574g (1962).
108. N. Clauson-Kaas and N. Elming, U.S. Patent 2,806,852 (1957); *Chem. Abstr.*, **52**, 10202b (1958).
109. J. B. Petersen, K. Norris, N. Clauson-Kaas, and K. Svanholt, *Acta Chem. Scand.*, **23**, 1785 (1969).
110. K. Undheim and T. Greibrokk, *Acta Chem. Scand.*, **23**, 2475 (1969).
111. K. Undheim, R. Johan, and T. Greibrokk, *Acta Chem. Scand.*, **23**, 2501 (1969).
112. K. Undheim and M. Gacek, *Acta Chem. Scand.*, **23**, 2488 (1969).
113. W. König, *J. Prakt. Chem.*, **70**, 19 (1904).
114. A. F. Vompe and N. F. Turitsyna, *Dokl. Akad. Nauk S.S.S.R.*, **114**, 1017 (1957); *Chem. Abstr.*, **52**, 464d (1958).
115. W. König, M. Coenen, W. Lorenz, F. Bahr, and A. Bassl, *J. Prakt. Chem.*, **30**, 96 (1965).
116. H. J. Bestmann, H. J. Lang, and W. Distler, *Angew. Chem.*, **84**, 65 (1972).
116a. G. G. Abott and D. Leaver, *Chem. Commun.*, 150 (1973).
117. A. I. Meyers and S. Singh, *Tetrahedron*, **25**, 4161 (1969).
118. N. Sugiyama, K. Kubota, G. Inouye, and T. Kubota, *Bull. Chem. Soc. Jap.*, **37**, 637 (1964).
119. K. Wallenfels and W. Hanstein, *Angew Chem.*, **77**, 861 (1965).
120. H. Sund, in "Biological Oxidations," T. P. Singer (Ed.), Wiley-Interscience, New York, 1968, pp. 621–624.
121. J. J. Steffens and D. M. Chipman, *J. Amer. Chem. Soc.*, **93**, 6694 (1971).
122. J. Rochlitz, *Tetrahedron*, **23**, 3043 (1967).
123. A. N. Nesmeyanov, V. A. Sazonova, V. N. Drozd, and N. A. Rodionova, *Dokl. Akad. Nauk S.S.S.R.*, **160**, 335 (1965); *Chem. Abstr.*, **62**, 14725b (1965).

123a. H. M. Relles, *J. Org. Chem.*, **38**, 1570 (1973).
124. A. Camerman, *Can. J. Chem.*, **48**, 179 (1970).
125. C. Bugg, R. Desiderato, and R. L. Sass, *J. Amer. Chem. Soc.*, **86**, 3157 (1964); C. Bugg and R. L. Sass, *Acta Crystallogr.*, **18**, 591 (1965).
126. R. J. Chuck and E. W. Randall, *Spectrochim. Acta*, **22**, 221 (1966).
127. A. K. Sheinkman, L. M. Kapkan, L. G. Gakh, E. V. Titov, S. N. Baranov, and A. N. Kost, *Dokl. Akad. Nauk S.S.S.R.*, **193**, 366 (1970); *Chem. Abstr.*, **73**, 135794q (1970).
127a. G. P. Schiemenz, *J. Molec. Struc.*, **16**, 99 (1973).
128. F. W. Wehrli, W. Giger, and W. Simon, *Helv. Chim. Acta*, **54**, 229 (1971).
129. W. Giger, P. Schauwecker, and W. Simon, *Helv. Chim. Acta*, **54**, 2488 (1971).
130. D. J. Elias, A. G. Moritz, and D. B. Paul, *Aust. J. Chem.*, **25**, 427 (1972).
131. J. A. Zoltewicz and L. S. Helmick, *J. Amer. Chem. Soc*, **92**, 7547 (1970); R. A. Abramovitch, G. M. Singer, and A. R. Vinutha, *Chem. Commun.*, 55 (1967).
132. J. A. Zoltewicz and P. E. Kandetzki, *J. Amer. Chem. Soc.*, **93**, 6562 (1971), footnote 2.
133. M. Katcka and T. Urbanski, *Bull. Acad. Polon. Sci., Ser. Sci. Chim.*, **12**, 615 (1964); *Chem. Abstr.*, **62**, 11662f (1965).
134. J. P. Shoffner, L. Bauer, and C. L. Bell, *J. Heterocycl. Chem.*, **7**, 479 (1970).
135. E. M. Kosower, in "The Enzymes," (P. D. Boyer, H. Lardy, and K. Myrbäck, (Eds.), 2nd ed., Vol. 3, Academic Press, New York, Chap. 13; E. M. Kosower, "Molecular Biochemistry," McGraw-Hill, New York, 1962, pp. 180ff.
136. A. Hantzsch, *Chem. Ber.*, **44**, 1783 (1911).
137. E. M. Kosower, J. A. Skorcz, W. M. Schwarz, Jr., and J. W. Patton *J. Amer. Chem. Soc.*, **82**, 2188 (1960).
138. S. F. Mason, *J. Chem. Soc.*, 2437 (1960).
139. E. M. Kosower and J. A. Skorcz, *J. Amer. Chem. Soc.*, **82**, 2195 (1960).
140. S. Shifrin, *Biochemistry*, **3**, 829 (1964); *Biochim. Biophys. Acta*, **96**, 173 (1965).
141. J. W. Verhoeven, I. P. Dirkx, and T. J. de Boer, *Tetrahedron Lett.*, **37**, 4399 (1966); *Tetrahedron*, **25**, 3395, 4037 (1969).
142. M. Itoh, *J. Amer. Chem. Soc.*, **93**, 4750 (1971).
143. J. P. Saxena, B. K. Tak, and T. N. Agrawal, *J. Indian Chem. Soc.*, **47**, 863 (1970).
144. M. Farcasin and D. Farcasin, *Tetrahedron Lett.*, **48**, 4833 (1967).
145. E. M. Kosower and B. G. Ramsey, *J. Amer. Chem. Soc.*, **81**, 856 (1959).
146. R. A. Mackay and E. J. Poziomek, *J. Amer. Chem. Soc.*, **92**, 2432 (1970).
147. R. F. Cozzens and T. A. Gover, *J. Phys. Chem.*, **74**, 3003 (1970).
148. A. J. de Gee, J. W. Verhoeven, I. P. Dirkx, and T. J. de Boer, *Tetrahedron*, 25, 3407 (1969).
149. S. Sakanoue, Y. Kai, N. Yosuoka, N. Kasai, M. Kakudo, and H. Mikawa, *Chem. Commun.*, 176, (1969); *Bull. Chem. Soc. Jap.*, **43**, 1306 (1970).
150. S. Sakanoue, Y. Yasuoka, N. Kasai, M. Kakudo, S. Kusabayaski, and H. Mikawa, *Bull. Chem. Soc. Jap.*, **42**, 2408 (1969).
151. K. Nakamura, N. Yasuoka, N. Kasai, H. Mikawa, and M. Kakudo, *Chem. Commun.*, 1135 (1970).
152. E. M. Kosower, *J. Amer. Chem. Soc.*, **80**, 3253 (1958).
153. A. H. Ewald and J. A. Scudder, *Aust. J. Chem.*, **23**, 1939 (1970).
154. U. Bem, K. Guminski, and T. Zyczkowska, *Rocz. Chem.*, **44**, 887 (1970); *Chem. Abstr.*, **73**, 103158u (1970).
155. D. Guerin-Ouler, C. Nicollin, and A. Olivier, *C. R. Acad. Sci., Ser. C, Paris,* **270**, 1500 (1970).

156. S. G. Mairanovsky, "Catalytic and Kinetic Waves in Polarography," Plenum Press, New York, 1968; V. Tamas and C. Bodea, *Rev. Roum. Chim.*, **15**, 655 (1970).
157. A. K. Scheinkman, S. L. Portnova, Y. N. Scheinker, and A. N. Kost, *Dokl. Akad. Nauk S.S.S.R.*, **157**, 1416 (1964); *Chem. Abstr.*, **61**, 14157h (1964).
158. G. K. Budnikov and N. V. Evdokimova, *J. Gen. Chem. U.S.S.R.*, **40**, 1909 (1970).
159. A. Calderbank, D. F. Charlton, J. A. Farrington, and R. James, *J. Chem. Soc. Perkin I*, 138 (1972).
160. D. J. McClemens, A. K. Garrison, and A. L. Underwood, *J. Org. Chem.*, **34**, 1867 (1969).
160a. R. M. Elofson and R. L. Edsberg, *Can. J. Chem.*, **35**, 646 (1957).
161. A. Calderbank, *Adv. Pest Contr. Res.*, **8**, 127 (1968).
162. P. Beak, J. Bonham, and J. T. Lee, Jr., *J. Amer. Chem. Soc.*, **90**, 1569 (1968).
163. P. Beak and J. T. Lee, Jr., *J. Org. Chem.*, **34**, 2125 (1969).
164. E. W. Anaker and H. M. Ghose, *J. Amer. Chem. Soc.*, **90**, 3161 (1968).
165. G. J. Papenmeier and J. M. Campagnoli, *J. Amer. Chem. Soc.*, **91**, 6579 (1969).
166. L. Kundu, *J. Indian Chem. Soc.*, **47**, 483 (1970).
166a. E. H. Cordes and C. Gitler, in "Progress in Bioorganic Chemistry," Vol.2, E. T. Kaiser and F. J. Kézdy (Eds.), Wiley-Interscience, New York, 1973, pp. 1ff.
167. D. M. Updegraff, D. C. Kvam, and J. E. Robertson, *J. Pharm. Sci.*, **59**, 188 (1970).
168. W. U. Malik and A. K. Jain, *J. Indian Chem. Soc.*, **48**, 217 (1971).
169. R. K. Joshi, L. Krasnec, and I. Lacko, *Helv. Chim. Acta*, **54**, 112 (1971).
170. M. Tomizawa and T. Kondo, *Chem. Pharm. Bull.* (Tokyo), **18**, 2158 (1970).
171. F. M. Plakogiannis, E. J. Lien, C. Harris, and J. A. Biles, *J. Pharm. Sci.*, **59**, 197 (1970).
172. A. Wu and W. T. Smith, Jr., *Anal. Chem.*, **40**, 1578 (1968).
173. R. Preussmann, H. Hengy, and H. Druckrey, *Ann. Chem.*, **684**, 57 (1965).
174. G. W. Fisher, *Z. Chem.*, **9**, 300 (1969).
174a. M. Yamamoto, S. Nakamura, K. Yoshimura, M. Yuge, S. Morosawa, and A. Yokoo, *Bull. Chem. Soc. Jap.*, **46**, 1509 (1973).
175. P. A. Claret and G. H. Williams, *J. Chem. Soc. C*, 146 (1969).
176. J. Kuthan, N. V. Koshmina, J. Palecek, and V. Skala, *Collect. Czech. Chem. Commun.*, **35**, 2787 (1970).
177. G. H. Schmid and A. Brown, *Tetrahedron Lett.*, **45**, 4695 (1968).
178. D. E. Guttman and D. Brooke, *J. Pharm. Sci.*, **57**, 1677 (1968).
178a. M. L. Tosato, L. Soccorsi, M. Cignitti, and L. Paoloni, *Tetrahedron*, **29**, 1339 (1973).
178b. K. Undheim and P. E. Hansen, *Org. Mass Spectrom.*, **7**, 635 (1973).
179. J. A. Berson and T. Cohen, *J. Amer. Chem. Soc.*, **78**, 416 (1956).
180. E. Matsumura, T. Nashima, and F. Ishibashi, *Bull. Chem. Soc. Jap.*, **43**, 3540 (1970).
181. H. Tomisawa and H. Hongo, *Chem. Pharm. Bull.* (Tokyo), **18**, 925 (1970); H. Hongo, *ibid.*, **20**, 226 (1972).
182. T. Robinson and C. Cepurnek, *Phytochemistry*, **4**, 75 (1965).
183. R. A. Abramovitch and A. R. Vinutha, *J. Chem. Soc. B*, 131 (1971).
184. H. Decker, *Ber.*, **25**, 443 (1892), *J. Prakt. Chem.*, **47**, 29 (1893); H. Decker and A. Kaufmann, *ibid.*, **84**, 425 (1911).
185. H. Möhrle and H. Weber, *Tetrahedron*, **26**, 2953 (1970).
186. H. Möhrle and H. Weber, *Chem. Ber.*, **104**, 1478 (1971).
187. H. Möhrle and H. Weber, *Tetrahedron*, **26**, 3779 (1970).
188. T. Severin, D. Bätz, and H. Lerch, *Chem. Ber.*, **103**, 1 (1970).
189. Y. Ban, R. Sakaguchi, and M. Nagai, *Chem. Pharm. Bull.* (Tokyo), **13**, 931 (1965).

190. E. M. Kosower and J. W. Patton, *Tetrahedron,* **22**, 2081 (1966).
191. G. R. Waller and L. M. Henderson, *J. Biol. Chem.,* **236**, 1186 (1961).
192. R. Mukherjee and A. Chatterjee, *Chem. Ind.* (London), 1524 (1964).
193. P. Fu and T. Robinson, *Phytochemistry,* **9**, 2443 (1970).
194. P. Fu, J. Kobus, and T. Robinson, *Phytochemistry,* **11**, 105 (1972). P. Fu and T. Robinson, *ibid.*, 95 (1972).
195. G. R. Waller, K. S. Yang, R. K. Gholson, L. A. Hadwiger, and S. Chaykin, *J. Biol. Chem.,* **241**, 4411 (1966).
196. T. C. Bruice and S. Benkovic, "Bioorganic Mechanisms," Vol. 2, Benjamin, New York, 1966, pp. 326 ff.
197. H. Sund. H. Diekmann, and K. Wallenfels, *Advan. Enzymol.,* **26**, 115 (1964).
198. R. E. Lyle and G. Gauthier, *Tetrahedron Lett.,* 4615 (1965).
199. P. S. Anderson, W. E. Krueger, and R. E. Lyle, *Tetrahedron Lett.,* 4011 (1965).
200. K. Wallenfels and H. Schüly, *Ann. Chem.,* **621**, 106, 215 (1959).
201. R. N. Lindquist and E. H. Cordes, *J. Amer. Chem. Soc.,* **90**, 1269 (1968).
201a. A. C. Lovesey, *J. Med. Chem.,* **13**, 693 (1970); **12**, 1018 (1969).
202. J. Baumrucker, M. Calzadilla, M. Centeno, G. Lehrmann, P. Lindquist, D. Dunham, M. Price, B. Sears, and E. H. Cordes, *J. Phys. Chem.,* **74**, 1152 (1970).
203. E. H. Cordes and R. B. Dunlap, *Accounts Chem. Res.,* **2**, 329 (1969).
203a. R. Foster and C. A. Fyfe, *Tetrahedron,* **25**, 1489 (1969).
203b. K. Wallenfels and W. Hanstein, *Ann. Chem..,* **709**, 151 (1967).
203c. K. Wallenfels and W. Hanstein, *Ann. Chem.,* **732**, 139 (1970).
204. K. Wallenfels and H. Diekmann, *Ann. Chem.,* **621**, 166 (1959).
205. K. Wallenfels and H. Schüly, *Ann. Chem.,* **621**, 86 (1959).
206. L. J. Winters, A. L. Borror, and N. Smith, *Tetrahedron Lett.,* **24**, 2313 (1967).
207. C. Schiele, D. Staudacher, D. Hendricks and G. Arnold, *Tetrahedron,* **24**, 5017 (1968); C. Schiele, M. Ruch, and D. Hendricks, *ibid.*, **23**, 3733 (1967).
208. C. Schiele, D. Staudacher, K. Halfar, and G. Arnold, *Tetrahedron*, **24**, 5023 (1968).
209. C. Schiele and H. O. Kalinowski, *Angew. Chem.,* **78**, 389 (1966); *Angew. Chem. Int. Ed.,* **5**, 416 (1966); *Ann. Chem.,* **696**, 81 (1966).
210. R. E. Lyle and E. White V, *J. Org. Chem.,* **36**, 772 (1971).
211. R. F. Stadnüchuk, G. T. Pilyugin, and O. E. Petrenki, *J. Gen. Chem. U.S.S.R.,* **40**, 1817 (1970).
211a. L. I. Zakharkin, L. E. Litovchenko, and A. V. Kazantsev, *J. Gen. Chem. U.S.S.R.,* **40**, 113 (1970).
212. K. Dimroth, K. Wolf, and H. Kroke, *Ann. Chem.,* **678**, 183 (1964).
213. M. Takeda, A. E. Jacobson, K. Kanematsu, and E. L. May, *J. Org. Chem.,* **34**, 4154, 4158 (1969); M. Takeda, A. E. Jacobson, and E. L. May, *ibid.*, 34, 4161 (1969); E. M. Fry, *ibid.*, **28**, 1869 (1963).
213a. G. V. Boyd, A. W. Ellis, and M. D. Harms, *J. Chem. Soc.C,* 800 (1970).
214. D. N. Kursanov and N. K. Baranetskaya, *Bull. Acad. Sci. U.S.S.R., Div. Chem. Sci.,* **341 (1958).**
215. D. N. Kursanov, N. K. Baranetskaya, and V. N. Setkina, *Proc. Acad. Sci. U.S.S.R., Chem. Sec.,* **113**, 191 (1957).
216. F. Kröhnke, K. Ellegast, and E. Bertram, *Ann. Chem.,* **600**, 176 (1956).
216a. *cf.* however, H. Prinzbach and H. Sauter, *Angew. Chem. Int. Ed.,* **11**, 133 (1972).
217. J. A. Berson, E. M. Evelth, Jr., and Z. Hamlet, *J. Amer. Chem. Soc.,* **87**, 2887 (1965).
218. G. V. Boyd and N. Singer, *J. Chem. Soc.B,* 1017 (1966).

219. G. V. Boyd and L. M. Jackman, *J. Chem. Soc.*, 548 (1963).
220. P. Blumbergo, A. B. Ash, F. A. Daniher, C. L. Stevens, H. O. Michel, B. E. Hackley, Jr., and J. Epstein, *J. Org. Chem.*, **34**, 4065 (1969).
221. H. Ahlbrect and F. Kröhnke, *Ann. Chem.*, **704**, 133 (1967); **717**, 96 (1968).
222. J. Ludowieg, N. Bhacca, and A. Levy, *Biochem. Biophys. Res. Commun.*, **14**, 431 (1964).
222a. I. H. Pitman, E. Shefter, and M. Ziser, *J. Amer. Chem. Soc.*, **92**, 3413 (1970).
223. G. Klopman, *J. Amer. Chem. Soc.,*, **90**, 223 (1968).
223a. J. W. Happ, M. T. McCall, and D. G. Whitten, *J. Amer. Chem. Soc.*, **93**, 5496 (1971).
224. V. J. Traynelis and A. I. Gallagher, *J. Org. Chem.*, **35**, 2792 (1970).
225. H. J. Roth and H. Möhrle, *Naturwissenschaften*, **51**, 107 (1964); *idem.*, *Arch. Pharm.* (Weinheim), **299**, 315 (1966).
226. H. J. Roth and S. Al Sarraj, *Arch. Pharm.* (Weinheim), **300**, 44 (1967).
227. H. J. Roth, S. Al Sarraj, and K. Jäger, *Arch. Pharm.* (Weinheim), **299**, 605 (1966).
228. H. Möhrle and D. Schittenhelm, *Arch. Pharm.* (Weinheim), **303**, 771 (1970).
229. T. Melton, J. Taylor, and D. G. Wibberley, *Chem. Commun.*, 151 (1965).
230. F. W. Kröch and F. Kröhnke, *Chem. Ber.*, **102**, 659, 669 (1969).
231. J. Taylor and D. G. Wibberley, *J. Chem. Soc.C*, 2693 (1968).
232. C. Schiele and D. Staudacher, *Tetrahedron*, **24**, 471 (1968).
233. J. A. Zoltewicz, G. M. Kauffmann, and C. L. Smith, *J. Amer. Chem. Soc.*, **90**, 5939 (1968).
234. P. Beak and E. M. Monroe, *J. Org. Chem.*, **34**, 589 (1969).
235. F. Kröhnke and W. Zecher, *Angew. Chem. Int. Ed.*, **1**, 626 (1962); F. Kröhnke, *ibid.*, **2**, 225 (1963).
236. A. W. Johnson, "Ylid Chemistry," Academic Press, New York, 1966, pp. 260ff.
237. C. A. Henrick, E. Ritchie, and W. C. Taylor, *Aust. J. Chem.*, **20**, 2441 (1967).
238. P. B. D. De la Mare, *Nature*, **195**, 441 (1962).
239. J. W. Cornforth, R. Gigg, and M. S. Tute, *Aust. J. Chem.*, **20**, 2479 (1967).
240. C. A. Henrick, E. Ritchie, and W. C. Taylor, *Aust. J. Chem.*, **20**, 2455 (1967).
241. T. Eicher, E. von Angerer, and A. Hansen, *Ann. Chem.*, **746**, 102 (1971).
242. T. Sasaki, K. Kanematsu, and A. Kakehi, *J. Org. Chem.*, **36**, 2451 (1971).
243. T. Eicher and E. von Angerer, *Ann. Chem.*, **746**, 120 (1971).
244. I. C. Calder, Q. N. Porter, and C. M. Richards, *Aust. J. Chem.*, **25**, 345 (1972).
245. M. Ikeda, N. Tsujimoto, and Y. Tamura, *Org. Mass Spectrom.*, **5**, 61 (1971).
246. R. J. Bass, *Chem. Commun.*, 322 (1970).
247. M. Yamazaki, K. Noda, and M. Hamana, *Chem. Pharm. Bull.* (Tokyo), **18**, 901, 908 (1970).
247a. M. Yamazaki, K. Noda, N. Honjo, and M. Hamana, *Chem. Pharm. Bull.* (Tokyo), **21**, 712 (1973).
247b. D. I. Schütze and F. Kröhnke, *Ann. Chem.*, **765**, 20 (1972).
248. J. E. Douglass and J. M. Wesolosky, *J. Org. Chem.*, **36**, 1165 (1971).
249. V. Boekelheide and K. Fahrenholtz, *J. Amer. Chem. Soc.*, **83** , 458 (1961).
250. R. Huisgen, R. Grashey, and E. Steingruber, *Tetrahedron Lett.*, 1441 (1963).
251. W. J. Linn, O. W. Webster, and R. E. Benson, *J. Amer. Chem. Soc.*, **87**, 3651 (1965).
252. C. A. Henrick, E. Ritchie, and W. C. Taylor, *Aust. J. Chem.*, **20**, 2467 (1967).
253. T. Sasaki, K. Kanematsu, Y. Yukimoto, and S. Ochiai, *J. Org. Chem.*, **36**, 813 (1971).
254. N. Basketter and A. O. Plunkett, *Chem. Commun.*, 1578 (1971).
255. Y. Kobayashi, T. Kutsuma, and K. Morinaga, *Chem. Pharm. Bull.* (Tokyo), **19**, 2106 (1971).

256. H. Wittmann, J. Kuhn-Kuhnenfeld, and E. Ziegler, *Monatsh. Chem.*, **102**, 1120 (1971).
257. H. Wittmann, P. Beutel, and E. Ziegler, *Monatsh. Chem.*, **100**, 1362 (1969).
258. J. Streith and J. M. Cassal, *C. R. Acad. Sci., Ser. C, Paris,* **264**, 1307 (1967).
259. T. Sasaki, K. Kanematsu, A. Kakehi, I. Ichikawa, and K. Hayakawa, *J. Org. Chem.*, **35**, 426 (1970).
260. A. Balasubramanian, J. M. McIntosh, and V. Snieckus, *J. Org. Chem.*, **35**, 433 (1970).
261. F. Kröhnke and I. Vogt, *Ann. Chem.*, **589**, 26 (1954).
262. H. Moureu, P. Chovin, R. Sabourin, and G. Flad, *Bull. Soc. Chim. Fr.*, 624 (1969).
263. F. Kröhnke and H. Ahlbrecht, *Chem. Ber.*, **100**, 1756 (1967).
264. Y. A. Zhdanov, G. V. Bogdanova, and N. N. Artamonova, *Khim. Geterotsikl. Soedin.*, 567 (1966); *Chem. Abstr.*, **66**, 46539y (1967).
265. F. Kröhnke, *Chem. Ber.*, **68**, 1177 (1935).
266. T. Melton and D. G. Wibberley, *J. Chem. Soc.*, 983 (1967).
267. C. H. Gaozza and S. Lamdan, *Tetrahedron Lett.*, 4945 (1969).
268. F. Kröhnke and H. H. Stevernagel, *Chem. Ber.*, **97**, 1118 (1964).
269. H. J. Roth and S. A. Sarraj, *Arch. Pharm.* (Weinheim), **299**, 385 (1966).
270. H. J. Roth and S. A. Sarraj, *Arch. Pharm.* (Weinheim), **299**, 394 (1966).
271. A. San Pietro, *J. Biol. Chem.*, **217**, 589 (1955).
272. S. Golding and A. R. Katritzky, *Can. J. Chem.*, **43**, 1250 (1965).
273. W. G. Phillips and K. W. Ratts, *J. Org. Chem.*, **35**, 3144 (1970).
274. J. E. Douglass, M. W. Tabor, and J. E. Spradling, III, *J. Heterocycl. Chem.*, **9**, 53 (1972).
275. J. H. Nelson, R. G. Garvey, and R. O. Ragsdale, *J. Heterocycl. Chem.*, **4**, 591 (1967).
276. K. Undheim and T. Greibrokk, *Acta Chem. Scand.*, **23**, 2501, 2505, 2509 (1969).
277. A. R. Katritzky and Y. Takeuchi, *J. Chem. Soc.C*, 874 (1971).
278. A. R. Katritzky and Y. Takeuchi, *J. Amer. Chem. Soc.*, **92**, 4134 (1970).
278a. T. Sasaki, K. Kanematsu, K. Hayakawa, and M. Uchide, *J. Chem. Soc. Perkin I*, 2750 (1972).
279. M. Tsuda and Y. Kawazoe, *Chem. Pharm. Bull.* (Tokyo), **18**, 2499 (1970).
280. M. Freifelder, *J. Pharm. Sci.*, **55**, 535 (1966).
281. For example, see J. Lee and W. Freudenberg, *J. Org. Chem.*, **9**, 537 (1944); K. Hohenlohe-Oehringen, *Monatsh Chem.*, **93**, 586 (1962).
282. R. E. Lyle, G. H. Warner, and D. A. Nelson, *Bol. Soc. Quim. Peru,* **31**, 89 (1965); *Chem. Abstr.*, **64**, 19548a (1966).
283. C. A. Grob and F. Ostermayer, *Helv. Chim. Acta,* **45**, 1119 (1962).
284. E. Wenkert and B. Wickberg, *J. Amer. Chem. Soc.*, **87**, 1580, 5810 (1965); E. Wenkert, K. G. Dave, and F. Haglid, *ibid.*, 5461 (1965).
285. E. Wenkert, K. G. Dave, F. Haglid, R. G. Lewis, T. Oishi, R. V. Stevens, and M. Terashima, *J. Org. Chem.*, **33**, 747 (1968).
286. M. Freifelder, *J. Org. Chem.*, **29**, 2895 (1964).
287. M. Ferles, *Collect. Czech. Chem. Commun.*, **24**, 2221 (1959).
288. M. Ferles, *Collect. Czech. Chem. Commun.*, **23**, 479 (1958).
289. M. Ferles, M. Kovarik, and Z. Vondrackova, *Collect. Czech. Chem. Commun.*, **31**, 1348 (1966).
290. M. Holik and M. Ferles, *Collect. Czech. Chem. Commun.*, **32**, 3067 (1967).
291. M. Jankovsky and M. Ferles, *Collect. Czech. Chem. Commun.*, **35**, 2802 (1970).
292. R. E. Lyle, D. A. Nelson, and P. S. Anderson, *Tetrahedron Lett.*, **13**, 553 (1962).
293. P. S. Anderson and R. E. Lyle, *Tetrahedron Lett.*, **3**, 153 (1964).
294. R. E. Lyle and P. S. Anderson, *Adv. Heterocycl. Chem.*, **6**, 46 (1966).

295. F. Liberatore, V. Carelli, and M. Cardellini, *Tetrahedron Lett.*, **46**, 4735 (1968).
296. F. Liberatore, V. Carelli, and M. Cardellini, *Chim. Ind.* (Milan), **51**, 55 (1969); *Chem. Abstr.*, **70**, 96577n (1969).
297. P. S. Anderson, W. E. Krueger, and R. E. Lyle, *Tetrahedron Lett.*, **45**, 4011 (1965).
297a. H. Oediger and N. Joop, *Ann. Chem.*, **764**, 21 (1972).
297b. P. P. Zarin, E. E. Liepin, E. S. Lavrinovich, and A. K. Aren, *Khim. Geterotsikl. Soedin.*, 115 (1974).
298. E. M. Fry, *J. Org. Chem.*, **29**, 1647 (1964).
299. E. M. Fry and J. A. Beisler, *J. Org. Chem.,*, **35**, 2809 (1970).
299a. R. T. Parfitt and S. M. Watters, *J. Med. Chem.*, **14**, 565 (1971).
300. W. Hanstein and K. Wallenfels, *Tetrahedron,* **23**, 585 (1967).
301. G. Büchi, D. L. Coffen, K. Kocsis, P. E. Sonnet, and F. E. Ziegler, *J. Amer. Chem. Soc.*, **87**, 2073 (1965).
302. D. L. Coffen, *J. Org. Chem.*, **33**, 137 (1968).
303. N. Kinoshita, M. Hamana, and T. Kawaski, *Chem. Pharm. Bull.* (Tokyo), **10**, 753 (1962).
304. N. Kinoshita, M. Hamana, and T. Kawasaki, *Yakugaku Zasshi*, **83**, 115, 120, 123, 126, (1963); *Chem. Abstr.*, **59**, 5126d (1963).
305. M. Saunders and E. H. Gold, *J. Org. Chem.*, **27**, 1439 (1962).
305a. U. Eisner and J. Kuthan, *Chem. Rev.*, **72**, 1 (1972).
306. K. Schenker and J. Druey, *Helv. Chim. Acta,* **42**, 1960 (1959).
307. E. Wenkert, R. A. Massy-Westropp, and R. G. Lewis, *J. Amer. Chem. Soc.*, **84**, 3732 (1962).
308. K. T. Potts and D. R. Liljegren, *J. Org. Chem.,*, **28**, 3066 (1963).
308a. J. A. Beisler, *Chem. Ber.*, **103**, 3360 (1970); *Tetrahedron,* **26**, 1961 (1970).
309. E. Wenkert, K. G. Dave, C. T. Gnewuch, and P. W. Sprague, *J. Amer. Chem. Soc.*, **90**, 5251 (1968).
310. E. Wenkert, K. G. Dave, R. G. Lewis, and P. W. Sprague, *J. Amer. Chem. Soc.*, **89** 6741 (1967); E. Wenkert and B. Wickberg, *ibid.*, **87**, 1580 (1965).
310a. M. S. Allen, A. J. Gaskell, and J. A. Joule, *J. Chem. Soc.C,* 736 (1971).
311. K. T. Potts and I. D. Nasri, *J. Org. Chem.*, **29**, 3407 (1964).
312. F. E. Ziegler and J. G. Sweeny, *J. Org. Chem.*, **32**, 3216 (1967).
313. M. Saunders and E. H. Gold, *J. Amer. Chem. Soc.* **88**, 3376 (1966).
314. F. Liberatore, A. Casini, V. Carelli, A. Arnone, and R. Mondelli, *Tetrahedron Lett.*, **26**, 2381 (1971).
315. F. Liberatore, A. Casini, V. Carelli, A. Arnone, and R. Mondelli, *Tetrahedron Lett.*, **41**, 3829 (1971).
316. R. E. Lyle and C. K. Spicer, *Tetrahedron Lett.*, **14**, 1133 (1970); R. E. Lyle, K. R. Carle, C. R. Ellefson, and C. K. Spicer, *J. Org. Chem.*, **35**, 802 (1970).
316a. P. Bichaut, G. Thuillier, and P. Rumpf, *Bull. Soc. Chim. Fr.*, 3322 (1971).
317. R. Lukes, J. N. Zvonkova, A. F. Mironov, and M. Ferles, *Collect. Czech. Chem. Commun.*, **25**, 2668 (1960).
318. O. Cervinka and O. Kruz, *Collect. Czech. Chem. Commun.*, **30**, 1700 (1965).
319. P. Karrer and O. Warburg, *Biochem. Z.*, **285**, 297 (1936); M. Pullman, A. San Pietro, and S. P. Colowick, *J. Biol. Chem.*, **206**, 129 (1954).
320. D. Manzerall and F. H. Westheimer, *J. Amer. Chem. Soc.*, **77**, 2261 (1955); R. F. Hutton and F. H. Westheimer, *Tetrahedron,* **3**, 73 (1958).
321. W. S. Caughey and K. A. Schellenberg, *J. Org. Chem.*, **31**, 1978 (1966).
322. J. F. Biellmann and H. J. Callot, *Tetrahedron Lett.*, 3991 (1966); *idem.*, *Bull. Soc. Chim. Fr.*, 1154, 1159 (1968); 1299 (1969).

322a. J. H. Supple, D. A. Nelson, and R. E. Lyle, *Tetrahedron Lett.*, **24**, 1645 (1963).
322b. G. Pfleiderer, E. Sann, and A. Stock, *Chem. Ber.*, **93**, 3083 (1960).
323. W. R. Boon, *Chem. Ind.* (London), 782 (1965).
324. R. F. Homer, G. C. Mees, and T. E. Tomlinson, *J. Sci. Food Agr.*, **11**, 309 (1960).
325. G. C. Mees, *Ann. Appl. Biol.*, **48**, 601 (1960).
326. A. Calderbank, *Biochem. J.*, **101**, 2P (1966).
327. L. A. Summers. *Tetrahedron*, **24**, 5433 (1968); A. L. Black and L. A. Summers, *ibid.*, 6453 (1968).
328. L. A. Summers, *Tetrahedron*, **24**, 2697 (1968).
329. J. E. Dickeson and L. A. Summers, *J. Heterocycl. Chem.*, **7**, 719 (1970).
330. A. L. Black and L. A. Summers, *J. Chem. Soc.C*, 2394 (1970).
331. J. E. Dickeson and L. A. Summers, *J. Heterocycl. Chem.*, **7**, 401 (1970).
332. J. E. Dickeson and L. A. Summers, *Experientia*, **25**, 1247 (1969).
333. J. E. Dickeson and L. A. Summers, *J. Chem. Soc.C*, 1643 (1969).
334. P. J. S. Wang and E. R. Thornton, *J. Amer. Chem. Soc.*, **90**, 1216 (1968).
335. G. Schroll, S. P. Nygaard, S. O. Lawesson, A. M. Duffield, and C. Djerassi, *Ark. Kemi*, **29**, 525 (1968); *Chem. Abstr.*, **70**, 28779e (1969).
335a. R. E. Lyle and E. White V, *Tetrahedron Lett.*, **22**, 1871 (1970).
336. M. B. Yarmolinski and S. P. Colowick, *Biochim. Biophys. Acta*, **20**, 177 (1956).
336a. A. C. Lovesey and W. C. J. Ross, *J. Chem. Soc.B*, 192 (1969).
337. E. M. Kosower and S. W. Bauer, *J. Amer. Chem. Soc.*, **82**, 2191 (1960).
338. K. A. Schellenberg and F. H. Westheimer, *J. Org. Chem.*, **30**, 1859 (1965) and references therein cited.
339. C. H. Wang, S. M. Linnell, and N. Wang, *J. Org. Chem.*, **36**, 525 (1971).
340. E. M. Kosower, *Prog. Phys. Org. Chem.*, **3**, 81 (1965).
341. J. J. Steffens and D. M. Chipman, *J. Amer. Chem. Soc.*, **93**, 6694 (1971).
342. J. N. Burnett and A. L. Underwood, *Biochemistry*, **4**, 2060 (1965).
343. A. M. Wilson and D. G. Epple, *Biochemistry*, **5**, 3170 (1966).
344. A. J. Cunningham and A. L. Underwood, *Arch. Biochem. Biophys.*, **117**, 88 (1966).
345. A. J. Cunningham and A. L. Underwood, *Biochemistry*, **6**, 266 (1967).
346. J. N. Burnett and A. L. Underwood, *J. Org. Chem.*, **30**, 1154 (1965).
346a. B. Janík and P. J. Elving, *Chem. Rev.*, **68**, 295 (1968).
346b. J. F. Biellmann, J. P. Samama, and A. D. Wrixon, *Biochimie*, **55**, 1469 (1973).
346c. L. E. Overman, *J. Org. Chem.*, **37**, 4214 (1972).
347. S. Chaykin, *Ann. Rev. Biochem.*, **36**, 149 (1967).
348. E. M. Kosower and J. L. Cotter, *J. Amer. Chem. Soc.*, **86**, 5524 (1964).
348a. F. E. Geiger, C. L. Trichilo, F. L. Minn, and N. Filipescu, *J. Org. Chem.*, **36**, *357* (1971).
348b. M. Frangopol, P. T. Frangopol, C. L. Trichilo, F. E. Geiger, and N. Filipescu, *J. Org. Chem.*, 38, 2355 (1973).
349. L. J. Winters, N. G. Smith, and M. I. Cohen, *Chem. Commun.*, 642 (1970).
350. J. G. Carey, J. F. Cairns, and J. E. Colchester, *Chem. Commun.*, 1280 (1969).
350a. J. A. Farrington, A. Ledwith, and M. F. Stam, *Chem. Commun.*, 259 (1969).
351. A. T. Nielsen, D. W. Moore, G. M. Muha, and K. H. Berry, *J. Org. Chem.*, **29**, 2175 (1964).
352. K. Wallenfels and M. Gellrich, *Chem. Ber.*, **92**, 1406 (1959).
353. H. Diekmann, G. Englert, and K. Wallenfels, *Tetrahedron*, **20**, 281 (1964).
354. T. Zincke, *Ann. Chem.*, **338**, 107 (1905); **339**, 193 (1905); *Ber.*, **38**, 3824 (1905).
355. W. König, *J. Prakt. Chem.*, **69**, (2) 105 (1904).
356. A. F. Vompe and N. F. Turitsyna, *Zhur. Obshch. Khim.*, **28**, 2864 (1958); *Chem. Abstr.*, **53**, 9207i (1959).

357. A. F. Vompe, I. I. Levkoev, N. F. Turitsyna, V. V. Durmashkina, and L. V. Ivanova, *Zhur. Obshch. Khim.*, **34**, 1758 (1964); *Chem. Abstr.*, **61**, 8269b (1964).
358. B. Lipke, *Z. Chem.*, **10**, 463 (1970); 11, 150 (1971).
359. E.. P. Lira, *J. Heterocycl. Chem.*, **9**, 713 (1972).
360. H. Beyer and E. Thieme, *J. Prakt. Chem.*, **31**, 293 (1966).
361. Y. Tamura, N. Tsujimoto, and M. Uchimura, *J. Pharm. Soc. Jap.*, **91**, 72(1971).
362. Y. Tamura, N. Tsujimoto, and M. Mano, *Chem. Pharm. Soc. Bull.* (Tokyo), **19**, 130 (1971); Y. Tamura and N. Tsujimoto, *Chem. Ind.* (London), 926 (1970); Y. Tamura, Y. Miki, T. Honda, and M. Ikeda, *J. Heterocycl. Chem.*, 9, 865 (1972).
363. P. Pfeiffer and E. Enders, *Chem. Ber.*, **84**, 313 (1951).
364. G. Schwarzenbach and R. Weber, *Helv. Chim. Acta*, **25**, 1628 (1942).
365. G. W. Fischer, *Z. Chem.*, **8**, 379 (1968).
366. J. Bartos, *Ann. Pharm. Fr.*, **29**, 221 (1971).
367. Y. Tamura, N. Tsujimoto, and Y. Hirano, *J. Pharm. Soc. Jap.*, **92**, 546 (1972).
368. R. Fusco, S. Rossi, and S. Maiorana, *Gazz. Chim. Ital.*, **95**, 1237 (1965).
369. J. Lee and W. W. Paudler, *Chem. Commun.*, 1636 (1971).
370. A. R. Fersht and W. P. Jencks, *J. Amer. Chem. Soc.*, **91**, 2125 (1969); **92**, 5432 (1970).
371. A. R. Fersht and W. P. Jencks, *J. Amer. Chem. Soc.*, **92**, 5442 (1970).
372. M. Wakselman and E. Guibé-Jampel, *Tetrahedron Lett.*, **18**, 1521 (1970).
373. T. Agawa and S. Miller, *J. Amer. Chem. Soc.*, **83**, 449 (1961).
374. G. Fraenkel, J. W. Cooper, and C. M. Fink, *Angew. Chem.*, **82**, 518 (1970).
374a. A. K. Sheinkman, A. N. Kost, A. N. Prilepskaya, and N. A. Klyuev, *Khim. Geterotsikl. Soedin.*, 1105 (1972).
375. H. V. Dobenek, H. Deubel, and F. Heichele, *Angew. Chem.*, **71**, 310 (1959).
376. J. Bergman, *J. Heterocycl. Chem.*, **7**, 1071 (1970).
377. A. N. Kost, A. K. Sheinkman, and N. F. Kazarinova, *J. Gen. Chem. U.S.S.R.*, **34**, 2059 (1964).
378. L. R. Kolomoitsev, N. F. Kazarinova, N. I. Geonya, and A. K. Sheinkman, *Mikrobiol. Zh. Akad. Nauk. Ukr. RSR*, **24** 23 (1962); *Chem. Abstr.*, **57**, 9958h (1962); L. R. Kolomoitsev, N. G. Geonya, N. V. Strangovs'ka, and A. K. Sheinkman, *Mikrobiol. Zh. Akad. Nauk, Ukr. RSR*, **27**, 56 (1965); *Chem. Abstr.*, **63**, 603lh (1965).
379. A. K. Sheinkman, G. V. Samoilenko, and S. N. Baranov, *J. Gen. Chem. U.S.S.R.*, **40**, 671 (1970).
380. E. Guibé-Jampel and M. Wakselman, *Chem. Commun.*, 267 (1971).
381. S. L. Johnson and K. A. Rumon, *J. Phys. Chem.*, **68**, 3149 (1964).
382. P. Baudet, C. Otten, and D. Rao, *Helv. Chim. Acta*, **53**, 859 (1970).
383. M. Wakselman and E. Guibé-Jampel, *Tetrahedron Lett.*, **54**, 4715 (1970).
384. R. Camps, *Arch. Pharm.* (Weinheim), **240**, 345 (1902); R. A. Jones and A. R. Katritzky, *J. Chem. Soc.*, 1317 (1959).
385. G. W. Fischer, *Chem. Ber.*, **103**, 3470 (1970).
386. G. W. Fischer, *Chem. Ber.*, **103**, 3489 (1970).
387. K. Undheim and P. O. Tveita, *Acta Chem. Scand.*, **25**, 5 (1971).
388. C. S. Giam and E. E. Knaus, *Tetrahedron Lett.*, **52**, 4961 (1971).
389. G. A. Olah and P. J. Szilagyi, *J. Amer. Chem. Soc.*, **91**, 2949 (1969).
390. I. Fleming and J. B. Mason, *J. Chem. Soc.C*, 2509 (1969).
391. R. E. Lyle, *Chem. Eng. News*, **44**, 72 (1966).
392. F. Kröhnke, *Angew. Chem. Int. Ed.*, **2**, 380 (1963).
393. A. B. Ash, P. Blumbergs, C. L. Stevens, H. O. Michel, B. E. Hackley, Jr., and J. Epstein, *J. Org. Chem.*, **34**, 4070 (1969).
394. L. Ruzicka, P. A. Plattner, and M. Furrer, *Helv. Chim. Acta*, **27**, 524 (1944).

395. H. R. Nace and D. H. Nelander, *J. Org. Chem.*, **29**, 1677 (1964).
396. K. Prezewowsky, R. Wiechert, and W. Hohlweg, *Ann, Chem.*, **752**, 68 (1971).
397. K. L. Williamson and W. S. Johnson, *J. Org. Chem.*, **26**, 4563 (1961).
398. E. W. Warnhoff, *J. Org. Chem.*, **27**, 4587 (1962).
399. K. Y. Novitskii, G. T. Khachaturova, and Y. K. Yur'ev, *Khim. Geterotsikl. Soedin.*, 406 (1969); *Chem Abstr.*, **71**, 124295k (1969).
400. W. Schulze and H. Willitzer, *J. Prakt. Chem.*, **31**, 131 (1966).
401. H. Rembges, F. Kröhnke, and I. Vogt, *Chem. Ber.*, **103**, 3427 (1970).
402. T. A. Nour and A. Salama, *J. Chem. Soc.C*, 2511 (1969).
403. B. Lipke, *Z. Chem.*, **8**, 23 (1968).
404. C. A. Cupas and R. L. Pearson, *J. Amer. Chem. Soc.*, **90**, 4742 (1968).
405. V. Boekelheide and W. G. Gall, *J. Amer. Chem. Soc.*, **76**, 1832 (1954).
406. E. E. Glover and G. H. Morris, *J. Chem. Soc.*, 3885 (1965).
407. D. H. Corr and E. E. Glover, *Chem. Ind.* (London), 2128 (1964); 847 (1965); *idem., J. Chem. Soc.*, 5816 (1965).
408. I. C. Calder and W. H. F. Sasse, *Tetrahedron Lett.*, 1465 (1965).
409. I. C. Calder and W. H. F. Sasse, *Aust. J. Chem.*, **21**, 2951 (1968).
410. I. C. Calder and W. H. F. Sasse, *Tetrahedron Lett.*, 3871 (1964).
411. E. E. Glover and G. Jones, *J. Chem. Soc.*, 3021 (1958).
412. A. Fozard and C. K. Bradsher, *Chem. Commun.*, 288 (1965); *idem., J. Org. Chem.*, **31**, 2346 (1966).
413. A. Fozard and C. K. Bradsher, *J. Org. Chem.*, **31**, 3683 (1966).
414. A. Fozard, L. S. Davies, and C. K. Bradsher, *J. Chem. Soc.C*, 3650 (1971).
415. K. Undheim and K. R. Reistad, *Acta Chem. Scand.*, **24**, 2949 (1970).
416. F. S. Babichev and V. N. Bubnovskaya, *Ukr. Khim. Zh.*, **30**, 848 (1964); *Chem. Abstr.*, **62**, 1766c (1965).
417. C. K. Bradsher and D. F. Lohr, Jr., *Chem. Ind.* (London), 1801 (1964); *idem., J. Heterocycl. Chem.*, **3**, 27 (1966).
418. K. Undheim and K. R. Reistad, *Acta Chem. Scand.*, **24**, 2956 (1970).
419. C. K. Bradsher and J. E. Boliek, *J. Org. Chem.*, **32**, 2409 (1967).
420. C. K. Bradsher and M. F. Zinn, *J. Heterocycl. Chem.*, **1**, 219 (1964); *idem.*, **4**, 66 (1967).
421. G. Jones and D. G. Jones, *J. Chem. Soc.C*, 515 (1967).
422. R. H. Good and G. Jones, *J. Chem. Soc.C*, 1938 (1970).
423. D. G. Jones and G. Jones, *J. Chem. Soc.C*, 707 (1969).
424. T. Greibrokk and K. Undheim, *Acta Chem. Scand.*, **25**, 2251 (1971).
425. V. Boekelheide and W. Feely, *J. Amer. Chem. Soc.*, **80**, 2217 (1958).
426. K. Winterfeld and W. Fahlisch, *Arch. Pharm.* (Weinheim), **304**, 923 (1971).
427. E. Schraufstätter, *Angew. Chem. Intern. Ed.*, **1**, 593 (1962).
428. C. K. Bradsher, *Accounts Chem. Res.*, **2**, 181 (1969).
429. J. Adamson and E. E. Glover, *J. Chem. Soc.C*, 861 (1971).
430. C. K. Bradsher and S. A. Telang, *J. Org. Chem.*, **31**, 941 (1966).
431. E. E. Glover and M. J. R.Loadman, *J. Chem. Soc.C*, 2391 (1967).
432. C. K. Bradsher, R. D. Brandau, J. E. Boliek, and T. L. Hough, *J. Org. Chem.*, **34**, 2129 (1969).
433. C. K. Bradsher, E. F. Litzinger, Jr., and M. F. Zinn, *J. Heterocycl. Chem.*, **2**, 331 (1965).
434. K. B. Moser and C. K. Bradsher, *J. Amer. Chem. Soc.*, **81**, 2547 (1959).
435. C. K. Bradsher, L. D. Quin, R. E. LeBleu, *J. Org. Chem.*, **26**, 3273 (1961).

436. C. K. Bradsher, L. D. Quin. R. E. LeBleu, and J. W. McDonald, *J. Org. Chem.*, **26**, 4944 (1961).
437. C. K. Bradsher and J. W. McDonald, *Chem. Ind.* (London), 1797 (1961); *idem., J. Org. Chem.*, **27**, 4475, 4478 (1962).
438. C. K. Bradsher and D. F. Lohr, Jr., *J. Org. Chem.*, **31**, 978 (1966).
439. M. Davis, P. Knowles, B. W. Sharp, R. J. A. Walsh, and K. R. H. Wooldridge, *J. Chem. Soc., C*, 2449 (1971).
440. Ya. L. Gol'dfarb and Ya. L. Danyshevski, *Dokl. Akad. Nauk S.S.S.R.*, **87**, 223 (1952); *Chem. Abstr.*, **48**, 679e (1954).
441. E. D. Rossiter and J. E. Saxton, *J. Chem. Soc.*, 3654 (1953).
442. O. G. Lowe and L. C. King, *J. Org. Chem.*, **24**, 1200 (1959).
443. M. Fraser, A. Melera, B. B. Molloy, and D. H. Reid, *J. Chem. Soc.*, 3288 (1962).
444. W. L. F. Armarego, *J. Chem. Soc.*, 4226 (1964).
445. M. Fraser, S. McKenzie, and D. H. Reid, *J. Chem. Soc.B*, 44 (1966).
446. P. J. Black, M. L. Hefferman, L. M. Jackman, Q. N. Porter, and G. R. Underwood, *Aust. J. Chem.*, **17**, 1128 (1964).
447. E. T. Borrows and D. O. Holland, *Chem. Rev.*, **42**, 611 (1948).
448. M. L. Mosby, "Heterocyclic Systems with Bridgehead Nitrogen Atoms," Part 1, Interscience, New York, 1961, pp. 239ff.
449. D. R. Bragg and D. G. Wibberley, *J. Chem. Soc.*, 2627 (1962).
450. W. Flitsch and E. Gerstmann, *Chem. Ber.*, **105**, 2344 (1972).
451. Y. Tamura, N. Tsujimoto, Y. Sumida, and M. Ikeda, *Tetrahedron*, **28**, 21 (1972).
452. V. Boekelheide and R. J. Windgassen, Jr., *J. Amer. Chem. Soc.*, **81**, 1456 (1959).
453. G. R. Cliff, G. Jones, and J. Stanyer, *J. Chem. Soc.C*, 3426 (1971).
454. F. B. Mallory, C. S. Wood, and J. T. Gordon, *J. Amer. Chem. Soc.*, **86**, 3094 (1964) and references cited therein.
455. R. E. Doolittle and C. K. Bradsher, *Chem. Ind.* (London), 1631 (1965); *idem.*, J. Org. Chem., **31**, 2616 (1966).
456. A. Fozard and C. K. Bradsher, *Tetrahedron Lett.*, 3341 (1966); *idem., J. Org. Chem.*, **32**, 2966 (1967).
456a. D. E. Portlock, M. J. Kane, J. A. Bristol, and R. E. Lyle, *J. Org. Chem.*, **38**, 2351 (1973).
457. R. M. Wilson and F. DiNinno, Jr., *Tetrahedron Lett.*, 289 (1970).
458. F. Kröhnke and D. Mörler, *Tetrahedron Lett.*, 3441 (1969).
459. W. Augstein and F. Kröhnke, *Ann. Chem.*, **697**, 158 (1966).
460. I. C. Calder, T. M. Spotswood, and W. H. F. Sasse, *Tetrahedron Lett.*, 95 (1963); I. C. Calder and W. H. F. Sasse, *Aust. J. Chem.*, **18**, 1819 (1965).
461. R. F. Homer and T. E. Tomlinson, *J. Chem. Soc.*, 2498 (1960).
462. W. P. Shyluk, *J. Polym. Sci., Part A*, **2**, 2191 (1964).
463. V. A. Kargin, V. A. Kabanov, K. V. Aliev, and E. F. Razvodovskii, *Proc. Acad. Sci. U.S.S.R.*, **160**, 604 (1965).
464. V. A. Kabanov, T. I. Patrikeeva, and V. A. Kargin, *Proc. Acad. Sci. U.S.S.R.*, **166**, 1350 (1966).
465. V. A. Kabanov, K. V. Aliev, O. V. Kargina, T. I. Patrikeeva, and V. A. Kargin, *J. Polym. Sci., Part C*, **16**, 1079 (1967).
466. T. I. Patrikeeva, T. E. Nechaeva, M. I. Mustafaev, V. A. Kabanov, and V. A. Kargin, *Vysokomol. Soedin., A*, **9**, 332 (1967); *Chem. Abstr.*, **66**, 86029p (1967).
467. V. A. Kabanov, T. I. Patrikeeva, O. V. Kargina, and V. A. Kargin, *J. Polym. Sci., Part C*, **23**, 357 (1968).

468. V. A. Kabanov, K. V. Aliev, and V. A. Kargin, *Vysokomol. Soedin., A,* **10**, 1618 (1968); *Chem. Abstr.*, **69**, 77774n (1968).
469. O. V. Kargina, M. V. Ul'yanova, V. A. Kabanov, and V. A. Kargin, *Vysokomol. Soedin., A,* **9**, 340 (1967); *Chem. Abstr.*, **66**, 86030g (1967).
470. V. A. Kabanov, V. A. Petrovskaya, and V. A. Kargin, *Vysokomol. Soedin., A,* **10**, 925 (1968); *Chem. Abstr.*, **69**, 19561e (1968).
471. V. A. Kabanov and V. A. Petrovskaya, *Vysokomol. Soedin., B,* **10**, 797 (1968); *Chem. Abstr.*, **70**, 78444n (1969).
472. A. V. Gvozdetzskii and V. A. Kobanov, *Vysokomol. Soedin., B,* **11**, 397 (1969); *Chem. Abstr.*, **71**, 81795d (1969).
473. S. Iwatsuki, T. Kokubo, K. Motomatsu, M. Tsuji, and Y. Yamashita, *Makromol. Chem.*, **120**, 154 (1968); *Chem. Abstr.*, **70**, 47940s (1969).
474. T. Otsu, M. Ko, and T. Sato, *J. Polym. Sci., Part A-1*, **8**, 791 (1970).
475. D. D. Reynolds and T. T. M. Laakso, U.S. Patent 2,725,381 (1955); *Chem. Abstr.*, **50**, 10797b (1956).
476. J. C. Salamone, B. Snider, and W. L. Fitch, *Macromolecules,* **3**, 707 (1970).
477. J. C. Salamone, B. Snider, and W. L. Fitch, *J. Polym. Sci., Part A-1,* **9**, 1493 (1971).
478. L. M. Richards, U.S. Patent 2,487,829 (1949); *Chem. Abstr.*, **44**, 1732b (1950).
479. A. Rembaum, W. Baumgartner, and A. Eisenberg, *J. Polym. Sci., Part B,* **6**, 159 (1968).
480. "Encyclopedia of Polymer Science and Technology," Vol. 14, Wiley-Interscience, New York, 1971, pp. 656–657.
481. R. Hart and D. Timmerman, *J. Polym. Sci.*, **28**, 638 (1958).
482. Yu. E. Kirsh, V. A. Kabanov, and V. A. Kargin, *Dokl. Akad. Nauk S.S.S.R.*, **177**, 112 (1967); *Chem. Abstr.*, **69**, 18351z (1968).
483. Yu. E. Kirsh, V. A. Kabanov, and V. A. Kargin, *Vysokomol. Soedin, A,* **10**, 349 (1968); *Chem. Abstr.*, **68**, 105580w (1968).
484. S. K. Pluzhnov, Yu. E. Kirsh, V. A. Kabanov, and V. A. Kargin, *Dokl. Akad. Nauk S.S.S.R.*, **185**, 843 (1969); *Chem. Abstr.*, **71**, 22442p. (1969).
485. Yu. E. Kirsh and V. A. Kabanov, *Dokl. Akad. Nauk S.S.S.R.*, **195**, 1109 (1970); *Chem. Abstr.*, **74**, 54331e (1971).
486. Yu. E. Kirsh, S. K. Pluzhnov, T. S. Shomina, V. A. Kabanov, and V. A. Kargin *Vysokomol. Soedin., A,* **12**, 186 (1970); *Chem. Abstr.*, **72**, 86675v (1970).
487. L. M. Minsk and W. O. Kenyon, U.S. Patent 2,484,420 (1949); *Chem. Abstr.*, **44**, 6197c (1950).
488. A. Ya. Chernikhov and S. S. Medvedev, *Dokl. Akad. Nauk S.S.S.R.*, **180**, 913 (1968); *Chem. Abstr.*, **69**, 44235j (1968).
489. A. A. Berlin, V. I. Ganina, V. A. Kargin, A. G. Kronman, and D. M. Yanovskii, *Vysokomol. Soedin.*, **6**, 1684 (1964); *Chem. Abstr.*, **62**, 2897d (1965).
490. V. V. Korshak, A. B. Davankov, L. B. Zubakova, and I. A. Plakunova, *Zh. Prikl. Khim.*, **42**, 1618 (1969); *Chem. Abstr.*, **71**, 113802x (1969).
491. J. H. Lupinski and K. D. Kopple, *Science,* **146**, 1038 (1964); J. H. Lupinski, K. D. Kopple, and J. J. Hertz, *J. Polym. Sci., Part C,* **16**, 1561 (1967).
492. J. M. Bruce and J. R. Herson, *Polymer,* **8**, 619 (1967).
493. C. Mercier and J. P. Dubosc, *Bull. Soc. Chim. Fr.*, 268 (1969).
494. Yu. E. Kirsh, S. G. Starodubtsev, Yu. B. Grebenshchikov, G. I. Liktenshtein, and V. A. Kabanov, *Dokl. Akad. Nauk S.S.S.R.*, **194**, 1357 (1970); *Chem. Abstr.*, **74**, 54294v (1971).
495. C. L. Arcus and W. A. Hall, *J. Chem. Soc.*, 5995 (1964).
496. P. Roy-Chowdhury, *J. Polym. Sci., Part A-2,* **7**, 1451 (1969).

497. M. I. Mustafaev, K. V. Aliev, and V. A. Kabanov, *Vyskomol. Soedin., A,* **12**, 855 (1970); *Chem. Abstr.,* 73, 25983t (1970).
498. V. A. Kabanov, *Pure Appl. Chem.,* **15**, 391 (1967).
499. V. A. Petrovskaya, V. A. Kabanov, and V. A. Kargin, *Vysokomol. Soedin., A,* **12**, 1645 (1970); *Chem. Abstr.,* **73**, 88249d (1970).
500. V. A. Kabanov, K. V. Aliev, and M. I. Mustafaev, *Azerb. Khim. Zh.,* **1-2**, 167 (1970); *Chem. Abstr.,* **74**, 32040g (1971).
501. V. A. Kabanov, V. G. Popov, and D. A. Topchiev, *Dokl. Akad. Nauk S.S.S.R.,* **188**, 1056 (1969); *Chem. Abstr.,* **72**, 32276x (1970).
502. A. I. Yurzhenko, I. N. Kirichenko, V. A. Vilshanskii, and N. N. Zayats, *Dokl. Akad. Nauk S.S.S.R.,* **190**, 616 (1970); *Chem. Abstr.,* **72**, 122000g (1970).
503. J. C. Salamone, B. Snider, and W. L. Fitch, *J. Polym. Sci., Part B,* **9**, 13 (1971).
504. I. Mielke and H. Ringsdorf, *Makromol. Chem.,* **142**, 319 (1971); *idem., J. Polym. Sci., Part B,* **9**, 1 (1971).
505. T. Alfrey and C. C. Price, *J. Polym. Sci.,* **2**, 101 (1947).
506. I. N. Duling and C. C. Price, *J. Amer. Chem. Soc.,* **84**, 578 (1962).
507. J. J. Monagle and W. Mosher, *Polym. Preprints,* **10**, 705 (1969).
508. J. Thomas, D. F. Elsden, and S. M. Partridge, *Nature,* **200**, 651 (1963).
509. N. R. Davis and R. A. Anwar, *J. Amer. Chem. Soc.,* **92**, 3778 (1970).
510. O. V. Kargina, V. A. Kabanov, and V. A. Kargin, *J. Polym. Sci., C,* **22**, 339 (1968).
511. L. D. Narkevich, O. V. Kargina, V. A. Kabanov, and V. A. Kargin, *Vysokomol. Soedin., A,* **12**, 1817 (1970); *Chem. Abstr.,* **73**, 99249e (1970).
512. V. A. Kabanov, O. V. Kargina, and V. A. Petrovskaya, *Vysokomol. Soedin., A,* **13**, 348 (1971); *Chem. Abstr.,* **74**, 112456n (1971).
513. V. A. Kargin, V. A. Kabanov, and O. V. Kargina, *Dokl. Akad. Nauk S.S.S.R.,* **161**, 1131 (1965); *Chem. Abstr.,* **63**, 3068h (1965).
514. M. Okawara, T. Sasaoka, and E. Imoto, *Kogyo Kagaku Zasshi,* **65**, 1652 (1962); *Chem. Abstr.,* **58**, 8051 (1963).
515. A. S. Lindsey, S. E. Hunt, and N. G. Savill, *Polymer,* **7**, 479 (1966).
516. Yu. E. Kirsh, L. Y. Bessmertnaya, V. P. Torchilin, I. M. Papisov, and A. V. Kabanov, *Dokl. Akad. Nauk S.S.S.R.,* **191**, 603 (1970); *Chem. Abstr.,* **73**, 15463j (1970).
517. N. A. Vengerova, Yu. E. Kirsh, V. A. Kabanov, and V. A. Kargin, *Dokl. Akad. Nauk S.S.S.R.,* **190**, 131 (1970); *Chem. Abstr.,* **72**, 107401v (1970).
518. K. Martinek, Yu. E. Kirsh, A. A. Strongina, V. N. Dorovska, A. K. Yatsimirskii, I. V. Berezin, and V. A. Kabanov, *Dokl. Akad. Nauk, S.S.S.R.,* **199**, 148 (1971); *Chem. Abstr.,* **75**, 109488q (1971).
519. C. J. Cavallito, H. S. Yun, J. C. Smith, and F. F. Foldes, *J. Med, Chem.,* **12**, 134 (1969); *ibid.,* **14**, 1251 (1971); J. C. Smith, C. J. Cavallito, and F. F. Foldes, *Biochem. Pharmacol.,* **16**, 2438 (1967).
520. C. J. Cavallito, H. S. Yun, T. Kaplan, J. C. Smith, and F. F. Foldes, *J. Med. Chem.,* **13**, 221 (1970).
521. R. C. Allen, G. L. Carlson, and C. J. Cavallito, *J. Med. Chem.,* **13**, 909 (1970).
522. C. J. Cavallito, H. S. Yun, M. L. Edwards, and F. F. Foldes, *J. Med. Chem.,* **14**, 130 (1971).
523. B. R. Baker and J. A. Hurlbut, *J. Med. Chem.,* **12**, 677 (1969).
524. B. R. Baker and J. A. Hurlbut, *J. Med. Chem.,* **12**, 902 (1969).
525. B. R. Baker and M. H. Doll, *J. Med. Chem.,,* **14**, 793 (1971).
526. B. R. Baker and M. Cory, *J. Med. Chem.,* **14**, 119 (1971).527.
527. B. R. Baker and J. A. Hurlbut, *J. Med. Chem.,* **12**, 221 (1969).

528. M. Leven, G. Pfleiderer, J. Berghauser, and C. Woenckhaus, *Z. Physiol. Chem.*, **350**, 1647 (1969).
529. J. R. Heitz and B. M. Anderson, *Mol. Pharmacol.*, **4**, 44(1968).
530. W. G. Bardsley, J. S. Ashford, and C. M. Hill, *Biochem. J.*, **122**, 557 (1971).
531. G. B. Koelle, "Handbuch der experimentelle Pharmakologie," Vol. 15, Springer-Verlag, Berlin, 1963, pp. 229, 921.
532. E. Usdin in "International Encyclopedia of Pharmacology and Therapeutics," Section 13, "Acetylcholinsterase Agents," Vol. 1, A. G. Karczmar (Ed.), Pergamon, New York, 1970, (a) pp. 151ff; (b) pp. 222ff.
533. R. I. Ellin and J. H. Wills, *J. Pharm. Sci.*, **53**, 995, 1143 (1964).
534. H. C. Froede and I. B. Wilson, in "The Enzymes," Vol. 5, 3rd. ed., P. D. Boyer (Ed.), Academic, New York, 1971, pp. 102ff.
535. D. F. Heath, "Organophosphorus Poisons," Pergamon, New York, 1961, pp. 124ff.
536. A. Lüttringhaus and I. Hagedorn, *Arzneim.-Forsch.*, **14**, 1 (1964).
537. Y. Ashani and S. Cohen, *J. Med Chem.*, **14**, 621 (1971); **13**, 471 (1970).
538. Y. Ashani and S. Cohen, *Israel J. Chem.*, **5**, 59 (1967).
539. C. F. Barfknecht, F. W. Benz, and J. P. Long, *J. Med. Chem.*, **14**, 1003 (1971).
540. J. C. Lamb, G. M. Steinberg, S. Solomon, and B. E. Hackley, Jr., *Biochemistry*, **4**, 2475 (1965).
541. P. Franchetti, M. Grifantini, and M. L. Stein, *J. Pharm. Sci.*, **59**, 710 (1970).
542. C. F. Barfknecht, J. P. Long, and F. W. Benz, *J. Pharm. Sci.*, **60**, 138 (1971).
543. J. Patocka, *Collect. Czech. Chem. Commun.*, **36**, 2677 (1971).
544. E. Dirks, A. Scherer, M. Schmidt, and G. Zimmer, *Arzneim.-Forsch.*, **20**, 197 (1970).
545. K. Schoene and E. M. Strake, *Biochem. Pharmacol.*, **20**, 1041 (1971).
546. H. Kuhnen, *Arzneim.-Forsch.*, **20**, 774 (1970).
547. T. Nishimura, C. Yamazaki, and T. Ishiura, *Bull. Chem. Soc. Jap.*, **40**, 2434 (1967).
547a. R. Reiner and A. Scherer, *Arch. Pharm.* (Weinheim), **306**, 424 (1973).
548. K. B. Augustinsson and H. Hasselquist, *Acta Chem. Scand.*, **18**, 1006 (1964).
549. K. B. Augustinsson, *Biochem. Biophys. Acta*, **128**, 351 (1966).
550. J. H. Blanch and O. T. Onsager, *J. Chem. Soc.*, 3729 (1965).
551. J. H. Blanch and O. T. Onsager, *J. Chem. Soc.*, 3734 (1965).
552. N. V. Khromov-Borisov, M. L. Indenbom, and A. F. Danilov, *Dokl. Akad. Nauk S.S.S.R.*, **183**, 134 (1968); *Chem. Abstr.*, **71**, 11449v (1969).
553. S. P. Pappas, B. C. Pappas, and K. A. Marchant, Jr., *Biochemistry*, **6**, 3264 (1967).
554. C. Zervos and E. H. Cordes, *J. Amer. Chem. Soc.*, **90**, 6892 (1968); *idem.*, *J. Org. Chem.*, **36**, 1661 (1971).
555. R. H. Hook and W. C. Robinson, *J. Biol. Chem.*, **239**, 4257 4263 (1964).
556. K. V. Thimann and S. Mahadevan, *Arch. Biochem. Biophys.*, **105**, 133 (1964); S. Mahadevan and K. V. Thimann, *Arch. Biochem. Biophys.*, **107**, 62 (1964).
557. W. C. J. Ross, *J. Chem. Soc.C*, 1816(1966).
558. F. A. Loewus, H. R. Levy, and B. Vannesland, *J. Biol. Chem.*, **223**, 589 (1965).
559. N. O. Kaplan, "The Enzymes," Vol. 3, Academic Press, New York, 1960, Chap. 12.
560. S. R. Humphreys, J. M. Vendetti, C. J. Ciotti, I. Kline, A. Goldin, and N. O. Kaplan, *Cancer Res.*, **22**, 483 (1962).
561. O. M. Friedman, K. Pollak, and E. Khedouri, *J. Med. Chem.*, **6**, 462 (1963).
562. W. C. J. Ross, "Biological Alkylating Agents," Butterworth, London, 1962, p. 11.
563. W. Schulze, W. Gutsche, and W. Jungstand, *Arzneim.-Forsch.*, **17**, 605 (1967).
564. W. Gruszecki and E. Borowski, *Rocz. Chem.*, **42**, 533 (1968); *Chem. Abstr.*, 69, 51963y (1968).

565. W. Gruszecki and E. Borowski, *Rocz. Chem.*, **42**, 733 (1968); *Chem. Abstr.*, 71, 38774j (1969); *Rocz. Chem.*, **41**, 1611 (1967); *Chem. Abstr.*, **68**, 95656j (1968).
566. B. F. Cain, G. J. Atwell, and R. N. Seeley, *J. Med. Chem.*, **12**, 199 (1969); *14*, 311 (1971).
567. G. J. Atwell, B. F. Cain, and R. N. Seeley, *J. Med. Chem.*, **15**, 611 (1972).
568. G. J. Atwell and B. F. Cain, *J. Med. Chem.*, **10**, 706 (1967); **11**, 295 (1968).
569. G. J. Atwell, B. F. Cain, and R. N. Seeley, *J. Med. Chem.*, **11**, 300, 690 (1968).
570. B. F. Cain, G. J. Atwell, and R. N. Seeley, *J. Med. Chem.*, **11**, 963 (1968).
571. H. Lund, *Acta Chem. Scand.*, **11**, 1323 (1957).
572. V. J. Bauer, H. P. Dalalian, W. J. Fanshawe,, S. R. Safir, E. C. Tocus, and C. R. Boshart, *J. Med. Chem.*, **11**, 981 (1968).
573. G. E. Wiegand, V. J. Bauer, S. R. Safir, D. A. Blickens, and S. J. Riggi, *J. Med. Chem.*, **12**, 891 (1969).
574. G. E. Wiegand, V. J. Bauer, S. R. Safir, D. A. Blickens, and S. J. Riggi, *J. Med. Chem.*, **12**, 943 (1969).
575. V. J. Bauer, W. J. Fanshawe, H. P. Dalalian, and S. R. Safir, *J. Med. Chem.*, **11**, 984 (1968).
576. S. J. Riggi, D. A. Blickens, and C. R. Boshart, *Diabetes,* **17**, 646 (1968).
577. D. A. Blickens and S. J. Riggi, *Toxicol. Appl. Pharmacol.*, **14**, 393 (1969); *idem.*, *Diabetes,* **18**, 612 (1969).
578. W. J. Fanshawe, V. J. Bauer, S. R. Safir, D. A. Blickens, and S. J. Riggi, *J. Med. Chem.*, **12**, 381 (1969).
579. G. E. Wiegand, V. J. Bauer, D. A. Blickens, and S. J. Riggi, *J. Med. Chem.*, **14**, 214 (1971).
580. W. J. Fanshawe, V. J. Bauer, S. R. Safir, D. A. Blickens, and S. J. Riggi, *J. Med. Chem.*, **13**, 993(1970).
581. V. J. Bauer, G. E. Wiegand, W. J. Fanshawe, and S. R. Safir, *J. Med. Chem.*, **12**, 944 (1969).
582. V. J. Bauer, H. P. Dalalian, and S. R. Safir, *J. Med. Chem.*, **11**, 1263 (1968).
583. V. J. Bauer, W. J. Fanshawe, G. E. Wiegand, and S. R. Safir, *J. Med. Chem.*, **12**, 945 (1969).
584. J. L. Archibald, B. J. Alps, J. F. Cavalla, and J. L. Jackson, *J. Med. Chem.*, **14**, 1054 (1971).
585. R. T. Brittain, J. B. Farmer, D. Jack, L. E. Martin, and A. C. Ritchie, *Nature,* **213**, 731 (1967).
586. V. A. Cullum, J. B. Farmer, and S. L. Handley, *Brit. J. Pharmacol.*, **31**, 435 (1968).
587. T. B. O'Dell, C. Luna, and M. D. Napoli, *J. Pharmacol. Exp. Ther.*, **114**, 317 (1955).
588. A. K. Sheinkman, A. N. Prilepskaya, L. R. Kolomoitsev, and A. N. Kost, *Vestn. Mosk. Univ. Ser. II Khim.*,6, 74 (1964); *Chem. Abstr.*, **62**, 10400f (1965).
589. A. Pedrazzoli, L. Dall'Asta, and G. M. Cipelletti, *Chim. Ther.*, **3**, 81 (1968).
590. R. Quisno and M. J. Foster, *J. Bact.*, **52**, 111 (1946).
591. C. A. Lawrence, "Surface-Active Quaternary Ammonium Germicides," Academic, New York, 1950; L. W. Hedgecock, "Antimicrobial Agents," Lea and Febiger, Philadelphia, 1967, pp. 67ff; C. A. Lawrence and S. S. Block, "Disinfection, Sterilization, and Preservation," Lea and Febiger, Philadelphia, 1968, pp. 430ff.
592. P. F. D'Arcy and E. P. Taylor, *J. Pharm. Pharmacol.*, **14**, 193 (1962).
593. H. N. Glassman, *Bact. Rev.*, **12**, 105 (1948).
594. W. E. Knox, V. H. Auerbach, K. Zarudnaya, and M. J. Spirtes, *J. Bact.*, **58**, 443 (1949).

594a. American Medical Association Council on Drugs, *J. Amer. Med. Assoc.*, **170**, 196 (1959).
595. E. F. Rogers, R. L. Clark, A. A. Pessolano, H. J. Becker, W. J. Leanza, L. H. Sarett, A. C. Cuckler, E. McManus, M. Garzillo, C. Malanga, W. H. Ott, A. M. Dickinson, and A. Van Iderstine, *J. Amer. Chem. Soc.*,, **82**, 2974 (1960).
596. E. F. Rogers, *Ann. N.Y. Acad. Sci.*, 98, 412 (1962).
597. R. H. Mizzoni, R. A. Lucas, R. Smith, J. Boxer, J. E. Brown, F. Goble, E. Konopka, J. Gelzer, J. Szanto, D. C. Maplesden, and G. deStevens, *J. Med. Chem.*, **13**, 878 (1970).
598. T. Singh, R. G. Stein, and J. H. Biel, *J. Med. Chem.*,**12**, 949 (1969).
599. J. W. McFarland and H. L. Howes, Jr., *J. Med. Chem.*, **12**, 1079 (1969).
600. I. B. Wood, J. A. Pankavich, and E. Waletzky, *J. Parasitol.*, **51**, (2), Sec. 2, 34 (1965).
601. I. B. Wood, J. A. Pankavich, and R. E. Bambury, U.S. Patent 3,177,166 (1965); *Chem. Abstr.*, **62**, 16204h (1965); I. B. Wood, R. E. Bambury, and H. Berger, U.S. Patent 3,179,559 (1965); *Chem. Abstr.*, **62**, 15998g (1965).
602. I. B. Wood, J. A. Pankavich, and R. E. Bambury, U.S. Patent 3,152,042 (1964); *Chem. Abstr.*, **62**, 4016c (1965).
603. G. Y. Paris, D. L. Garmaise, J. Komlossy, and R. C. McCrae, *J. Med. Chem.*, **13**, 122 (1970).
604. G. Y. Paris, D. L. Garmaise, and J. Komlossy, *J. Heterocycl. Chem.*, **8**, 169 (1971).
605. A. L. Black and L. A. Summers, *J. Chem. Soc.C*, 610 (1969).
606. N. N. Mel'nikov, B. A. Khaskin, and I. V. Sablina, *J. Gen. Chem. U.S.S.R.*, **38**, 1509 (1968).
607. W. O. Foye, Y. J. Cho, and K. H. Oh, *J. Pharm. Sci.*, **59**, 114 (1970).
608. C. Ainsworth, D. N. Benslay, J. Davenport, J. L. Hudson, D. Kau, T. M. Lin, and R. R. Pfeiffer, *J. Med. Chem.*, **10**, 158 (1967).
609. J. R. Maley and T. C. Bruice, *J. Amer. Chem. Soc.*, **90**, 2843 (1968); *Arch. Biochem. Biophys.*, **136**, 187 (1970).
610. D. T. Browne, S. S. Hixson, and F. H. Westheimer, *J. Biol. Chem.*, **246**, 4477 (1971).
611. D. A. Deranleau and R. Schwyzer, *Biochemistry*, **9**, 126 (1970).
612. R. A. Bradshaw and D. A. Deranleau, *Biochemistry*, **9**, 3310(1970).
613. P. Mohr and W. Scheler, *Eur. J. Biochem.*, **8**, 444 (1969).
614. D. Manson, *Biochem. J.*, **119**, 541 (1970).
615. F. M. Plakogiannis, E. J. Lien, C. Harris, and J. A. Biles, *J. Pharm. Sci.*, **59**, 197 (1970).
616. J. T. MacGregor and T. W. Clarkson, *Biochem. Pharmacol.*, **20**, 2833 (1971), and references therein cited.
617. D. Gross, *Prog. Chem. Org. Nat. Prod.*, **28**, 109 (1970).
618. K. Torssell and K. Wahlberg, *Tetrahedron Lett.*, **445** (1966); *idem., Acta Chem. Scand.*, **21**, 53 (1967).
619. M. Noguchi, H. Sakuma, and E. Tamaki, *Arch. Biochem. Biophys.*, **125**, 1017 (1968); *idem., Phytochemistry*, **7**, 1861 (1968).
620. S. Mizusaki, Y. Tanabe, T. Kisaki, and E. Tamaki, *Phytochemistry*, **9**, 549 (1970).

SUBJECT INDEX

Italicized page numbers refer to tables; individual compounds from tables are not indexed. In some cases nomenclature in the index may differ from that in text.